CHROMIUM IN THE NATURAL AND HUMAN ENVIRONMENTS

Volume

20

in the Wiley Series in

**Advances in Environmental
Science and Technology**

JEROME O. NRIAGU, Series Editor

CHROMIUM IN THE NATURAL AND HUMAN ENVIRONMENTS

Edited by
Jerome O. Nriagu
National Water Research Institute
Burlington, Ontario, Canada

Evert Nieboer
Department of Biochemistry and
Occupational Health Program
McMaster University
Hamilton, Ontario, Canada

WILEY

A WILEY-INTERSCIENCE PUBLICATION
JOHN WILEY & SONS
New York · **Chichester** · **Brisbane** · **Toronto** · **Singapore**

Library of Congress Cataloging-in-Publication Data:

Chromium in the natural and human environments / edited by Jerome O.
    Nriagu, Evert Nieboer.
        p. cm. — (Advances in environmental science and technology :
    v. 20)
        "A Wiley-Interscience publication."
        Includes bibliographies and index.
        ISBN 0-471-85643-6
        1. Chromium—Toxicology. 2. Chromium—Environmental aspects.
    3. Chromium. I. Nriagu, Jerome O. II. Nieboer, Evert.
    III. Series.
        [DNLM: 1. Chromium—adverse effects. 2. Chromium—analysis.
    3. Environmental Pollutants—adverse effects. 4. Environmental
    Pollutants—analysis. W1 AD552T v. 20 / QV 290 C557]
    TD180.A38 vol. 20
    [RA1231.C5]
    628 s—dc19
    [363.1'79]
    DNLM/DLC   87-27303
    for Library of Congress   CIP

Printed in the United States of America

10 9 8 7 6 5 4 3 2 1

CONTRIBUTORS

AISLABIE, JACKIE, Microbiology Department, University of Otago, Dunedin, New Zealand

BARTLETT, RICHARD J., Department of Plant and Soil Science, The University of Vermont, Burlington, Vermont

BESZEDITS, STEPHEN, B & L Information Services, Toronto, Ontario, Canada

BREMER, PHILIP, Microbiology Department, University of Otago, Dunedin, New Zealand

CALDER, LYNN M., Golden Associates, Mississauga, Ontario, Canada

COLEMAN, RICHARD N., Microbiology Group, Alberta Environmental Center, Vegreville, Alberta, Canada

DAVIDSON, CLIFF I., Departments of Civil Engineering and Public Policy, Carnegie-Mellon University, Pittsburgh, Pennsylvania

HAINES, A. TED, Department of Clinical Epidemiology and Biostatistics, and Occupational Health Program, McMaster University, Hamilton, Ontario, Canada

HANDA, B. K., Central Chemical Laboratory, Lucknow, India

HOLDWAY, DOUGLAS A., Aquatic Ecology Division, National Water Research Institute, Canada Center for Inland Waters, Burlington, Ontario, Canada

JAMES, BRUCE R., Department of Plant and Soil Science, The University of Vermont, Burlington, Vermont

JUSYS, ALGIS A., Department of Biochemistry and Occupational Health Program, McMaster University, Hamilton, Ontario, Canada

LOUTIT, MARGARET, Microbiology Department, University of Otago, Dunedin, New Zealand

MAYER, LAURENCE M., Program in Oceanography, Center for Marine Studies, Ira C. Darling Center, University of Maine, Walpole, Maine

MILFORD, JANA B., Department of Engineering and Public Policy, Carnegie-Mellon University, Pittsburgh, Pennsylvania

NIEBOER, EVERT, Department of Biochemistry and Occupational Health Program, McMaster University, Hamilton, Ontario, Canada

NRIAGU, JEROME O., National Water Research Institute, Burlington, Ontario, Canada

ONO, BUN-ICHIRO, Laboratory of Environmental Hygiene Chemistry, Faculty of Pharmaceutical Sciences, Okayama University, Okayama, Japan

PACYNA, JOZEF M., Norwegian Institute for Air Research, Lillestrøm, Norway

SHAW, SUSAN L., Department of Biochemistry and Occupational Health Program, McMaster University, Hamilton, Ontario, Canada

TREVORS, JACK T., Department of Environmental Biology, Ontario Agricultural College, University of Guelph, Guelph, Ontario, Canada

WONG, PAUL T. S., Fisheries and Oceans, Great Lakes Fisheries Research Branch, Canada Center for Inland Waters, Burlington, Ontario, Canada

YASSI, ANNALEE, Occupational Health Program, Department of Community Health Sciences, University of Manitoba, Winnipeg, Manitoba, Canada

INTRODUCTION
TO THE SERIES

The deterioration of environmental quality, which began when mankind first congregated into villages, has existed as a serious problem since the industrial revolution. In the second half of the twentieth century, under the ever-increasing impacts of exponentially growing population and of industrializing society, environmental contamination of the air, water, soil, and food has become a threat to the continued existence of many plant and animal communities of various ecosystems and may ultimately threaten the very survival of the human race. Understandably, many scientific, industrial, and governmental communities have recently committed large resources of money and human power to the problems of environmental pollution and pollution abatement by effective control measures.

Advances in Environmental Sciences and Technology deals with creative reviews and critical assessments of all studies pertaining to the quality of the environment and to the technology of its conservation. The volumes published in the series are expected to service several objectives: (1) stimulate interdisciplinary cooperation and understanding among the environmental scientists; (2) provide the scientists with a periodic overview of environmental developments that are of general concern or that are of relevance to their own work or interests; (3) provide the graduate student with a critical assessment of past accomplishment which may help stimulate him or her toward the career opportunities in this vital area; and (4) provide the research manager and the legislative or administrative official with an assured awareness of newly developing research work on the critical pollutants, and with the background information important to their responsibility.

As the skills and techniques of many scientific disciplines are brought to bear on the fundamental and applied aspects of the environmental issues, there is a heightened need to draw together the numerous threads and to present a coherent picture of the various research endeavors. This need and the recent tremendous growth in the field of environmental studies have clearly made some editorial adjustments necessary. Apart from the changes in style and format, each future volume in the series will focus on one particular theme or timely topic, starting with Volume 12. The author(s) of each pertinent section will be expected to critically review the literature and the most important recent developments in the particular field; to critically evaluate new concepts, methods, and data; and to focus attention on important unresolved or controversial questions and on probable future trends. Monographs embodying the results of unusually extensive and well-rounded investigations will also be published in the series. The net result of the new editorial policy should be more integrative and comprehensive volumes on key environmental issues and pollutants. Indeed, the development of realistic standards of environmental quality for many pollutants often entails such a holistic treatment.

JEROME O. NRIAGU, Series Editor

PREFACE

Chromium and its compounds are widely used by modern industries, resulting in large quantities of this element being discharged into the environment. In view of its paradoxical roles as an essential micronutrient in human and animal nutrition at low concentrations and a known carcinogen at elevated levels (as chromate), there is now a growing concern about the fate and effects of chromium in the environment. This volume provides a comprehensive assessment of the uses of chromium, its sources (natural and anthropogenic), and occurrence in the air, water, and soil. Its toxicity to terrestrial and aquatic biota and humans is examined in relation to its chemistry and biochemistry, and the role of chromium as an essential element is properly evaluated. The critical assessment of the published literature and the large ensemble of references to chromium in diverse disciplines should make the volume an essential reading for researchers and serious students of environmental and occupational health effects of metals. Some aspects of our knowledge about the relative risks and benefits of chromium to human health remain unresolved, and it is hoped that this timely volume will stimulate further research on the environmental toxicology, hypersensitivity, geno-toxicity, and carcinogenicity of chromium compounds. Special emphasis is warranted on documenting the adverse effects on the liver, kidney and reproduction.

We thank the staff at John Wiley & Sons, Inc., for invaluable editorial assistance. Any success of this volume belongs to our distinguished group of contributors.

JEROME O. NRIAGU
EVERT NIEBOER

January 1988

CONTENTS

CHROMIUM IN THE
NATURAL AND HUMAN
ENVIRONMENTS

1

HISTORICAL PERSPECTIVES

Jerome O. Nriagu

National Water Research Institute
Burlington, Ontario, Canada

There is no hard evidence to suggest that metallic chromium was known to the ancients. It is, however, possible that the Hittites used chromium minerals in the fabrication of their hand weapons of "stainless steel"; these artifacts have yet to be analyzed for chromium (Langård, 1982). Medieval and Renaissance artists used many yellow and orange minerals that were called *ochres* or *giallulinum* (Thompson, 1956). Whether chromium compounds were actually included among such pigments remains open to question. Such an early use of a chromium compound would have been accidental, however. It is certainly easy to understand why this particular metal was late in making its debut in the industrial stage. The only mineral (chromite) from which chromium is recovered economically is uncommon, extremely refractory, unattractive in appearance, and has few desirable properties. It is not surprising that, ultimately, the metal was first isolated not from chromite but from crocoite, a rare but exceedingly attractive mineral found in a Siberian gold mine.

Nicolas-Louis Vauquelin, the Frenchman who discovered chromium, made the serendipitous observation that the new mineral would find few uses because of its brittleness and infusibility. Less than 150 years later, the unglamorous metal had become crucial to the military survival and economic well-being of all the industrialized nations. It is now the most important alloying element in modern cast iron, steel, and nonferrous metallurgy and has played a key role in all the major technologic developments of the twentieth century. Without chromium, the development of high-speed trains, the automobile, satellites and the associated space industry, electrical heating appliances, the chemical, plating, and machine-tool industries may well not have been economically feasible. The metallurgical applications of the unique properties of chromium have succeeded in turning steel, a very base material, into a noble metal that does not rust at the seashore or in highly polluted environments, does not dissolve in hot strong acids, and does not scale at 2,000°F. The consumption of chromite ore and concentrates by the world's metallurgical, chemical, and refractory industries increased from about 10,000 in 1850 to about 100,000 in 1904 and now stands at nearly 10 million tonnes per year (see Chapter 2). Each day, new processes and consumer products are being discovered that depend on chromium, a truly remarkable metal of our technologic culture.

1. DISCOVERY

A new mineral, *plomb rouge de Siberie* from the Beresof gold mine in Siberia, was first described and analyzed in 1766 by Johann Gotlob Lehmann, a former Prussian mining official and then professor of chemistry in St. Petersburg. The bright red-orange mineral (natural lead chromate, now known as *crocoite*) was subsequently analyzed by various chemists, with different results, and thus remained a matter of much curiosity and speculation for nearly three decades. In 1794, the French chemists Nicolas-Louis Vauquelin and L.C.H. Macquart inferred that the mineral contained lead oxide, iron, aluminum, and a large fraction (approximately 38%) of oxygen. Three years later, Vauquelin was able to demonstrate that the lead was united to a peculiar acid (chromic acid) with the following general characteristics (Vauquelin, 1809):

I observed that when the powdered mineral is boiled with a solution of two parts of potassium carbonate, the lead combines with the carbonic acid, and the alkali with the peculiar acid to form a yellow solution which furnishes a crystalline salt of the same color. The mineral is decomposed by mineral acids and when the solution is evaporated, it furnishes a lead salt of the mineral acid and *l'acide du plomb rouge* in long prisms the same color as ruby. When the compound of *l'acide du plomb rouge* with potash is treated with mercury nitrate, it gives a red precipitate the color of cinnabar; with lead nitrate, an orange-yellow precipitate; with copper nitrate, a maroon-red, etc. *L'acide du plomb rouge,* free or in combination, dissolves in fused borax, microcosmic salt, or glass to which it communicates a beautiful emerald green color. . . .

Later in the same year, Vauquelin succeeded in isolating the chromium metal by heating a mixture of the chromic acid and carbon in a graphite crucible. After consultation with his friends (notably Anton Francois Fourcroy, his mentor and Rene-Just Hauy, a crystallographer), Vauquelin named the new mineral *chrom,* from the Greek word χρωμα, because of the brilliant hues of its compounds. At about the same time, Martin Heinrich Klaproth (1798) was also able to isolate the new metal by dissolving the red Siberian ore in hydrochloric acid (HCl), precipitating out the lead chloride and saturating the supernatant liquid with sodium carbonate. In 1798, both Klaproth and Tobias Lowitz (a German chemist serving as an apothecary in the Imperial Court at St. Petersburg) successfully isolated the metal from chromite ore samples from the northern Urals, and in the following year Tassaert (a German chemist at the Paris School of Mines) isolated chromium in a sample from the chrome iron ore deposits at Gassin, France, which were discovered in 1798 by Pierre-Henri Pontier (Rollinson, 1973). Since then, chromite has remained the only commercial source of chromium (see Chapter 2).

Vauquelin himself showed that chromium was the coloring material in emeralds and some varieties of ruby. The preferential accumulation of chromium in serpentine rocks of Saxony was established in 1800 by Valentine Rose; Isaac Tyson later used this geochemical association in prospecting for chromite in the United States. Around 1818, J.J. Berzelius, the famous Swedish chemist, determined the various degrees of oxidation of chromium and reported, among other things, that chromic trioxide contain twice as many oxygen atoms as the green chromic oxide (*Ciba Reviews,* 1953).

Russia was the principal source of chrome ore until Isaac Tyson discovered additional supplies near Baltimore, Maryland, in 1811. Using the inference that chromium was often associated with serpentine rocks that were comparatively barren of vegetation, Tyson was also able to discover other chromite deposits in Maryland, Virginia, and Pennsylvania. The mines in Maryland attained prominence as the world's leading source of chromite from 1828 to 1850. With the discovery of the large Turkish deposits in 1848, production from the mines in Maryland was drastically reduced, and the United States has never again dominated the chromite export market. The Turkish monopoly lasted until 1897, when there was increased production from the mines in Russia and New Caledonia. The continuing search for chromium to meet the ever expanding demand subsequently led to the discovery of other major chromite deposits in Cuba (1840–1850), Greece (1890), Brazil (1907), South Africa (1908), Zimbabwe (1917), the Philippines (1925), Albania (1937), and Madagascar (1954) (EPA, 1984).

2. COMMERCIAL PRODUCTION

Even after its discovery, commercial production of pure chromium remained technologically unfeasible for several decades. The first successful attempt was made in 1821 by Berthier, who produced metallic chromium by reducing the oxides of the metal with carbon in a crucible, and in 1857 Fremy succeeded in reducing chromium chloride with sodium vapor. The early attempts yielded small quantities of very impure metal (alloy), and the production of pure chromium had to await the development of the aluminothermic reaction process by H. Sainte-Claire Deville in 1856. The first pyrometallurgical reduction of chromium oxides by means of finely divided aluminum was achieved in 1859 by Wöhler, who is also credited with determining most of the properties of chromium as we know them today. Development of the technique that eventually led to industrial scale production of chromium by the aluminothermic process, however, was by H. Goldschmidt (1895, 1897). The Thermit process yields 97–99% pure chromium metal that has a low carbon content, with aluminum and iron as the principal impurities. The silicon-reduction process in an electric-arc furnace for the production of pure chromium was developed between 1906 and 1908 by F.M. Becket (1907; 1908); this process is still in use.

The winning of chromium by an electrolytic process was first proposed by Antoine Cesar Becquerel in his classic book on electrochemistry published in 1843. Five years later, C. Junot (1848) followed Becquerel's suggestions and was able to obtain several patents covering chromium plating on iron objects. In 1854, R.W. von Bunsen (inventor of the Bunsen burner) succeeded in obtaining a metallic deposit by electrolysing a hot mixture of chromous and chromic chloride solutions in a porous pot at a high current density. For the next 70 years, varied attempts were made (Carveth and Curry, 1905; Sargent, 1920) to electrodeposit chromium using an assortment of compounds under different physicochemical conditions from all kinds of baths, but no one was able to develop an efficient and commercially profitable process (see Sully and Brandes, 1967). These early pioneering efforts were aimed basically at electroplating of chromium, and the preparation of commercial quantities of chromium by the electrolytic process was not achieved until 1927, when F. Adcock, C.G. Fink, M.J. Udy, and E. Liebreich concurrently did the needed breakthrough work and established the fundamentals of the reaction processes actually involved. Electrowinning of chromium from hexavalent baths was not developed until the 1950s (Sully and Brandes, 1967).

3. USES

3.1 Chemicals

From the time of discovery of crocoite, chromium salts were used primarily as pigments, and the metal itself had very limited industrial application until around the end of the nineteenth century. The process of manufacturing chromates by roasting chromite with lime and soda ash was discovered by 1800, and with it the chromium compounds became increasingly available commercially. Apollo Count von Mussin-Pushkin discovered the important violet-colored chromium alum (chromium potassium sulfate) in 1805. By 1812, chromium red (basic lead chromate) was already being produced commercially by boiling potassium chromate with white lead, and in 1816, Andreas Kurtz (a pupil of Vauquelin) started manufacturing chromium yellow and other chromium chemicals in London. Princess Charlotte, the daughter of King George IV, is reputed to have had her coach painted with the brilliant hue of chromium yellow (the yellow cabs presumably can be traced back to this popular princess). The discovery of large quantities of chromium ore in the United States at the turn of the last century led to more widespread commercialization of the chromium pigments (Harley, 1970). A succinct account of the discovery, development, and use of chromium chemicals can be found in MCC (1941).

It is believed that it was Vauquelin who encouraged Alexander Brongniart to try the green chromic oxide as a coloring material for porcelain. By 1802, Brongniart had successfully developed a pigment that formed the most attractive green glaze then available. Two years later, in 1804, C. Gordon de Saint-Memin produced a fast green resistant to light and atmospheric exposure using a mixture of mercury chromate and alumina. Soon after that, chromium salts became increasingly popular as the colorant in oil and watercolor painting, porcelain, glass, and enamel.

The excellent dyeing properties of chromium salts for fabrics of vegetable and animal origin were independently determined in 1819 by Ernst Ausust Geitner and by Jean-Louis Lassaigne. Lassaigne experimented with silk, linen, wool, and cotton and obtained fast shades of color using lead acetate and potassium chromate. In 1820, Daniel Koechlin-Schouch introduced chromium yellow into calico printing, and soon after that Walter Crum, a Glasgow chemist, invented the "orange resist" process of drawing fabric dyed chromium yellow through a bath of caustic soda. In 1831, James Thomson, an English calico printer, developed the process for discharging indigo blue cloths to white by means of chromic acid, and a year later Camille Koechlin successfully used the chromic acid oxidation process to produce fast blacks and grays with logwood and brown colors with cutch.

The first patent on chrome mordants (the combination of chromium with dyestuff to form lakes) was filed in 1840 by Charles Kober, a textile manufacturer in Leeds. At about the same time, textile printers started to fix chrome pigments onto fabric with the aid of albumin or other protein, a process still being used to a limited extent for printing wallpaper. Guignet's green quickly became a very popular textile dye after the process for its manufacture was established in 1858. Thus, chrome salts had become the most widely used metallic dyes and mordants worldwide, as noted by Camille Koechlin in 1853 (*CIBA Review,* 1953, p. 3620):

Our industries have come to regard chromium as the metal for color in the same way that to the world at large iron is the metal of civilization. The properties peculiar to chromium are of great importance in painting, in the structure of precious stones, in enamelling, in all colors and coloring arts. No other substance in any branch of applied chemistry has found and continues daily to find such varied use as does chromium in the dyeing of fabric. Every oxygen molecule combining with this metal opens up a new field of applications and these applications are already so numerous that dyes produced with chromium comprise all the colors of the rainbow and have a part in every one of our processes. By its nature, a coloring matter in two respects, chromium also serves to produce, fix, and destroy organic dyes; combined with oxygen, it is able to exercise all of these functions in succession or sometimes simultaneously. In the dyer's language, then, chromium is dye, mordant, discharge color and oxidation agent all in one. . . . This substance, which is now at the center of the largest number and the most ingenious of our processes and forms a resource we cannot dispense with, has an interesting history and stands as one of the finest examples of what the art of calico printers may achieve with one single element.

Between 1830 and 1900, the available records show clearly that textiles were colored either with minerals or with natural dyes of a mordant type fixed on the fiber by metallic compounds. During this period, chromium compounds were paramount in such processes. With the introduction of coal tar and petroleum-derived dyes, which give fast colors without mordants, the role of chrome compounds in the cotton textile industry gradually declined, and today their use is limited to oxidation processes and as after-treatment dyes. At the same time, the use of chromium compounds as mordants for fixing synthetic dyes on wool increased markedly, with many chromium compounds still being used in these processes (see Chapter 2).

Chromium also played an important role in the early synthesis of colored dyes and other organic reagents of commercial interest. Turkey red dye, with alizarin as the most important ingredient, was prepared as early as 1820 by Daniel Koechlin, who used potassium as a mordant in dyeing wool with the natural extract of madder root. The synthesis of nepthazarin, or alizarin black, in 1861 and anthragallol, or alizarin brown, in 1877 presumably would not have been accomplished without chromium. Familiar dyes that were mordanted with chromium include alizarin, gallocyanine, azorubine, and diazo compounds, resulting in the widely popular chrome wool dyes in various shades of brown, red, blue, yellow, olive, violet, and so on. In 1912, there was another milestone in the history of mordant chromium complexes when René Bohn, a Swiss chemist, synthesized chromium complexes of azo dyes, which are soluble in water and hence can be applied even more simply to fabric. Currently, chromium compounds find application in the industrial synthesis of saccharin, benzoic acid, anthraquinone, hydroquinone, camphor, and synthetic fibers. A detailed historical outline of chromium mordant and chromium complex dyes can be found in the *CIBA Reviews* (1953).

Friedrich Ludwig Knapp was the first person to describe chromium tanning in his *Textbook of Chemical Technology* published in 1847. In 1853, both Hylten Cavalin of Sweden and René de Kercado of France obtained patents covering chromium tanning processes using iron salts sweetened with chromium. In a later compendium on "Natural and Essential Character of the Tanning Process and Leather" published in 1858, Knapp discussed his detailed studies on single-batch tanning and emphasized the use of alkaline solution for precipitating the oxides on the leather surface (*CIBA Reviews,* 1953, p. 3642):

> Owing to the acid reaction of the iron and chrome salts, even light skins become hard, especially the grain, which tends to break easily. If, however, enough soda or caustic soda is added to the hydrochloric acid solution of the iron or chrome oxide, short of causing it to form a permanent precipitate, the oxide compound will precipitate on the skin more easily and in greater quantity.

Although clearly Knapp invented chromium tanning, the commercial use of the process was developed much later—in 1884 by Augustus Schultz. His was a two-bath process, in which the pelt was treated with sodium or potassium dichromate and the chromate ion was subsequently reduced on the skin with sodium thiosulfate. The commercially feasible one-bath process that had eluded Knapp was patented in 1893 by Martin Dennis, an American tanning specialist. Further refinements of the chromium tanning process came from Henry Richardson Proctor, an English leather chemist, who introduced chromium as the tanning agent and the reduction of the bichromate ion by glucose; his basic process is still in common use.

The use of chromium yellow and chromium green for printing wallpapers was introduced in 1818 by the Frenchman Zuber. The light-sensitive properties of chromium compounds were subsequently reported in 1839 by Mungo Ponton, who noted that paper impregnated with dichromate turned to a brown color on exposure to strong light. The following year, E. Becquerel observed that when a dichromated starch coating on paper was exposed to light, it became insoluble and that, after washing, the design could be colored blue with iodine. These observations led to the photoengraving by gravure process invented by Fox Talbot in 1852. He found that dichromated gelatin on steel became light-hardened, and that after etching with acid and removing the light-hardened stencil, an intaglio image was formed that accepted printing ink. Soon after that, in 1855, Louis Poitevin developed the photolithographic transfer process using dichromated gelatin to produce images on lithographic stones. These early photomechanical processes pioneered by Talbot and Poitevin formed the backbone of contemporary graphic arts (Jorgensen and Bruno, 1956).

3.2   Refractories

The use of chromite as refractory material in open hearth steel furnaces began in 1879 in France. It comes as no surprise that this application coincided with the development of the basic steel process by Thomas and Gilchrist (Udy, 1956). The first molded, tar-bonded, but unfired bricks of chromite were fashioned in England in 1886. Although the manufacture of fired bricks began in the United States in 1896, the procedure did not gain worldwide acceptance until the 1930s. During this period, innovations in the use of chromite for refractory purposes included (1) chromite-magnesite bricks with chromite dominant, (2) magnesite-chromite bricks, with magnesite dominant, (3) chromite-magnesite bricks containing equal amounts of the two materials; an English patent for this type of brick was first issued in 1915 (Searle, 1924), (4) chromite bricks with enhanced physical and chemical features. Special iron oxide-resistant chromite-magnesite bricks were developed in 1939, and the improved steel-plated bricks were introduced into the market in 1941 (Heuer et al., 1956).

3.3 Alloys

Electroplating of objects with chromium had a long and tortuous history starting with the pioneering work of Bunsen in 1854. The foundation for modern commercial electroplating was eventually laid by J. Sigrist, P. Winkler and M. Wantz in 1924 and refined by C.G. Fink in 1926. These pioneers were able to overcome the difficulties encountered by previous workers by using baths containing chromic acid and catalytic free acid radicals (typically sulfate or fluoride) maintained at a fixed ratio of 100:1. Chromizing is a related process at an elevated temperature for applying protective coatings of chromium to a metallic base. It was first demonstrated in 1923 by F.C. Kelley, although actual commercial development came much later (Sully and Brandes, 1967).

A method for making small quantities of high-carbon ferrochromium was developed by P. Berthier in 1821 by heating a mixture of iron and chromium oxides with powdered charcoal in a crucible. This method was modified in 1857 by E. Fremy, who reduced the oxides with coal or coke in a blast furnace. The resulting alloy, which typically contained 7–8% chromium, was often referred to as *Tasmanian pig iron* because it could be produced by the direct reduction of chromium-rich iron ores found in Tasmania. Alloys containing 74% chromium were made successfully in a crucible in 1875 by S. Kern, and in the same year A. Brustlein (1889) achieved an 85% chromium content in an alloy. The regular use of very high blast furnace temperatures and excess coke for the commercial production of ferrochromium containing 30–40% chromium began between 1870 and 1880 to meet the increasing demand by the chromium steels works. Ferrochromium, as we now know it, containing 70% chromium was first reported in Sweden in 1886 by E.G. Odelstjerna, who used a specially designed blast furnace. The eventual large-scale production of ferrochromium containing 67–71% chromium and 4–6% carbon by reducing chromite ore in an electric arc furnace was pioneered by the Frenchman, H. Moissan (1893). Large-scale production of electric furnace, high-carbon ferrochromium was initiated by Moorehead at Halcomb Rock, Virginia, in 1897.

The first large-scale production of low-carbon ferrochromium and chromium metal by the aluminothermic process was by H. Goldschmidt in 1895. By 1905, the reduction of chromite by aluminum or silicon in the electric furnace was yielding ferrochromium with less than 1% carbon, and by 1909 the average carbon content had been reduced to under 0.5%. By 1937, ferrochromium with less than 0.06% maximum carbon content had become a standard product (Kinzel and Crafts, 1937). The widely popular Perrin process for making low-carbon ferrochromium was patented in 1939 (see Udy, 1956, p. 37).

The history of chromium-containing steels is long and convoluted. The French mineralogist, P. Berthier, who was the first to make ferrochromium is also credited with being the first to make chromium–iron alloys in 1821. He found that his alloys containing 1–1.5% chromium had increased hardness, improved corrosion resistance, and altered magnetic properties. He made a knife and a razor blade of fine quality from his alloy, and his statement that chromium steels were excellent for cutlery was ignored for over a century. In 1822, J. Stodart and M. Faraday made a steel containing about 3% chromium and likewise observed that (Faraday, 1859, p. 79):

> The blade of such a sabre, or some such instrument made from this alloy, and heat treated in this way, would assuredly be beautiful, whatever its other properties might be; for the value of the chrome alloy for edge-tools we are not prepared to speak, not having made trials of its cutting power. The sabre blade, thus colored, would amount to a proof of its being well tempered; the bluish-black would indicate the temper of a watchspring, while the straw color towards the edge would announce the requisite degree of hardness.

The enhanced corrosion resistance of the chromium steels was subsequently confirmed by many others, including Mallet (1838) and Frémy (1857).

Small quantities of chromium steel became commercially available when J. Baur established the Chrome Steel Works in Brooklyn in 1869. These steels, which contained up to 2% chromium, were made by adding about 2% ferrochromium with 30% chromium and 3% carbon to wrought iron and melting the charge in a crucible. The first major structural application of the Bauer alloy was the famous bridge built between 1867 and 1874 by Captain J.D. Eads across the Mississippi River at St. Louis. Soon after that, production of similar steels was started at the Jacob Holzer Works at Unieux, France (Brustlein, 1886) and at other places (Boussingault, 1878). The Tasmanian Iron Company of Sheffield used a blast furnace to make alloys containing 7% chromium and 4–5% carbon (Riley, 1877; Howe, 1891), and in 1866–1867 a forge was set up near Medellin in Colombia, South America, to produce hard, chrome-bearing pig iron. Because of their intense hardness, these low-chromium alloys were widely used in crusher parts and bulletproof safes and cells, as well as in munitions and armaments. Arnold (1889) drew attention to the desirable properties of chromium in Bessemer steel tires, which were immensely popular at the end of the last century.

In 1864, J. Percy, the famous British metallurgist, apparently became the first person to produce various grades of chromium-rich alloys using ferrochromium in a crucible. His alloys contained 4%, 27%, 54%, and 76% chromium, and he made the very important observation that the magnetic properties were marked at 4% and 27% chromium and absent at the higher chromium concentrations. The relationship between magnetism and microstructure, however, stayed unexplored for many decades. The introduction of the blast furnace technique allowed the Terre Noire Company in France to start manufacturing chromiferous Spiegeleisen, containing 25% chromium and 13% manganese, in 1874 (Snelus, 1874; Busek, 1889; Eckardt, 1889). Hadfield (1892) studied a number of alloys containing up to 16% chromium, but is remembered for reaching the fatal conclusion that chromium was inimical to the corrosion resistance of steels in sulfuric acid.

The historical records thus show that up to 1900, chromium had been demonstrated conclusively to be beneficial in steels where corrosion resistance, strength, and hardness were required. Industrial applications of the chromium steels, however, were sporadic, and very little attempt was made to investigate the exact role of chromium in steels. The damaging effects of the high carbon concentration in steel, with particular reference to corrosion resistance, had been recognized at the turn of the century (Carnot and Goutral, 1898). One cannot fail to mention the eclectic patent filed in England in 1896 by E. Placet, in which he claimed to have succeeded in producing pure electrolytic chromium and to have invented a method of introducing it into molten metals and alloys. He further asserted that the chromium improved any metal to which it was added and that it particularly made the metals stable at high temperatures and raised the electrical resistance of the iron–nickel alloys and other metals. These claims have overshadowed most of the subsequent major developments in chromium metallurgy (Schneidewind, 1927, 1928).

The major breakthroughs in the development and understanding of the metallurgy of stainless steels came in quick succession between 1900 and 1915 and were predicated on the availability of large quantities of low-carbon chromium produced by the aluminothermit process. It became possible for W. Hittorf (1898, 1899) and P. Monnartz (1911) to undertake their detailed studies that firmly established the passivation phenomenon and, hence, the stainlessness of steels. The actual development of the commercially important stainless steel compositions and the unraveling of their metallographic features, however, were left to three pioneers: Leon Guillet (1904a,b,c, 1905, 1906) of France, who was the first person to produce and define the very important groups of alloys now known as martensitic, ferritic, and austenitic steels, W. Giesen (1909) of Britain, who provided the critical studies of chromium–nickel austenitic steels as well as other steels containing 8–18% chromium and 0.3% carbon, and A.M. Portevin (1909, 1911–12), also a Frenchman, who is generally credited with inventing the most widely used alloy containing 17.38% chromium and 0.2% carbon (Zapffe, 1949).

Development of the major industrial applications for these stainless steels occurred mainly between 1910 and 1930. The first of these materials to be popularized was the stainless cutlery developed by Harry Brearley in 1912. His patent covered a heat-treated alloy containing 9–16% chromium and less than 0.7% carbon (Brearley, 1915, 1919). Although Woods and Clark (1872) had recognized the commercial value of chromium steels much earlier and Portevin of France had patented in 1909 a method for softening chrome steels by annealing and very slow cooling, the first commercial-scale production of low (less than 0.07%) carbon, high (13–17%) chromium ferroalloys was made in 1920 at the Brown Bayley's Steel Works Ltd., Sheffield (Monypenny, 1951). In fact, Brearley takes credit for assigning the name *stainless steels* to these ferritic iron–chromium alloys.

Much of the credit for developing the austenitic chromium–nickel–iron alloys in 1912 certainly belongs to Strauss and Maurer of the Krupp Laboratories, Essen. Strauss (1924) gave the following account of their invention (Thum, 1935, p. 5):

> Metallurgists for a long time have been accustomed to alloy iron with nickel with a view to the improvement of the iron, whereas the discovery of the valuable properties of chromium is of comparatively recent date. In seeking for a suitable material from which to make protective tubes for thermo-electric couples, I started on the supposition that it ought to be possible to make use of the valuable properties of chromium, that is, its resistance to the action of oxygen, for the purpose of increasing the resistance of iron against attack of hot gases. In 1910, five experimental alloys high in chromium were prepared for this purpose. Polished specimens taken from them, after having been exposed for some time to the atmosphere of the laboratory, revealed the curious fact particularly that those steels which contained 20% chromium had remained perfectly bright while other bars, even those containing up to 25% nickel had become rusty. . . . From this work two brands of high chromium-nickel steel were first applied to practical uses. The first of these belongs to the martensite group (0.15% carbon, 14% chromium and 1.8% nickel) and the second to the austenite group (0.25% carbon, 29% chromium and 7% nickel). In the autumn of 1912, patents were applied for both groups of alloys and they were soon afterward placed on the market. Goods made from both groups were first shown at the Malmo Exhibition in April, 1914.

By 1914, large quantities of the 20% chromium steels were being supplied to the Fritz Haber's ammonia synthesis plants in Europe. Continued work soon led to the famous "18–8" (18% chromium and 8% nickel) alloy, which came into the market during the early 1920s.

The commercialization of the ferritic steels is the work of two American pioneers, Dantsizen, who introduced the nonhardening stainless steels containing 14–16% chromium and 0.07–0.15% carbon that are unsurpassed for turbine blades, and Becket (1920, 1924, 1926), who popularized the highly heat-resistant and corrosion-resistant chromium–irons containing over 20% chromium and before that was the inventor of the silicon-reduction process for producing chromium metal with very low carbon content (Becket, 1907, 1908, 1926). The important group of ternary alloys containing iron, silicon, and chromium (the so-called silcrome or valve steel) came about in 1914 when Armstrong accidentally introduced silicon into his crucible melt of chromium steel. A related group of silicon-rich chromium steels (the rezistals) was discovered by Johnson in 1917 while he was developing alloys to resist the oxyacetylene cutting flame.

Chromium also forms many commercially important alloys with the nonferrous metals. As early as 1895, Haynes invented a class of exceedingly hard and abrasion-resistant alloys containing chromium, cobalt, and nickel. In a later quest (between 1907 and 1912) for a suitable material for gas engine sparkplugs, this pioneer American automobile maker developed the famous stellite alloys containing cobalt, chromium, and molybdenum or tungsten. These alloys are now widely used in rock-cutting tools (Haynes, 1917, 1920). Another unique group of alloys is the electrical resistors containing nickel and chromium that are indispensable in all electrical heating devices. Their discovery dates back to 1903, when the search for a good thermocouple material led Marsh to melt nickel and chromium together in a crucible. By 1916, they had been introduced into carburizing boxes and retorts. One also cannot fail to mention the alloys known as "illium" first reported by Parr. They contain nickel and chromium as well as significant amounts of copper, molybdenum, and tungsten and are renowned for their superior electrical and heat resistance properties. Reference must also be made to the affinity of chromium for nitrogen and the formation of chromium nitride, which was reported by Ufer as early as 1859. A nitriding process was later patented by Fry (1923).

Since 1930, there have been numerous accomplishments in the practical and theoretical metallurgy of chromium and its alloys. Some of the recent advances include the rustless process for melting stainless steels by A.L. Field, stabilized austenitic stainless steels by E. Houdremeont, free-machining stainless steels by F.R. Palmer, precipitation hardening of steels by E. Wyche, and, obviously, the theoretical framework established by R. Franks, E.C. Bain, V.N. Krivobok, and A.K. Kinzel, among many others. Since the 1960s, the carbon content of the chromium alloys has been reduced to extremely low (less than 0.01%) levels using argon–oxygen decarburization, vacuum–oxygen decarburization, or electron-beam refining. These decarburization processes have produced a whole new group of ferritic stainless steels (or superalloys) containing primarily chromium (18–29%), molybdenum (1–4%), and iron (Streicher, 1977). It is clear that attempts to develop new chromium alloys or to improve the old ones have been a never-ending quest, as it should be considering the enormous commercial potential of each new material.

This historical outline is based on the following publications to which an interested reader should refer: Percy (1864), Howe (1891), Hadfield (1892), Schneidewind (1927), Thum (1935), Kinzel and Franks (1940), Zapffe (1949), Monypenny (1951), *CIBA Reviews* (1953), *Vancoram Review* (1959), Udy (1956), Krainer (1962), Bain and Paxton (1966), Sully and Brandes (1967), Klopp (1972), and Streicher (1977). A selected list of references on the history of chromium follows for the benefit of the more serious students.

REFERENCES

Adcock, F. (1927). "Alloys of iron research. Part V—Preparation of pure chromium." *J. Iron Steel Inst.* **115**, 369–392, 435–442.

Adcock, F. (1931). "Alloys of iron research. Part X–The chromium-iron constitutional diagram," with an appendix by Preston, G.D. "X-ray examination of chromium-iron alloys." *J. Iron Steel Inst.* **124**, 99–149.

Aitchison, L. (1921–1922). "Chromium steels and irons." *Proc. Inst. Automobile Eng.* **16** (part 1), 183–217. *Engineering* **112**, 771–772, 805–807.

Aitchison, L. (1915). "Experiments on the influence of composition upon the corrosion of steel." *Trans. Faraday Soc.* **11**, 212–234.

Arnold, J.O. (1889). "On the influence of chemical composition on the strength of Bessemer-steel tires." *Min. Proc. Inst. Civil Eng.* **95**, 115–166.

Austin, C.R. (1923). "Some mechanical properties of a series of chromium steels." *J. Iron Steel Inst.* **107**, 419–438.

Bain, E.C., and Paxton, H.W. (1966). *Alloying Elements in Steel,* 2nd ed. American Society Metallurgy, Cleveland, Ohio.

Baur, J. (Aug. 22, 1865). "Manufacture of steel," U.S. Patent 49,495.

Becket, F.M. (1928, 1929). "Chromiium alloys." *Mining Metallurg.* **9**, 551, **10**, 10–24.

Becket, F.M. (1920). "Iron alloy resistant to oxidation at high temperatures." U.S. Patent 1,333,151. (1924). "Rust-resisting chromium alloys." U.S. Patent 1,508,211. (1926). "Rustless iron." U.S. Patent 1,567,898.

Becket, F.M., (1907). "Reducing metallic oxides." U.S. Patent 854,018. (1908). "Making low-carbon ferro-alloys." U.S. Patent 891,898. (1908). "Producing low-carbon ferro-alloys." U.S. Patent 892,211. (1908). "Reducing metals and obtaining alloys." U.S. Patent 866,421. (1910). "Reducing Ores." U.S. Patent 967,159. (1924). "Rust-resisting chromium alloys." Patent 1,508,211.

Behrens, H., and Van Linge, A.R. (1894). "Sur l'acier cémenté, ferrochrome, le ferro-tungstène, l'acier chromé et l'acier tungstaté (Cemented steel, ferrochromium, ferrotungsten, chromium steel, and tungsten steel)." *Recueil Travaux Chim. Pays-Bas* **13**, 155–181.

Benedicks, C. (1904). "Recherches physiques et physico-chimiques sur l'acier au carbone (Physical and physico-chemical investigations on carbon steel)." Dissertation, Uppsala, 1904, 215 pp. Preliminary publication (1902). *Jernk. Ann.* **57**, 124–133. Condensed: (1906). *Jernk. Ann.* **61**, 1–231.

Berthier, P. (1821). "Sur les alliages du chrôme avec le fer et avec l'acier (Chromium irons and chromium steels)." *Ann. Chim. Phys. Paris* **17**, 55–64.

Boussingault. (1878). "Sur la production, la constitution et les propriétés des aciers chromés (Production, constitution, and properties of chromium steels)." *Ann. Chim. Phys. Paris,* ser. 5, **15**, 91–126.

Brearley, H. (1917). "Cutlery", U.S. Patent 1,197,256, 1916; French Patents 483,152 and 484,693.

Brearley, H. (1915, 1919). "Steel production." Canadian Patent 164,622, Canadian Patent 193,550.

Brearley, H. (1921). *The Case Hardening of Steel.* Longmans, Green & Company, New York, 207 pp.

Brustlein, M. (1886). "On chrome pig iron and steel." *J. Iron Steel Inst.* **II**, 770–778.

Brustlein, A. (1889). "Le ferro-chrome" (International Congress of Mining and Metallurgy). *Societé l'industrie minerale bulletin (Saint-Etienne),* Ser. 3, No. 3, pp. 975–983.

Bunsen, R.W. (1854). "Über die Darstellung von metallischem Chrom auf galvanischem Wege (The preparation of metallic chromium by the galvanic method)." *Ann. Physik Chemie,* ser. 4, **1**, 619–625. Abst.: *Ann. Chim. Phys. Paris,* ser. 3, **41**, 354–357.

Busek, R. (1889). "Remarks on ferrochromium and chromium steel." *Stahl Eisen,* **9**, 727; *J. Iron Steel Inst.* **21**, 409–421.

Carnot, A., and Goutal, E. (1898). "Recherches sur l'état où se trouvent le silicium et le chrome dans les produits sidérurgiques (Investigations on the state of silicon and chromium in ferrous products)." *C.R. Acad. Sci. Paris* **126**, 1240–1245.

Carpenter, H.C.H. (1905). "The types of structure and the critical ranges on heating and cooling of high-speed tool steels under varying thermal treatment." *J. Iron Steel Inst.* **67**, 433–473.

Carveth, H.R., and Curry, B.E. (1905) "Chromium and the electrolysis of chromic acid." *Trans. Am. Electrochem. Soc.* **7**, 115–142.

CIBA Review. (December 1953). No. 101.

Corbin, J.H.H. (1869). "Contributions from the laboratory of the Lehigh University. No. 2. On certain compounds of chromium with iron." *Am. J. Sci. Arts* **48**, 346; *Silliman's J.* **48**, 348–360.

Deville Sainte-Claire, H. (1856). "Mémoire sur la production des températures très élevées (Memoir on the production of very high temperatures)." *Ann. Chim. Phys. Paris*, ser. 3, **46**, 182–203 (200–201).

Deville Sainte-Claire, H., and Debray, H. (1859). "Du platine et des métaux qui l'accompagnent (Platinum and metals which accompany it)." *Ann. Chim. Phys. Paris*, ser. 3, **56**, 385–496.

DeYoung, J.H., Lee, M.P., and Lipin, B.R. (1984). *International Strategic Minerals Inventory Summary Report—Chromium,* U.S. Geological Survey Circular 930-B, Washington, D.C.

Dickenson, J.H.S. (1918–1919; 1919–1920). "Nickel-chromium steels." *J. West of Scotland Iron Steel Inst.* **26**, 110–125; **27**, 1–5.

Dupuy, E.L., and Portevin A.M. (1915). "La thermo-électricité des aciers spéciaux (The thermo-electric properties of special steels)." *Rev. de Mét., Mém.* **12**, 657–679. *J. Iron Steel Inst.* **91**, 306–335.

Eckardt, A. (1888; 1889). "Production of chromiferous ferromanganese." *Berg- und Hutten-mannisches Zeitschrift* **48**, 10; *J. Iron Steel Inst.* (No. 1), 311–337.

Edwards, C. (1908). "Function of chromium and tungsten in high-speed tool steel." *J. Iron Steel Inst.* **77**, 104–132.

Edwards, C.A., and Kikkawa, H. (1915). "The effect of chromium and tungsten upon the hardening and tempering of high-speed tool steel. *J. Iron Steel Inst.* **92**, 6–46.

Ehrensberger, E. (1922). "Aus der Geschichte der Herstellung der Panzerplatten in Deutschland (From the history of the manufacture of armor plate in Germany)." *Stahl Eisen* **42**, 1229–1236, 1276–1282, 1320–1330.

EPA (1984). *Health Assessment Document for Chromium.* Report No. EPA-600/8-83-014F. Office of Health and Environmental Assessment, U.S. Environmental Protection Agency, Research Triangle Park, NC.

Faraday, M. (1859). *Experimental Researches in Chemistry and Physics.* Richard Taylor & Henry Francis, London.

Fink, C.G. (1926). U.S. Patent 1,581,188. (1931). U.S. Patent 1,802,463. Assigned to the Chemical Treatment Co.

Forbes, D. (1875). "Report on the progress of the iron and steel industries in foreign countries." *J. Iron Steel Inst.* **II**, 581–676 (654).

Frémy, E. (1857). "Note sur le chrome cristallisé et sur ses alliages (Crystallized chromium and its alloys)." *C. R. Acad. Sci. Paris* **44**, 632–634.

French, H.J. (1922). "Tensile properties of some structural alloy steels at high temperatures." *Trans. Am. Soc. Steel Treat.* **2**, 409–422.

French, H.J. (1922). "Effect of heat treatment on mechanical properties of a carbon-molybdenum and a chromium-molybdenum steel." *Trans. Am. Soc. Steel Treat.* **2**, 769–798.

Friend, J.N., West, W., and Bentley, J.L. (1913). "The corrodibility of nickel, chromium, and nickel-chromium steels." *J. Iron Steel Inst.* **87**, 388–398.

Fry, A. (1923). "Stickstoff in Eisen, Stahl und Sonderstahl. Ein neues Oberflächenhärtungsverfahren

(Nitrogen in iron, steel, and special steel. A new method of surface hardening)." *Kruppsche Monatshefte* **4**, 137–151. Condensed: *Stahl Eisen,* **43**, 1271–1279.

Giesen, W. (1909). "The special steels in theory and practice. *Iron Steel Inst. Carnegie Schol. Mem.* **1**, 1–59.

Goerens, P., and Stadeler, A. (1907). "Über den Einfluss des Chroms auf die Lösungsfähigkeit des Eisens für Kohlenstoff und die Graphitbildung (The effect of chromium on the solubility of carbon in iron and on graphite formation)." *Metallurgie* **4**, 18–24.

Goldschmidt, H. (1897). German Patent 96,317. (1897). U.S. Patent 578,868.

Greaves, R.H. (1935). *Chromium Steels.* Her Majesty's Stationery Office, London.

Grenet, L. (1917). "The penetration of the hardening effect in chromium and copper steels."*J. Iron Steel Inst.* **95**, 107–117.

Griffiths, F.J. (1917). "The rôle of chrome vanadium. *Proc. Am. Soc. Test. Mat.* **17** (part 2), 33–44, 55–56. Condensed: *Iron Age* **100**, 266–268.

Guillet, L. (1904a). "La cémentation des aciers au carbone et des aciers spéciaux (Cementation of carbon and special steels)." *Mém. C. R. Soc. Ing. Civils de France* (part 1), 177–207.

Guillet, L. (1904b). "Les aciers au chrome (Chromium steels)." *Rev. Mét. Mém.* **1**, 155–183. Also: (1905). "Les aciers spéciaux (Special steels)." Dunod, Paris, **2**, 132 pp.

Guillet, L. (1904c). "Aciers au tungstène (Tungsten steels)." *Rev. Mét. Mém.* **1**, 263–283. Also: (1905). "Les aciers spéciaux (Special steels)." Dunod, Paris, **2**, 132 pp.

Guillet, L. (1905). "Comparaison des propriétés et classification des aciers ternaires (Comparison of the properties and classification of the ternary steels)." *Rev. Mét. Mém.* **2**, 350–367.

Guillet, L. (1906). "Quaternary steels." *J. Iron Steel Inst.* **70**, 1–141.

Guillet, L. (1922). "Les aciers au chrome et leurs récentes applications (Chromium steels and their recent uses)." *Rev. Mét. Mém.* **19**, 499–504.

Hadfield, R.A. (1915). "Introductory address on the corrosion of steel alloys." *Trans. Faraday Soc.* **11**, 183–197.

Hadfield, R.A. (1892). "Alloys of iron and chromium, including a report by F. Osmond." *J. Iron Steel Inst.* **II**, 49–175.

Hadfield, R.A. (1931). "A research on Faraday's 'steel and alloys.' " *Philosophical Trans. Royal Soc.,* ser. A, **230**, 221–242.

Hadfield, R.A. (1922). "The corrosion of iron and steel." *Proc. Royal Soc.,* ser. A, **101**, 472–486.

Hall, R.G. (1921–1922). "Stainless steels and the making of cutlery." *Trans. Am. Soc. Steel Treat.* **2**, 561–568.

Harley, R.D. (1970). *Artists' Pigments,* c. 1600–1835. Butterworths, London.

Hart, B.F., and Calish, J. (1889; 1889; 1892). "Analysis of chrome steel."*Iron* **33**, 78.*J. Iron Steel Inst.* No. 1, 376. *Stevens Indicator* **9**, 49.

Hatfield, W.H. (1925). "Modern developments in steels resistant to corrosion." *Engineering* **120**, 657–660.

Hatfield, W.H. (1919). "Cutlery, stainless and otherwise, from the scientific point of view, paper before Sheffield Cutlery Trades Tech. Soc., December 1919. Abst.: *Ironmonger* **169**, 91.

Haynes, E. (1920). "Stellite and stainless steels." *Proc. Eng. Soc. Western Pennsylvania* **35**, 467–474.

Haynes, E. (1917). "Wrought metal for tools." U.S. Patent 1,299,404.

Heuer, R.P., Trostel, L.J., and Grisby, C.E. (1956). "Chromium in refractories." In M.J. Udy, Ed., *Chromium.* Reinhold, New York, Vol. 2, pp. 327–390.

Hittorf, W. (1898; 1899). *Z. Physik. Chemie* **25**, 729; **30**, 481. Cited by M.A. Streicher (1977).

Howe, H.M. (1891). *The Metallurgy of Steel,* 2nd ed. Scientific Publishing Company, New York, 385 pp.

Jänecke, E. (1917). "Über die Konstitution der Eisen-Chromlegierungen (The constitution of iron-chromium alloys)." *Z. Elektrochemie* **23**, 49–55.

Johnson, C.M. (1920–1921). "Properties and microstructure of heat-treated, non-magnetic flame-, acids-, and rust-resisting steel." *Trans. Am. Soc. Steel Treat.* **1**, 554–575.

Johnson, C.M. (1924). "General discussion (of symposium on corrosion-resistant, heat-resistant, and electrical-resistance alloys)." *Proc. Am. Soc. Test. Mat.* **24** (part 2), 423–426.

Johnson, J.B., and Christiansen, S.A. (1924). "Characteristics of material for valves operating at high temperatures. *Proc. Am. Soc. Test. Mat.* **24** (part 2), 383–400.

Jörgensen, G.W., and Bruno, M.H. (1956). "Chromium chemicals in the graphic art." In M.J. Udy, Ed., *Chromium*. Reinhold, New York, Vol. 1, pp. 385–405.

Junot, C.J.E. (1848). French Patent 3564. (1854). U.K. Patent 1183.

Kayser, J.F. (1923–1924). "Heat- and acid-resisting alloys (Ni-Cr-Fe). *Trans. Faraday Soc.* **19**, 184–195.

Kelley, F.C. (1923). *Trans. Am. Electrochem. Soc.* **43**, 35.

Kern, S. (1875). "Alloys of iron and chromium." *J. Iron Steel Inst.*, No. 2, 654–673.

Kinzel, A.B., and Crafts, W. (1937). *The Alloys of Iron and Chromium.* McGraw-Hill, New York, Vols. 1 and 2.

Kinzel, A.B., and Franks, R. (1940). *The Alloys of Iron and Chromium.* New York, Vol. 2.

Klaproth, M.H. (1798). "Nachricht von einem neu entdeckten Metall aus dem rothen Sibirischen Bleyspathe (Report on a newly discovered metal from the red Siberian lead)." *(Crell's Annalen) Chemische Ann. Freunde Naturlehre*, part 1, 80.

Klopp, W.D. (1972). "Chromium-base alloys." In C. Sims and W. Hagel, Eds., *The Superalloys.* John Wiley & Sons, New York.

Krainer, H. (1962). *Techn. Mitt. Krupp-Werksberichte* **20**, 165–179.

Langård S. (1982). *Environmental and Biological Aspects of Chromium.* Elsevier, Amsterdam.

Lebasteur, H. (1880). *The Metals at the Paris International Exhibition of 1878.* Dulan and Co., London, pp. 152–162.

Mallet, R. (1838). "The action of water on iron." Report of the British Association for the Advancement of Science, 7, 253–261.

Marble, W.H. (1920–1921). "Stainless steel: Its treatment, properties, and applications." *Trans. Am. Soc. Steel Treat.* **1**, 170–179.

Mars, G. (1909). "Magnetstahl und permanenter Magnetismus (Magnet steel and permanent magnetism)." *Stahl Eisen* **29**, 1673–1678, 1769–1781.

Mathews, J.A. (1925). "Retained austenite—A contribution to the metallurgy of magnetism." *Trans. Am. Soc. Steel Treat.* **8**, 565–588.

Mathews, J.A. (1925). "Austenite and austenitic steels." *Trans. Am. Inst. Min. Met. Eng.* **71**, 568–596.

Mathews, J.A. (1925). "Retained austenite." *J. Iron Steel Inst.* **112**, 299–312.

Maurer, E., and Hohage, R. (1921). "Über die Wärmebehandlung der Spezialstähle im allgemeinen und der Chromstähle im besonderen (Heat treatment of special steels in general and of chromium steels in particular)." *Mitt. K.-W. Inst. Eisenforschung* **2**, 91–105.

MCC (1941). *Chromium Chemicals, Their Discovery, Development and Use.* Mutual Chemical Company, New York.

McWilliam, A., and Barnes, E.J. (1910). "Some physical properties of two per cent chromium steels." *J. Iron Steel Inst.* **81**, 246–267, 276–286.

Mellor, J.W. (1931). *A Comprehensive Treatise on Inorganic and Theoretical Chemistry.* Longmans, Green & Co., London, Vol. 11.

Moissan, H. (1893; 1894). "Préparation rapide du chrome et du manganèse à haute température

(Rapid preparation of chromium and manganese at high temperature)." *C. R. Acad. Sci. Paris* **116**, 349–351; *Bull. Soc. Chim. Paris,* ser. 3, **11**, 13–15.

Moissan, H. (1894). "Nouvelles recherches sur le chrome (New investigations on chromium)." *C. R. Acad. Sci. Paris* **119**, 185–191; *Bull. Soc. Chim. Paris,* ser. 3, **11**, 1014–1020.

Moissan, H. (1897). "Le four électrique." Steinheil, Paris, 1897; translated by Lenher, V. (1904). "The Electric Furnace." Chemical Publishing Company, Easton, Pa., p. 148.

Moissan, H. (1897). "Nouvelle méthode de préparation des carbures par l'action du carbure de calcium sur les oxydes (New method of preparing carbides by the reaction of calcium carbide with oxides). *C. R. Acad. Sci. Paris* **125**, 839–844.

Monnartz, P. (1909). "Verschmelzen von Chromeisenstein im Schachtofen mit sauerstoffreichem Winde (Smelting chromium iron ore in the blast furnace with an oxygen-enriched blast)." *Metallurgie* **6**, 160–167.

Monnartz, P. (1911). "Beitrag zum Studium der Eisen-Chromlegierungen unter besonderer Berücksichtigung der Säurebeständigkeit (Iron-chromium alloys with special consideration of resistance to acids)." *Metallurgie* **8**, 161–176.

Monypenny, J.H.G. (1920). "The structure of some chromium steels." *J. Iron Steel Inst.* **101**, 493–525.

Monypenny, J.H.G. (1920). "Stainless steel." *J. Soc. Chem. Ind.* **39**, 000–000.

Monypenny, J.H.G. (1951). *Stainless Iron and Steel,* 3rd ed. John Wiley & Sons, New York.

Moore, H. (1910). "The A_2 point in chromium steel." *J. Iron Steel Inst.* **81**, 268–286.

Murakami, T. (1918). "On the structure of iron-carbon-chromium alloys." *Sci. Rep., Sendai,* ser. 1, **7**, 217–276.

Oberhoffer, P., and Esser, H. (1927). "Zur Kenntnis des Zustandsdiagramms Eisen-Chrom (The iron-chromium diagram)." *Stahl Eisen* **47**, 2021–2031.

Odelstjerna, E.G. (1887). "Ingeniör E. G.:son Odelstjernas tjensteberättelse för år 1886 (Engineer Odelstjerna's report for 1886)." *Jernk. Ann.* 297–305.

Oertel, W., and Wurth, K. (1927). "Über den Einfluss des Molybdäns und Siliziums auf die Eigenschaften eines nichtrostenden Chromstahls (The effect of molybdenum and silicon on the properties of corrosion-resistant chromium steels)." *Stahl Eisen* **47**, 742–753.

Osmond, F. (1890). "On the critical points of iron and steel. *J. Iron Steel Inst.* **I**, 38–80.

Osmond, F. (1904). "Contribution à la théorie des aciers rapides (Contribution to the theory of high-speed steels)." *Rev. Mét. Mém.* **1**, 348–352.

Osmond, F. (1905). "À propos du rôle du chrome dans les aciers (Role of chromium in steels)." *Rev. Mét. Mém.* **2**, 798–799.

Percy, J. (1864). *Metallurgy—Iron and Steel.* John Murray, London, 1864, pp. 185–188.

Piwowarsky, E. (1925). "Die Gusseisenveredelung durch Legierungszusätze (The improvement of cast iron by alloy additions)." *Stahl Eisen* **45**, 289–297; *Foundry Tr. J.* **31**, 331–334, 345–346.

Portevin, A.M. (1909). "Contribution to the study of the special ternary steels." *Iron Steel Inst. Carnegie Schol. Mem.* **1**, 230–364.

Portevin, A.M. (1911). "Chromium steels." *Comptes rendus* **153**, 64; *Iron Age,* July 3, 1911, and Feb. 8, 1912.

Rapatz, F. (1925). *Die Edelstähle, ihre metallurgischen Grundlagen* (High-quality steels). Julius Springer, Berlin, 219 pp.

Riley, E. (1877). "On chromium pig iron made by the Tasmanian Iron Company." *J. Iron Steel Inst.* No. 1, 104–115.

Rollinson, C.L. (1973). "Chromium." In J.C. Baillar, Ed., *Comprehensive Inorganic Chemistry.* Pergamon, Oxford, pp. 623–629.

Rosenberg, S.J., Darr, J.H. (1948). "Stabilization of austenitic stainless steel." *J. Research,* National Bureau of Standards, R. P. 1878, pp. 321–346.

Russell, T.F. (1921). "On the Constitution of chromium steels." *J. Iron Steel Inst.* **104**, 247–295.

Sargent, G.J. (1920). "Electrolytic chromium." *Trans. Am. Electrochem. Soc.* **37**, 479–497.

Schneidewind, R. (Nov., 1927). "A study of patents dealing with the electrodeposition of chromium." Dept. of Eng. Res., Univ. of Michigan, *Bull.* **8**, 49 pp.

Schneidewind, R. (1928). "A study of chromium plating." Dept. of Eng. Res., Univ. of Michigan, *Bull.* **10**, 141 pp.

Searle, A.B. (1924). *Refractory Materials, Their Manufacture and Use.* Griffin, London.

Seidell, L.R., and Horvitz, G.J. (1919). "Physical qualities of high-chrome steel." *Iron age* **103**, 291–294.

Sigrist, J., Winkler, P., and Wantz, M. (1924). *Helv. Chim. Acta* **7**, 968.

Sklodowska-Curie, M. (1898). "Propriétés magnétiques des aciers trempés (Magnetic properties of quenched steels)." *Bull. Soc. d'Encouragement pour l'Ind. Nat.* **97**, 36–76.

Snelus, G.J. (1974). "On the manufacture and use of spiegeleisen." *J. Iron Steel Inst.* No. 1, 68, discussion p. 82.

Stodart, J., and Faraday, M. (1820). "Experiments on the alloys of steel, made with a view to its improvements." *Q. J. Science, Literature and the Arts* **9**, p. 319–332.

Stodart, J., and Faraday, M. (1822). "On the alloys of steel." *Philosophical Trans. Royal Soc.* **112**, 253–270.

Strauss, B. (1914). "Microscopic investigation of steel." *Zeitschrift Ange. Chemie* **27**, 633–639.

Strauss, B. (1928). Presentation at the Inst. Metals Div., American Inst. Mining Met. Engineers, Oct. 10. Cited by Thum (1935), p. 5.

Struass, B., and Maurer, E. (1920). "Die hochlegierten Chromnickelstähle als nichtrostende Stähle (Stainless high chromium-nickel steels)." *Kruppsche Monatschefte* **1**, 126–146.

Streicher, M.A. (1977). "Stainless steels: Past, present and future." In *Stainless Steel '77.* Climax Molybdenum Inc., New York.

Sully, A.H., and Brandes, E.A. (1967). *Chromium,* 2nd ed. Butterworths, London.

Thompson, D.V. (1956). *The Materials and Techniques of Medieval Painting.* Dover Publications, New York.

Thum, E.E. (1935). *The Book of Stainless Steels.* American Society for Metals, Cleveland, Ohio.

Treitschke, W., and Tammann, G. (1907). "Über die Legierungen des Eisens mit Chrom (The alloys of iron and chromium)." *Z. Anorg. Chemie* **55**, 402–411.

Udy, M.J. (1956). *Chromium.* Rheinhold, New York, 2 vols.

Ufer, C.E. (1850). "Über das Stickstoffchrom (Chromium nitride)." *Ann. Chemie Physik* **112** (new ser. **36**), 281–303.

Vancoram Review. (1959). **14** (2), 14.

Vanick, J.S. (1922). "The mechanical properties of some chrome-vanadium steels." *Trans. Am. Soc. Steel Treat.* **3**, 196–217.

Vauquelin, L.N. (1797). "Analyse du plomb rouge de Sibérie et expériences sur le nouveau metal qu'il contient (Analysis of the red lead of Siberia and investigations of the new metal which it contains)." *J. Mines,* ser. 1, **6**, 737–760.

Vauquelin, L.N. (1809). "Memoire sur la meilleure methode pour decomposer le chromate de fer, obtenir l'oxide de chrome, preparer l'acide chromique, et sur quelques combinaisons de ce dernier." *Ann. Chimie* **70**, 70–94.

Voss, G. (1908). "Die Legierungen: Nickel-Zinn, Nickel-Blei, Nickel-Thallium, Nickel-Wismut, Nickel-Chrom, Nickel-Magnesium, Nickel-Zink und Nickel-Cadmium (Nickel-tin, nickel-lead, nickel-thallium, nickel-bismuth, nickel-chromium, nickel-magnesium, nickel-zinc, and nickel-cadmium alloys)." *Z. Anorg. Allgem. Chemie* **57**, 34–71.

Whitman, W.G., and Chappell, E.L. (1926). "Corrosion of steels in the atmosphere." *J. Ind. Eng. Chem.* **18**, 533–535.

Williams, P. (1898). "Sur un carbure double de fer et de tungstène (A double carbide of iron and tungsten)." *C. R. Acad. Sci. Paris* **127**, 410–412.

Wöhler, F. (1859). "Leichte Darstellungsweise des metallischen Chroms (An easy method of producing metallic chromium)." *Ann. Chemie Pharmacie* **111** (new ser. **35**), 230–234.

Woods, J.E.T., and Clark, J. (1872). "Weather resisting alloy." British Patent No. 1923.

Wyche, E.H., and Smith, R. (April 9, 1946). "Precipitation hardenable austenitic stainless steels." U.S. Patent 2,397,997.

Zapffe, C.A. (1949). *Stainless Steels.* American Society for Metals, Cleveland, Ohio, pp. 1–27.

2

BIOLOGIC CHEMISTRY OF CHROMIUM

E. Nieboer

A.A. Jusys

Department of Biochemistry and Occupational Health Program
McMaster University
Hamilton, Ontario, Canada

1.   INTRODUCTION

Chromium toxicity is dependent on chemical speciation, and thus associated health effects are influenced by the chemical form of exposures. The flow chart provided in Figure 1 indicates that, in industry, workers are exposed to a limited number of chromium compounds. Included in this list, or implied by it, are the following hexavalent compounds: sodium chromate (Na_2CrO_4) and sodium dichromate ($Na_2Cr_2O_7 \cdot 2H_2O$), calcium chromate ($CaCrO_4 \cdot 2H_2O$), the pigment chromates including those of zinc (e.g., $ZnO \cdot ZnCrO_4 \cdot xH_2O$, $3ZnCrO_4 \cdot K_2CrO_4 \cdot Zn(OH)_2 \cdot 2H_2O$) and lead (e.g., $PbCrO_4$, $PbO \cdot PbCrO_4$; see Table 1), and the chromium(VI) oxide (CrO_3) that colloquially is called *chromic acid*.

PRODUCTION OF CHROMIUM METAL ALLOYS
AND COMPOUNDS FROM CHROMITE ORE

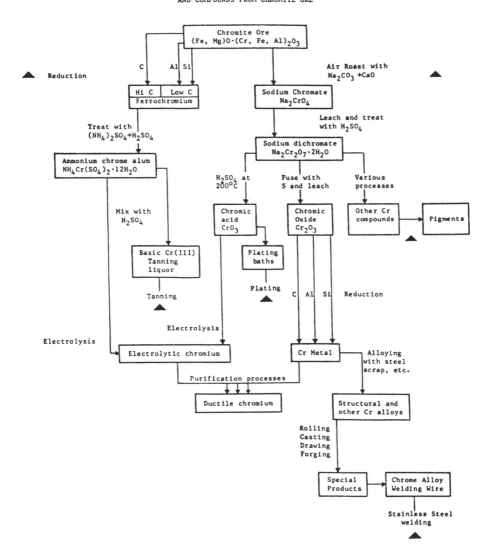

Figure 1. A simplified flow chart summarizing the steps in the production of metallic chromium and some of the important commercial chromium compounds. Processes for which occupational exposure data are available are denoted by the symbol ▲. Basic tanning liquor is often derived from sodium dichromate, sulfuric acid, and a reducing agent. (*Source:* Stern, 1982.)

Table 1 Chemical Composition and Analytic Specifications for Lead Chromate Color Pigments

	$PbCrO_4$, %	$PbSO_4$, %	Other components, %
Primrose chrome yellow			
Theoretical	77.3	22.7	—
ASTM[a] spec, minimum	50	—	—
Actual			
Minimum	52.0	4.2	—
Maximum	82.7	25.9	—
Lemon chrome yellow			
Theoretical	72.7	27.3	—
ASTM spec, minimum	65	—	—
Actual			
Minimum	52.4	17.4	—
Maximum	68.8	39.0	—
Medium chrome yellow			
Theoretical	100.0	—	—
ASTM spec, minimum	87	—	—
Actual			
Minimum	82.4	nd[b]	—
Maximum	98.2	nd	—
Chrome orange			
Theoretical	59.2	—	PbO, 40.8
ASTM spec, minimum	55	—	—
Actual	58.2	—	PbO, 39.4
Molybdate orange			
Theoretical	82.3	2.8	$PbMoO_4$, 14.9
ASTM spec, minimum	70	—	—
Lead Silicochromate			
Theoretical	50.0	—	SiO_2, 50.0

[a] ASTM, American Society for Testing and Materials.
[b] nd = not determined.
Source: Hartford (1979).

Exposure to trivalent or elemental chromium is in the following forms: chromite ore ($FeO \cdot Cr_2O_3$) and calcium chromite ($CaCr_2O_4$), chromium(III) oxide (Cr_2O_3), basic chromium sulfates [e.g., $Cr(OH)(SO_4)$, $Cr_2(OH)_4(SO_4)$], and chromium metal and alloys (Gafafer, 1953; Foley, 1978; Morgan, 1978; Hartford, 1979; Stern, 1982).

Fundamental to a study of chromium toxicology is an understanding of the chemistry, biochemistry, pharmacodynamics, and biologic function of chromium compounds. This is the goal of this chapter. In addition, the analytic technology required for the determination of ultratrace concentrations of chromium in body fluids and tissues is discussed in the context of biological monitoring of exposed people.

2. CHROMIUM CHEMISTRY

2.1 Stable Oxidation States

Chromium exists in a number of oxidation states, not all of which are of equal stability (Fig. 2). As indicated in the legend for Figure 2, thermodynamic energy considerations show that positive values of the standard electrode potential ($E°$) indicate that the reduced form is favored, whereas negative $E°$ values denote that the oxidized species is relatively more stable.* Consequently, it is clear from this

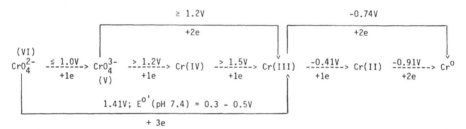

Figure 2. Reduction potential diagram for chromium. The standard electrode potential in volts ($E°$) is indicated above the arrow linking two species and refers to rather concentrated solutions (standard state, of unit activity; ≥ 1 M); for more dilute solutions, the E values are more negative. Of the two species linked by an arrow, the reduced form at the right of the arrow is favored when $E°$ is positive because of energy considerations (see Table 2, footnote a, for additional explanation). Conversely, that at its left is more stable when $E°$ is negative. For CrO_4^{2-} and CrO_4^{3-}, the $E°$ value is pH dependent because protons are consumed in the reaction; as shown for CrO_4^{2-}, its value (given the symbol $E°'$ for pH $\neq 0$) is ~0.5V at pH 7.4 (see Table 2, footnote a, for supplemental rationale). Diagram was constructed from data reported in Westheimer (1949), Beattie and Haight (1972), and Cotton and Wilkinson (1980).

*Reduction corresponds to a gain of electrons, whereas oxidation is accompanied by a loss of electrons. In the absence of kinetic limitations, the oxidizing power increases with the numerical (positive) value of $E°$ (see Table 2 and Fig. 2 for additional explanation).

reduction potential diagram that Cr(III) is the most stable form of chromium in solution, since considerable energy would be required to convert it to lower or higher oxidation states. Although CrO_4^{2-} is relatively stable, its high positive reduction potential denotes that it is strongly oxidizing and is unstable in acid solution in the presence of electron donors, such as Fe^{2+}, H_3AsO_3, HSO_3^-, and all organic molecules with oxidizable groups (alkanes, alkenes, alcohols, aldehydes, ketones, carboxylic acids, mercaptans, etc.) (Wiberg, 1965; Beattie and Haight, 1972). At physiologic pH values or higher, CrO_4^{2-} is less reactive, as indicated by the relatively low $E^{\circ\prime}$ values in Table 2 (see also Fig. 2)[†].

Table 2   Oxidizing Power of Chromate in Relation to pH[a]

	$E^{\circ\prime}$	pH
Weakly	− 0.13	14
oxidizing	− 0.04	13
	+ 0.16	11
	+ 0.36	9
	+ 0.46	8
	+ 0.52 $(0.31)^b$	7.4 (physiologic pH)
	+ 0.56 $(0.36)^b$	7
	+ 0.66 $(0.54)^b$	6
	+ 0.76 $(0.66)^b$	5

Basic solution:
$$CrO_4^{2-} + 4H_2O + 3\ e^- \rightleftharpoons Cr(OH)_3(s) + 5\ OH^-\ (E^\circ = -0.13V)$$

	$E^{\circ\prime}$	pH
	+ 0.79	4
	+ 0.93	3
	+ 1.07	2
Strongly	+ 1.21	1
oxidizing	+ 1.35	0

Acidic solution:
$$HCrO_4^- + 7\ H^+ + 3\ e^- \rightleftharpoons Cr^{3+} + 4H_2O\ (E^\circ = +1.35V)$$

[a]The formal potential $E^{\circ\prime}$ (which refers to nonstandard conditions) was calculated employing the Nernst equation for the reactions shown. In basic solution, the reduction of CrO_4^{2-} generates OH^- ions against a gradient. This is costly from an energy point of view, thus destabilizing the Cr(VI) state relative to the Cr(VI) state (as denoted by the negative $E^{\circ\prime}$ values). In acid solution, the reduction of $HCrO_4^-$ is accompanied by the consumption of H^+ ions, and therefore increased acidity pushes the reaction from left to right as written (as denoted by the positive $E^{\circ\prime}$ values). Supplemental explanatory information is provided in the legend to Figure 2.

[b]Potential-pH equilibrium diagrams show that the hydrated Cr^{3+} ion cannot exist above pH 4 because of hydrolysis (Cary, 1982). In the absence of chelating agents, $Cr(OH)^{2+}$ is the predominant species at pH 5 and 6, $Cr(OH)_2^+$ at physiologic pH values around 7, and $Cr(OH)_3$ at pH ≥ 8. For the $E^{\circ\prime}$ values in parentheses, this speciation and the presence of $HCrO_4^-$ were taken into consideration.

[†]$E^{\circ\prime}$ is the *formal* electrode potential and refers to nonstandard conditions, whereas E°, the *standard* electrode potential, corresponds to solutions of unit activity (concentrations ≥ 1 mol L^{-1} for most species).

Presumably, it is for this reason that gross cellular destruction does not occur in bioassays. Chromium(V) species are derived from the anion CrO_4^{3-} and are long-lived enough to be observed directly. However, there are relatively few stable compounds containing Cr(V). Chromium(IV) compounds are even less common. The Cr(IV) ion and its compounds are not very stable and, because of short half-lives, defy detection as reaction intermediates between CrO_4^{2-} and Cr(III). It should be emphasized that CrO_4^{2-} and CrO_4^{3-} are anions (like carbonate, CO_3^{2-}, and sulfate, SO_4^{2-}) and, in fact, can behave as ligands (e.g., like ammonia, NH_3, EDTA, CO_3^{2-}) toward metal ions [e.g., Cu(II), Ni(II), Fe(III), or even the Cr(III) ion]. Chromium(IV) appears to react as a metal ion. This distinction between anionic and cationic forms of chromium is essential to the understanding of their divergent biologic effects.

The negative $E°$ value for the Cr(III)/Cr(II) metal-ion couple (see Fig. 2) signifies that Cr(II) is a strong reducing agent [it readily gives up electrons and thus is itself oxidized to Cr(III)]. Only under stringent anaerobic conditions is it stable in aqueous solution. Even then, it slowly decomposes by reducing water, with liberation of hydrogen gas (Cotton and Wilkinson, 1980, p. 724).

2.2 Chromium(VI) Chemistry

2.2.1 Chromate Speciation

Chromium(VI) is strongly oxidizing and exists only in oxo species, such as CrO_3 [Cr(VI) oxide, chromium trioxide, colloquially referred to as chromic acid], CrO_2Cl_2 (chromyl chloride), and CrO_4^{2-} (chromate ion). CrO_3 is highly soluble in water (625.3 g L^{-1} or 6.25 mol L^{-1} at 20°C) and, in basic solution, forms the yellow chromate ion; dissolution in acidic medium yields a number of pH- and concentration-dependent species. As may be deduced from the acid dissociation constant in Equation (1), H_2CrO_4 is a strong acid ($pK_{a1} = -0.6$) and is the main species below pH -0.6 (Cotton and Wilkinson, 1980, p. 733).

$$H_2CrO_4 \rightleftharpoons H^+ + HCrO_4^- \qquad K_{a1} = 10^{0.6} \qquad (1)$$

$$HCrO_4^- \rightleftharpoons H^+ + CrO_4^{2-} \qquad K_{a2} = 10^{-5.9} \qquad (2)$$

By contrast, the monohydrogen chromate ion ($HCrO_4^-$) is a weak acid ($pK_{a2} = 5.9$; Equation 2) and is the primary species between pH 1 and pH 6, with CrO_4^{2-} predominating at pH greater than 6. At concentrations of Cr(VI) above $10^{-2} M$, the $HCrO_4^-$ ion dimerizes to give the dichromate ion (Baes and Mesmer, 1976; Connett and Wetterhahn, 1983; see Equation 3).

$$Cr_2O_7^{2-} + H_2O \rightleftharpoons 2HCrO_4^- \qquad K = 10^{-2.2} \qquad (3)$$

In biologic systems, the chromate concentrations encountered are likely to be considerably below $10^{-2} M$, and thus the existence of dichromate is not expected to be significant, especially at physiologic pH values.

2.2.2 Water Solubility of Metal Chromates

Since the carcinogenic potency of chromates has been linked to their water solubility, a brief review of the available data is presented. Metal chromates may be divided into three arbitrary categories (Table 3). The first group, designated as sparingly soluble to insoluble, includes chromates of zinc, lead, barium, and strontium. Of intermediate solubility are calcium chromate, potassium dichromate, and apparently industrial strontium chromate pigments. It should be emphasized that solutions of calcium chromate and potassium dichromate of significant concentration (e.g., 0.2–1.0 mol L^{-1}) are possible at ambient temperature. The third solubility category includes the highly soluble potassium and sodium chromates, as well as sodium and calcium dichromates.

Biologic media, such as serum and cytosol, have a high affinity for metal ions (e.g., Zn^{2+}, Pb^{2+}, Ba^{2+}) and contain reducing agents that can react with chromate. For these reasons, the solubility of metal chromates may be expected to be augmented in biologic solutions. The enhancement of solubility by complex formation of the released metal ion has been demonstrated clearly for the dissolution of nickel compounds in rat serum and renal cytosol (Kuehn and Sunderman, 1982). No similar studies have been reported for metal chromates.

2.2.3 Chromium(VI) Oxide

It is evident from the flow chart in Figure 1 that Cr(VI) oxide (CrO_3, chromic acid) is an important industrial intermediate (see also IARC, 1980). As mentioned in Section 2.2.1, the solid CrO_3 (mp 197°C) dissolves readily in water, yielding chromic acid (H_2CrO_4) and its deprotonated forms ($HCrO_4^-$, CrO_4^{2-}), depending on pH. It is the acid anhydride of H_2CrO_4 and is prepared industrially by the dehydration with sulfuric acid of sodium dichromate (Fig. 1) (Brown, 1974). Because of its high solubility and because the dissolved forms are powerful oxidizing agents (see Section 2.2.4), CrO_3 is a primary irritant. Contact with combustible material may cause fire (Windholz et al., 1983). The special precautions required in handling this material have been outlined by the National Institute for Occupational Safety and Health (NIOSH, 1973, 1975).

2.2.4 Chemical Reactivity of the Chromate Ion

Extensive use has been made of chromic acid as an oxidizing agent in both organic and inorganic chemistry. Comprehensive reviews on these subjects are available (Wiberg, 1965; Espenson, 1970; Beattie and Haight, 1972). A perusal of these reviews suggests that strongly acidic chromate solutions are powerful oxidants. As illustrated in Section 2.1, the oxidizing power of chromate is considerably diminished at physiologic pH values. Consequently, a legitimate question is whether the versatility demonstrated in acidic solution applies to

Table 3 Water Solubility of Metal Chromates[a]

Compound	Molecular Weight (amu)	g · L^{-1}	Moles · L^{-1} (of CrO$_4^{2-}$)	Temperature (°C)
		Insoluble to Sparingly Soluble		
PbCrO$_4$ · PbO	546.4	Insoluble	Insoluble	25
PbCrO$_4$	323.2	$5.8 \cdot 10^{-4}$	$1.8 \cdot 10^{-6}$	25
Pb$_2$(OH)$_2$CrO$_4$	564.4	0.14	$2.5 \cdot 10^{-4}$	17
Lead chromate pigments	Variable	$2\text{--}45 \cdot 10^{-3}$ (as CrO$_4^{2-}$)[b]	$2\text{--}39 \cdot 10^{-5}$	—
BaCrO$_4$	253.3	$4.4 \cdot 10^{-3}$[b]	$1.7 \cdot 10^{-3}$	28
4ZnO · ZnCrO$_4$ · XH$_2$O (tetroxychromate)	507 (anhydrous)	2.2[b]	$4.3 \cdot 10^{-3}$	—
SrCrO$_4$	203.6	1.2	$5.9 \cdot 10^{-3}$	15
3ZnCrO$_4$ · K$_2$CrO$_4$ · Zn(OH)$_2$ · 2H$_2$O (zinc yellow)	873.8	1.8[b]	$8.2 \cdot 10^{-3}$	—
		Intermediate Solubility		
CaCrO$_4$	156.1	22.3[c]	0.14	20
K$_2$Cr$_2$O$_7$	294.2	49	0.17	0
		(1020)	(3.5)	(100)
CaCrO$_4$ · 2H$_2$O	192.1	163	0.85	20
Strontium chromate pigments	~204	$250\text{--}810$[b]	$1.2\text{--}3.9$	—
		Highly Soluble		
K$_2$CrO$_4$	194.2	629	3.2	20
Na$_2$CrO$_4$	162.0	873	5.4	30
Na$_2$Cr$_2$O$_7$ · 2H$_2$O	298.0	940[b]	6.3	—
CaCr$_2$O$_7$ · 4.5 H$_2$O	337.1	1939[c]	11.5	30

[a]When values are calculated from the solubility products quoted in Sillén and Martell (1964), there are some discrepancies; note that the unit moles of CrO$_4^{2-}$ per liter of H$_2$O does not correspond to molarity (moles of solute per liter of solution) for concentrated solutions.

[b]From Levy et al., 1986.

[c]From Hartford et al., 1950.

Source: Weast, 1973, and IARC, 1980.

biologic systems. Data have indicated that, at pH 7.4, chromate is indeed capable of oxidizing both low-molecular-weight reductants (e.g., cysteine, ascorbate, glutathione, lipoic acid) and proteins. These reactions adhere to second order kinetics and thus are dependent on the concentration of both the chromate and the reductant (Connett and Wetterhahn, 1983). Since the concentrations of the mentioned low-molecular-weight reductants are often in the range 10^{-5} to 10^{-1} M in biologic fluids, the published rate constants predict reaction half-times in the range of minutes to hours when chromate is present in comparable concentrations. Connett and Wetterhahn (1983) speculate that because of the high intracellular levels of glutathione ($0.8–8$ mmol L^{-1}), it may be the most effective reagent for converting chromate to Cr(III). We may conclude, therefore, that chromate is chemically unstable at physiologic pH values but does not decompose as easily as expected on the basis of the highly positive standard reduction potential so frequently quoted in health assessment documents ($E° = 1.33V$). Consequently, a biologic half-life in hours for the chromate ion is not unrealistic.

Chemically, it is difficult to displace the oxygen atoms from the chromate ion, and it is said to be nonlabile kinetically. In contrast, the hydroxyl group of the monoprotonated form of chromate ($HCrO_4^-$; Fig. 3) is more readily replaced. In fact, $HCrO_4^-$ forms esters with other species containing –OH or –SH groups; formation of the dichromate ion is a prime example. It can be seen from Equation (3) that this involves the elimination of water from the two units being linked. Adducts are known to form with oxyacids (e.g., HSO_4^-, HSO_3^-, $HS_2O_3^-$, $H_2PO_4^-$), alcohols (R–OH), thiols (R–SH), carboxylic acids (R–COOH), and other anions (e.g., Cl^-, SCN^-) (Wiberg, 1965; Beattie and Haight, 1972; Connett and Wetterhahn, 1983). Ester formation is favored in acid solution but is less important at physiologic pH values, presumably because of the low concentration of the monoprotonated form of chromate. Nevertheless, recent kinetic studies suggest that the rate-determining step in the oxidation of thiols by chromate at physiologic pH is the rate of thioester formation. Consequently, chromate esters may occur as reaction intermediates and, in this manner, could be of toxicologic consequence.

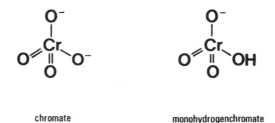

chromate   monohydrogenchromate

Figure 3. Structural formulae for the biologically important forms of Cr(VI) at the physiologic pH of 7.4; $HCrO_4^-$ constitutes 3% of the total, and the rest exists as CrO_4^{2-}.

2.2.5   Chromyl Chloride [Dichlorodioxochromium(VI)]

Chromyl chloride, CrO_2Cl_2, is used extensively in organic synthesis (Wiberg, 1965). It is prepared by the action of hydrogen chloride on chromic(VI) oxide or by warming dichromate with an alkali metal chloride in concentrated sulfuric acid (Cotton and Wilkinson, 1980, p. 734). Because of these reactions, the use of chromic–sulfuric acid cleaning solution may yield chromyl chloride (King, 1978). This compound is an unusual Cr(VI) derivative in that it is a liquid at room temperature (bp 116°C) and is soluble in organic solvents (Hartford and Darrin, 1958). It undergoes hydrolysis on contact with water, yielding H_2CrO_4 (see Equation 4) or its dissociation products, $HCrO_4^-$ and CrO_4^{2-}.

$$CrO_2Cl_2(1) + 2H_2O(1) \rightleftharpoons H_2CrO_4(aq) + 2HCl(g) \qquad (4)$$

This reaction is exothermic and explains the fuming of CrO_2Cl_2 in moist air. Furthermore, as an oxidant, it can react explosively with combustible substances. Hydrolysis and oxidation probably are the cause of the burns and blisters on contact with skin (or other tissues).

Chromyl compounds of other halides (especially fluoride) and of other anions (e.g., nitrate, cyanate, acetate) are known (Hartford and Darrin, 1958; Wasson, 1976). These chromyl compounds, like the chloride derivative, are volatile materials, although there are wide differences in their boiling points and volatilities. Chromyl chloride serves as a prototype of the chemical activity of these compounds. They are important synthetically, although it is difficult to assess the extent of their employment in industry. Extreme care must be exercised when working with chromyl compounds (e.g., containment in a high performance fumehood) because of their high volatility and chemical reactivity.

2.3   Chromium(III) Chemistry

2.3.1   Kinetics of Chromium(III) Complex Formation

As described in Section 2.2, the chemistry of Cr(VI) is essentially that of the chromate anion (CrO_4^{2-}). Other than in acidic solutions, it is difficult to displace the oxygen atoms from this anion. Consequently, the immediate environment around Cr(VI) (referred to as the *coordination polyhedron*) is fixed, that is, occupied by oxide or hydroxide groups. The affinity of Cr(III) for the oxide and hydroxide ions is considerably lower, and numerous complexes with other ligands are possible. Complexes are known with ligands for which the point of attachment is the oxygen atom (e.g., water, hydroxyl ion, carboxylic acids, phenols, phosphates), the nitrogen atom (e.g., ammonia, amines, pyridines, purines, pyrimidines, imidazoles), or the sulfur atom (e.g., sulfhydryl groups). It is largely because of kinetic inertness that so many complex species of Cr(III) can be isolated or are stable in solution. In fact, Cr(III) species are considered nonlabile, and ligand displacement is slow (hours to days at room temperature) compared to most metal ions ($10^{-9}-10^{-3}$ seconds at room temperature) (Cotton and Wilkinson, 1980, pp. 727, 1188).

One should not conclude, however, that reactions of the hydrated Cr(III) ion, $Cr(OH_2)_6^{3+}$, or of other complex species of Cr(III) do not occur. Under selected conditions (e.g., in the presence of a catalyst, by ligand-assisted labilization, or at elevated temperatures), even substitution-inert complexes may be induced to exchange their ligands somewhat more rapidly, and this characteristic has been used synthetically for Cr(III) (Garner and House, 1970; De Pamphilis and Cleland, 1973; also see Sulfab et al., 1976). At physiologic temperature (37°C), Cr(III) reactions are faster by a factor of 4–5 than at ambient temperatures (20–25°C). Because of this and because suitable catalysts are available (e.g., oxidizing–reducing agents), reactions of Cr(III) are plausible in mammalian systems. As explored in Section 3, it may be surmised that the nonlability of Cr(III) complexes is of special biologic significance.

2.3.2 Types of Chromium(III) Complexes

Like other trivalent metal ions [e.g., Fe(III), Al(III)], the hydrated Cr(III) ion, $Cr(OH_2)_6^{3+}$, has a tendency to hydrolyze, which is often accompanied by polymerization. Hydrolysis involves the conversion of a bound water molecule to the hydroxide ion and results in the release of a proton. Equilibrium measurements have identified, among others, the existence of the following species: $Cr(OH)^{2+}$, $Cr(OH)_2^+$, $Cr(OH)_3(s)$, $Cr_2(OH)_2^{4+}$, $Cr_3(OH)_4^{5+}$, and $Cr_4(OH)_6^{6+}$ (Smith and Martell, 1976; Baes and Mesmer, 1976; Connett and Wetterhahn, 1983). In the polymeric species, the chromium ions are linked by hydroxyl bridges. At pH 4, about one-half of the hydrated Cr(III) ions are present as the monohydroxo complex, $Cr(OH)^{2+}$. To prevent this hydrolysis process, Cr(III) solutions must be stored under acidic conditions (pH $\leq$ 3).

Chromium(III), like other first-row transition metal ions, may be designated as a borderline ion that possesses considerable class A character. The borderline classification denotes that this ion has the ability to combine with O-, N-, and S-containing ligands, in which the designated donor atoms are the point of attachment (Nieboer and Richardson, 1980; Nieboer and Sanford, 1985). This catholic affinity is the basis for the multiplicity of possible complexes already alluded to (see Section 2.3.1). The significant class A character of Cr(III), like that of Fe(III), signifies that the stability of its complexes follows in the order F^- $> Cl^- > Br^-$ and O ligands $>$ S ligands (Earley and Cannon, 1965). Class B ions, such as Hg(II), have the opposite ligand preference. Thus, Cr(III) and Fe(III) often behave like Al(III), Ca(II), and Mg(II), which, in biologic systems, predominantly seek out charged oxygen centers (Nieboer and Richardson, 1980; Nieboer and Sanford, 1985). The high affinity for the OH^- and PO_4^{3-} ions is consistent with this (Rollinson et al., 1967; Mertz, 1969).

From a biochemical viewpoint (see Section 3), the similarities and differences in atomic parameters between Cr(III) and Fe(III) are important. Chromium(III) exists exclusively with octahedral geometry (i.e., is surrounded by six donor atoms), whereas Fe(III) complexes can have four, five, or six coordinates. The octahedral Cr(III) and Fe(III) ions are of comparable size, with ionic radii of 76 picometers (pm) and 79 pm (high-spin state), respectively. These similarities in charge and size account for the observation that Cr(III) replaces Fe(III) in a number of biomolecules (see Sections 3.2 and 3.4.2).

2.3.3 Solubility of Chromium(III) Complexes

Chromium(III) oxide (Cr_2O_3), even when hydrated ($Cr_2O_3 \cdot xH_2O$), is insoluble in water. Chromium(III) phosphate complexes are also virtually insoluble. Anhydrous complexes of common anions have low water solubility [e.g., $CrBr_3$, $CrCl_3$, $Cr_2(SO_4)_3$],* whereas corresponding hydrated forms have considerable water solubility [e.g., $(Cr[H_2O]_6)Br_3$, $(Cr[H_2O]_4Cl_2)Cl \cdot 2H_2O$, $Cr_2(SO_4)_3 \cdot 15H_2O$]. Salts of acetate, oxalate, and nitrate, which occur only in hydrated forms, are also soluble (Weast, 1973; IARC, 1980).

2.4 Chemistry of Less Stable Oxidation States

Judging by the lack of published documentation, industrial exposure to Cr(II) compounds appears not to be a serious concern. Anhydrous Cr(II) salts (e.g., chloride, acetate, formate) are relatively stable (Wasson, 1976; Hartford, 1979; Windholz et al., 1983). They are powerful reducing agents (see Section 2.1) and may be oxidized to Cr(III) by water, with the liberation of hydrogen gas (H_2), depending on the acidity and the anion present. In addition, reaction with molecular oxygen (O_2) is rapid, and in fact Cr(II) solutions are convenient O_2 absorbants in the purification of laboratory gases. It is obvious that because of this reactivity, Cr(II) solutions must be maintained anaerobically to minimize exposure to these strong irritants. Recent equilibrium studies indicate that Cr(II) complexes of bidentate ligands have formation constants similar to Zn(II), whereas with ligands of higher denticity, the magnitudes are comparable to those for Mn(II) (Micskei et al., 1983).

As indicated in Section 2.1, Cr(V) and Cr(IV) are relatively unstable and usually require special handling procedures (Nag and Bose, 1985). Although few complex compounds of chromium in these oxidation states have been isolated and fully characterized (Wasson, 1976; Cotton and Wilkinson, 1980, p. 731), there has been a steady growth in their known chemistry (Nag and Bose, 1985; Mitewa and Bontchev, 1985). Only a few such complexes are stable in air and dissolve in water without disproportionation to Cr(III) and Cr(VI) (Gould, 1986). Other than Cr(IV) oxide, CrO_2, which is used in magnetic recording and storage devices, very few compounds have industrial use. Considerable evidence has accumulated that both Cr(V) and Cr(IV) are formed as transient inter-

---

*Some of these chromic salts exist in two forms, a soluble and an insoluble modification.

mediates in the reduction of Cr(VI) (i.e., chromates) to Cr(III). Of these intermediates, those of Cr(V) are somewhat more stable. Diol ligands are known to stabilize this oxidation state most effectively (Mitewa and Bontchev, 1985). The magnitude of the half-lives amounts to minutes or longer, allowing characterization by spectrometric techniques (Wiberg, 1965; Beattie and Haight, 1972; Wetterhahn Jennette, 1982). Because of the inherent instability of these oxidation states, it is conceivable that they are able to generate cytotoxic and genotoxic radicals, such as those derived from peroxide (e.g., the hydroxyl radical, $\cdot OH$, and superoxide anion, O_2^-). Active oxygen species of this type have been incriminated in tumor promotion (Ames, 1983; Marx, 1983).

Numerous organochromium compounds have been synthesized and characterized (Sneeden, 1975; Wasson, 1976; Hartford, 1979). (Organometallic compounds contain a distinct metal–carbon bond.) In many of the compounds, chromium has a low valency (-2, -1, 0, $+1$), although derivatives of Cr(II), Cr(III), and Cr(IV) are also common. The majority of organochromium compounds are prepared under stringent conditions: in organic solvents and under anaerobic and anhydrous conditions. Many of the substances are insoluble in water and are air and moisture sensitive, although a small number are water soluble and stable in aqueous media. Chromium hexacarbonyl $[Cr(CO)_6]$ is considerably less volatile than nickel carbonyl $[Ni(CO)_4]$ and iron pentacarbonyl $[Fe(CO)_5]$, even though it sublimes at room temperature (Wender and Pino, 1968; King, 1976; Stokinger, 1981). Brief et al. (1971) suggest that because of this low volatility, chromium carbonyl poses less of an occupational hazard than its nickel and iron counterparts. Although metal carbonyls are used in the preparation of metals of high purity, as catalysts for organic reactions, and as gasoline additives, the specific industrial use of chromium carbonyl appears to be low compared to that of manganese carbonyl, iron carbonyl, cobalt carbonyl, and nickel carbonyl (Wender and Pino, 1968; Brief et al., 1971; Windholz et al., 1983; Stokinger, 1981). A chromium carbonyl LD_{50} in mice of $30 \, mg \, kg^{-1}$ (IV) has been reported (NIOSH, 1977), compared to about $50 \, mg \, kg^{-1}$ (IV) for nickel carbonyl (NIOSH, 1977; Stokinger, 1981).

3. CHROMIUM BIOCHEMISTRY

3.1 Introductory Remarks

It may be said with some certainty that chromium is an element of which trace levels are required but which, when accumulated in excess, becomes toxic. Because it is an essential metal in animals, metabolic pathways for chromium must have evolved. In this section, a brief description is given of what is known about the interaction of chromium with low-molecular-weight biomolecules, proteins, and polynucleotides and about its biologic function.

3.2 Interaction with Low-Molecular-Weight Biomolecules

The strong affinity of Cr(III) for charged oxygen centers has been mentioned (Section 2.3.2). Complexes with formation constants comparable in magnitude to those recorded for Fe(III) are known, for example, with the following ligands: hydroxide, phosphate, acetate, oxalate, malonate, phenolate, the iron chelating agents (the siderophores or siderochromes) involving the hydroxamato ($\overset{O}{\underset{C}{}}\overset{O}{\underset{N}{}}$) and catecholato ($\overset{O}{\underset{C}{}}\overset{O}{\underset{C}{}}$) functional groups (Martell and Smith, 1977; Smith and Martell, 1976). There is now general agreement that siderochromes are responsible for microbial iron transport and perhaps for the solubilization of extracellular iron (Neilands, 1979; Hughes, 1981). The Cr(III) analogs also are known to be taken up by cell suspensions and are thus biologically active (Emery, 1978). Other ligands are known to affect the transport of Cr(III) across absorptive tissue. Oxalate administered concurrently with Cr(III) enhanced this ion's transport through rat intestine in both in vitro and in vivo (oral feeding) experiments. This ligand not only enhanced absorption but also increased urinary excretion of chromium. Phytate (inositol hexaphosphate, a fully phosphorylated six-carbon cyclic polyhydric alcohol) reduced both transport and absorption, whereas EDTA and citrate had little effect (Chen et al., 1973). No doubt these ligand effects are the result of complex formation and are related to the permeability of the resulting Cr(III) complexes (i.e., their molecular weight, charge, and lipid solubility). Complex formation is also known to enhance or prevent the polymerization of Cr(III) that accompanies hydrolysis. Rollinson et al. (1967) have demonstrated that hydrolysis in the presence of phosphate at pH 7.0 [Cr(III), $10^{-4} M$; HPO_4^{2-}, $0.02 M$] was prevented by pyrophosphate ($10^{-3} M$) and enhanced by oleate ($10^{-3} M$). EDTA and citrate also have such inhibitive capability (Sanderson, 1976, 1982). This was tested by dialysis and thus by the ability to diffuse as low-molecular-weight species across a semi-permeable membrane (4.8 nm pore size).

Chromium(III) is known to form complexes with amino acids. Those of aspartic acid and glutamic acid have been characterized (Evans et al., 1979), and crystallographic data are available for $Na[Cr(L\text{-}cys)_2] \cdot 2H_2O$, $Cr(gly)_3$, and $Cr(L\text{-}his)(D\text{-}pen)$.* As expected for a borderline metal ion, these amino acid ligands are attached to Cr(III) through the N atom of the α-amino group and an O atom of the α-carboxylate, as well as through the S atom of the sidechain sulfhydryls (L-cys, D-pen), the imidazole heterocyclic-N of histidine, and presumably the O atom of the β-carboxylate of aspartic acid (de Meester et al., 1977; de Meester and Hodgson, 1977; Evans et al., 1979; Gergeley and Sóvágó, 1979; Connett and Wetterhahn, 1983). The class A character of Cr(III) is characterized by the $Cr(L\text{-}cys)_2^-$ complex anion. In acid solution, this anion is subject to acid hydrolysis, which cleaves the Cr(III)–S bond and replaces it with water (the α-amino and α-carboxylate groups remain attached) (Connett and

*L-cys, L-cysteine; gly, glycine; L-his, L-histidine; D-pen, D-penicillamine (β,β-dimethylcysteine).

Wetterhahn, 1983). By contrast, metal–sulfur bonds of class B ions, such as Hg(II), are extremely stable, even in very acid solution (Section 2.3.2; Nieboer and Richardson, 1980).

The oxygen-seeking ability that characterizes chromium is specifically highlighted in its interaction with nucleotides. Unlike borderline metal ions with considerable class B character [e.g., Cu(II), Cd(II)] and class B ions [e.g., Hg(II), Pd(II)], Cr(III) does not exhibit much affinity for the N-donor centers provided by the nucleotide bases (Martin and Mariam, 1979; Spiro, 1980). Instead, it is attached through the phosphate moieties of the nucleotide molecules (DePamphilis and Cleland, 1973; Legg, 1978; Campomar et al., 1986), and in this it resembles Mg(II) and Ca(II). As might be expected from Section 2.3.1, the Cr(III)–nucleotide complexes are kinetically inert, and this appears to be of biologic consequence in relation to Mg(II) biochemistry. Magnesium(II) plays a major role in living organisms because it is a cofactor in the production, storage, and use of the high-energy compound adenosine triphosphate (ATP) and because it is essential for the stabilization of polynucleotides, such as DNA and RNA, by acting as a counterion for the negatively charged phosphate moieties in these macromolecules. As explained more fully in Section 3.7, the replacement of Mg(II) by Cr(III) usually results in inhibition of critical enzymes and, furthermore, may be expected to alter the three-dimensional geometry of DNA and RNA so essential to their role in genetic processes. Finally, Cr(III) has considerably greater affinity for N-containing ligands than has Mg(II) (e.g., Warren et al., 1981), which creates additional modes of interaction with biomolecules, potentially increasing its toxicity.

In Section 2.2.4, the chemical reactivity of the chromate ion was discussed. Because this Cr(VI) species is an anion, the central Cr(VI) ion does not interact directly with biologic ligands at physiologic pH. However, because chromate can oxidize intracellular reductants, the generated Cr(III) ion is available for such complex-formation reactions, often with unreacted substrate. At physiologic pH, chromate attacks most low-molecular-weight intracellular reducing agents. Of these, thiol-containing compounds are most prominent and include cysteine, glutathione, and lipoic acid. They are oxidized to the corresponding disulfide (Gergely and Sóvágó, 1979; Rabenstein et al., 1979; Connett and Wetterhahn, 1983). Although Samitz and Katz (1964) reported that methionine was oxidized by chromate at pH 7, more recent work has shown this reaction to be slow compared to thiols (Connett and Wetterhahn, 1983). Ascorbic acid and lactic acid also are capable of reducing chromate at physiologic pH values (Samitz and Katz, 1964; Connett and Wetterhahn, 1983). In fact, chromate is potentially capable of oxidizing most intracellular reducing agents because of its significant positive formal potential ($E^{\circ\prime}$) at physiologic pH (see Section 2.1). Other possibilities include the reduced forms of nicotinamide adenine dinucleotide, flavin adenine dinucleotide, and coenzyme Q (a quinone), all of which have $E^{\circ\prime}$ values less positive than chromate (Connett and Wetterhahn, 1983). Since a number of these reductants are present in significant concentrations in cells, the chromate ion will eventually be reduced. The magnitude of the biologic half-life of chromate is predicted to be minutes to hours (see Section 2.2.4).

3.3   Interaction with Polynucleotides

One of the earliest indications that chromium can interact with nucleic acids was reported by Herrmann and Speck (1954). They observed that chromate-treated tissue homogenates decreased the extractability of nucleic acid into hot trichloro-acetic acid. These authors correctly concluded (based on solution color changes) that chromate is reduced to Cr(III) by tissues (see Section 3.6) and that the reduced form interacts with the nucleic acids. Laboratory studies have indicated that Cr(III) binds to nucleic acids and most likely is bound to exterior phosphate groups (Eisinger et al., 1962; Danchin, 1975). Chromium(III):DNA-P ratios up to 1 have been observed (Okada et al., 1981). Furthermore, Cr(III) has a tendency to precipitate polynucleotides and, like Ni(II) and Zn(II), increases their thermal stability (Fuwa et al., 1960). Significant amounts of chromium (20–400 μg g^{-1} of RNA or DNA) have been found in nucleic acids from a number of sources (Wacker and Vallee, 1959).

Ono et al. (1981) examined the distribution of several metals in nuclei and nucleoli from rats fed a conventional diet containing 0.3 μg g^{-1} of chromium. They found that, in contrast to Ca, Zn, Cu, and Mn, the ratio of Ni and Cr in nuclei to that in whole cells was relatively high (20% compared to 1–3%). Concentrations of Ni and Cr in nucleoli (relative to the nuclear content) were considerably enhanced compared to the other metals and were more resistant to release on treatment with nucleases. Additional evidence that Cr(III) accumulates in the nucleus because of its high affinity for nucleic acids may be derived from cell culture studies. Levis et al. (1978a,b) observed that DNA isolated from BHK fibroblasts treated with potassium dichromate contained large amounts (> 1%) of chromium measured as Cr(III). Chromate-induced cellular DNA-protein crosslinks have been observed in vivo and in vitro and attest further to the strong interaction of Cr(III) with DNA and other polynucleotides (see Section 3.4.1).

3.4   Interaction with Proteins and Enzymes

3.4.1   Chromium(III) Crosslinks

Because of its application in the tanning of leather, the interaction of Cr(III) with collagen is undoubtedly the most comprehensively studied reaction with proteins. Chemical modification studies have revealed that Cr(III) crosslinks predominantly involve the carboxylate groups of the polypeptide chains of collagen, although the polymerization of the hydrated Cr(III) ion by hydroxide ion is also required (Gustavson, 1958; Kuehn and Gebhardt, 1960; Mertz, 1969). Other protein systems have been examined. The dimerization of conarachin II (a groundnut protein) at pH 5.3 occurred more slowly with Cr(III) than with Al(III) and Fe(III), and the product was more stable to dialysis against phosphate buffer (Naismith, 1958). The dialysis step reversed the Al(III)- and Fe(III)-induced dimerization by removal of the metal ion. Chromium(III) salts have produced crosslinking between DNA and protein in vitro when the constituents were mixed in solution or when isolated nuclei were treated with Cr(III). Hexavalent chromate was ineffective in these experiments (Fornace et al., 1981). However, chromate-treated rats (Tsapakos et al., 1981, 1983) and chromate-treated cultured mammalian cells (Fornace et al., 1981) were found to have similar cellular DNA–protein crosslinks. It is now widely accepted that chromate is reduced intracellularly to Cr(III) (see Section 3.6). Fornace et al. (1981) suggest that, in vivo, the linking of DNA and protein or perhaps other macromolecules by the Cr(III) ion may affect polymerase activity during both DNA replication and repair, with resultant mutagenic and carcinogenic consequences. Recent observation and characterization of DNA–interstrand and DNA–protein crosslinks in kidney and liver of rats treated with Cr(III) salts or Cr(VI) salts are described in detail in Chapter 16.

3.4.2 Chromium(III) Binding and Its Consequences

There is solid evidence that Cr(III) binds to the Fe(III) transport protein, transferrin. The bulk of the Cr(III) administered to rats and occurring in serum is bound to transferrin, whereas secondary binding occurs to other proteins (e.g., albumin) (Hopkins and Schwarz, 1964; Jett et al., 1968). In vitro studies have shown that Cr(III) competes with Fe(III) for the same binding sites of transferrin (Aisen et al., 1969; Harris, 1977; Emery, 1978). Since iron rarely occupies more than one-third of the available binding sites in transferrin, the small amount of plasma chromium ($\sim 1\ \mu g\ L^{-1}$) can be easily accommodated. It is considered that transferrin is the main serum transport protein for Cr(III). Serum transferrin carries iron from the breakdown sites of hemoglobin in the reticuloendothelial cells of spleen and liver back to the sites in bone marrow for the synthesis of hemoglobin (Hughes, 1981). Interestingly, the intravenous administration of Cr(III) in humans has established that Cr(III) is initially localized primarily in the liver, spleen, body soft tissues, and bone (Lim et al., 1983). Studies with reticulocytes have demonstrated that transferrin is transported into cells by endocytosis (Light and Morgan, 1982). In addition, transferrin is involved in the transport of iron to cells after the ferritin-mediated absorption of iron from the gastrointestinal tract. Transferrin uptake, and thus iron uptake, by other cells also appears to occur by receptor-mediated endocytosis (Emery, 1978; Dautry-Varsat and Lodish, 1984). The possibility exists that Cr(III) may be taken up in a similar manner.

In vitro experiments indicate that Cr(III) binding to enzymes is accompanied by the enhancement or inhibition of catalytic activity. Chromium(III) is known to restore the activity of apocarboxypeptidase (Vallee et al., 1958), demonstrating that Cr(III) can replace Zn(II), since carboxypeptidase is a zinc metalloenzyme.

Chromium(III) also satisfies the second-metal requirement of phosphogluco-mutase. The greatest activity of this enzyme occurs in the presence of three activators: hexosediphosphate, Mg(II), and a second metal (Stickland, 1949). Consideration of affinities and the relative concentrations required suggests that if this two-metal activation has any physiologic analog, Cr(III) and Mg(II) are the metals concerned. Concentrations as low as $10^{-6} M$ of Cr(III) were effective. In in vitro experiments, nonphysiologic concentrations of the metal ion are often required, making extrapolation to in vivo conditions difficult. Nevertheless, in vitro experiments often demonstrate potential effects. This is illustrated by a recent example involving the alteration by Cr(III) of RNA synthesis directed by DNA and chromatin isolated from mouse liver. Maximum enhancement (sevenfold) occurred when Cr(III) was pre-incubated with DNA or chromatin, with a Cr(III):DNA-P molar ratio near 1 (Okada et al., 1981, 1982). Although such high Cr(III):phosphate ratios are unlikely to occur in vivo, experiments with mice demonstrated that the enhancement did occur in vivo after intraperitoneal administration of Cr(III) chloride (≥ 0.05 mg Cr/kg) (Okada et al., 1983).

Chromium(III) has been shown to inhibit blood clotting in vitro (Chargaff and Green, 1948), and more recently, selective inhibition of the electron transfer functions of cytochrome c and plastocyanin have been reported (Connett and Wetterhahn, 1983). The large fraction of erythrocyte chromium derived from chromate uptake is bound to hemoglobin, although its effect on oxygen binding has not been demonstrated (Kitagawa et al., 1982; Sanderson, 1982). In a detailed study of soluble skin proteins, Samitz et al. (1969) illustrated that more chromium bound to protein (20-fold) in Cr(III) sulfate solutions than in equimolar potassium dichromate solutions. Blocking of amino groups by acetylation and carboxylate groups by methylation decreased the binding of chromium. The decrease was much more marked for the latter, again strongly suggesting the prominence of carboxylate groups in the interaction of Cr(III) with protein. Such interactions may be significant in the etiology of chromium dermatitis (see Chapter 18).

Inhibition with potentially serious physiologic consequences involves the replacement of Mg(II), and perhaps Mn(II) in those instances where it is the native ion (Ochiai, 1977), in reactions where ATP, ADP, or similar nucleotides are substrates. Chromium(III) forms nonlabile complexes with these mono-nucleotides (see Section 3.2). The Cr(III) nucleoside diphosphates and tri-phosphates inhibit kinases (which involve the transfer of phosphate from ATP to a substrate; DePamphilis and Cleland, 1973; Janson and Cleland, 1974; Gupta et al., 1976), pyruvate carboxylase (Armbruster and Rudolph, 1976), and ATPases (Schuster et al., 1975). For example, Cr(III) inhibition of Mg(II)-ATPase activity has been reported for plasma membranes of intact hamster fibroblasts in culture (Luciani et al., 1979). Inhibition of nucleoside permeases in this system has also been observed (Levis et al., 1978a).

The relevance of DNA-interstrand and DNA-protein crosslinks in the patho-genesis of cancer is not clearly understood (see Chapter 16).

Finally, when protein systems become saturated with Cr(III), denaturation occurs (Clark, 1959). Bianchi et al. (1979) postulate that denaturation of membrane structures may contribute to the cytotoxicity of Cr(III).

3.4.3   Chromate as Oxidant and Substrate

Connett and Wetterhahn (1983) summarized the evidence for the conclusion that not many purified proteins reduce chromate at physiologic pH values. However, this statement needs to be qualified because the analytic procedures used often precluded detection of the chromate consumed by the sulfhydryl equivalents available in the proteins studied. Unequivocal evidence for the reduction of chromate by hemoglobin is available (Samitz and Katz, 1964; Connett and Wetterhahn, 1983), and the resultant Cr(III) appears to be bound to hemoglobin (Gray and Sterling, 1950; Kitagawa et al., 1982). Glutathione reductase is also inhibited by chromate, and it is surmised that this involves the oxidation of the active site sulfhydryl groups (Connett and Wetterhahn, 1983). In vitro studies with mitochondria have shown that respiratory dehydrogenases are also inactivated (Broughall and Reid, 1974). Since sulfhydryl groups are critical to the catalytic function of a number of other enzymes (e.g., δ-aminolevulinic dehydrase, urease, adenylate kinase), chromate oxidation of such enzymes should be investigated. Preliminary inhibition data for urease have been reported (Henry and Smith, 1946).

Evidence is mounting that chromate is reduced by microsomal electron-transport cytochrome P-450 systems. The slow rate of reduction in control samples (microsomes without NADPH; NADPH in the absence of microsomes) and the inhibition by P-450-specific inhibitors confirm that this reaction is enzymatic (Gruber and Jennette, 1978; Garcia and Wetterhahn Jennette, 1981). A practical application of this observation is the metabolic deactivation of chromate mutagenicity in the *Salmonella*-microsome test. The resultant Cr(III) is nonmutagenic (Petrilli and DeFlora, 1978; DeFlora, 1978; Levis and Bianchi, 1982). An interesting development in relation to the microsomal reduction of Cr(VI) is the spectrometric characterization of relatively long-lived (hours) Cr(V) intermediates (Polnaszek, 1981; Wetterhahn Jennette, 1982). Goodgame et al. (1982) were able to generate Cr(V) in the absence of microsomes when chromate was mixed with ribonucleotides. Interestingly, the reaction did not occur with the corresponding deoxyribonucleotides, suggesting that oxidation of the ribose ring is involved. The toxicologic implication of unstable Cr(V) intermediates has been addressed (Section 2.4; see also Chapter 16).

3.5   Cellular Association and Uptake

Numerous cell culture studies have shown that cellular uptake of chromate is at least 10 times greater than that of Cr(III) from equimolar solutions (Gray and Sterling, 1950; Levis et al., 1978a). Speciation studies have revealed that intracellular reduction of chromate to Cr(III) occurs (Levis et al., 1978a,b; Kitagawa et al., 1982; Levis and Bianchi, 1982). At least in the case of erythrocytes, the cell-associated chromium is primarily cytosolic. Since the resultant Cr(III) has a high affinity for intracellular binding sites (see Sections 3.2 to 3.4), it is not surprising that little of the chromate accumulated by cells can be removed by washing with isotonic buffer (e.g., 6% at $t=0$ and 1% at $t=6$ hours after exposure to chromate; Kitagawa et al., 1982). The sensitivity of chromate uptake by erythrocytes to stilbene inhibitors suggests that it uses an anion channel, by analogy with sulfate and phosphate uptake (Kitagawa et al., 1982). However, measurement of activation energies and other uptake characteristics implicates diffusion-controlled uptake (Ormos and Mányai, 1974). In agreement with this are the findings of Sanderson (1976), who showed that metabolic inhibitors did not inhibit uptake by mastocytoma cells.

It appears that chromate uptake is not a prerequisite for all cytotoxic effects. Studies with intact cultured hamster fibroblasts demonstrated that chromate, on initial exposure, enhances the cell membrane permeability to nucleosides. This is followed by a second phase in which inhibition of uptake occurs. Structural modification of cell membranes, presumably by the oxidative action of chromate, is believed to stimulate simple inward diffusion of the nucleosides (Bianchi et al., 1979). Subsequently, when Cr(III) accumulates at the membrane surface, nucleoside permeases are thought to be inactivated and perhaps even denatured. Separate administration of Cr(III) salts does indeed inhibit nucleoside uptake (Levis et al., 1978a).

3.6 Interaction with Tissues

Chapter 18 deals with the hypersensitivity responses to chromium compounds. Allergic contact dermatitis is the most prominent reaction. Epicutaneous applications (patch tests) and percutaneous absorption measurements have shown the importance of valency, type of salt, pH, and such factors as the presence of reducing agents. In this section, we briefly outline some basic concepts that help to rationalize the often conflicting results reported for the transport of chromium compounds into and through the skin. Comprehensive reviews summarizing the empirical data are provided by Pedersen (1982) and Polak (1983) (see also Chapter 18). In brief, the major findings are as follows:

1. Chromium(VI) compounds penetrate the skin more readily than do Cr(III) compounds.
2. The penetration of chromate increases with increasing pH.
3. The penetration of Cr(III) salts depends on the nature of the anion.
4. The difference in penetration of Cr(VI) and Cr(III) compounds appears to be governed by the epidermal layer, which forms the main barrier to uptake.
5. Skin and skin components reduce chromate.
6. Chromium(III) binds strongly to skin and skin proteins.

Before these results are interpreted and rationalized, a brief review of the structure and composition of human skin is presented.

A cross-section of human skin is shown in Figure 4. The first phase of percutaneous absorption is diffusion of the toxicant through the epidermis (Klaassen, 1980). The stratum corneum comprises a dead surface layer of the skin, and movement through it is by passive diffusion. Water-soluble (polar) substances appear to diffuse through the hydrated component of the stratum corneum, whereas nonpolar molecules dissolve in and diffuse through the lipid matrix situated between the protein filaments that make up this layer. After diffusion through the epidermis, the toxicant passes through the dermis, which corresponds to a porous, nonselective, aqueous diffusion medium. Movement is again by simple diffusion, and the toxicant now has easy access to the systemic circulation (Fig. 4). Several additional points need to be made. Because it acts as a barrier, removal of the stratum corneum (e.g., by abrasion, injurious agents, such as acids and bases, and burns) causes an abrupt increase in the permeability of the epidermis. Hydration of the stratum corneum is also known to enhance absorption, as does the removal of its lipid components on contact with organic solvents. Furthermore, penetration increases with the diffusivity and thickness of the stratum corneum. The hair follicle also appears to be a route of entry.

A guideline to the composition of the stratum corneum is as follows: 52–55% water-insoluble material, 19–27% water soluble, 5–10% water, and 10–22% ether-soluble constituents. The water-soluble materials consists of 40% free

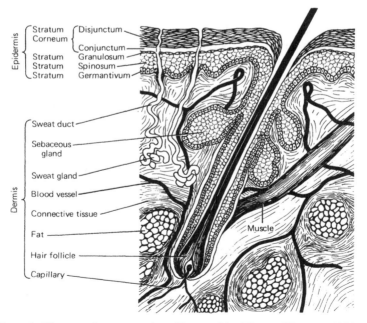

Figure 4. Diagram of a cross-section of human skin. (*Source:* Klaassen, 1980.)

amino acids, 10% carbohydrates, 1% citric acid, and 12% lactic acid, among other constituents (Flesch, 1958; Samitz and Katz, 1964).

There is considerable evidence that the chromate ion diffuses freely through the epidermis, whereas the movement of the Cr(III) ion is inhibited (Mali et al., 1963). Diffusion of the Cr(III) ion is retarded for a number of reasons. First, Cr(III), in contrast to chromate, binds strongly with dermal tissue and its low-molecular-weight constituents (Mali et al., 1963; Samitz and Katz, 1963, 1964; Samitz et al., 1969). Second, the species present in solution is highly dependent on the Cr(III) salt dissolved. For example, solutions of $[Cr(H_2O)_6]Cl_3$, like those of $KCr(SO_4)_2 \cdot 12H_2O$, generate the fully hydrated ion $Cr(H_2O)_6^{3+}$, whereas the normal, commercially available salt $[CrCl_2(H_2O)_4]Cl \cdot 2H_2O$ yields the singly charged ion $CrCl_2(H_2O)_4^+$. The net charge on the Cr(III) species influences both the diffusion rate and the ability to bind to tissues. The degree of hydrolysis, and thus polymerization with a concomitant reduction in diffusion, is also highly dependent on the Cr(III) salt selected. Of course, the latter process increases with increasing pH of the vehicle. These differences adequately explain that undamaged human skin is poorly penetrated by the commercially available sulfate and nitrate of Cr(III), whereas the chloride penetrates almost as well as potassium dichromate (Polak, 1983).

The observation that diffusion of chromate increases with pH is in agreement with the concomitant decrease in its oxidizing power (see Sections 2.1 and 2.2.4). Consequently, less of it is converted to Cr(III) because of reaction with skin proteins and water-soluble constituents (i.e., lactic acid, amino acids; see Section 3.2).

The observation that the removal or damage of the stratum corneum permits the uptake of Cr(III) salts may have serious toxicologic consequences. In a case report, significant quantities of chromium were absorbed by a man who became accidentally immersed in hot acidic Cr(III) sulfate solution (Kelly et al., 1982). Although no oral ingestion of chromium occurred, the chromium content of the tissues at necropsy was severely elevated, as was the content in blood and urine before death. Burnt skin also accumulated extremely high Cr levels. The clinical course was dominated by burns, intravascular hemolysis, and acute renal failure. Other cases of severe nephritis resulting from the cutaneous absorption of chromate are known (see references in Stokinger, 1981; Kelly et al., 1982). Severe kidney toxicity has been observed in acute chromate poisoning (Goldman and Karotkin, 1935; Sharma et al., 1978; Kelly et al., 1982).

3.7 Biologic Functions

In Section 3.4.2, we reviewed the consequences of Cr(III) binding to enzymes. Both inhibition and stimulation of catalysis occurred. Although the exact physiologic importance of this type of role is hard to assess, it illustrates the potential of Cr(III) to regulate and alter physiologic processes. This was most vividly illustrated by the discovery of the chromium glucose tolerance factor (CrGTF).

There seems not much doubt that chromium is an essential metal, since its deficiency has serious consequences in animals and humans. Signs of chromium deficiency in humans include glucose intolerance, elevated circulating insulin, elevated serum cholesterol and triglycerides, and peripheral neuropathy and metabolic encephalopathy (Jeejeebhoy et al., 1977; Freund et al., 1979; Anderson, 1981). An increased risk of coronary disease may occur (Mertz, 1982b; Wallach, 1985). The CrGTF has been identified in brewers' yeast as a Cr(III)–dinicotinic acid–glutathione complex (Mertz, 1975, 1982a, 1983). In vitro studies with rat tissues indicate insulin-potentiating capability, perhaps by forming a complex with insulin or acting independently at the membrane level. Less insulin is required in the presence of the CrGTF (Mertz, 1969). Biosynthesis of cholesterol by rat liver microsomes in vitro can be enhanced or inhibited by addition of Cr(III), depending on concentration (Deliconstantinos et al., 1984). Quite a few Cr(III) complexes can mimic the yeast isolate in its hypoglycemic action and appear to be further characterized by enhanced intestinal absorption as well as enhanced placental transfer (Mertz et al., 1974; Mertz, 1975; Anderson et al., 1980). Glucose tolerance factor activity of Cr(III) compounds with nonphysiologic ligands, for example, potassium trioxalato-chromium(III), has been demonstrated in rats when given by stomach tube supplementation (Schwarz and Mertz, 1959). It may be recalled that oxalate enhanced the gastrointestinal absorption of Cr(III) in rats (see Section 3.2), which is characteristic of GTF-active compounds.

Synthetic analogs of brewers' yeast CrGTF are as active as the natural product. These compounds are prepared by refluxing a 1:2:2:1:1 mixture of, respectively, Cr(III) sulfate, nicotinic acid, glycine, glutamic acid, and cysteine in aqueous ethanol (Mertz et al., 1974; Toepfer et al., 1977; Haylock et al., 1983a). Gonzalez-Vergara et al. (1982) claim to have isolated a CrGTF from yeast that appeared to contain a tryptophanlike ligand rather than nicotinic acid and glutathione. Since all of these ligands are physiologic, it seems possible that GTFs can be synthesized in vivo from Cr(III) and that there are a number of active forms. The work of Haylock et al. (1983b) and Cooper et al. (1984) suggests that Cr(III)-containing GTFs and Cr(III)-free GTFs are cations with geometries in which the orientation and separation of two protonated nitrogen centers are similar. Such compounds, on binding to membranes, are believed to increase glucose permeability.

Results of studies of the effect of chromium supplementation in the improvement of glucose tolerance in humans are conflicting. Polansky et al. (1981, 1982; also see Anderson et al., 1982a, 1983a,b), in controlled experiments in free-living human subjects, found a significant reduction in glucose levels when subjects with somewhat elevated glucose levels were given chromium supplementation. The chromium was administered orally as Cr(III) chloride or as brewers' yeast. Similar results were obtained in other studies (Offenbacher and Pi-Sunyer, 1980; Mertz, 1982b; Mossop, 1983; Saner et al., 1983), although some investigators have reported negative results (Rabinowitz et al., 1983; Offenbacher et al., 1985). A similar ambivalence exists concerning increases in the fraction of high-density lipoprotein (HDL) cholesterol (Donaldson et al., 1985). In spite of the inconsistencies in the human response to chromium supplements, Wallach (1985) concluded that chromium supplementation improves human glucose tolerance and can decrease the total cholesterol to HDL cholesterol ratio. However, good evidence is lacking that chromium deficiency occurs in glucose-intolerant and overtly diabetic subjects. Elias et al. (1984) showed in stable maturity onset diabetics that supplementation with brewers' yeast results in a significant decrease ($P < 0.02$) in insulin requirement during a glucose challenge. There is also evidence that chromium supplementation in humans promotes higher insulin levels (Rabinowitz et al., 1983; Potter et al., 1985). Collectively, these observations leave unanswered the question whether the observed CrGTF and chromium supplementation phenomena reflect an essential biologic function of chromium or are simply a manifestation of a hypoglycemic drug effect.

4. ABSORPTION, DISTRIBUTION, AND EXCRETION

4.1 Intake and Absorption

As for most toxicants, there are three primary modes of chromium absorption: via the airways, the skin, and the gastrointestinal tract. We have already considered the dermal route (see Section 3.6), and it is more appropriate to treat uptake by the respiratory tract in relation to biological monitoring (see Section 5). Gastrointestinal absorption is considered in this section.

Chromium is poorly absorbed, and often a figure of 0.5–3% is quoted (Doisy et al., 1969; Underwood, 1977; Kumpulainen et al., 1979; Offenbacher et al., 1986). Chromium absorption has been shown to be inversely related to dietary intake (Anderson and Kozlovsky, 1985). However, as noted in Section 3.7, the uptake may also be dependent on the chemical form of the chromium. For example, preliminary animal experiments suggest that 10–25% of the chromium in brewers' yeast is absorbed by rats (Mertz et al., 1974; Underwood, 1977). Furthermore, Underwood (1977) points out that the presence of other metals, such as zinc, may have a marked effect, as is the case for the absorption of most metals. Rat studies indicate that fasting also enhances uptake (MacKenzie et al., 1959). Even though Cr(VI) compounds are somewhat better absorbed than Cr(III) compounds (Donaldson and Barreras, 1966; Underwood, 1977), the difference is less than would be expected from the much greater ability of Cr(VI) to cross cell membranes (see Section 3.5). Donaldson and Barreras (1966) have demonstrated unequivocally that human gastric fluids effectively reduce chromate to Cr(III), and in this way absorption is diminished. Absorption after duodenal administration in humans showed that 6% of Cr(III) chloride was taken up (compared to 0.5% for oral administration), whereas about 40% of Na_2CrO_4 was absorbed after duodenal administration (compared to 10% for oral administration) (Donaldson and Barreras, 1966).

In a very careful analytic study, Kumpulainen et al. (1979) determined that in representative American diets the daily chromium intake was 62 ± 28 μg (high fat diets) and 89 ± 56 μg (low fat diets). Similar results have been reported in other studies (Guthrie, 1982; Smart and Sherlock, 1985; Gibson et al., 1985). In relation to CrGTF function, a provisional recommendation for the dietary intake of trivalent chromium of 50–200 μg day^{-1} has been made (Mertz, 1979, 1980). However, recent metabolic balance studies suggest that intakes <50 μg · day^{-1} may be adequate (Bunker et al., 1984; Anderson and Kozlovsky, 1985; Offenbacher et al., 1986).

4.2   Distribution

The models depicted in Figures 5 and 6 help to organize our discussion of the distribution and excretion of chromium. The individual functional compartments or components depicted in these figures are based mostly on animal studies, but the overall model was proposed and verified for the kinetics of Cr(III) in the human body (Lim et al., 1983). Lim et al. studied the distribution and kinetics of IV injected ^{51}Cr(III) in six human subjects using a whole-body scintillation scanner, a whole-body counter, and plasma counting. Since chromium in the diet is most likely in the Cr(III) state, Figures 5 and 6 should reflect the normal chromium distribution patterns. The special cases of enteral and parenteral administration of chromates are considered separately.

After absorption, Cr(III) is transported bound primarily to plasma transferrin

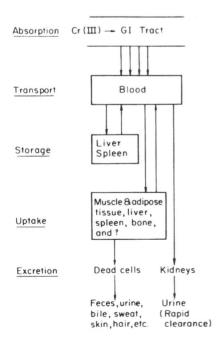

Figure 5. Proposed pathways for Cr(III) in the human body, based on human and animal data. Chromium(III) is transported bound mainly to plasma transferrin. (*Source:* Lim et al., 1983.)

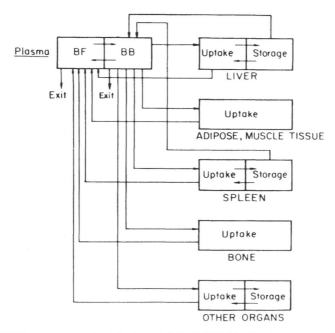

Figure 6. Multicompartment physiologic model for Cr(III) kinetics in the human body. Plasma chromium is the central compartment, consisting of two subcompartments, the plasma protein-bound (BB; i.e., to transferrin) and the plasma protein-unbound (BF) chromium. Return from the organs is shown to occur in the unbound form but may well also take place as BB. For a mathematical solution, the uptake compartments take up and release Cr(III) relatively rapidly (half-lives, $t_{1/2}$, of <1 hour to 2 days), whereas the uptake in the storage compartments occurs more slowly ($t_{1/2}$, 4 days), with extremely slow release ($t_{1/2}$, 315 days). Excretion occurs via kidney clearance ($t_{1/2}$, 3.5 hours) and other slower routes (e.g., desquamated cells and feces; $t_{1/2} = 7$ days). OTHER ORGANS denote those that could not be resolved by the whole-body scintillation scanning technique, including kidney, heart, and CNS. (*Source:* Lim et al., 1983.)

(70–90%) (Hopkins and Schwarz, 1964; Jett et al., 1968; Sayato et al., 1980; all studies with rats). In humans, as in animals, Cr(III) is localized primarily in the blood, liver, spleen, kidney, body soft tissues, and bone (Figs. 5 and 6) (Hopkins, 1965; Onkelinx, 1977; Jain et al., 1981; Langård, 1982; Lim et al., 1983). The pool of protein-bound chromium in the plasma presumably corresponds to transferrin. Long-term storage appears to occur in certain tissues, especially in the liver and the spleen, suggesting that the deposition of chromium in these tissues is related to the function and processing of transferrin rather than of red blood cells, since very little Cr(III) administered by stomach tube to rats, or introduced into the intestine, becomes associated with erythrocytes (MacKenzie et al., 1959). The association in humans of hemosiderin with ^{51}Cr(III) introduced in labeled erythrocytes (Kurzweil et al., 1984) suggests that Cr(III), like Fe(III), is stored as ferritin in tissues (hemosiderin is believed to be a degradation product of ferritin).

Distribution studies for the IV injection of Na_2CrO_4 in rats yield similar results to those discussed for Cr(III) (accumulation in blood, lungs, liver, kidney, and spleen) (Langård, 1979). The work of MacKenzie et al. (1959) with rats has shown that oral (stomach tube) or intestinal administration of chromate results in the strong labeling of both erythrocytes and plasma. Sayato et al. (1980) illustrated that the bulk of erythrocyte chromium was bound to hemoglobin, as expected from in vitro erythrocyte studies (see Section 3.5). Shih et al. (1972) have studied erythrocyte survival in man by injecting ^{51}Cr-labeled erythrocytes. The half-life of the label, measured by urine excretion of ^{51}Cr, was 50–60 days, whereas that of the erythrocytes was 25–35 days. Presumably, this observation is related to the chromium accumulation with time in rat spleen seen in Langård's sodium chromate work (1979), since the spleen disintegrates old red blood cells.

In applying the model of Figure 6, accumulations from the central plasma compartment with short (in minutes) and intermediate (1–24 hours) half-lives ($t_{1/2}$) were considered simply as *uptake*. Release from such uptake compartments occurred with slightly longer $t_{1/2}$ values (≤ 2.2 days). Deposition in the storage compartments took considerably longer (~ 4 days), with very slow release ($t_{1/2} = 315$ days). Interestingly, the storage compartments accumulated by far the most chromium (96%). Note that the kidney and lungs, which animal data suggest initially also accumulate Cr(III), are designated OTHER ORGANS in Figure 6 because they could not be resolved individually by the whole-body scan technique.

4.3 Excretion

In the model (Figs. 5 and 6), two excretion pathways are noted: that cleared from the unbound (plasma) pool is excreted by the kidney with $t_{1/2}$ of 3.5 hours, and that removed by desquamated cells and feces with $t_{1/2}$ of 7 days (Lim et al., 1983). This dual excretion pathway is supported strongly by animal data, even though the earlier conclusion that urinary excretion is the only route (Visek et al., 1953; Mertz, 1969) is often quoted without question. The animal data summarized by Langård (1982) and Sayato et al. (1980) clearly show that 5–20% of chromium parenterally administered as Cr(III) is excreted in the feces, whereas as much as 10–30% of the injected Cr(VI) is excreted in this way. It is interesting that in a study of welders exposed primarily to Cr(VI), Tossavainen et al. (1980) assigned a $t_{1/2}$ value of 15–41 hours to the urinary excretion of chromium, which is comparable to the assignment of 2.2 days for the efflux from the uptake compartments (see Section 4.2). Nomiyama et al. (1980) provide evidence that chromium occurs in the urine of rabbits as Cr(III). Urine itself was found to be reducing toward chromate. More recent work has shown that Cr(VI) administered IV in rats is reduced quickly (minutes) to Cr(III) in blood in vivo (Cavalleri et al., 1985). Furthermore, erythrocyte-bound Cr(III), as encountered in exposure to chromate, appears to be an additional long-term storage pool, with a release time

$(t_{1/2})$ of 50–60 days; see earlier comments in Section 4.2 on the work by Shih et al. (1972).

Donaldson et al. (1982) concluded from serum ultrafiltration and urinary excretion studies of rats and dogs that "there may not be a strong reabsorptive mechanism for Cr(III) in the kidney, as has been previously reported." In fact, in a detailed study with dogs, Donaldson et al. (1984) found that the mean ratio of the clearance of ultrafiltrable plasma ^{51}Cr(III) (given as CrCl$_3$) to that of endogenous creatinine was near unity (0.97 ± 0.11 to 1.14 ± 0.10 by gavage, $n = 5$; 0.84 ± 0.05 to 0.97 ± 0.05, IV injection, $n = 4$). They concluded that "regardless of the route of administration, ultrafiltrable plasma chromium-51 concentration and glomerular filtration rate appeared to be primary determinants of renal chromium-51 excretion." Collins et al. (1961) concluded that 63% of the glomerular filtrate was reabsorbed. Perhaps the data reported for dogs by these workers did not refer to an equilibrium situation, since the dialyzable fraction of Cr(III) in plasma was time-dependent (9% at 2.5 hours and 3% at 10 hours after injection). Onkelinx (1977) found that the ultrafiltrable fraction of Cr(III) in rats remained constant for times ≥ 3 hours after IV injection (at $7 \pm 1.5\%$). Mutti et al. (1979) concluded from ultrafiltration studies on serum from welders that about 90% reabsorption occurred. They reported variable chromium clearance values (volume of plasma cleared per unit time of nonprotein-bound chromium), which is an unlikely situation and suggests that methodologic errors occurred. As summarized in Chapter 16, considerable evidence now exists from animal studies that both urinary and biliary excretion of chromium is facilitated by its association [as Cr(III)] with an endogenous ligand consisting of glutamic acid, glycine, and cysteine as the major amino acids and forming a low-molecular-weight ($\sim 1,500$) anionic complex (Manzo et al., 1983; Wada et al., 1983; Yamamoto et al., 1984). These observations add plausibility to the views that renal excretion of chromium is similar to that of creatinine and that some biliary excretion occurs.

Three lines of evidence indicate that the clearance of chromium in humans is similar to that of creatinine and thus occurs by glomerular filtration *without* tubular reabsorption or excretion (Faulkner and King, 1976). First, this is implied by the reduction in variability of urinary chromium determinations when creatinine normalization or corrections are applied for differences in urinary flow rates (see Section 5.3; Tola et al., 1977; Rahkonen et al., 1983). In a careful and definitive study of nonexposed people, Anderson et al. (1982a) found that creatinine adjustment reduced the standard error by a factor of 2. Second, the linear relationships observed in humans (nonexposed) between urinary chromium and urinary creatinine (Saner, 1980; Anderson et al., 1983a) denote that Cr(III) and creatinine have the same functional dependence on urinary flow rate and, thus, by the basic mathematics of clearance (Faulkner and King, 1976; Nieboer et al., 1984), must have the same excretion mechanism. [Anderson et al. (1982a) have commented that Saner's chromium data are high because of analytic problems; also see Section 5.2.]. Third, when the apparent reabsorption fraction

(Equation 5) reported for controls and insulin-dependent diabetics by Vanderlinde et al. (1979) is recalculated to incorporate the estimate that in humans only 5–6% of the plasma chromium is available for glomerular filtration (Lim et al., 1983), no reabsorption results.

$$\text{Reabsorption} = 1 - \frac{\text{serum creatinine}}{\text{urine creatinine}} \cdot \frac{\text{urine chromium}}{\text{serum chromium}} \qquad (5)$$

Interestingly, Vanderlinde et al. (1979) found urinary chromium to be at least twofold higher in insulin-dependent diabetics than in healthy normal people. (The analytic work in this study is of the highest quality.) Severe exercise also appears to induce chromium uresis (Anderson et al., 1982b).

5. BIOLOGICAL MONITORING

5.1 Status of Chromium Concentrations in Body Fluids and Tissues of Normal Adults

The general availability of methods using sensitive instruments for the routine analysis of elements at the ng g^{-1} (ultratrace) and μg g^{-1} (trace) levels in environmental and biologic samples has revealed that contamination by containers, analytic reagents, and airborne dust constitutes a serious source of both systematic and random errors. Interlaboratory comparisons (Parr, 1977; Brown et al., 1981; Sunderman et al., 1982) and the use of standard reference materials (Parr, 1977; Kosta and Byrne, 1982) clearly show contamination as a limiting parameter in trace and ultratrace analyses (Kosta, 1982). This realization has resulted in severe downward adjustments in the normal levels of elements in human tissues (American Institute of Nutrition, 1981; Versieck, 1985) and body fluids (Versieck and Cornelis, 1980; Versieck, 1985). Recent compilations of chromium levels in human tissues and body fluids provide a vivid example. Normal serum and urine chromium concentrations have decreased markedly with time: from 200 μg L^{-1} to 0.1–0.2 μg L^{-1}, and from 150 μg L^{-1} to < 1 μg L^{-1}, respectively (Versieck and Cornelis, 1980; Guthrie, 1982; Versieck, 1985). A graphic illustration of this trend is provided in Figures 7 and 8. Gross contamination during collection, handling, storage, and analysis can be incriminated (Versieck and Cornelis, 1980; Torgrimsen, 1982). Chromium, like nickel, is ubiquitous because of its presence in stainless steels. The uncertainties in normal body fluid chromium levels have made it difficult to assess the nutritional role of chromium (the chromium–glucose–insulin interrelationships; Veillon et al., 1982a) and to evaluate the extent of exposure in the workplace by biological monitoring. As with most other trace and ultratrace metals, the normal chromium concentrations reported for human tissues are likely in gross error because of extraneous additions during sampling and analysis (American Institute of Nutrition, 1981; Guthrie, 1982; Versieck, 1985). Recent develop-

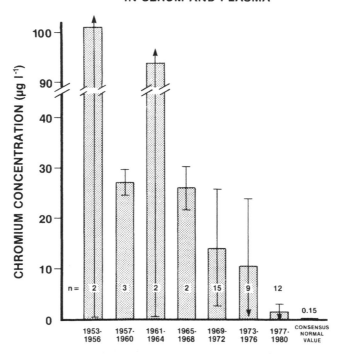

Figure 7. Chronologic decrease in normal chromium concentrations reported for sera (plasma) of unexposed humans. Contamination control during collection, handling, and analysis accounts for both the drastic downward trend and the improvement in precision. The error bars denote the standard deviation, and the number of data points contributing to the mean are shown for each histogram. The consensus value is $\sim 0.15 \, \mu g \, L^{-1}$. (*Source:* Compilation reported by Guthrie, 1982.)

ments in the analytic chemistry of chromium now permit the routine determination of nanogram quantities of this element in biologic samples.

5.2 Technical Feasibility

5.2.1 Available Instrumental Methods

The extremely low natural levels of chromium in body fluids limit the number of analytic methods. For routine analysis, electrothermal atomic absorption spectrometry (EASS) is the preferred technique (Fishbein, 1984). It is the least labor-intensive of the available methods, yet it exhibits suitable analytic sensitivity and precision. Techniques appropriate for intermethod comparisons are neutron activation and isotope dilution mass spectrometry. Since phosphorus-32 interferes with the analytic peak of chromium-51, chromium needs to be

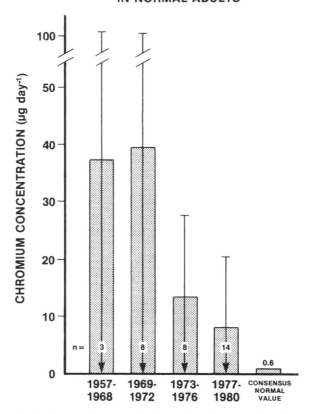

Figure 8. Chronologic decrease in normal urinary chromium excretion in unexposed adults. Contamination control during collection, handling, and analysis adequately explains the drastic downward trend. The error bars denote the standard deviation, and the number of data points averaged are shown for each histogram. The consensus value is ∿0.6 μg dayL^{-1}. (*Source:* Compilation reported by Guthrie, 1982.)

separated in neutron activation analysis from the irradiated sample (distilled as chromyl chloride; Versieck et al., 1978, 1979). A similar complication occurs with the isotope dilution technique because a volatile chelate suitable for gas chromatographic processing must be isolated (Veillon et al., 1979). Because of elaborate sample treatment and handling, clean air facilities are required with both techniques to prevent adventitious contamination. Another drawback circumventing routine uses of these techniques is that the highly specialized instrumentation required is not generally available. Nevertheless, both neutron activation and isotope dilution mass spectrometry serve as suitable references in intermethod comparisons (Harnly et al., 1983).

5.2.2   Contamination Control

In a review of contamination control in routine ultratrace analysis of toxic metals, Nieboer and Jusys (1983) suggested that the rigorous application of measures 1 to 9 in Table 4 will assure adequate blank levels and acceptable precision in most situations. However, they emphasize that if ultratrace analyses are to be carried out in a laboratory known to be subject to ambient metal fallout, as is often encountered in industry, clean-air facilities are warranted. In their review, Nieboer and Jusys consider the major sources of contamination, recommend a systematic approach to dealing with and evaluating adventitious additions, and provide appropriate references for cleaning laboratory ware and other de-contamination procedures. In the approach recommended by Nieboer and Jusys (1983) for the avoidance of contamination, the sophistication of the control measures required is linked to analytic need.

5.2.3   Determination of Chromium in Urine by EAAS

There is no doubt that the main deterrent to reliable chromium determinations is contamination (Versieck, 1985). Direct analysis of neat urine by EAAS is the most common practice, and this avoids many of the hindrances to attainment of low blank values (measures 6–9 in Table 4). Proper control of contamination is readily achieved when collection and storage procedures known to minimize extraneous additions are followed, including the use of thoroughly acid-washed containers and other laboratory utensils (measures 1–5 in Table 4; Brown et al., 1981; Nieboer and Jusys, 1983, 1988).

Table 4   Measures to Assure Low Blank Values

Measure	Rationale
1. Assure cleanliness of work area	To avoid obvious contamination
2. Wear talc-free disposable gloves	Sweat and talc are known sources of metals
3. Acid-wash glass and plastic ware	To prevent leaching and adsorption of metals
4. Contamination-free specimen collection	To eliminate additions during handling
5. Sealed specimen storage	To avoid contact with ambient air
6. Minimum sample size	To minimize additions and contact with laboratory ware
7. Use high-purity reagents	To reduce addition of contaminants
8. Minimize reagent additions	To reduce addition of contaminants
9. Direct, small-scale analytic procedures	To avoid excessive handling and transfers; to minimize surface area contact with containers and ambient air
10. If required, work in clean-air installations	To reduce blank values to acceptable levels

Source: Nieboer and Jusys, 1983.

Some technical complications in the EAAS determination step are evident (see Torgrimsen, 1982, for review). In determination of normal urinary chromium levels, uncompensated, nonspecific spectrometer absorption (uncorrected background) no doubt has contributed to artificially elevated concentrations (Guthrie et al., 1978; Veillon et al., 1982a,b). These difficulties arise because of the complexity of the urine medium matrix (Harnly and O'Haver, 1977) and are accentuated by the need to operate the spectrometer near the detection limit. Nise and Vesterberg (1979) clearly illustrate that for elevated chromium levels (such as observed in industrially exposed humans), this technical difficulty is less important (also see Aitio et al., 1984). A perusal of available reports makes it clear that the use of modern spectrometers virtually eliminates the background correction problem (Veillon et al., 1982a,b; Anderson and Bryden, 1983) and circumvents the need to modify commercially available equipment (Harnly and O'Haver, 1977; Kayne et al., 1978; Kumpulainen et al., 1979). Nevertheless, the method of standard additions is still recommended. Because background interference is highly dependent on the spectrometer and graphite furnace characteristics, it behooves the analyst to assess quantitatively its dependence on such factors as temperature programming (Guthrie et al., 1978; Nise and Vesterberg, 1979; Veillon et al., 1982b). For example, Schaller et al. (1973) used a charring temperature of $1,600°C$, which, according to Veillon et al. (1982b), causes premature loss of chromium before atomization. However, Nise and Vesterberg (1979) found only a small dissipation at this temperature. Similarly, Halls and Fell (1983) found, with conventional equipment and deuterium background correction, that reducing the atomization temperature from $2,700°C$ to $2,400°C$ with ashing at $1,000°C$ permitted the direct determination of chromium in 1:1 diluted urine without anomalous background interference.

5.2.4 Determination of Chromium in Serum by EAAS

Although serum chromium concentrations have not been extensively used to assess exposure, their determination is technically feasible. To date, only results obtained by neutron activation, EAAS, and gas chromatography–mass spectrometry have yielded reliable values (Versieck and Cornelis, 1980; Versieck, 1985). As indicated in Section 5.2.1, EAAS is most suitable for routine work. It is now widely accepted that almost all of the published normal serum chromium levels are in error, as is the case for urinary chromium (see Figs. 7 and 8). Two main reasons have been postulated to explain the observed discrepancies: volatilization during sample drying or dry ashing and contamination during sample collection and handling (Versieck and Cornelis, 1980). Considerable evidence is now available that losses during dry ashing procedures do not occur (Versieck et al., 1978, 1979; Versieck and Cornelis, 1980), leaving systematic and random errors due to contamination as explanation of the divergent reports. [Contrary to the findings of Shapcott (1979), our experience is that chromium volatilization *does* occur during perchloric acid digestion.] Versieck et al. (1982) have demonstrated that stainless steel syringe needles contribute significantly to the chromium concentration in blood samples. Consequently, use of the IUPAC-recommended blood collection procedure for nickel with a Teflon IV catheter needle seems prudent (Brown et al., 1981).

Dry ashing (Versieck et al., 1978; Anderson et al., 1985) or wet digestion (Kayne et al., 1978; Vanderlinde et al., 1979) techniques have proven satisfactory in the oxidative destruction of the serum matrix. It is our experience that dry ashing is more susceptible to extraneous additions or losses (by adsorption). The wet digestion techniques inherently involve less handling and minimize the surface area contact of the sample with containers and ambient air (see Table 4). Low-temperature (overnight at 80°C) HNO_3/H_2O_2 digestion has been used successfully. The chromium in the digested mixture is then determined after a fivefold dilution with ultrapure water (Kayne et al., 1978; Vanderlinde et al., 1979). Direct serum analysis after 1:1 dilution with H_2O also appears feasible (Kumpulainen et al., 1983).

The reliability and performance of methods for the determination of chromium in whole blood have not been extensively evaluated (Rahkonen et al., 1983).

5.2.5   Determination of Chromium in Tissues

In the regulatory framework, the measurement of chromium in tissues is not an option, and thus a detailed review of the techniques used is not warranted in the context of biological monitoring. Neither has the chromium content of hair been used extensively as an index of industrial exposure (see Section 5.4). Nevertheless, two comments about interpretation of the chromium content of autopsy samples are pertinent. First, the lack of a reliable database on the mineral content of human tissues (see Sections 5.1 and 5.4) is not a serious problem for assessing the accumulation of chromium in exposed organs, such as the lung, small intestine, and skin, but may well prevent the detection of marginal elevations in other organs. Second, the absence of evidence in many studies that rigorous protocols were followed to eliminate contamination during collection, handling, and analysis has cast doubt on the reported levels of chromium in organs not exposed to exterior factors.

5.3   Chromium in Body Fluids of Exposed People

Alsbirk et al. (1981) have used serum chromium convincingly as an index of exposure. These authors found in welders that serum chromium increased from 0.3 $\mu g\ L^{-1}$ to values near 2 $\mu g\ L^{-1}$ during 2.5 days of work. More recently, Rahkonen et al. (1983) compared the daily mean increase in chromium levels in whole blood, plasma, and urine of manual stainless steel welders. Although all three concentrations correlated with ambient chromium [as total chromium or Cr(VI)], the most significant relationship was found for whole blood (see Section 5.5.3 for further comments). Unfortunately, the authors do not elaborate on the analytic methods they used for blood and serum determinations nor on their reliability and performance. Other than that the levels reported for nonexposed people are high, the data presented appear to be reasonable. A report of average plasma levels of 35 ± 20 $\mu g\ L^{-1}$ for chromium manufacturing plant workers seems suspect (Wieser et al., 1982). These extremely high chromium concentrations are suggestive of contamination errors and need confirmation. Although there are some earlier reports of plasma and whole blood chromium for workers (Mancuso, 1951; Gafafer, 1953; Baetjer et al., 1959), the studies were completed before the awareness of the devastating effects of contamination on ultratrace analysis. Because of limited data, the present discussion is restricted to urinary chromium as an index of exposure.

Typical concentrations of chromium observed in the urine of exposed humans are depicted in Figures 9, 10, and 11, where they are plotted as a function of water-soluble chromium in the working environment. Values up to 150 μg L^{-1} have been reported. The parameters describing the relationships between urinary and ambient chromium concentrations in welders are summarized in Table 5.

Several conclusions may be derived from the data depicted in Figures 9, 10, and 11 and compiled in Table 5. First, a substantial fraction of the ambient chromium in manual metal arc welding of stainless steel is present as water-soluble Cr(VI), in agreement with Stern's conclusions. Stern (1982) also points out that welding fume is almost entirely respirable. Consequently, it may be deduced from the good relationships observed with the water-soluble fraction that chromate is readily absorbed by the respiratory tract. Second, since ΔCr yielded the best correlations with ambient Cr(VI), a rapid response of the worker to his environment must have occurred (half-time in terms of hours). This is in

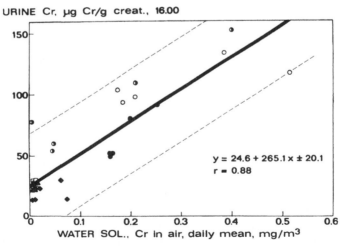

Figure 9. Relationship between the water-soluble chromium concentration, primarily Cr(VI), in the air and the creatinine-corrected urinary chromium for welders. Urine specimens were collected at the end of the work shift (at 4 PM) after showering. Six welders were monitored daily for 1 week; four of them did manual stainless steel welding, one did inert gas stainless steel welding, and one was involved only in the welding of mild steel. See Table 5 for complementary data on the cohort. (*Source:* Tola et al., 1977.)

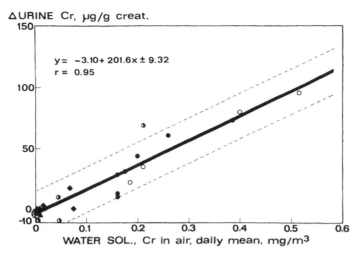

Figure 10. Relationship between the water-soluble chromium concentration in air and the increase in creatinine-corrected urinary chromium concentration (ΔCr) during the workday for the group of welders referred to in Figure 9. The preshift morning measurement (7 AM) was subtracted from that obtained for the afternoon collection (4 PM). (*Source:* Tola et al., 1977.)

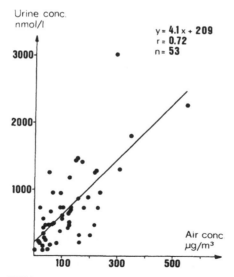

Figure 11. Eight-hour TWA air concentrations and unadjusted postshift (end-of-day) urine concentrations of chromium for stainless steel welders. To express the chromium concentration in MG L^{-1}, multiply by 0.052. (*Source:* Sjögren et al., 1983.)

57

Table 5 Relationship Between Ambient and Urinary Chromium (for Welders) (Urinary chromium concentrations, y, are in $\mu g\ g^{-1}$ creatinine or $\mu g\ L^{-1}$, and air chromium levels, x, are in mg Cr m^{-3}.)

Relationship	Equation	Correlation Coefficient	n
A. ΔCr(urine) versus Cr in air (water soluble)			
Creatinine corrected	$y = -3 + 202\,x \pm 9$	0.95	27[a,b]
Creatinine corrected	$y = 2 + 198\,x$	0.92	6[b,c,d]
Specific gravity corrected	$y = -11 + 236\,x \pm 23$	0.82	27[a]
Neat	$y = -21 + 335\,x \pm 56$	0.65	27[a]
B. ΔCr(urine) versus Cr in air (total)			
Creatinine corrected	$y = -12 + 179\,x \pm 13$	0.90	27[a]
Creatinine corrected	$y = 2 + 140\,x$	0.89	19[e]
C. ΔCr(urine) versus Cr in air (water insoluble)			
Creatinine corrected	$y = 15 + 51\,x \pm 30$	0.11	27[a]
D. Cr(urine) versus Cr in air (water soluble)			
Creatinine corrected (end of shift)	$y = 25 + 265\,x \pm 20$	0.88	27[a,b]
Creatinine corrected (end of shift)	$y = 6 + 466\,x$	0.96	6[b,c,f]
Creatinine corrected (end of shift)	$(y = 26 + 100\,x)$	0.91	5[b,g]
Neat (end of shift)	$y = 25 + 486\,x \pm 42$	0.85	27[a]
E. Cr(urine) versus Cr in air (total)			
specific gravity corrected (end of shift)	$y = 14 + 198\,x$	0.72	53[h]
Neat (end of shift)	$y = 11 + 213\,x$	0.68	53[h]
Neat (end of shift)	$(y = 30 + 400\,x)$		23[i]

[a] From Tola et al., 1977. ΔCr(urine) refers to difference in urinary concentration between afternoon (4 PM) and morning (7 AM) specimens. Nearly all the water-soluble chromium was Cr(VI). Reference specific gravity was 1.018.

[b] Recently, Mutti et al. (1984) reported the following relationships for 9 workers (6 welders, 2 chrome platers, and 1 chromate worker): $y = 4 + 167x$, $r = 0.88$, $n = 8$, $y = \Delta$Cr(urine, creat. corr.), $x = $ water-soluble Cr(VI); and $y = 11 + 384x$, $r = 0.88$, $n = 9$, $y = $ Cr(urine, creat. corr.), $x = $ water-soluble Cr(VI). The end-of-shift value and the daily increase in urinary chromium corresponding to the 50 $\mu g \cdot m^{-3}$ TLV were estimated as 30 and 12 $\mu g \cdot g^{-1}$ creatinine, respectively (predicted value for incremental increase is 10 $\mu g \cdot g^{-1}$, Table 6).

[c] From Mutti et al., 1979.

58

agreement with the accumulation and release kinetics of the uptake compartments depicted in Figure 6. Mathematical modeling determined that elimination half-times of 15–41 hours ($n = 4$) occurred for manual stainless steel welders in Tola et al.'s study (Tossavainen et al., 1980), and 5–35 hours ($n = 12$) for those studied by Welinder et al. (1983). Third, the nonzero y-intercept value for plots of urinary chromium versus soluble ambient chromium implies the existence of a compartment that decays with a longer half-time (see item D in Table 5). Welinder et al. (1983) assigned half-times of 14 days to infinity for the slowly decaying stores in welders and estimated that, on average, the fraction of chromium originating from the slow compartment was 31% (range 11–82%; $n = 12$). Again, this observation is in agreement with the model discussed in Sections 4.2 and 4.3. Finally, the strongest correlations were observed when creatinine-corrected urinary chromium values were employed, relative to specific gravity adjusted and neat concentrations. In Section 4.3, we interpreted this result as evidence that the dependence of chromium excretion on urine flow is similar to that known for creatinine (see Section 5.5.2 for additional explanation).

A preliminary estimate of the scaling factor relating urinary chromium concentrations to ambient water-soluble Cr(VI) concentrations is obtained from the ΔCr increment data in Table 5 (item A). Although additional confirmation is required, this evaluation permits the prediction of the daily increase in urinary chromium levels equivalent to a range of water-soluble ambient Cr(VI) values. Results for some ambient concentrations are given in Table 6, including the TWAs* recommended by ACGIH[†] and NIOSH. Both of these latter ambient concentrations would result in mild elevation of chromium excretion.

The data in Tables 5 and 7 suggest that the relationships found between ambient chromium and urinary levels seem to hold for chromeplating workers as well. This group is also exposed to water-soluble chromates. The reports by Guillemin and Berode (1978), Bovet et al. (1977), and Franchini et al. (1983) indicate that urinary chromium concentrations during the mid-1970s in chrome-plating shops were similar in magnitude to those reported in Figures 9 and 11 for

*TWA, time-weighted average threshold limit value (TLV).

[†]ACGIH, American Conference of Governmental Industrial Hygienists.

---

[d]In another group of workers ($n = 8$) the relationship was $y = 0.03 + 86 x$. This group was said to have lower chromium clearance, but see earlier comments in Section 4.3.

[e]From Welinder et al., 1983.

[f]In another group of workers ($n = 8$) the relationship was $y = 3 + 98 x$. See footnote d for additional comment.

[g]From Rahkonen et al., 1983. These data exhibit wide scatter.

[h]From Sjögren et al., 1983. End-of-shift, afternoon urine specimens were collected. Reference specific gravity was 1.024.

[i]From Gylseth et al., 1977. The equation was estimated from a nonlinear log plot in the original publication. Because of the large volume of urine samples analyzed, the large amounts of reagent addition, and the prolonged treatment of the sample, these results are judged to be unreliable.

Table 6 Predicted End-of-Shift Urinary Chromium Concentration Increase in Relation to Water-soluble Ambient Cr(VI) Concentrations

| Ambient Cr(VI) Concentration (mg Cr m^{-3}) | Predicted Incremental Increase in Urine Chromium Levels | |
	Creatinine corrected[a,b,c] (μg g^{-1} creatinine)	Neat[b,c] (μg L^{-1})
1.00	200	300
0.50	100	150
0.10	20	30
0.05[d]	10[e]	15
0.025[f]	5	7.5
0.02	4	6

[a] ΔCr(urine; μg/g of creatinine) = 200 Cr (air; mg m^{-3}).
[b] ΔCr(urine; μg L^{-1}) = 300 Cr (air; mg m^{-3}), based on the expression in footnote[a] and a urinary creatinine concentration of 1.5 g L^{-1} at an excretion rate of 1 ml minute^{-1} (Nieboer et al. 1984).
[c] To convert μg Cr to μmol Cr divide by 52.0.
[d] ACGIH recommended TWA.
[e] See footnote [b], Table 5.
[f] NIOSH recommended TWA.

Table 7 Urinary Chromium as an Index of Exposure for Chromeplating Workers

| Type of Plating | Water-soluble Ambient Cr(VI) (mg Cr m^{-3}) | n^a | Urinary Chromium (μg g^{-1} creatinine) | | |
			Observed[b]	n^a	Predicted[c]
Hard	0.06 $\pm$ 0.09	17	18 $\pm$ 16	21	18
Bright	0.006 $\pm$ 0.006	9	5 $\pm$ 3	13	4.5

[a] n, number of data points averaged (separate specimen collections).
[b] End-of-shift specimens.
[c] Calculated from: Urine (μg Cr g^{-1} creatinine) = 250 air (mg Cr m^{-3}) + 3. The relationship reported by Lindberg and Vesterberg (1983) appears unlikely, because it implies that daily urinary excretion exceeds uptake by a factor of about 4, namely, Urine (μg Cr L^{-1}) = 4,000 air (mg Cr m^{-3}) − 2.
Source: Guillemin and Berode, 1978.

welders. Current exposure appears to have been reduced considerably by improved industrial hygiene measures (Franchini et al., 1983).

Tandon et al. (1977) have reported extremely high urinary chromium levels for electroplaters and paint-pigment workers (up to 1,000 μg L^{-1}). Judging by the high value of 25 μg L^{-1} found for control subjects, the analytic methods employed appear suspect, especially since no information about them is provided. As explained in footnote c to Table 7, the recent report by Lindberg and Vesterberg (1983) for electroplaters also seems to contain technical deficiencies.

5.4   Chromium in Human Tissues and Hair

As pointed out in Section 5.1, the reference values for chromium in tissues of normal people are very dubious because of contamination during collection, handling, and analysis. Recently, background chromium levels in lung that appear to be reliable have been reported (Versieck, 1985). Unfortunately, authors reporting tissue chromium levels in exposed people fail to describe the analytic details of the methods employed, nor is care exercised to prevent contamination. Consequently, only those data that show a substantial increase over control values ($>$ 10-fold excess) can be considered as indicative of accumulation. A summary of the available data is provided in Table 8. It is clear that the lung is the main target organ. This is not surprising, since direct deposition occurs there. The occurrence of large amounts of chromium in lymph tissue also is found regularly. Accumulations in some of the other tissues may

Table 8   Chromium Contents of Tissues at Necropsy of Occupationally Exposed People[a]

Type of Worker	Tissue with Elevated Chromium Concentrations[b]	Other Tissues Examined[c]	Reference
Chromate manufacture	Lung	More than 20: liver, kidney, lymph nodes, etc.	Mancuso and Hueper, 1951
Chromate manufacture	Lung and other respiratory tract components, lymph nodes	—	Baetjer et al., 1959
Chromate manufacture	Lung	—	Mancuso, 1975
Chrome plater	Lung	—	Mancuso, 1975
Chromate-manufacture	Lung, adrenal	10 others: brain, liver, kidney, etc.	Hyodo et al., 1980
Chromate manufacture	Lung, lymph node	—	Tsuneta et al., 1980; Tsuneta, 1982
Chromate manufacture[d]	Lung, lymph node	Liver, spleen, kidney, heart	Teraoka, 1981
Chromeplater[e]	Lung, lymph node	Liver, spleen, kidney, heart	Teraoka, 1981

[a] All died of lung cancer, unless otherwise specified.
[b] A finding was considered reliable if the chromium concentration exceeded that of the control by at least a factor of 10.
[c] Accumulations in some of these tissues appeared to have occurred, but the analytic protocol described does not permit a firm conclusion (see text).
[d] Three workers: one died of stomach cancer, one of esophagus cancer, and one of lung cancer.
[e] Two workers: one died of lung cancer, and the other of pancreas cancer and metastatic lung cancer.

have occurred, but increases are not large enough relative to controls to conclude with confidence that this is so.

The data in Table 8 constitute good evidence for the presence of high chromium levels in the lung tissues of both chromate manufacture workers and chromeplaters, most of whom died of lung cancer. These pools appear to increase with the duration of exposure (Tsuneta et al., 1980; Tsuneta, 1982) and decay at a very slow rate ($t_{1/2}$ in years), since elevated levels still occurred 16 years (Mancuso, 1975) and 30 years (Teraoka, 1981) after last exposure. This finding is in agreement with the non-zero y-intercept values observed for plots of urinary chromium versus water-soluble chromium for welders (see Section 5.3), which implied such slowly decaying compartments.

Hair chromium levels are generally between 0.10 and 1.0 $\mu g\,g^{-1}$ (Pankhurst and Pate, 1979; Guthrie, 1982). Apparently, the significance of chromium concentrations in hair as an indicator of body stores is uncertain, although there is some evidence that diabetics have reduced amounts in hair. Indeed, elevated serum glucose levels have been correlated with increases in chromium excretion (Anderson et al., 1982a,b). Chromium in hair has not been used extensively as an index to exposure (Jenkins, 1979). Saner et al. (1984) have reported chromium levels in tannery workers of 17 ± 4 $\mu g\,g^{-1}$ in hair and of 7 ± 1 $\mu g\,L^{-1}$ in urine (compared to 0.6 ± 0.1 $\mu g\,g^{-1}$ and 0.22 ± 0.03 $\mu g\,L^{-1}$ for unexposed controls).

5.5   Urinary Chromium as an Exposure Index

5.5.1   Short-Term Versus Long-Term Excretion

We conclude that urinary chromium is a suitable index of exposure to water-soluble chromates [i.e., Cr(VI)]. The incremental increase in urine chromium during the working day (ΔCr) appears to be proportional to the ambient Cr(VI) level (Fig. 10, Table 5). Strong evidence exists that in vivo Cr(VI) is reduced to Cr(III). With reference to the model in Figure 6, release of Cr(III) from the uptake compartments is characterized by a half-time of 2.2 days or less, and this is in agreement with the value of 1–2 days established for the elimination half-time of chromium excretion in welders exposed to Cr(VI) (Tossavainen et al., 1980). This suggests that the Cr(VI)-reduction hypothesis in man is correct. Because the storage compartments in Figure 6 release Cr(III) slowly and because slowly decaying pools have been identified in the respiratory tract of chromium workers, a residual chromium excretion in urine should persist after the initial decay of newly absorbed chromium. As already indicated, the nonzero intercepts in plots such as in Figures 9 and 11 may be taken as evidence of such residual (basal) release in welders.

On the basis of the necropsy data reviewed in Table 8, one might suspect that the deposits in the respiratory tract of chromium workers are significant and, together with the naturally slow-releasing storage compartments, might therefore be expected to determine the long-term decay kinetics. As already indicated, the size of such pools appears to increase with the duration of exposure (Tsuneta et al., 1980; Tsuneta, 1982). Furthermore, the decay half-time may be expected to depend strongly on the oxidation state and solubility in biologic fluids of the deposits and other properties of solids, such as crystallinity, surface morphology, and surface reactivity (e.g., adsorption of protein and cell lysis). There is some experimental evidence for this conclusion. Bragt and van Dura (1983) administered three chromate salts by intratracheal injection to rats. They estimated that the release into the blood was 17 μg Cr $L^{-1}day^{-1}$ for sodium chromate (water soluble, see Table 3), 11 μg Cr $L^{-1}day^{-1}$ for zinc chromate (water insoluble), and 0.2 μg Cr $L^{-1}day^{-1}$ for lead chromate (water insoluble). (The zinc and lead salts were prepared by precipitation and most likely were amorphous rather than crystalline; the injected material had maximal diameters of 2.0 ± 0.9 μm and 3 ± 2 μm, respectively.) All three compounds were removed from the lung with half-times ($t_{1/2}$) of around 2 days and from the whole body with $t_{1/2}$ of 2–5 days (fast initial phase in first 4 days) and $t_{1/2} = 31$–95 days (slow subsequent phase). After the initial fluxes following the administration of the dose had settled down, all three chromates were lost from the blood and excreted into the urine with $t_{1/2} = 2$–6 days. Loss in the feces was monophasic ($t_{1/2} \sim 1$ day). At 10 days after the injection, the amount of the chromium dose excreted in the feces varied from 75% (Pb), to 45% (Zn) and 21% (Na), compared to total excretion in the urine of 20% (Na and Zn) and 2% (Pb). On day 51, the residual chromium content of the body was 8.3% of the initial sodium chromate dose, whereas it was 5.3% and 12.5% for the zinc and lead compounds, respectively. Langård et al. (1978) have also found in inhalation studies with rats that zinc chromate was absorbed quickly and excreted slowly. From these animal studies one concludes that water solubility is not a good index to the rate of internal mobilization and pathways of metal chromates. The similarity for all three chromates in the half-time values for whole-body chromium elimination, for the decay of blood chromium levels, and for the urinary excretion rates is strongly suggestive that absorbed chromate is handled in the same manner in the rat regardless of the source. The low urinary and high fecal excretion in the case of lead chromate most likely reflects a different clearance mechanism. Interestingly, retention of stainless steel manual-welding fumes in the lungs of man and rat have been shown to be significantly higher and clearance lower than of mild steel manual-welding fumes (Kalliomäki et al., 1983a,b). For nickel workers, the scaling factor that relates urinary nickel (in μg L^{-1}) to ambient nickel levels (in mg m^{-3}) had values of 700 for exposures to aerosols of dissolved nickel salts encountered in electrolytic refining and 20–85 for exposures to particulate nickel compounds encountered in pyrometallurgical refining (Nieboer et al., 1984). The magnitude of such linear scaling factors is, therefore, diagnostic of a particular working environment and likely reflects the speciation, composition, surface morphology and reactivity, solubility, and particle size of the ambient particulates.

Chromium(III) compounds appear generally to be taken up less readily than Cr(VI) compounds (see Section 4.1). Like welders (Tola et al., 1977; Stern, 1982) and ferrochromium workers (Langård et al., 1980), chromate manufacturing workers (Bourne and Yee, 1950; Buckell and Harvey, 1951; Gafafer,

1953) are known to be exposed to chromium in more than one oxidation state, namely, Cr(III) and Cr(VI) compounds. In addition, welders and ferrochromium workers are known to be exposed also to particulates of chromium metal or its alloys. Not surprisingly, therefore, Kalliomäki et al. (1981) found a good correlation ($r = 0.85$) between urinary chromium levels and the cumulative exposure (exposure time in years $\times$ % of time performing manual metal arc stainless steel welding), as well as between the average measured remanent magnetic fields of the chest and the cumulative exposure ($r = 0.82$). (The remanent magnetic field of the chest area is a direct measure of the metal contaminants that have accumulated in the lungs.) In follow-up studies, other correlations have been reported. Among manual stainless steel welders, the retention rate of magnetic dust in the lungs (the remanent magnetic field divided by exposure period in years) showed a relationship ($P < 0.01$) with ambient chromium levels [total Cr or Cr(VI)]. A strong correlation with the daily mean increase in whole blood chromium also existed (Rahkonen et al., 1983). For manual mild-steel welders, Kalliomäki et al. (1983a) estimated that about 1% of the dust deposited in the lung was retained ($\sim$70 mg year^{-1} at current exposures) and that the average clearance rate of retained material was 20% per year ($t_{1/2} =$ 3.5 years). This long half-time corresponds to a biologic residence time of 5 years and confirms the existence of long-term deposits.

To reiterate, the daily incremental increase in urine chromium (ΔCr) appears to be a better index of current exposure to water-soluble Cr(VI) than the end-of-shift value because of the large contribution of long-term excretion. This conclusion follows clearly from the analysis by Sjögren et al. (1983), as well as from the data in Table 5, although it was not realized by these researchers. A small contribution of water-soluble Cr(III) to ΔCr might also occur, although such compounds have low abundance in most industrial airborne particulates (Gafafer, 1953; Stern, 1982) and might be expected to be absorbed less readily. Consequently, the water-soluble ambient chromium concentration is a good measure of the short-term release of chromium in urine. Water extraction of air samples from locations in chromate production plants (Gafafer, 1953) and from airborne welding fumes (Thomsen and Stern, 1979; Stern, 1982) yield only Cr(VI). Mild alkali elution (with 1% carbonate) appears to dissolve the same fraction, whereas leaching with hot $NaOH-Na_2CO_3$ is effective in removing water-insoluble chromate constituents as well [total Cr(VI)] (Thomsen and Stern, 1979; Bhargava et al., 1983; Moreton et al., 1983). In view of the animal work by Bragt and van Dura (1983) reviewed in detail previously, the measurement of total Cr(VI) seems inherently of greater biologic significance than the evaluation of water-soluble Cr(VI). Supplementary to total Cr(VI) levels, the total ambient chromium level should be an index to the contaminant pool building up in the body and thus be determinative in the long-term basal urinary excretion. As suggested, the naturally slow-releasing storage compartments will necessarily contribute to this residual excretion. The in vivo mobilization of chromium deposits of exogenous origin may be expected to depend on the composition of the material to which people were exposed, and thus the work category, as already discussed. End-of-shift urinary chromium action levels (sum of short-term and long-term excretion) of $20–30 \mu g L^{-1}$ are recommended in some countries (Tola et al., 1977; Kalliomäki et al., 1981; Sjögren et al., 1983). Of course, as pointed out by Rantanen et al. (1982), action levels for biologic monitoring are arbitrary and serve only as an adjunct to environmental monitoring.

5.5.2 Creatinine Adjustment of Urinary Chromium

In Section 4.3, the evidence was reviewed that suggests that Cr(III) is largely excreted by glomerular filtration without significant tubular reabsorption. If confirmed, this indicates that creatinine adjustments do correct appropriately for differences in urine flow rates. On the other hand, if tubular reabsorption were to occur to a significant degree, as appears to be the case for Ni(II) and solutes (Nieboer et al., 1984; Sanford et al., 1984), creatinine adjustments would overcompensate for dilution.

For a single person, Araki (1980) has shown that the relationship in Equation (6) has general application for a number of analytes, including lead, creatinine, and total solutes:

$$U_i V^{b_i} = U_i^{\circ} \qquad (6)$$

In Equation (6), U_i is the concentration of i in the urine, V is the urine flow (ml minute^{-1} or L day^{-1}), U_i° is the urinary concentration of i at an excretion rate of 1 ml minute^{-1} or 1 L day^{-1} (and is arbitrarily designated the standard analyte excretion rate), and b_i is the volume exponent or power coefficient. For creatinine-like substances that are largely excreted by glomerular filtration, $b_i = 1.0$, whereas for substances that are subject to tubular reabsorption $b_i < 1.0$, such as appears to be the case for lead ($b_{Pb} = 0.75$), solutes ($b_{solute} = 0.67$), and nickel ($b_{Ni} = 0.67$) (Araki, 1980; Sanford et al., 1984). When $b_i = 1.0$, the excretion rate of an analyte (amount per unit time) is independent of urine flow, and simple dilution effects regulate its concentration. In instances when $b_i \neq 1.0$, the analyte excretion rate (i.e., UV; weight per unit time) has an implied dependence on urine flow.

Equation (7) may be derived from Equation 6, providing $b_{creat} = b_{Cr} = 1.0$.

$$U_{Cr} = \frac{U_{cr}^{\circ}}{U_{creat}^{\circ}} U_{creat} \qquad (7)$$

Since U_{Cr}° and U_{creat}° are constants, a linear relationship might be expected between U_{Cr} and U_{creat}, providing substances are excreted at a constant rate (amount per unit time). The latter condition is approximated by creatinine and may only be expected for chromium for healthy people who are not occupationally exposed and are on a normal diet. Since such relationships are observed (see Section 4.3; Saner, 1980; Anderson et al., 1983a), their existence may be correctly interpreted as indicating that chromium is excreted in a creatinine-like manner (i.e., by glomerular filtration). Creatinine correction of urinary chromium, therefore, appears justified.

5.5.3   Proposed Relationship Between Chromium in Plasma and Urine

When it is assumed that chromium is excreted primarily by glomerular filtration, it is possible to predict a relationship between plasma (serum), P_{Cr}, and urine chromium (U_{Cr}) concentrations. Equation (8) follows directly from the conventional clearance formula after correction for the fraction of chromium (α_{Cr}) not bound to protein (i.e., that available for filtration). Because chromium is an essential metal, reabsorption might be postulated for very low plasma concentrations, and this explains the intercept b in Equation 8:

$$P_{Cr} = \frac{V}{\alpha_{Cr}C_{Cr}} U_{Cr} + b \qquad (8)$$

C_{Cr} is the clearance of chromium (volume of plasma cleared per minute; units of ml minute^{-1}), and all concentrations are in $\mu g\,L^{-1}$. In a group of people, such as workers, V averages near 1.0 ml minute^{-1} (Nieboer et al., 1984), α_{Cr} is known to be 0.055 (Lim et al., 1983), and C_{Cr} would have the value 117 $\pm$ 20 ml minute^{-1} if glomerular filtration dominates and if there is normal kidney performance (Faulkner and King, 1976; Nieboer et al., 1984). Plugging these values into Equation 8 and considering the consensus normal value of P_{Cr} as 0.15 $\mu g\,L^{-1}$ and of U_{Cr} as 0.4 $\mu g\,L^{-1}$, Equation 9a follows:

$$P_{Cr} = 0.16\ U_{Cr} + 0.09\ \text{(neat urine)} \qquad (9a)$$

$$P_{Cr} = 0.27\ U_{Cr} + 0.09 \qquad (9b)$$

(creatinine-adjusted urine, $\mu g\ Cr\ g^{-1}$)

The slope in Equation 9b was estimated by multiplying that in Equation 9a by the urine/ambient chromium scaling factor ratio for neat and creatinine-corrected urines (see Table 5, entry A). The recent data reported by Rahkonen et al. (1983) for welders fit Equation 9b extremely well. Equation 9a implies that for $U_{Cr} > 5$ $\mu g\,L^{-1}$, P_{Cr} is about 15% of the value of U_{Cr}. This makes the analysis of urine more attractive than analysis of plasma, as is readily illustrated by the following examples. The action reference U_{Cr} value mentioned earlier of 30 $\mu g\,L^{-1}$ corresponds to 4.9 $\mu g\,L^{-1}$ of chromium in plasma, whereas a value of $U_{Cr} = 100$ $\mu g\,L^{-1}$ (Figs. 9, 11) yields $P_{Cr} = 16\ \mu g\,L^{-1}$. Additional measurements are required to confirm the relationships in Equation 9a and 9b.

Since absorption of chromate results in the accumulation of chromium in erythrocytes and absorption of Cr(III) compounds does not (see Section 4.2), erythrocyte chromium concentration (or perhaps whole blood chromium) would appear to be a specific index of exposure to Cr(VI). Indeed, the daily mean increase in whole blood chromium has been strongly correlated with ambient Cr(VI) for manual stainless steel welders (Rahkonen et al., 1983). Plasma chromium concentrations were about 80% of whole blood values, indicating a preference for erythrocytes.

ACKNOWLEDGMENT

Financial support for the preparation of this review was received from the Occupational Health and Safety Division of the Ontario Ministry of Labour and is gratefully acknowledged.

REFERENCES

Aisen, P., Aasa, R., and Redfield, A.G. (1969). "The chromium, manganese, and cobalt complexes of transferrin." *J. Biol. Chem.* **244**, 4628–4633.

Aitio, A., Järvisalo, J., Kiilunen, M., et al. (1984). "Urinary excretion of chromium as an indicator of exposure to trivalent chromium sulphate in leather tanning." *Int. Arch. Occup. Environ. Health* **54**, 241–249.

Alsbirk, K.E., Mogensen, C.E., Husted, S.E., and Geday, E. (1981). "Lever-og nyrefunktion ved svejsning i rustfrit stal." *Ugeskr. Laeg.* **143**, 112–116.

American Institute of Nutrition. (1981). "Research needed to improve data on mineral content of human tissues." *Fed. Proc.* **40**, 2111–2158. (Report on workshop sponsored by the American Institute of Nutrition, May 28–30, 1980. University of Maryland, College Park.)

Ames, B.N. (1983). "Dietary carcinogens and anticarcinogens." *Science* **221**, 1256–1264.

Anderson, M., Riley, D., and Rotruck, J. (1980). "Chromium(III) tris-acetylacetonate: An absorbable, bioactive source of chromium." *Fed. Proc.* **39**, 787.

Anderson, R.A. (1981). "Nutritional role of chromium." *Sci. Total Environ.* **17**, 13–29.

Anderson, R.A., and Bryden, N.A. (1983). "Concentration, insulin potentiation, and absorption of chromium in beer." *J. Agric. Food Chem.* **31**, 308–311.

Anderson, R.A., and Kozlovsky, A.S. (1985). "Chromium intake, absorption and excretion of subjects consuming self-selected diets." *Am. J. Clin. Nutr.* **41**, 1177–1183.

Anderson, R.A., Bryden, N.A., and Polansky, M. (1985). "Serum chromium of human subjects: Effects of chromium supplementation and glucose." *Am. J. Clin. Nutr.* **41**, 571–577.

Anderson, R.A., Polansky, M.M., Bryden, N.A., et al. (1983a). "Effects of chromium supplementation on urinary chromium excretion of human subjects and correlation of chromium excretion with selected clinical parameters." *J. Nutr.* **113**, 276–281.

Anderson, R.A., Polansky, M.M., Bryden, N.A., et al. (1983b). "Chromium supplementation of human subjects: Effects on glucose, insulin and lipid variables." *Metabolism* **32**, 894–899.

Anderson, R.A., Polansky, M.M., Bryden, N.A., et al. (1982a). "Urinary chromium excretion of human subjects: Effects of chromium supplementation and glucose loading." *Am. J. Clin. Nutr.* **36**, 1184–1193.

Anderson, R.A., Polansky, M.M., Bryden, N.A., et al. (1982b). "Effect of exercise (running) on serum glucose, insulin, glucagon, and chromium excretion." *Diabetes* **31**, 212–216.

Araki, S. (1980). "Effects of urinary volume on urinary concentrations of lead, δ-aminolaevulinic acid, coproporphyrin, creatinine, and total solutes." *Br. J. Ind. Med.* **37**, 50–54.

Armbruster, D.A., and Rudolph, F.B. (1976). "Rat liver pyruvate carboxylase. Inhibition by chromium nucleotide complexes." *J. Biol. Chem.* **251**, 320–323.

Baes, C.F. Jr., and Mesmer, R.E. (1976). *The Hydrolysis of Cations.* John Wiley & Sons, New York.

Baetjer, A.M., Damron, C., and Budacz, V. (1959). "The distribution and retention of chromium in men and animals." *A.M.A. Arch. Ind. Health* **20**, 136–150.

Beattie, J.K., and Haight, G.P. Jr. (1972). "Chromium(VI) oxidations of inorganic substrates." In J.O. Edwards, Ed., *Progress in Inorganic Chemistry Series, Inorganic Reaction Mechanisms.* Interscience Publishers (John Wiley & Sons), New York, Part II, pp. 93–145.

Bhargava, O.P., Bumsted, H.E., Grunder, F.I., et al. (1983). "Study of an analytical method for hexavalent chromium." *Am. Ind. Hyg. Assoc. J.* **44**, 433–436.

Bianchi, V., Levis, A.G., and Saggioro, D. (1979). "Differential cytotoxic activity of potassium dichromate on nucleoside uptake in BHK fibroblasts." *Chem. Biol. Interact.* **24**, 137–151.

Bourne, H.G. Jr., and Yee, H.T. (1950). "Occupational cancer in a chromate plant—An environmental appraisal." *Ind. Med. Surg.* **19**, 563–567.

Bovet, P., Lob, M., and Grandjean, M. (1977). "Spirometric alterations in workers in the chromium electroplating industry." *Int. Arch. Occup. Environ. Health* **40**, 25–32.

Bragt, P.C., and van Dura, E.A. (1983). "Toxicokinetics of hexavalent chromium in the rat after intratracheal administration of chromates of different solubilities." *Ann. Occup. Hyg.* **27**, 315–322.

Brief, R.S., Blanchard, J.W., Scala, R.A., and Blacker, J.H. (1971). "Metal carbonyls in the petroleum industry." *Arch. Environ. Health* **23**, 373–384.

Broughall, J.M., and Reid, R.A. (1974). "Some effects of potassium dichromate on mitochondrial adenosine triphosphate and respiratory activity." *Biochem. Soc. Trans.* **2**, 498–501.

Brown, G.I. (1974). *Introduction to Inorganic Chemistry.* Longman Group Ltd., London, p. 335.

Brown, S.S., Nomoto, S., Stoeppler, M., and Sunderman, F.W. Jr. (1981). "IUPAC reference method for analysis of nickel in serum and urine by electrothermal atomic absorption spectrometry." *Pure Appl. Chem.* **53**, 773–781.

Buckell, M., and Harvey, D.G. (1951). "An environmental study of the chromate industry." *Br. J. Ind. Med.* **8**, 298–301.

Bunker, V.W., Lawson, M.S., Delves, H.T., and Clayton, B.E. (1984). "The uptake and excretion of chromium by the elderly." *Am. J. Clin. Nutr.* **39**, 797–802.

Campomar, J.A., Fiol, J.J., Terron, A., and Moreno, V. (1986). "Chromium(III) interactions with nucleotides. II." *Inorg. Chim. Acta* **124**, 75–81.

Cary, E.E. (1982). "Chromium in air, soil and natural waters." In S. Langård, Ed., *Biological and Environmental Aspects of Chromium, Topics in Environmental Health.* Elsevier Biomedical Press, Amsterdam, Vol. 5, pp. 49–64.

Cavalleri, A., Minoia, C., Richelmi, P., et al. (1985). "Determination of total and hexavalent chromium in bile after intravenous administration of potassium dichromate in rats." *Environ. Res.* **37**, 490–496.

Chargaff, E., and Green, C. (1948). "On the inhibition of the thromboplastic effect." *J. Biol. Chem.* **173**, 263–270.

Chen, N.S.C., Tsai, A., and Dyer, I.A. (1973). "Effect of chelating agents on chromium absorption in rats." *J. Nutr.* **103**, 1182–1186.

Clark, J.H. (1959). "The denaturation of proteins by chromium salts." *A.M.A. Arch. Ind. Health* **20**, 117–123.

Collins, R.J., Fromm, P.O., and Collings, W.D. (1961). "Chromium excretion in the dog." *Am. J. Physiol.* **201**, 795–798.

Connett, P.H., and Wetterhahn, K.E. (1983). "Metabolism of the carcinogen chromate by cellular constituents." In *Structure and Bonding. Inorganic Elements in Biochemistry*. Springer-Verlag, Berlin, Vol. 54, pp. 93–124.

Cooper, J.A., Blackwell, L.F., and Buckley, P.D. (1984). "Chromium(III) complexes and their relationship to the glucose tolerance factor. Part II. Structure and activity of amino acid complexes." *Inorg. Chim. Acta* **92**, 23–31.

Cotton, F.A., and Wilkinson, G. (1980). *Advanced Inorganic Chemistry, A Comprehensive Text*, 4th ed. John Wiley & Sons, New York.

Danchin, A. (1975). "Labelling of biological macromolecules with covalent analogs of magnesium. II. Features of the chromic Cr(III) ion." *Biochimie* **57**, 875–880.

Dautry-Varsat, A., and Lodish, H.F. (1984). "How receptors bring proteins and particles into cells." *Sci. Am.* **250** (5), 52–58.

DeFlora, S. (1978). "Metabolic deactivation of mutagens in the *Salmonella*-microsome test." *Nature* **271**, 455–456.

Deliconstantinos, G., Trichopoulou, A., and Kapoulas, V.M. (1984). "Rat liver microsomal cholesterol biosynthesis and drug oxidase activity are affected by chromium (Cr^{3+}) in vitro." *Int. J. Biochem.* **16**, 935–938.

DePamphilis, M.L., and Cleland, W.W. (1973). "Preparation and properties of chromium(III)-nucleotide complexes for use in the study of enzyme mechanisms." *Biochemistry* **12**, 3714–3724.

Doisy, R.J., Streeten, D.H.P., Levine, R.A., and Chodos, R.B. (1969). "Effects and metabolism of chromium in normals, elderly subjects, and diabetics." In D.D. Hemphill, Ed., *Trace Substances in Environmental Health*. University of Missouri, Columbia, MO, Vol. 2, pp. 75–82.

Donaldson, D.L., Smith, C.C., and Yunice, A.A. (1984). "Renal excretion of chromium-51 chloride in the dog." *Am. J. Physiol.* **246** (Renal Fluid Electrolyte Physiol., 15), F870–878.

Donaldson, D.L., Anderson, R.A., Veillon, C., and Mertz, W.E. (1982). "Renal excretion of orally and parenterally administered chromium-51." *Fed. Proc.* **41**, 391.

Donaldson, D.L., Lee, D.M., Smith, C.C., and Rennert, O.M. (1985). "Glucose tolerance and plasmid lipid distributions in rats fed a high-sucrose, high-cholesterol, low-chromium diet." *Metabolism* **34**, 1086–1093.

Donaldson, R.M. Jr., and Barreras, R.F. (1966). "Intestinal absorption of trace quantities of chromium." *J. Lab. Clin. Med.* **68**, 484–493.

Earley, J.E., and Cannon, R.D. (1965). "Aqueous chemistry of chromium(III)." In R.L. Carlin, Ed., *Transition Metal Chemistry*. Marcel Dekker, Inc., New York, Vol. 1, pp. 33–109.

Eisinger, J., Shulman, R.G., and Szymanski, B.M. (1962). "Transition metal binding in DNA solutions." *J. Chem. Phys.* **36**, 1721–1729.

Elias, A.N., Grossman, M.K., and Valenta, L.J. (1984). "Use of the artificial beta cell (ABC) in the assessment of peripheral insulin sensitivity: Effect of chromium supplementation in diabetic patients." *Gen. Pharmacol.* **15**, 535–539.

Emery, T. (1978). "The storage and transport of iron." In H. Sigel, Ed., *Metal Ions in Biological Systems*. Marcel Dekker, Inc., New York, Vol. 7, pp. 77–126.

Espenson, J.H. (1970). "Oxidation of transition metal complexes by chromium(VI)." *Accounts Chem. Res.* **3**, 347–353.

Evans, C.A., Guevremont, R., and Rabenstein, D.L. (1979). "Metal complexes of aspartic acid and glutamic acid." In H. Sigel, Ed., *Metal Ions in Biological Systems*. Marcel Dekker, Inc., New York, Vol. 9, pp. 41–75.

Faulkner, W.R., and King, J.W. (1976). "Renal function." In N.W. Tietz, Ed., *Fundamentals of Clinical Chemistry*. W.B. Saunders, Philadelphia, pp. 975–1014.

Fishbein, L. (1984). "Overview of analysis of carcinogenic and/or mutagenic metals in biological and environmental samples. 1. Arsenic, beryllium, cadmium, chromium and selenium." *Int. J. Environ. Anal. Chem.* **17**, 113–170.

Flesch, P. (1958). "Chemical data on human epidermal keratinization and differentiation." *J. Invest. Derm.* **31**, 63–73.

Foley, E.F. Jr. (1978). "Chromium chemicals manufacture." In *Proceedings of the Symposium on Health Aspects of Chromium Containing Materials.* September 15, 1977. Baltimore, MD, pp. 1–11.

Fornace, A.J. Jr., Seres, D.S., Lechner, J.F., and Harris, C.C. (1981). "DNA-protein cross-linking by chromium salts." *Chem. Biol. Interact.* **36**, 345–354.

Franchini, I., Magnani, F., and Mutti, A. (1983). "Mortality experience among chromeplating workers." *Scand. J. Work Environ. Health* **9**, 247–252.

Freund, H., Atamian, S., and Fischer, J.E. (1979). "Chromium deficiency during total parenteral nutrition." *JAMA* **241**, 496–498.

Fuwa, K., Wacker, W.E.C., Druyan, R., et al. (1960). "Nucleic acids and metals. II: Transition metals as determinants of the conformation of ribonucleic acids." *Proc. Natl. Acad. Sci. USA* **46**, 1298–1307.

Gafafer, W.M. (Ed.) (1953). *Health of Workers in Chromate Producing Industry: A Study.* U.S. Public Health Service, Division of Occupational Health, Pub. No. 192, Washington, D.C.

Garcia, J.D., and Wetterhahn Jennette, K. (1981). "Electron-transport cytochrome P-450 system is involved in the microsomal metabolism of the carcinogen chromate." *J. Inorg. Biochem.* **14**, 281–295.

Garner, C.S., and House, D.A. (1970). "Amine complexes of chromium(III)." In R.L. Carlin, Ed., *Transition Metal Chemistry.* Marcel Dekker, Inc., New York, Vol. 6, pp. 59–295.

Gergely, A., and Sóvágó, I. (1979). "The coordination chemistry of L-cysteine and D-penicillamine." In H. Sigel, Ed., *Metal Ions in Biological Systems.* Marcel Dekker, Inc., New York, Vol. 9, pp. 77–102.

Gibson, R.A., MacDonald, A.C., and Martinez, O.B. (1985). "Dietary chromium and manganese intakes of a selected sample of Canadian elderly woman." *Hum. Nutr. Appl. Nutr.* **39A**, 43–52.

Goldman, M., and Karotkin, R.H. (1935). "Acute potassium bichromate poisoning." *Am. J. Med. Sci.* **189**, 400–403.

Gonzalez-Vergara, E., Hegenauer, J., and Saltman, P. (1982). "Biological complexes of chromium: A second look at glucose tolerance factor." *Fed. Proc.* **41**, 286.

Goodgame, D.M.L., Hayman, P.B., and Hathway, D.E. (1982). "Carcinogenic chromium(VI) forms chromium(V) with ribonucleotides but not with deoxyribonucleotides." *Polyhedron* **1**, 497–499.

Gould, E.S. (1986). "Reductions of carboxylate-bound chromium(V)." *Accounts Chem. Res.* **19**, 66–72.

Gray, S.J., and Sterling, K. (1950). "The tagging of red cells and plasma proteins with radioactive chromium." *J. Clin. Invest.* **29**, 1604–1613.

Gruber, J.E., and Jennette, K.W. (1978). "Metabolism of the carcinogen chromate by rat liver microsomes." *Biochem. Biophys. Res. Commun.* **82**, 700–706.

Guillemin, M.P., and Berode, M. (1978). "A study of the difference in chromium exposure in workers in two types of electroplating process." *Ann. Occup. Hyg.* **21**, 105–112.

Gupta, R.K., Oesterling, R.M., and Mildvan, A.S. (1976). "Dual divalent cation requirement for activation of pyruvate kinase: Essential roles of both enzyme- and nucleotide-bound metal ions." *Biochemistry* **15**, 2881–2887.

Gustavson, K.H. (1958). "A novel type of metal-protein compounds." *Nature* **182**, 1125–1128.

Guthrie, B.E. (1982). "The nutritional role of chromium." In S. Langård, Ed., *Biological and Environmental Aspects of Chromium, Topics in Environmental Health.* Elsevier Biomedical Press, Amsterdam, Vol. 5, pp. 117–148.

Guthrie, B.E., Wolf, W.R., and Veillon, C. (1978). "Background correction and related problems in the determination of chromium in urine by graphite furnace atomic absorption spectrometry." *Anal. Chem.* **50**, 1900–1902.

Gylseth, B., Gundersen, N., and Langård, S. (1977). "Evaluation of chromium exposure based on a simplified method for urinary chromium determination." *Scand. J. Work Environ. Health* **3**, 28–31.

Halls, D.J., and Fell, G.S. (1983). "Determination of chromium in urine by graphite furnace atomic absorption spectrometry." In P. Brätter and P. Schramel, Eds., *Trace Element-Analytical Chemistry in Medicine and Biology.* Walter de Gruyter, Berlin, Vol. 2, pp. 667–673.

Harnly, J.M., and O'Haver, T.C. (1977). "Background correction for the analysis of high-solids samples by grpahite furnace atomic absorption." *Anal. Chem.* **49**, 2187–2193.

Harnly, J.M., Patterson, K.Y., Veillon, C., et al. (1983). "Comparison of electrothermal atomic absorption spectrometry and atomic emission spectrometry for determination of chromium in urine." *Anal. Chem.* **55**, 1417–1419.

Harris, D.C. (1977). "Different metal-binding properties of the two sites of human transferrin." *Biochemistry* **16**, 560–564.

Hartford, W.H. (1979). "Chromium compounds." In *Kirk-Othmer Encyclopedia of Chemical Technology,* 3rd ed. John Wiley & Sons, New York, Vol. 6, pp. 82–120.

Hartford, W.H., and Darrin, M. (1958). "The chemistry of chromyl compounds." *Chem. Rev.* **58**, 1–61.

Hartford, W.H., Lane, K.A., and Meyer, W.A. (1950). "Calcium dichromate." *J. Am. Chem. Soc.* **72**, 3353–3356.

Haylock, S.J., Buckley, P.D., and Blackwell, L.F. (1983a). "Separation of biologically active chromium-containing complexes from yeast extracts and other sources of glucose tolerance factor (GTF) activity." *J. Inorg. Biochem.* **18**, 195–211.

Haylock, S.J., Buckley, P.D., and Blackwell, L.F. (1983b). "The relationship of chromium to the glucose tolerance factor. II." *J. Inorg. Biochem.* **19**, 105–117.

Henry, R.J., and Smith, E.C. (1946). "Use of sulfuric acid-dichromate mixture in cleaning glassware." *Science* **104**, 426–427.

Herrmann, H., and Speck, L.B. (1954). "Interaction of chromate with nucleic acids in tissues." *Science* **119**, 221.

Hopkins, L.L. Jr. (1965). "Distribution in the rat of physiological amounts of injected ^{51}Cr(III) with time." *Am. J. Physiol.* **209**, 731–735.

Hopkins, L.L. Jr., and Schwarz, K. (1964). "Chromium(III) binding to serum proteins, specifically siderophilin." *Biochim. Biophys. Acta* **90**, 484–491.

Hughes, M.N. (1981). *The Inorganic Chemistry of Biological Processes,* 2nd ed. John Wiley & Sons, Chichester, UK, pp. 243–255.

Hyodo, K., Suzuki, S., Furuya, N., and Meshizuka, K. (1980). "An analysis of chromium, copper, and zinc in organs of a chromate worker." *Int. Arch. Occup. Environ. Health* **46**, 141–150.

IARC (1980). "Chromium and chromium compounds." *IARC Monogr. Eval. Carcinog. Risk Chem. Humans* **23**, 205–323.

Jain, R., Verch, R.L., Wallach, S., and Peabody, R.A. (1981). "Tissue chromium exchange in the rat." *Am. J. Clin. Nutr.* **34**, 2199–2204.

Janson, C.A., and Cleland, W.W. (1974). "The specificity of chromium nucleotides as inhibitors of selected kinases." *J. Biol. Chem.* **249**, 2572–2574.

Jeejeebhoy, K.N., Chu, R.C., Marliss, E.B., et al. (1977). "Chromium deficiency, glucose intolerance, and neuropathy reversed by chromium supplementation, in a patient receiving long-term total parenteral nutrition." *Am. J. Clin. Nutr.* **30**, 531–538.

Jenkins, D.W. (1979). *Toxic Trace Metals in Mammalian Hair and Nails.* EPA Report 600/4-79-049, U.S. Environmental Protection Agency, Las Vegas, Nevada, 184 pp.

Jett, R. Jr., Pierce, J.O., and Stemmer, K.L. (1968). "Toxicity of alloys of ferrochromium. III. Transport of chromium(III) by rat serum protein studied by immunoelectrophoretic analysis and autoradiography." *Arch. Environ. Health* **17**, 29–34.

Kalliomäki, P.-L., Kalliomäki, K., Rahkonen, E., and Aittoniemi, K. (1983a). "Follow-up study on the lung retention of welding fumes among shipyard welders." *Ann. Occup. Hyg.* **27**, 449–452.

Kalliomäki, P.-L., Junttila, M.-L., Kalliomäki, K.K., et al. (1983b). "Comparison of the behaviour of stainless and mild steel manual metal arc welding fumes in rat lung. *Scand. J. Work Environ. Health* **9**, 176–180.

Kalliomäki, P.-L., Rahkonen, E. Vaaranen, V., et al. (1981). "Lung-retained contaminants, urinary chromium and nickel among stainless steel welders." *Int. Arch. Occup. Environ. Health* **49**, 67–75.

Kayne, F.J., Komar, G., Laboda, H., and Vanderlinde, R.E. (1978). "Atomic absorption spectrophotometry of chromium in serum and urine with a modified Perkin-Elmer 603 atomic absorption spectrophotometer." *Clin. Chem.* **24**, 2151–2154.

Kelly, W.F., Ackrill, P., Day, J.P., et al. (1982). "Cutaneous absorption of trivalent chromium: Tissue levels and treatment by exchange transfusion." *Br. J. Ind. Med.* **39**, 397–400.

King, M.V. (1978). "Toxic hazards arising from contamination during use of chromic-sulfuric acid cleaning solution." *Am. Lab.* **10**, 68.

King, R.B. (1976). "Transition metal carbonyls." In K. Niedenzu and H. Zimmer, Eds., *Methodicum Chimicum. Preparation of Transition Metal Derivatives.* Academic Press, New York, Vol. 8, pp. 421–468.

Kitagawa, S., Seki, H., Kametani, F., and Sakurai, H. (1982). "Uptake of hexavalent chromium by bovine erythrocytes and its interaction with cytoplasmic components; The role of glutathione." *Chem. Biol. Interact.* **40**, 265–274.

Klaassen, C.D. (1980). "Adsorption, distribution, and excretion of toxicants." In J. Doull et al., Eds., *Casarett and Doull's Toxicology,* 2nd ed. Macmillan Publishing Co., New York, pp. 35–37.

Kosta, L. (1982). "Contamination as a limiting parameter in trace analysis." *Talanta,* **29**, 985–992.

Kosta, L., and Byrne, A.R. (1982). "Analytical evaluation of comparative data on trace elements in biological materials." *J. Radioanal. Chem.* **69**, 117–129.

Kuehn, K., and Gebhardt, E. (1960). "Chemische und elektronenoptische untersuchungen uber die keaktion von chrom(III)-komplexen mit kollagen." *Z. Naturforsch.* **156**, 23–30.

Kuehn, K., and Sunderman, F.W. Jr. (1982). "Dissolution half-times of nickel compounds in water, rat serum, and renal cytosol." *J. Inorg. Biochem.* **17**, 26–39.

Kumpulainen, J.T., Wolf, W.R., Veillon, C., and Mertz, W. (1979). "Determination of chromium in selected United States diets." *J. Agric. Food Chem.* **27**, 490–494.

Kumpulainen, J.T., Lehto, J., Koivistoinen, P., et al. (1983). "Determination of chromium in human milk, serum and urine by electrothermal atomic absorption spectrometry without preliminary ashing." *Sci. Total Environ.* **31**, 71–80.

Kurzweil, P.R., Miller, D.R., Freeman, J.E., et al. (1984) "Use of sodium chromate Cr51 in diagnosing childhood idiopathic pulmonary hemosiderosis." *Am. J. Dis. Child.* **138**, 746–748.

Langård S. (1979). "The time-related subcellular distribution of chromium in the rat liver cell after intravenous administration of $Na_2{}^{51}CrO_4$." *Biol. Trace Elem. Res.* **1**, 45–54.

Langård, S. (1982). "Absorption, transport and excretion of chromium in man and animals." In S. Langård, Ed., *Biological and Environmental Aspects of Chromium, Topics in Environmental Health.* Elsevier Biomedical Press, Amsterdam, Vol. 5, pp. 149–169.

Langård, S., Andersen, Aa., and Gylseth, B. (1980). "Incidence of cancer among ferrochromium and ferrosilicon workers." *Br. J. Ind. Med.* **37**, 114–120.

Langård, S., Gundersen, N., Tsalev, D.L., and Gylseth, B. (1978). "Whole blood chromium level and chromium excretion in the rat after zinc chromate inhalation." *Acta Pharmacol. Toxicol.* **42**, 142–149.

Legg, J.I. (1978). "Substitution-inert metal ions as probes of biological function." *Coordination Chem. Rev.* **25**, 103–132.

Levis, A.G., and Bianchi, V. (1982). "Mutagenic and cytogenetic effects of chromium compounds." In S. Langård, Ed., *Biological and Environmental Aspects of Chromium, Topics in Environmental Health.* Elsevier Biomedical Press, Amsterdam, Vol. 5, pp. 171–208.

Levis, A.G., Bianchi, V., Tamino, G., and Pegoraro, B. (1978a). "Cytotoxic effects of hexavalent and trivalent chromium on mammalian cells in vitro." *Br. J. Cancer* **37**, 386–396.

Levis, A.G., Buttignol, M., Bianchi, V., and Sponza, G. (1978b). "Effects of potassium dichromate on nucleic acid and protein syntheses and on precursor uptake in BHK fibroblasts." *Cancer Res.* **38**, 110–116.

Levy, L.S., Martin, P.A., and Bidstrup, P.L. (1986). "Investigation of the potential carcinogenicity of a range of chromium-containing materials on rat lung." *Br. J. Ind. Med.* **43**, 243–256.

Light, A., and Morgan, E.H. (1982). "Transferrin endocytosis in reticulocytes: An electron microscope study using colloidal gold. *Scand. J. Haematol.* **28**, 205–214.

Lim, T.H., Sargent, T., and Kusubov, N. (1983). "Kinetics of trace element chromium(III) in the human body." *Am. J. Physiol.* **244**, (Regulatory Integrative Comp. Physiol., 13), R445–R454.

Lindberg, E., and Vesterberg, O. (1983). "Monitoring exposure to chromic acid in chromeplating by measuring chromium in urine." *Scand. J. Work Environ. Health* **9**, 333–340.

Luciani, S., DalToso, R., Rebellato, A.M., and Levis, A.G. (1979). "Effects of chromium compounds on plasma membrane Mg^{2+}-ATPase activity of BHK cells." *Chem. Biol. Interact.* **27**, 59–67.

MacKenzie, R.D., Anwar, R.A., Byerrum, R.U., and Hoppert, C.A. (1959). "Absorption and distribution of ^{51}Cr in the albino rat." *Arch. Biochem. Biophys.* **79**, 200–205.

Mali, J.W.H., van Kooten, W.J., and van Neer, F.C.J. (1963). "Some aspects of the behaviour of chromium compounds in the skin." *J. Invest. Derm.* **41**, 111–122.

Mancuso, T.F. (1951). "Occupational cancer and other health hazards in a chromate plant: A medical appraisal. II. Clinical and toxicologic aspects." *Ind. Med. Surg.* **20**, 393–407.

Mancuso, T.F. (1975). "Consideration of chromium as an industrial carcinogen." *Symposium Proceedings. International Conference on Heavy Metals in the Environment,* Toronto, Ontario, Canada, Oct. 27–31, Vol. 3, pp. 343–356.

Mancuso, T.F., and Hueper, W.C. (1951). "Occupational cancer and other health hazards in a chromate plant: A medical appraisal. I. Lung cancers in chromate workers." *Ind. Med. Surg.* **20**, 358–363.

Manzo, L., DiNucci, A., Edel, J., et al. (1983). "Biliary and gastrointestinal excretion of chromium after administration of Cr-III and Cr-VI in rats." *Res. Commun. Chem. Pathol. Pharmacol.* **42**, 113–125.

Martell, A.E., and Smith, R.M. (1977). *Critical Stability Constants. Other Organic Ligands.* Plenum Press, New York, Vol. 3, 495 pp.

Martin, R.B., and Mariam, Y.H. (1979). "Interactions between metal ions and nucleic bases, nucleosides, and nucleotides in solution." In H. Sigel, Ed., *Metal Ions in Biological Systems.* Marcel Dekker, Inc., New York, Vol. 8, pp. 57–116.

Marx, J.L. (1983). "Do tumor promoters affect DNA after all?" *Science* **219**, 158–159.

de Meester, P., and Hodgson, D.J. (1977). "Synthesis and structural characterization and L-histidinato-D-penicillaminatochromium(III) monohydrate." *J. Chem. Soc. Dalton Trans.* 1604–1607.

de Meester, P., Hodgson, D.J., Freeman, H.C., and Moore, C.J. (1977). "Tridentate coordination by the L-cysteine dianion. Crystal and molecular structure of sodium bis(L-cysteinato)chromate(III) dihydrate." *Inorg. Chem.* **16**, 1494–1498.

Mertz, W. (1969). "Chromium occurrence and function in biological systems." *Physiol. Rev.* **49**, 163–239.

Mertz, W. (1975). "Effects and metabolism of glucose tolerance factor." *Nutr. Rev.* **33**, 129–135.

Mertz, W. (1979). "Chromium—An overview." D. Shapcott and J. Hubert, Eds., *Developments in Nutrition and Metabolism. Chromium in Nutrition and Metabolism.* Elsevier/North Holland Biomedical Press, Amsterdam, Vol. 2, pp. 1–14.

Mertz, W. (1980). "Mineral elements: New perspectives." *J. Am. Diet. Assoc.* **77**, 258–263.

Mertz, W. (1982a). "Clinical and public health significance of chromium." In A.S. Prasad, Ed., *Clinical, Biochemical, and Nutritional Aspects of Trace Elements.* Alan R. Liss, Inc., New York, pp. 315–323.

Mertz, W. (1982b). "Chromium: An essential micronutrient." *Contemp. Nutr.* **7**(3), 2 pp.

Mertz, W. (1983). "Chromium: An ultra-trace element." *Chem. Scripta* **21**, 145–150.

Mertz, W., Toepfer, E.W., Roginski, E.E., and Polansky, M.M. (1974). Present knowledge of the role of chromium. *Fed. Proc.* **33**, 2275–2280.

Micskei, K., Debreczeni, F., and Nagypal, I. (1983). "Equilibria in aqueous solutions of some chromium (2+) complexes." *J. Chem. Soc. Dalton Trans.* 1335–1338.

Mitewa, M., and Bontchev, P.R. (1985). "Chromium(V) coordination chemistry." *Coord. Chem. Rev.* **61**, 241–272.

Moreton, J., Bettelley, J., Mathers, H., et al. (1983). "Investigation of techniques for the analysis of hexavalent chromium, total chromium and total nickel in welding fume: A co-operative study." *Ann. Occup. Hyg.* **27**, 137–156.

Morgan, J.F. (1978). "Use of chromates in pigments." In *Proceedings of the Symposium on Health Aspects of Chromium Containing Materials.* September 15, 1977. Baltimore, MD. Industrial Health Foundation, Pittsburgh, PA, pp. 39–48.

Mossop, R.T. (1983). "Effects of chromium(III) on fasting blood glucose, cholesterol and cholesterol HDL levels in diabetics." *Cent. Afr. J. Med.* **29**, 80–82.

Mutti, A., Cavatorta, A., Pedroni, C., et al. (1979). "The role of chromium accumulation in the relationship between airborne and urinary chromium in welders." *Int. Arch. Occup. Environ. Health* **43**, 123–133.

Mutti, A., Pedroni, C., Arfini, G., et al. (1984). "Biological monitoring of occupational exposure to different chromium compounds at various valency states." *Int. J. Environ. Anal. Chem.* **17**, 35–41.

Nag, K., and Bose, S.N. (1985). "Chemistry of tetra- and pentavalent chromium." *Struct. Bond. (Berl.)* **63**, 153–197.

Naismith, W.E.F. (1958). "The cross-linking of conarachin II with metal salts." *Arch. Biochem. Biophys.* **73**, 255–261.

Neilands, J.B. (1979). "Biomedical and environmental significance of siderophores." In N. Kharasch, Ed., *Trace Metals in Health and Disease.* Raven Press, New York, pp. 27–41.

Nieboer, E., and Jusys, A.A. (1983). "Contamination control in routine ultratrace analysis of toxic metals." In S.S. Brown and J. Savory, Eds., *Clinical Chemistry and Clinical Toxicology of Metals.* Academic Press, London, pp. 3–16.

Nieboer, E., and Jusys, A.A. (1988). "Collection, storage and shipment of urine, serum, plasma and blood specimens for nickel analysis." *Pure Appl. Chem.* in press.

Nieboer, E., and Richardson, D.H.S. (1980). "The replacement of the nondescript term 'heavy metals' by a biologically and chemically significant classification of metal ions." *Environ. Pollut. (B)* **1**, 3–26.

Nieboer, E., and Sanford, W.E. (1985). "Essential, toxic and therapeutic functions of metals (including determinants of reactivity)." In E. Hodgson, J.R. Bend and R.M. Philpot, Eds., *Reviews in Biochemical Toxicology,* Elsevier, New York, Vol. 7, pp. 205–245.

Nieboer, E., Yassi, A., Jusys, A.A., and Muir, D.C.F. (1984). *Technical Feasibility and Usefulness of Biological Monitoring in the Nickel-Producing Industry.* Special Report, McMaster University, Hamilton, Canada, 285 pp.

NIOSH. (1973). *Criteria for a Recommended Standard: Occupational Exposure to Chromic Acid.* DHEW (NIOSH), Pub. 73-11021, National Institute for Occupational Safety and Health, Cincinnati, Ohio.

NIOSH. (1975). *Criteria for a Recommended Standard: Occupational Exposure to Chromium(VI).* DHEW (NIOSH), Pub. 76-129, National Institute for Occupational Safety and Health, Cincinnati, Ohio.

NIOSH. (1977). *Registry of Toxic Effects of Chemical Substances.* DHEW (NIOSH), Pub. 78-104-B. National Institute for Occupational Safety and Health, Cincinnati, Ohio, Vol. 2, pp. 296, 590.

Nise, G., and Vesterberg, O. (1979). "Direct determination of chromium in urine by electrothermal atomic absorption spectrometry." *Scand. J. Work Environ. Health* 5, 404–410.

Nomiyama, H., Yotoriyama, M., and Nomiyama, K. (1980). "Normal chromium levels in urine and blood of Japanese subjects determined by direct flameless atomic absorption spectrophotometry, and valency of chromium in urine after exposure to hexavalent chromium." *Am. Ind. Hyg. Assoc. J.* 41, 98–102.

Ochiai, E. (1977). *Bioinorganic Chemistry: An Introduction.* Allyn and Bacon, Boston, pp. 388–413.

Offenbacher, E.G., and Pi-Sunyer, F.X. (1980). "Beneficial effect of chromium-rich yeast on glucose tolerance and blood lipids in elderly subjects." *Diabetes* 29, 919–925.

Offenbacher, E.G., Rinko, C.J., and Pi-Sunyer, F.X. (1985). "The effects of inorganic chromium and brewers' yeast on glucose tolerance, plasma lipids, and plasma chromium in elderly subjects." *Am. J. Clin. Nutr.* 42, 454–461.

Offenbacher, E.G., Spencer, H., Dowling, H.J., and Pi-Sunyer, F.X. (1986). "Metabolic chromium balances in men." *Am. J. Clin. Nutr.* 44, 77–82.

Okada, S., Ohba, H., and Taniyama, M. (1981). "Alterations in ribonucleic acid synthesis by chromium(III)." *J. Inorg. Biochem.* 15, 223–231.

Okada, S., Suzuki, M., and Ohba, H. (1983). "Enhancement of ribonucleic acid synthesis by chromium(III) in mouse liver." *J. Inorg. Biochem.* 19, 95–103.

Okada, S., Taniyama, M., and Ohba, H. (1982). "Mode of enhancement in ribonucleic acid synthesis directed by chromium(III)-bound deoxyribonucleic acid." *J. Inorg. Biochem.* 17, 41–49.

Onkelinx, C. (1977). "Compartment analysis of metabolism of chromium(III) in rats of various ages." *Am. J. Physiol.* 232, E478–E484.

Ono, H., Wada, O., and Ono, T. (1981). "Distribution of trace metals in nuclei and nucleoli of normal and regenerating rat liver with special reference to the different behaviour of nickel and chromium." *J. Toxicol. Environ. Health* 8, 947–957.

Ormos, G., and Mányai, S. (1974). "Chromate uptake by human red blood cells: Comparison of permeability for different divalent anions." *Acta Biochim. Biophys. Acad. Sci. Hung.* 9, 197–207.

Pankhurst, C.A., and Pate, B.D. (1979). "Trace elements in hair." In *Reviews in Analytical Chemistry.* Freund Publishing Co., Tel Aviv, Israel, Bo. IV, No. 2, 3, pp. 111–235.

Parr, R.M. (1977). "Problems of chromium analysis in biological materials: An international perspective with special reference to results for analytical quality control samples." *J. Radioanal. Chem.* 39, 421–433.

Pedersen, N.B. (1982). "The effects of chromium on the skin." In S. Langård, Ed., *Biological and Environmental Aspects of Chromium, Topics in Environmental Health.* Elsevier Biomedical Press, Amsterdam, Vol. 5, pp. 249–275.

Petrilli, F.L., and DeFlora, S. (1978). "Metabolic deactivation of hexavalent chromium mutagenicity." *Mutat. Res.* **54**, 139–147.

Polak, L. (1983). "Immunology of chromium." In D. Burrows, Ed., *Chromium: Metabolism and Toxicity.* CRC Press, Boca Raton, FL, pp. 51–136.

Polansky, M.M., and Anderson, R.A. (1980). "Role of inorganic chromium in placental transport." *Fed. Proc.* **39**, 903.

Polansky, M.M., Anderson, R.A., Bryden, N.A., and Glinsmann, W.H. (1982). "Chromium (Cr) and brewers' yeast supplementation of human subjects: Effect on glucose tolerance, serum glucose, insulin and lipid parameters." *Fed. Proc.* **41**, 391.

Polansky, M.M., Anderson, R.A., Bryden, N.A., et al. (1981). "Chromium supplementation of free-living subjects—Effect on glucose tolerance and insulin." *Fed. Proc.* **40**, 885.

Polnaszek, C.F. (1981). "Stable chromium(V) free radical species formed by the enzymatic reduction of chromate." *Fed. Proc.* **40**, 715.

Potter, J.F., Levin, P., Anderson, R.A., et al. (1985). "Glucose metabolism in glucose-intolerant older people during chromium supplementation." *Metabolism* **34**, 199–204.

Rabenstein, D.L., Guevremont, R., and Evans, C.A. (1979). "Glutathione and its metal complexes." In H. Sigel, Ed., *Metal Ions in Biological Systems.* Marcel Dekker, Inc., New York, Vol. 9, pp. 103–138.

Rabinowitz, M.B., Gonick, H.C., Levin, S.R., and Davidson, M.B. (1983). "Effects of chromium and yeast supplements on carbohydrate and lipid metabolism in diabetic men." *Diabetes Care* **6**, 319–327.

Rahkonen, E., Junttilla, M.-L., Kalliomäki, P.L., et al. (1983). "Evaluation of biological monitoring among stainless steel welders." *Int. Arch. Occup. Environ. Health* **52**, 243–255.

Rantanen, J., Aitio, A., Hemminki, K., et al. (1982). "Exposure limits and medical surveillance in occupational health." *Am. J. Ind. Med.* **3**, 363–371.

Rollinson, C.L., Rosenbloom, E., and Lindsay, J. (1967). "Reactions of chromium(III) with biological substances." In *Proceedings, International Congress of Nutrition, 7th Meeting, Hamburg.* Pergamon Press, New York, pp. 692–697.

Samitz, M.H., and Katz, S. (1963). "Preliminary studies on the reduction and binding of chromium with skin." *Arch. Dermatol.* **88**, 184–187.

Samitz, M.H., and Katz, S. (1964). "A study of the chemical reactions between chromium and skin." *J. Invest. Derm.* **43**, 35–43.

Samitz, M.H., Katz, S.A., Scheiner, D.M., and Gross, P.R. (1969). "Chromium-protein interactions." *Acta Dermato-Vener.* **49**, 142–146.

Sanderson, C.J. (1976). "The uptake and retention of chromium by cells." *Transplantation* **21**, 526–529.

Sanderson, C.J. (1982). "Applications of [51]chromium in cell biology and medicine." In S. Langård, Ed., *Biological and Environmental Aspects of Chromium, Topics in Environmental Health.* Elsevier Biomedical Press, Amsterdam, Vol. 5, pp. 101–116.

Saner, G. (1980). *Current Topics in Nutrition and Disease. Chromium in Nutrition and Disease.* Alan R. Liss, Inc., New York, Vol. 2, 135 pp.

Saner, G., Yüzbasiyan, V., and Cigdem, S. (1984). "Hair chromium concentration and chromium excretion in tannery workers." *Br. J. Ind. Med.* **41**, 263–266.

Saner, G., Yüzbasiyan, V., Neyzi, O., et al. (1983). "Alterations of chromium metabolism and effect of chromium supplementation in Turner's syndrome patients." *Am. J. Clin. Nutr.* **38**, 574–578.

Sanford, W.E., Jusys, A.A., Stetsko, P.I., and Nieboer, E. (1984). "Adjustment of urinary nickel concentrations for rates of urine flow." *Ann. Clin. Lab. Sci.* **14**, 423.

Sayato, Y., Nakamuro, K., Matsui, S., and Ando, M. (1980). "Metabolic fate of chromium compounds. I. Comparative behavior of chromium in rat administered with $Na_2{}^{51}CrO_4$ and $^{51}CrCl_3$." *J. Pharm. Dyn.* **3**, 17–23.

Schaller, K.H., Essing, H.-G., Valentin, H., and Schacke, G. (1973). "The quantitative determination of chromium in urine by flameless atomic absorption spectroscopy." *Atomic Absorption Newslett.* **12**, 147–150.

Schuster, S.M., Ebel, R.E., and Lardy, H.A. (1975). "Kinetic studies on rat liver and beef heart mitochondrial adenosine triphosphatase: The effects of the chromium complexes of adenosine triphosphate and adenosine diphosphate on the kinetic properties." *Arch. Biochem. Biophys.* **171**, 656–661.

Schwarz, K., and Mertz, W. (1959). "Chromium(III) and the glucose tolerance factor." *Arch. Biochem. Biophys.* **85**, 292–295.

Shapcott, D. (1979). "Preparation of samples for analysis." In D. Shapcott and J. Hubert, Eds., *Developments in Nutrition and Metabolism. Chromium in Nutrition and Metabolism.* Elsevier/North-Holland Biomedical Press, Amsterdam, Vol. 2, pp. 43–48.

Sharma, B.K., Singhal, P.C., and Chugh, K.S. (1978). "Intravascular haemolysis and acute renal failure following potassium dichromate poisoning." *Postgrad. Med. J.* **54**, 414–415.

Shih, S.C., Tauxe, W.N., Fairbanks, V.F., and Taswell, H.F. (1972). "Urinary excretion of ^{51}Cr from labeled erythrocytes." *JAMA* **220**, 814–817.

Sillén, L.G., and Martell, A.E., Eds. (1964). *Stability Constants of Metal-Ion Complexes, Section I: Inorganic Ligands. Section II: Organic Ligands.* The Chemical Society, London, pp. 89–91.

Sjögren, B., Hedström, L., and Ulfvarson, U. (1983). "Urine chromium as an estimator of air exposure to stainless steel welding fumes." *Int. Arch. Occup. Environ. Health* **51**, 347–354.

Smart, G.A., and Sherlock, J.C. (1985). "Chromium in foods and the diet." *Food Addit. Contam.* **2**, 139–147.

Smith, R.M., and Martell, A.E. (1976). *Critical Stability Constants. Inorganic Complexes.* Plenum Press, New York, Vol. 4, 257 pp.

Sneeden, R.P.A. (1975). *Organometallic Chemistry—A Series of Monographs. Organochromium Compounds.* Academic Press, New York, 327 pp.

Spiro, T.G., Ed. (1980). *Nucleic Acid-Metal Ion Interactions. Metal Ions in Biology.* John Wiley & Sons, New York, Vol. 1, 256 pp.

Stern, R.M. (1982). "Chromium compounds: Production and occupational exposure." In S. Langård, Ed., *Biological and Environmental Aspects of Chromium. Topics in Environmental Health.* Elsevier Biomedical Press, Amsterdam, Vol. 5, pp. 5–47.

Stickland, L.H. (1949). "The activation of phosphoglucomutase by metal ions." *Biochem. J.* **44**, 190–197.

Stokinger, H.E. (1981). "The metals." In G.D. Clayton and F.E. Clayton, Eds., *Patty's Industrial Hygiene and Toxicology. Toxicology,* 3rd ed. John Wiley & Sons, New York, Vol. 2A, pp. 1589–1605, 1792–1807.

Sulfab, Y., Taylor, R.S., and Sykes, A.G. (1976). "Rapid equilibration of the ethylenediamine-N,N,N′,N′-tetraacetatoaquochromate(III) complex with chromate(VI), molybdate(VI), tungstate(VI), and azide. Labilization of the aquo ligand by the free carboxylate and substitution at chromium(III)." *Inorg. Chem.* **15**, 2388–2393.

Sunderman, F.W. Jr., Brown, S.S., Stoepler, M., and Tonks, D.B. (1982). "Interlaboratory evaluations of nickel and cadmium analyses in body fluids." In H. Egan and T.S. West, Eds., *IUPAC Collaborative Interlaboratory Studies in Chemical Analysis.* Pergamon Press, Oxford, pp. 25–35.

Tandon, S.K., Mathur, A.K., and Gaur, J.S. (1977). "Urinary excretion of nickel and chromium among electroplaters and pigment industry workers." *Int. Arch. Occup. Environ. Health* **40**, 71–76.

Teraoka, H. (1981). "Distribution of 24 elements in the internal organs of normal males and the metallic workers in Japan." *Arch. Environ. Health* **36**, 155–165.

Thomsen, E., and Stern, R.M. (1979). "A simple analytical technique for the determination of hexavalent chromium in welding fumes and other complex matrices." *Scand. J. Work Environ. Health* **5**, 386–403.

Toepfer, E.W., Mertz, W., Polansky, M.M., et al. (1977). "Preparation of chromium-containing material of glucose tolerance factor activity from brewers' yeast extracts and by synthesis." *J. Agric. Food Chem.* **25**, 162–166.

Tola, S., Kilpiö, J., Virtamo, M., and Haapa, K. (1977). "Urinary chromium as an indicator of the exposure of welders to chromium." *Scand. J. Work Environ. Health* **3**, 192–202.

Torgrimsen, T. (1982). "Analysis of chromium." In S. Langård, Ed., *Biological and Environmental Aspects of Chromium. Topics in Environmental Health.* Elsevier Biomedical Press, Amsterdam, Vol. 5, pp. 65–99.

Tossavainen, A., Nurminen, M., Mutanen, P., and Tola, S. (1980). "Application of mathematical modelling for assessing the biological half-times of chromium and nickel in field studies." *Br. J. Ind. Med.* **37**, 285–291.

Tsapakos, M.J., Hampton, T.H., and Wetterhahn, K.E. (1983). "Chromium(VI)—Induced DNA lesions and chromium distribution in rat kidney, liver, and lung." *Cancer Res.* **43**, 5662–5667.

Tsapakos, M.J., Hampton, T.H., and Wetterhahn Jennette, K. (1981). "The carcinogen chromate induces DNA cross-links in rat liver and kidney." *J. Biol. Chem.* **256**, 3623–3626.

Tsuneta, Y. (1982). "Investigations of the pathogenesis of lung cancer observed among chromate factory workers." *Hokkaido J. Med. Sci.* **57**, 175–187.

Tsuneta, Y., Ohsaki, Y., Kimura, K., et al. (1980). "Chromium content of lungs of chromate workers with lung cancer." *Thorax* **35**, 294–297.

Underwood, E.J. (1977). *Trace Elements in Human and Animal Nutrition,* 4th ed. Academic Press, New York, pp. 258–270.

Vallee, B.L., Rupley, J.A., Coombs, T.L., and Neurath, H. (1958). "The release of zinc from carboxypeptidase and its replacement." *J. Am. Chem. Soc.* **80**, 4750–4751.

Vanderlinde, R.E., Kayne, F.J., Komar, G., et al. (1979). "Serum and urine levels of chromium. In D. Shapcott and J. Hubert, Eds., *Developments in Nutrition and Metabolism. Chromium in Nutrition and Metabolism.* Elsevier/North-Holland Biomedical Press, Amsterdam, Vol. 2, pp. 49–57.

Veillon, C., Patterson, K.Y., and Bryden, N.A. (1982a). "Chromium in urine as measured by atomic absorption spectrometry." *Clin. Chem.* **28**, 2309–2311.

Veillon, C., Patterson, K.Y., and Bryden, N.A. (1982b). "Direct determination of chromium in human urine by electrothermal atomic absorption spectrometry." *Anal. Chim. Acta* **136**, 233–241.

Veillon, C., Wolf, W.R., and Guthrie, B.E. (1979). "Determination of chromium in biological materials by stable isotope dilution." *Anal. Chem.* **51**, 1022–1024.

Versieck, J. (1985). "Trace elements in human body fluids and tissues." *CRC Crit. Rev. Clin. Lab. Sci.* **22**, 97–184.

Versieck, J., and Cornelis, R. (1980). "Normal levels of trace elements in human blood plasma or serum." *Anal. Chim. Acta* **116**, 217–254.

Versieck, J., Barbier, F., Cornelis, R., and Hoste, J. (1982). "Sample contamination as a source of error in trace-element analysis of biological samples." *Talanta* **29**, 973–984.

Versieck, J., Hoste, J., Barbier, F., et al. (1978). "Determination of chromium and cobalt in human serum by neutron activation analysis." *Clin. Chem.* **24**, 303–308.

Versieck, J., de Rudder, J., Hoste, J., et al. (1979). "Determination of the serum chromium concentration in healthy individuals by neutron activation analysis." In D. Shapcott and

J. Hubert, Eds., *Developments in Nutrition and Metabolism. Chromium in Nutrition and Metabolism.* Elsevier/North-Holland Biomedical Press, Amsterdam, Vol. 2, pp. 59–68.

Visek, W.J., Whitney, I.B., Kühn, U.S.G., and Comar, C.L. (1953). "Metabolism of ^{51}Cr by animals as influenced by chemical state." *Proc. Soc. Exp. Biol. Med.* **84**, 610–615.

Wacker, W.E.C., and Vallee, B.L. (1959). "Nucleic acids and metals. I. Chromium, manganese, nickel, iron, and other metals in ribonucleic acid from diverse biological sources." *J. Biol. Chem.* **234**, 3257–3262.

Wada, O., Wu, G.Y., Yamamoto, A., et al. (1983). "Purification and chromium-excretory function of low-molecular-weight, chromium-binding substances from dog liver." *Environ. Res.* **32**, 228–239.

Wallach, S. (1985). "Clinical and biochemical aspects of chromium deficiency." *J. Am. Coll. Nutr.* **4**, 107–120.

Warren, G., Schultz, P., Bancroft, D., et al. (1981). "Mutagenicity of a series of hexacoordinate chromium(III) compounds." *Mut. Res.* **90**, 111–118.

Wasson, J.R. (1976). "Chromium." In K. Niedenzu and H. Zimmer, Eds., *Methodicum Chimicum. Preparation of Transition Metal Derivatives.* Academic Press, New York, Vol. 8, pp. 161–180.

Weast, R.C. (1973). *Handbook of Chemistry and Physics,* 54th ed. 1973–1974. CRC Press, Cleveland, Ohio.

Welinder, H., Littorin, M., Gullberg, B., and Skerfving, S. (1983). "Elimination of chromium in urine after stainless steel welding." *Scand. J. Work Environ. Health* **9**, 397–403.

Wender, I., and Pino, P., Eds. (1968). *Organic Syntheses via Metal Carbonyls.* Interscience Publishers (John Wiley & Sons), New York, Vol. 1, 517 pp.

Westheimer, F.H. (1949). "The mechanisms of chromic acid oxidations." *Chem. Rev.* **45**, 419–451.

Wetterhahn Jennette, K. (1982). "Microsomal reduction of the carcinogen chromate produces chromium(V)." *J. Am. Chem. Soc.* **104**, 874–875.

Wiberg, K.B. (1965). "Oxidation by chromic acid and chromyl compounds." In K.B. Wiberg, Ed., *Oxidation in Organic Chemistry, Part A.* Academic Press, New York, pp. 69–184.

Wieser, O., Grünbacher, G., Prügger, R., et al. (1982). "Spastische bronchitis bei chromarbeitern." *Wien. Med. Wochenschr.* **132**, 59–62.

Windholz, M., et al., Eds. (1983). *The Merck Index,* 10th ed. Merck & Co., Inc., Rahway, NJ, pp. 315–319.

Yamamoto, A., Wada, O., and Ono, T. (1984). "Distribution and chromium-binding capacity of a low-molecular-weight chromium-binding substance in mice." *J. Inorg. Biochem.* **22**, 91–102.

3

PRODUCTION AND USES OF CHROMIUM

Jerome O. Nriagu

National Water Research Institute
Burlington, Ontario, Canada

1. OCCURRENCE

Although chromium is the seventh most abundant element on earth, most of it resides in the core and mantle (Table 1). In fact, it is the 21st most abundant element in the earth's crust, the average concentration being 100 $\mu g/g$ compared to 3,700 $\mu g/g$ for the earth as a whole. In its crustal distribution, it shows a particular preference for ultrabasic and basic rocks and specifically eschews the feldspar minerals (Matzat and Shiraki, 1978). It ranks fourth among the 29 elements of biologic importance.

Even though about 40 chromium-containing minerals are known (Table 2), the only one of economic importance is chromite. In nature, this mineral has a highly variable chemical composition, which can be represented by $(Me^{2+})O \cdot (Me^{3+})_2(O)_3$, with Me^{2+} being Mg or Fe^{2+}, and Me^{3+} being Cr, Fe^{3+}, or Al. The ideal (end-member) chemical formula for chromite is $FeO \cdot Cr_2O_3$ and should contain about 46% chromium. The chromium ores are often classified on the basis of their composition and industrial usage into high-chromium (46–55% Cr_2O_3, Cr:Fe ratio over 2.1) used primarily for metallurgical purposes, high-iron (40–46% Cr_2O_3, Cr:Fe 1.5–2.1) used mostly for chemical and metallurgic purposes, and high-aluminum (33–38% Cr_2O_3, 22–34% Al_2O_3) used widely as refractory material.

Most of the major chromium deposits known occur in three principal geologic settings: (1) Stratiform-type deposits, such as those of the Bushveld Complex of South Africa, the Great Dyke in Zimbabwe, and the Kemi intrusion of Finland. These deposits account for over 90% of the identified chrome ore resources and

Table 1   Average Chromium Concentrations in Geologic Material

Material	Concentration ($\mu g/g$)	Reference
Ultramafic rocks	2,400	Matzat and Shiraki, 1974
Basalts	200	Matzat and Shiraki, 1974
Granites	10	Matzat and Shiraki, 1974
Metamorphic rocks	90	Matzat and Shiraki, 1974
Pelagic sediments	80	Matzat and Shiraki, 1974
Shales	83	Matzat and Shiraki, 1974
Sandstones	27	Matzat and Shiraki, 1974
Limestones	9	Matzat and Shiraki, 1974
Weathered continental crust	55	Matzat and Shiraki, 1974
Continental crust (unweathered)	100	Liu, 1982
Oceanic crust	300	Liu, 1982
Upper mantle	3,100	Liu, 1982
Lower mantle	2,700	Liu, 1982
Outer core	5,200	Liu, 1982
Inner core	12,100	
Average, earth	3,700	Liu, 1982

Table 2   Chromium Minerals

Oxides, hydroxides, carbonates

Chromite	$FeCr_2O_4$
Donathite	$(Fe^{2+}, Mg, Zn)(Cr, Fe^{3+}, Al)_2O_4$
Eskolaite	Cr_2O_3
Guyanaite	$CrOOH$
Bracewellite	$CrOOH$
Grimaldiite	$CrOOH$
Mccommellite	$CrOOCu$
Stichtite	$Mg_{18}Cr_6(OH)_{48}[CO_3]_3 \cdot 12H_2O$
Barbertonite	$Mg_6Cr_2(OH)_{16}[CO_3] \cdot 4H_2O$

Sulfides

Daubréelite	$FeCr_2S_4$
Brezinaite	Cr_3S_4
Heideite	$(Fe, Cr)_{1+x}(Ti, Fe)_2S_4$

Nitrides

Carlsbergite	CrN

Chromates

Tarapacaite	$K_2[CrO_4]$
Lopezite	$K_2[Cr_2O_7]$
Chromatite	$Ca[CrO_4]$
Crocoite	$Pb[CrO_4]$
Phoenicochroite	$Pb_2[CrO_4]O$
Santanaite	$Pb_{11}[CrO_4]O_{12}$
Iranite	$Pb[CrO_4] \cdot H_2O$
Vauquelinite	$Pb_2Cu[CrO_4][PO_4](OH)$
Beresovite	$Pb_6[CrO_4]_3CO_3]O_2$
Dietzeite	$Ca_2[IO_3]_2[CrO_4]$
Fornacite	$(Pb,Cu)_3[(Cr,As)O_4]_2(OH)$
Hemihedrite	$ZnF_2Pb_{10}[CrO_4]_6[SiO_4]_2$
Embreyite	$Pb_5[CrO_4]_2[PO_4]_2 \cdot H_2O$

Silicates

Knorringite	$Mg_3Cr_2[SiO_4]_3$
Uvarovite	$Ca_3Cr_2[SiO_4]_3$
Tawmawite	$Ca_2(Al,Cr)_3[SiO_4]_3OH$
Ureyite (kosmochlor)	$NaCr[Si_2O_6]$
Krinovite	$NaMg_2Cr[Si_3O_{10}]$
Wolchonskoite	$(Cr_{0.33},Fe^{+3}_{0.6},Al_{0.4},Mg_1)[Al_2Si_{3.8}O_{10}](OH)_2$
Fuchsite	$K_2(Al,Cr)_4[Al_2Si_6O_{20}](OH)_4$
Kämmererite	$(Mg,Al,Fe,Cr)_6[(Si,Al)_4O_{10}](OH)_8$
Kotchubeite	$(Mg,Al,Fe)_6[(Si,Al,Cr)_4O_{10}](OH)_8$

Source: After Matzat and Shiraki, 1974.

include about 90% and 50%, respectively, of the high-chromium and high-iron ores (DeYoung et al., 1984). Only layered igneous rock complexes older than 1.9 billion years are known to host economic chromite deposits of the stratiform-type. (2) Podiform-type or Alpine-type deposits generally associated with the island-arcs (Philippine and Cuban deposits) and the major tectonic belts, such as the Appalachian Mountain (eastern North American occurrences), Tethyan mountain chains (Greek, Turkish, and Albanian deposits), and Ural Mountains (deposits in the Soviet Union). Podiform-type deposits account for about 10% of the world's chromium ore resources. Host rocks for this type of deposit typically are peridotite, and the deposits are seldom more than 1.1 billion years old (DeYoung et al., 1984). (3) Lateritic (placer) deposits, which are generally derived from the weathering of chromium-bearing peridotites. The ore tenor tends to be low, and few of these deposits have been exploited profitably.

The global distribution of the major chromite deposits and districts is shown in Figure 1.

1.1   Resources and Reserves

Most of the world's chromite reserves (economically exploitable deposits) are located in southern Africa. The reserves of chromite ore in South Africa exceed 1.0 billion t, or over 430 million t of chromium metal, and account for over 70% of the known world's reserves. The Great Dyke region of southern Zimbabwe contain over 130 million t of exploitable ore equivalent to about 62 million t of chromium metal. Other countries with significant (over 50 million t) chromite reserves include the Soviet Union, India, and Finland (Table 3). The worldwide reserves of chromium ores have been estimated to be over 1.5 billion t, equivalent to over 600 million t of the chromium metal (DeYoung et al., 1984).

Estimated world's resources of chromium ores exceed 5.5 billion t, or about 2.1 billion t of the metal (Table 3). This figure excludes the 5.4 billion t of the marginally economic deposit (chromite content of 38%) of the UG2 chromite layer of the Bushveld. Thus, over 90% of the world's chromite resources are located in the Bushveld complex of South Africa. The chromite resources of Zimbabwe probably exceed 750 million t, implying that for many years to come, chromium will likely remain an African metal. Significant resources (of over 50 million t) also occur in Greenland, Iran, Papau New Guinea, and the Soviet Union. Sizable (over 10 million t) chromite resources are known to occur in Albania, Brazil, Canada, India, and Turkey (Table 3).

Only 10 million t or so of the chromite reserves now known were discovered before 1900. The ore reserves were increased by about 550 million t by discoveries made between 1900 and 1919, and with the discoveries of the giant deposits of southern Africa during the early 1920s, further quest for chromite resources became almost redundant. Since then, only minor additions to the known reserves have been made by new finds in the 1930s (Soviet Union), 1940s (Canada), 1950s (India and Finland), and 1960s (Greenland and Papau New Guinea). The poor discovery record of recent times may be attributed to

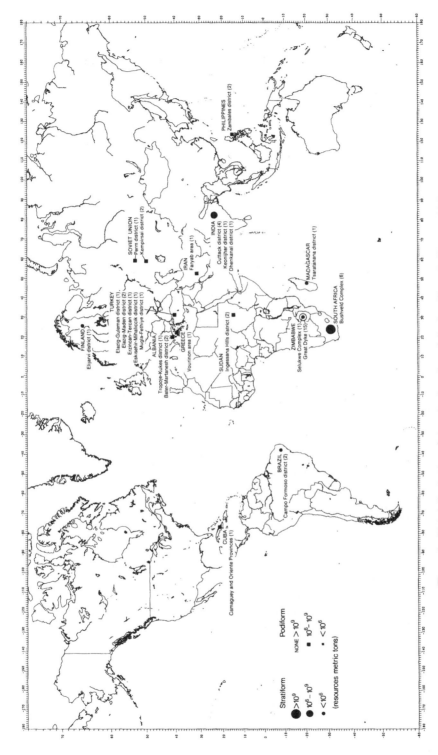

Figure 1. Global distribution of chromite deposits and districts (From DeYoung, 1984).

85

Table 3   World's Reserves and Resources of Chromium (in million t)

Country	Reserves		Resources	
	Chromite ore	Cr metal	Chromite ore	Cr metal
Albania	6.5	2.7	15	6.5
Brazil	17	3.4	20	4.0
Canada	—	—	19	0.88
Cuba	3.0	1.1	5.0	1.8
Finland	50	13	—	—
Greece	3.0	0.54	3.2	0.58
Greenland	—	—	100	38
India	79	26	26	8.4
Iran	2.2	1.1	50	18
Madagascar	10	3.1	—	—
Papau New Guinea	—	—	128	38
Philippines	12	3.3	—	—
South Africa	1,014	430	4620[a]	1632[a]
Soviet Union	>116	>58	?	?
Turkey	10	3.6	16	6.2
Zimbabwe	133	62	748	354
Global total	>1456	>608	>5750	>2108

Source: DeYoung et al., 1984.

[a]These figures do not include the 5,420 million t of ore (chromium content of 38%) known as the UG2 chromitite layer in eastern and western Bushveld (DeYoung et al., 1984).

curtailed exploration engendered by low chromium demand and may not reflect the nonexistence of additional resources.

2. RECOVERY

The mining of chromite ores uses both open-pit and underground methods. The latter technique is prevalent in South Africa, Zimbabwe, Turkey, Greece, and Yugoslavia, whereas the surface method is favored by some developing countries and the Soviet Union (DeYoung et al., 1984).

Most of the ores now being mined are of sufficiently high grade to be marketed without much dressing. The ores are often screened into three sizes: lumps, intermediates, and fines. The higher-grade lump ores are further sorted on sorting belts and cleaned by hand-cobbing, heavy-liquid separation, or by screening before they are shipped. The intermediates, fines, and size-reduced lower-grade ores are concentrated by gravity separation techniques and then agglomerated by briquetting with a binder or pelletizing by kiln firing.

Figure 1 in Chapter 2 outlines the process used in the production of chromium metal and its compounds. Since the industrial demand mostly calls for the intermediate forms of chromium, very little of the chromite is actually processed all the way to ductile chromium. Ferrochromium, chrome alum, chromic acid, and sodium chromate all find extensive industrial applications. Detailed discussions of the industrial production of chromium and its compounds are given by Udy (1956), Sully and Brandes (1967), Sittig (1978), and Westbrook (1979).

3. PRODUCTION AND CONSUMPTION

The Ural Mountains of Russia remained the principal source of the world's chromite until around 1827, when production from the deposits in Maryland came into prominence. During the 1860s, the United States was supplanted by Turkey as the leading producer of chromite. Since the early 1920s, the outputs from the deposits of southern Africa have accounted for 30–50% of the world's chromite market, although the Soviet Union regained a major share (12–30%) when the huge Kazakh deposits (Kempirsai area) came into production during the 1930s. Figure 2 shows the proportion of the world's chromite production derived from the various countries between 1920 and 1980.

Chromium is one of the key metals whose use is nurtured by our modern technologic society. From the total world production of 100,000 t in 1904, the consumption reached 1 million t in 1936 and exceeded 2 million t during the World War (II) but declined after that conflict. Chromite consumption reached 5 million t in 1965, exceeded 10.5 million t in 1980, and dropped to just over 9 million t in 1984. The historical trend in worldwide production of chromite (Fig. 3) shows pronounced plateaus corresponding to World War I and World War II and the Viet Nam War of the 1960s. A marked downturn in output corresponds to the great depression of the early 1930s, the post-World War II period, and the recent economic recession of the early 1980s. The wide fluctuations between the early 1950s and late 1960s may be attributed to the conflicts in southeast Asia. Between 1968 and 1980, chromite production increased at a sharp but steady rate of about 10% per year (Fig. 3).

On a tonnage basis, chromium now ranks 4th among the metals and 13th among all the mineral commodities. In 1984, chromium was produced in 19 different countries from over 50 different chromite deposits. Only 9 countries (South Africa, Soviet Union, Albania, Zimbabwe, Turkey, India, Finland, Philippines, and Brazil) produced over 300,000 t of chromite, and collectively these countries accounted for over 95% of all the current chromite production. About 41% of the total production in 1983 came from the Soviet Union and Albania and 33% from southern Africa (Table 4).

The cumulative production of chromite by each country between 1901 and 1983 is also shown in Table 4. The outputs from the Soviet Union and Albania exceed 74 million t, or about 34% of all the chromite consumed in this century. The production from South Africa and Zimbabwe totals about 76 million t, or 35% of the cumulative production. The total (all-time) production figure of about 320 million t is roughly equivalent to 22% of the chromite reserves but less than

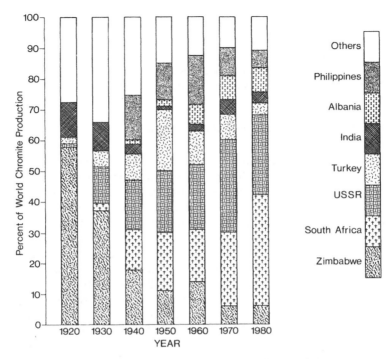

Figure 2. Historical changes in the regional production of chromite.

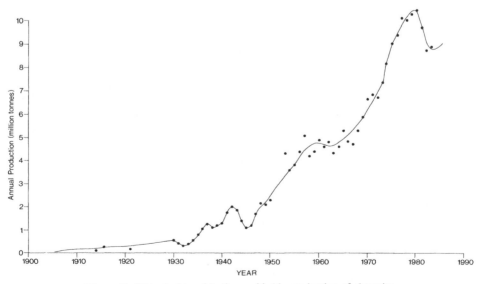

Figure 3. Historical trend in the worldwide production of chromite.

Table 4   Cumulative and Annual Production of Chromite by the Major Producing Countries

Country	Cumulative production[a] 1901–1983 (t)	Annual production[b] 1980 (t)	1983 (t)
Soviet Union	58	2.46	2.46
South Africa	53	3.42	2.24
Zimbabwe	23	0.56	0.43
Turkey	22	0.39	0.40
Philippines	18	0.50	0.33
Albania	16	0.76	0.90
India	7.3	0.32	0.36
Brazil	4.1	0.31	0.28
Finland	3.7	0.36	0.34
Cuba	3.7	0.03	0.03
Iran	3.4	0.08	0.05
Greece	2.0	0.04	0.04
Madagascar	2.0	0.18	0.09
Sudan	0.47	0.02	0.02
Others	20 (estimated)	0.21	0.25
Total	237	9.50	8.12

[a]*Source:* DeYoung et al., 1984.
[b]*Source:* Papp, 1983.

5% of the potential resources. At the present consumption rate of 10 million t per year, the available resources will certainly be adequate to meet the demand for many centuries to come. It should be obvious that in the foreseeable future, countries of southern Africa and the Soviet Union will likely remain the principal producers of chromite ores.

4.   SUPPLY–DEMAND RELATIONSHIPS

Chromium is a vital mineral commodity for which there is no ready substitute. As noted previously, well over 95% of the chromite reserves are located in the developing countries and the Soviet Block countries, and the principal consumer nations have few large and producing chromite deposits. Chromium is, therefore, a classic example of a strategic mineral, in that disruptions or loss of its supply would place the defense industry of the Western nations at some risk. The supply–demand relationships for chromium have been influenced by both political and economic factors. Notable developments and events that have controlled the production and consumption of chromium are summarized below.

Because of its strategic importance, chromium was one of the first metals to be designated for stockpiling by the industrialized countries of the West. In the

United States alone, the stockpiled chromium reached 8 million t in 1965 and currently is about 5 million t, equivalent to the industrial requirements for about 2–3 years (Wolfe, 1984). Many countries of western Europe also maintain large stockpiles of chromium for strategic or economic reasons.

Many countries have marginal (low-grade) or small chromite deposits that can be brought into production when there are restrictions on imports. For example, the production of chromite in Canada rose during World Wars I and II to peaks of 32,300 and 26,850 t in 1917 and 1943, respectively, and has been insignificant since 1950. The United States has over 2,000 known chromite deposits, and production since the 1800s has totalled about 2 million (Brantley, 1970). Most of this production was achieved under governmental assistance during the emergency periods of World War I (82,400 t in 1918 alone), World War II (from 800 t in 1938 to 145,000 t in 1943), and the Korean conflict (163,000 t in 1954). No domestic chromite ore has been mined in the United States since the early 1960s (Brantley, 1970).

Chromite supplies tend to be linked to geopolitics and to political upheavals. For example, the sanctions imposed by the United Nations inhibited the purchase of chromite from Rhodesia (now Zimbabwe) during that country's civil war of 1966–1980. Because of the Korean conflict, the Soviet Union stopped shipment of her chromite to the United States from 1950 to 1960. The need to safeguard supplies at critical periods led to the establishment of captive mines by United States companies, especially in South Africa. The captive mine strategy curtailed the development of chromite deposits in countries that gave less encouragement to foreign control of their mining operations (DeYoung, 1984). In recent times, the emphasis has been on diversification of supply sources in view of the political unrest in South Africa.

For technologic and economic reasons, plants for converting the chromite ore to ferrochromium traditionally had been sited near the steel-making facilities of the industrialized nations. Since the 1970s, the major chromite-producing nations have been expanding their ferrochromium production, thereby increasing the value added to the commodity at its source. As a result of this development, the import of chromite by the United States has declined from a little over 2 million t in 1979 to just over 300,000 t in 1983; this decline is closely matched by an increase in ferrochromium import.

The demand for chromium is generally dictated by the economic conditions of the steel industry, the dominant user of the metal.

The recovery of chromium from scraps, solid, and liquid wastes only returns small quantities of secondary chromium to the market. In the United States, for example, only 15% of the chromium consumed is recycled, mostly from stainless steel scrap (Westbrook, 1979). Chromium may be replaced by other material in limited instances, such as in some alloys and in refractory and pigment applications. In general, however, there are no satisfactory substitutes for chromium, and the mines will likely remain the principal source of this vital metal for many years to come.

Although chromite deposits often contain other minerals and metals, such as platinum, gold, nickel, vanadium, talc, and titaniferous magnetite, the chromite ore is hardly ever treated for the recovery of any associated metals. The lack of valuable byproducts means that the chromite deposits are developed based on specific profit-motive or political-motive criteria.

Environmental pollution control regulations have made an impact on the chrome industry in terms of financial investments on (1) the pollution control equipment and (2) the research and development on control technology to meet the restrictions for air and water discharges. It has been estimated that between 1970 and 1978, about 60% of the new investment dollars in the ferroalloy industry of the United States went for pollution control equipment rather than toward increasing the capacity of new furnaces (Westbrook, 1979).

The supply–demand relationships for chromium in the economy of the United States, by far the largest consumer of the metal, are depicted in Figure 4.

5. USES

Different industrial applications call for different forms of chromium, such as chromite, ferrochromium, chromium metal, chromite refractory bricks, chromite foundry sands, chromic acid, and other chromium compounds. The 10 million t of chromite consumed globally each year are distributed among the three principal industrial end uses as follows: 76% for metallurgical, 13% for refractory, and 11% for chemical applications (DeYoung et al., 1984). In the United States, the major uses of chromite between 1977 and 1983 were 58% metallurgical, 23% chemical, and 18% refractory (Papp, 1983). The use patterns in Japan, the European Economic Communities, and the Eastern Block countries are also different from the worldwide averages.

Chromium has become a part of our daily lives in very many ways (Udy, 1956, Vol. 1, p. 13):

> Chromium metal gives brightness to our kitchens, pots, pans, and furniture. Chromium steels give protection to our health and safety. At high temperatures chromium is becoming more important in the production of metals and alloys, making possible the jet turbine engines for industrial use and particularly for aviation. In fact, chromium is proving to be the wonder metal of them all.

The uses of chromium chemicals for our safety and welfare are even more pervasive (Darrin, 1956, p. 251):

> [They] are essential to the oil and gas industries for corrosion control and the preparation of catalysts; the food and beverage industries, in refrigerating brines and cleansing compounds; the transportation industries, in diesel locomotives and automobiles; the iron and steel industries, in stainless metal and chromium plate; the aircraft industry, for anodizing aluminum and pickling magnesium; the copper industry, for descaling brass and stripping copper; the electrical industry, in

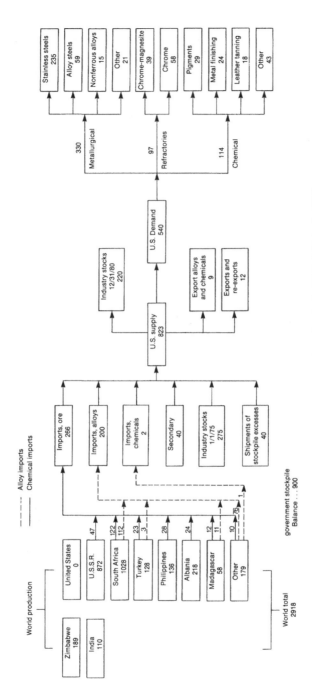

Figure 4. Supply–demand relationships for chromium in the United States in 1975.

mercury-arc rectifiers and dry cells; the pyrotechnical industries, in matches and fireworks; and photographic industries, in lithography and engraving.

Only the principal industrial applications of chromium and its compounds are discussed in this chapter.

5.1 Ferroalloys

The metallurgical chrome for steel production is expected to meet the following specifications: $>45\%$ Cr_2O_3, Cr:Fe ratio >3.1, and SiO_2 less than 8% (Wolfe, 1984). The addition of chromium to steel or wrought iron (1) enhances its passivity toward many corroding materials and (2) improves its mechanical properties in beneficial ways. Chromium-containing steels can be divided into three arbitrary groups based on the chromium content: low-chromium engineering steels with less than 3% chromium, the tool steels with 3–12% chromium, and the stainless steels with over 12% chromium (Sully and Brandes, 1967; Kinzel and Crafts, 1937). Of the 329,000 t of chromium consumed in 1979 by the metallurgical industries of the United States, 71% was used for stainless steels, 13% for wrought-alloy steels, and 5% each for nonferrous alloys and cast-alloy steels (Table 5). By 1983, the use pattern had shifted to 81% for stainless steels, 9% for full-alloys steels, and 3% for superalloys (Papp, 1983).

Small amounts of chromium (less than 3%) significantly increase the tempering resistance, hardness, and strength of steels by delaying the transformation that takes place during heat treatment, thereby increasing the capacity of the steel to harden (i.e., its hardenability). These low-chromium steels, which account for 20–25% of the annual chromium consumption in the United States, are extensively used in the manufacture of automobiles, trucks, buses, locomotives, aircraft, ships, oil and gas well equipment, mining, quarrying and lumbering machinery, ordinance material, and agricultural equipment.

Steels containing 3–12% chromium also show enhanced corrosion and oxidation resistance that increases with the chromium content, but passivation against corrosion is not generally achieved with less than 12% chromium (Udy, 1956). Low-carbon steels in this category find wide application in petroleum and chemical industries. Steels with medium carbon are used in castings and tools for light service, such as wrenches, screwdrivers, and pliers. High-carbon steels in this category arused primarily as magnetic steels and for certain valves (Zapffe, 1949).

The most important industrial application for chromium is in the making of stainless steels, which consumes about 75% of all the ferrochromium produced each year. Stainless steels are classified into three broad groups on the basis of their microstructures: austenitic, martensitic, and ferritic. In general, the quality of stainlessness (especially resistance to corrosion and oxidation) increases with the chromium content, and by changing the carbon content and the heat treatment, increased tensile and yield strengths and increased hardness can be attained. These steels can also be modified for special metallurgical service or

Table 5   Consumption of Chromium by Industrial Sectors of the United States in 1979

	Quantity Consumed 10^3	Percentage of U.S. Consumption
Metallurgical		
Wrought stainless and heat-resisting steels	235	44
Tool steels	6	1.1
Wrought-alloy steels	44	8.1
Cast-alloy steels	15	2.7
Alloy castirons	8	1.4
Nonferrous alloys	15	2.7
Other	6	1.1
Subtotal	329	61.0
Refractories		
Chrome and chrome-magnesite	16	3.0
Magnesite-chrome brick	23	4.2
Granular chrome-bearing	42	7.8
Granular chromite	16	3.0
Subtotal	97	18.0
Chemicals		
Pigments	29	5.4
Metal finishing	24	4.4
Leather tanning	18	3.3
Drilling muds	5	0.9
Wood treatment	7	1.3
Water treatment	7	1.3
Chemical manufacture	9	1.7
Textiles	4	0.7
Catalysts	<2	0.3
Other	9	1.7
Subtotal	114	21.0
Total	540	100

Source: EPA, 1984.

fabrication by adding other elements, such as Ni, Mn, Mg, Ti, Se, S, Mo, P, Zr, and rare earths (see, for example, Udy, 1956; Kinzel and Crafts, 1937; Kinzel and Franks, 1940).

Ferritic stainless steels with 15–30% chromium and 0.08–0.2% carbon are widely used in architecture, in automotive and railroad car trims, and for some equipment used in the chemical and food industries. The low-carbon variety is notably resistant to oxidation at high temperatures and thus is used in heat exchangers, furnace parts, and other high-temperature equipment. The so-called superferritics have recently appeared in the market and boast of superior toughness, weldability, and stress corrosion resistance (Streicher, 1977).

Martensitic stainless steels with 10–18% chromium and moderate (0.08–1.1%) carbon contents are the choice for mechanical parts where strength and hardness are desirable. Historically, this has been the type steel used in cutlery (see Chapter 1) but now also is widely used for valves, shafts, and bearings and in steam turbines, furnace parts, and other equipment subject to combined abrasion and corrosion. In the oil industry, they are often used for bubble trays, bubble caps, and liners.

Austenitic stainless steels with 16–26% chromium and 6–22% nickel and/or manganese account for 60–70% of all the stainless steel products (Westbrook, 1979; River et al., 1970). They are characterized by superior chemical inertness, shock resistance, and formability and hence are popularly used in the automotive, kitchen equipment, building, dairy, pulp and paper, petrochemical, and aircraft industries. Unique properties for specialized applications can be conferred on the renowned 18–8 (18% chromium and 8% nickel) alloy by additions of molybdenum or manganese (Kinzel and Franks, 1940).

Additional information and literature on stainless steels are given in Chapter 1.

5.2   Nonferrous Alloys

Metallic chromium is used primarily in nonferrous alloys where use of the less expensive ferrochromium alloys can introduce undesirable amounts of iron into the system. Although these alloys account for less than 5% of the chromium used by industries, they represent a vital role for chromium in that some of the applications are unique. Familiar applications include the superalloys used in jet engines and gas turbines, the cutting-tool and hard-facing stellite alloy, the electric-resistance heating elements of nickel–chromium type, the heat-treatable chromium bronzes, and the chromium-containing alloys used in nuclear reactors. Chromium basically confers corrosion and oxidation resistance properties on these alloys.

5.3   Refractories

The second most important use for the chrome ore is for refractory purposes. Such application stems from the fact that chromite has a high melting point (2,040°C) and is almost chemically inert and, therefore, resistant to attack by acids and bases at high temperatures. The minimum specifications for chromium ore designated for refractories include $Cr_2O_3 > 31\%$, $(Al_2O_3 + Cr_2O_3) > 58\%$, $Fe > 12\%$, and $SiO_2 > 6\%$ (Wolfe, 1984).

Chromite is used in the manufacture of refractory bricks, in making mortars, castables, and ramming gunning mixes, in joining of bricks within furnaces, for repairs, and in the bonding and coating of the basic bricks to close pores. Refractories containing chromite and magnesite are used in furnaces wherever basic slags and dust are encountered (Heuer et al., 1956). In the nonferrous industry, chrome-magnesite bricks are used mostly in converters. As the open hearth steel furnaces, where chromite bricks traditionally have been employed, are replaced by electric-arc furnaces, the use of chromium for refractories in the steel industry has been declining (Udy, 1956; Westbrook, 1979). The Kraft paper industry uses chromite bricks in the recovery furnace to reduce chemical attack by the spent liquors, and the glass industry uses chrome-magnesite bricks in the reheating chambers of the glass furnace (Heuer et al., 1956).

5.4 Chemicals

The first commercial application for chromium ore was in the manufacture of chromium compounds for use as pigments and in mordant dyeing (see Chapter 1). Today, over 70 chromium compounds are in commercial use (Table 6). However, only a few of these compounds are produced in large quantities, notably, sodium chromate, potassium chromate, potassium dichromate, ammonium dichromate, chromic acid, and the basic chromic sulfate used primarily for leather tanning (Darrin, 1956; Sittig, 1978; Hartford, 1979; Pedersen, 1982; Yassi and Nieboer, 1988).

5.4.1 Metal Finishing and Corrosion Control

Chromium compounds are used in metal surface treatments to prevent corrosion, improve durability of product, and enhance the retention of paints and other final finishes. The application techniques are diverse and expanding rapidly and include chromium plating, anodizing and sealing treatments for aluminum, the dips and phosphate-coating treatments for iron, steel, brass, and tin, and immersion and electrolytic surface treatments for zinc and magnesium (the Parkerizing, Bonderizing, Granodizing, Cronak, Iridite processes). Chromium compounds have been used also to produce water-resistant and oil-resistant coatings on plastics, textiles, and fiberglass under the trade names of Volan, Quilon, and Scotchgard (Hartford, 1979).

   The electroplating applications can be divided into decorative plating and hard (wear-resistant) plating. Decorative plating is used primarily in consumer goods industries, in making rail coach and automobile body parts (bumpers, radiator shells, headlights, handrails, decorative hardware), tableware, plumbing fixtures, and appliances. Typically, the thin chromium plate of 0.2–0.5 μm is applied over a nonferrous (Cu or Ni) base. It is of heuristic interest that an average 1968 vintage car contained about 4 kg of chromium, 30% of which went into the chromium plates or stainless steel trims, 15–20% in the engine and transmission, and the remainder distributed in the body frame and other parts (Brantley, 1970).

   Hard plating generally is reserved for industrial purposes, especially in chemical, food, printing, and paper industries. Gauges, gears, cylinder liners, and piston rings of rolling equipment and internal combustion engines are often chrome plated. The resistance of chromium to many types of corrosives has made

Table 6   Properties and Uses of Miscellaneous Chromium Compounds

Name	Formula	Uses
Oxidation state +6		
Aluminum chromate	Variable	In ceramics
Ammonium chromate	$(NH_4)CrO_4$	Flexible printing, photosensitization
Barium chromate	$BaCrO_4$	Pyrotechnics, high-temperature, batteries
Barium dichromate	$BaCr_2O_7 \cdot 2H_2O$	In ceramics
Barium potassium chromate	$K_2Ba(CrO_4)_2$	Corrosion-inhibiting pigment
Cadmium chromate	$CdCrO_4$	In catalysts, pigment
Cadmium dichromate	$CdCr_2O_7 \cdot H_2O$	Metal finishing
Calcium chromate	$CaCrO_4$	Metal Primers, corrosion inhibitor, high-temperature batteries
Calcium dichromate	$CaCr_2O_7 \cdot 4.5H_2O$	Metal finishing
Cesium chromate	Cs_2CrO_4	Electronics
Chromic chromate	Variable	In catalysts, mordants
Chromyl chloride	CrO_2Cl_2	Organic oxidation, Etard reaction
Cobalt chromate	$CoCrO_4$	In ceramics
Copper chromate, basic	$4CuO \cdot CrO_3 \cdot xH_2O$	Fungicides, in catalysts
Copper dichromate	$CuCr_2O_7 \cdot 2H_2O$	In catalysts, wood preservatives
Copper sodium chromate	$Na_2O \cdot 4CuO \cdot 4CrO_2 \cdot 3H_2O$	Antifouling pigment
Lithium chromate	$Li_2CrO_4 \cdot 2H_2O$	Corrosion inhibitor, especially in air-conditioning and nuclear reactors
Lithium dichromate	$Li_2Cr_2O_7 \cdot 2H_2O$	Corrosion inhibitor
Magnesium chromate	$MgCrO_4 \cdot 5H_2O$	Corrosion inhibitor for gas turbines; refractories
Magnesium dichromate	$MgCr_2O_7 \cdot 6H_2O$	In catalysts, refractories
Mercuric chromate	$HgCrO_4$	Antifouling formulations
Mercurous chromate	Hg_2CrO_4	Antifouling formulations
Morpholine chromate	$(OC_4H_8NH_2)_2CrO_4$	Vapor phase corrosion inhibitor
Nickel chromate	$NiCrO_4$	In catalysts

(Continued)

Table 6 (*Continued*)

Name	Formula	Uses
Pyridine–chromic acid	$CrO_3 \cdot 2C_5H_5N$	Research oxidant
Pyridine dichromate	$(C_5H_5NH)_2Cr_2O_7$	Photosensitizer in photoengraving, in ceramics
Nickel potassium chromate	$K_2Ni(CrO_4)_2 \cdot 6H_2O$	
Silver chromate	Ag_2CrO_4	In catalysts
Strontium chromate	$SrCrO_4$	Corrosion-inhibiting pigment, plating additive
Tetramino copper chromate	$Cu(NH_3)_4CrO_4$	Catalysts, gas absorbent
Zinc sodium chromate	$Na_2O \cdot 4ZnO \cdot 4CrO_3 \cdot 3H_2O$	Corrosion-inhibitive pigment

Oxidation state +3

Name	Formula	Uses
Ammonium tetrathiocyanato diamino chromate(III)	$NH_4(NH_3)_2Cr(SCN)_4$	Identification amines and alkaloids
Basic chromic chloride	$Cr(OH)Cl(H_2O)_4 \ Cl_2$	Manufacturing Quilon, Volan, Scotchgard, all trademarks
Chromic acetate	$Cr(OCOCH_3)_3 \cdot xH_2O$	Printing and dyeing textiles
Chromic acetylacetonate	$Cr(CH_3COCHCOCH_3)_3$	Preparation of Cr complexes, in catalysts, antiknock compounds
Chromic ammonium sulfate	$NH_4Cr(SO_4)_2 \cdot 12H_2O$	Electrolyte for manufacturing Cr metal
Chromic chloride	$CrCl_3$	Chromizing, Cr metal, organochromium compounds
Chromic chloride (hydrated)	$CrCl_3 \cdot 6H_2O$	Mordant, tanning, chromium complexes
Chromic fluoborate	$Cr(BF_4)_3$	Chromium plating, in catalysts
Chromic fluoride	$CrF_39(4)H_2O$	Mordants, in catalysts
Chromic fluoride	CrF_3	Chromizing

Table 6 (*Continued*)

Name	Formula	Uses
Chromic formate, basic	Cr(OH)(HCOO)	Skein printing of cotton tanning
	$(H_2O)_4(HCOO)$	
Chromic hydroxide	$Cr_2O_3xH_2O$	See Guignet's Green under Uses (in text)
Chromic lactate	$Cr(CH_3CHOHCOO)_3 \cdot xH_2O$	Mordant ingredient of Universal mordant 9333
Chromic naphthenate	No definite formula	Textile preservative
Chromic nitrate	$Cr(NO_3)_3 \cdot 9H_2O$	In catalysts, textiles, manufacturing CrO_2
Chromic phosphate	$CrPO_4$	Pigments, phosphate coating, wash primers
Chromic potassium oxalate	$K_3[(Cr(C_2O_4)_3] \cdot 3H_2O$	Mordant Eriochromal Mordant, and Chromosol
Chromic potassium sulfate	$KCr(SO_4)_2 \cdot 12H_2O$	Hardening photographic emulsions
Chromic sulfate	$Cr_2(SO_4)_3 \cdot xH_2O$	Insolubilizing gelatin
Cobalt chromite	$CoCr_2O_4$	Pigment, ceramics, in catalysts
Copper chromite	$CuCr_2O_4$	In catalysts, especially for automobile exhaust
Ferrous chromite	$FeCr_2O_4$	Impure (fe,Mg) $(Cr,AlFe)_2O_4$ is chrome ore of commerce
Magnesium chromite	$MgCr_2O_4$	Refractory
Triphenyl chromium tetrahydrofuranate	$(C_6H_5)_3Cr \cdot 3THF$	Unstable
Zinc chromite	$ZnCr_2O_4$	In catalyst

Other oxidation states

Oxidation state 0

Chromium carbonyl	$Cr(CO)_6$	Synthesis of "sandwich" compounds
Dicumene chromium	$[(CH_3)_2CHC_6H_5]_2Cr$	Preparation of Cr carbides by thermal vapor deposition

(*Continued*)

Table 6   (*Continued*)

Name	Formula	Uses
Oxidation state +2		
Chromous chloride	$CrCl_2$	Chromizing, preparation of Cr metal
Oxidation state +4		
Chromium(IV) oxide	CrO_2	Manufacturing magnetic tapes
Manganese(III) chromate(IV)	$Mn_2CrO_5 \cdot xH_2O$	Catalysts, in ceramics
Oxidation state +5		
Calcium chromate(V)	$Ca_3(CrO_4)_2$	Corrosion inhibitive pigment, may be carcinogenic

Source: After Hartford, 1979.

chromium plating almost indispensable in chemical apparatuses. The plate is generally applied over a hard substrate, such as steel, in thicknesses of $1-300\,\mu m$ (Hartford, 1979).

Chromium plates have been applied on steel beverage cans and as chemical conversion coatings on nonferrous metals, especially Mg, Al, Zn, and Cd. Chromium-plated plastics now are used widely in the automotive industry at the expense of plated die-cast zinc parts.

5.4.2 Leather Tanning and Finishing

Chromium tanning of leather was introduced in 1858 (see Chapter 1) and depends on the propensity of Cr(III) to form stable complexes with proteins, cellulosic material, dyestuff, and synthetic polymers. The reaction renders the leather resistant to bacterial attack and enhances its stability in the ambient environment. All hides, with the exception of heavy cattle hides, still receive chromium tannage. In the United States alone, the leather industry consumed over 18,000 t, or about 15% of the chromium chemical sales, in 1980 (Papp, 1983). The basic chromic sulfates used in the tanning are produced directly from sodium dichromate.

5.4.3 Pigments

Pigments and allied products account for a rapidly increasing share of the chromite ores and concentrates consumed in the United States in the 1980s (Papp, 1983). These pigments fall into two broad categories: chromate color pigments, which are mostly lead chromates, and corrosion inhibition pigments based on zinc chromates (Yassi and Nieboer, 1988). Familiar colors of lead chromates, usually with additives to improve their working properties, include the yellows (lemon, primrose, and medium yellows), chrome orange, molybdate orange, chrome green, and Guignet's green. They are used in residential and industrial structures, machinery and industrial equipment, bridges, automobiles, printing inks, rubber, paper, traffic markings, crayons, chalk, shoe polish, ceramics, cement, granulated rocks for asphalt roofing. Chrome oxide green, the most stable green pigment known, is used in coloring cements, plasters, and roofing granules. It is an ingredient of most camouflage colors because its reflection of infrared light is very similar to that of natural vegetation. It is sometimes referred to as "green rouge" on account of its use as a mild abrasive for polishing jewelry and fine metal parts. Its use in coloring artificial gem stones stems from the fact that emerald, ruby, and dichroic alexandrite owe their hues to chromic acid.

Zinc yellow was an extensively used pigment during World War II (production in the United States alone exceeded 11,000 t) for aircrafts and still remains unequaled as a corrosion-inhibiting primer on aircraft parts made of aluminum or magnesium. Zinc tetroxychromate has been used in wash primers, and the sodium salt of chromium sulfate finds occasional use as an antifouling marine pigment and also is an ingredient of dips for automobile bodies (Darrin, 1956; Hartford, 1979).

5.4.4 Miscellaneous Uses

Chromated copper sulfate has become extremely popular as a wood preservative; the treatment of utility poles, wood foundations, and building timber prevents them from bleeding, makes them more paintable and resistant to fungi and termite attack, and gives them an attractive olive-green hue. The fungicidal properties of chromium compounds have led to their use in agriculture for potato blight prevention, seed conditioning, and in weed killers, defoliants, and tobacco-bed cloths. Chromium compounds in fire-retardant formulations reduce the corrosion of the equipment treated and prevent the leaching of the retardant from the wood.

Chromium compounds are used in recirculating water systems in cooling towers and automobiles (antifreeze) to inhibit corrosion and stop the growth of slime and algae. Since 1941, chromium compounds have been added to drilling muds to reduce corrosion cracking of the drill strings. Many oils used in metal cutting and grinding contain chromium compounds as an anticorrosive. With the phaseout of lead tetralkyls, chromium compounds are being used increasingly as antiknock agents in unleaded gasolines.

Chromium compounds react with gelatin to produce hydrophilic gels used in treating paper and cloth, conditioning soils, and solidifying road surfaces. Their reaction with glue, gelatin, casein, and lignin provides the waterproof, grease-proof, and oil-repellent properties of wallpapers and printing rolls. Potassium dichromate is used in hardening photographic films, in developing blueprints, and in coating blueprint papers. Chromic acid and ammonium dichromate find interesting application in photoengraving and offset printing processes. Silkscreen devices containing light-sensitive chromium compounds have been used for producing printed electric circuits and the heating thread patterns on automobile rear windows. Because of their peculiar properties, chromium compounds are added to printing inks, and the addition of methacrylato chromic chloride makes it possible to form objects from fiberglass and plastic cements.

The use of chromium compounds in the textile industry as oxidants to improve washfastness of cotton fabric and in mordant dyeing of wool and synthetic fabric has a long history (see Chapter 1).

Chromium is an important colorant in porcelain enameling and green glass formulations. Insoluble chromium salts act as activators and depolarizers in fused-salt dry batteries, whereas chromium dioxide is a common ferromagnetic material in high-fidelity magnetic tapes. Match heads and fireworks often contain salts of chromium as an oxidizing agent. Other products that are fabricated with chromium include rail-joint packing and gasket materials, arc-welding rods, plasters, chromium catgut, and metal inplants (pacemakers and osteosynthetic material). Tattoo green colors often contain trivalent chromium compounds.

Chromium compounds have been used as catalysts for a large variety of chemical reactions, including alkylation, oxidation, isomerization, hydrogenation, polymerization, and dehydrogenations. They are, therefore, used by the petrochemical industry in the synthesis of saccharin, benzoic acid, anthraquinone, hydroquinone, camphor, and synthetic fibers. Chromic acid is a well-known bleaching agent for oils, fats, and waxes. Many chromium compounds are familiar reagent chemicals in industrial and research laboratories.

REFERENCES

Brantley, F.E. (1970). "Chromium." In *Mineral Facts and Problems*. Bureau of Mines Bulletin 650, U.S. Department of the Interior, Washington, DC, pp. 247–262.

Darrin, M. (1956). "Chromium compounds—Their industrial use. In M.J. Udy, Ed., *Chromium*. Reinhold, New York, Vol. 1, pp. 251–262.

DeYoung, J.H., Lee, M.P. and Lipin, B.R. (1984). *International Strategic Minerals Inventory Summary Report—Chromium*. U.S. Geological Survey Circular 930-B, U.S. Department of the Interior, Washington, DC, 41 pp.

EPA (1984). *Health Assessment Document for Chromiuim*. Report No. EPA-600/8-83-014F. Office of Health and Environmental Assessment, U.S. Environmental Protection Agency, Research Triangle Park, NC.

Hartford, W.H. (1979). "Chromium compounds." In *Encyclopedia of Chemical Technology*. John Wiley & Sons, New York, Vol. 6, pp. 82–120.

Heuer, R.P., Trostel, L.J. and Grisby, C.E. (1956). "Chromium in refractories." In M.J. Udy, Ed., *Chromium*. Reinhold, New York, Vol. 2, pp. 327–390.

Kinzel, A.B. and Crafts, W. (1937). *The Alloys of Iron and Chromium*. McGraw-Hill, New York, Vol. 1.

Kinzel, A.B. and Franks, R. (1940). *The Alloys of Iron and Chromium*. McGraw-Hill, New York, Vol. 2.

Liu, L.G. (1982). "Speculations on the composition and origin of the earth." *Gechem. J.* **16**, 287–310.

Matzat, E. and Shiraki, K. (1974). "Chromium." In K. Wedepohl et al., Eds., *Handbook of Geochemistry*. Springer-Verlag, Heidelberg.

Papp, J.F. (1980–1983). "Chromium." In *U.S. Bureau of Mines, Minerals Yearbooks, 1980–1983*, U.S. Government Printing Office, Washington, DC.

Pedersen, N.B. (1982). "The effects of chromium on the skin." In S. Langård, Ed., *Biological and Environmental Aspects of Chromium*. Elsevier, Amsterdam, pp. 249–275.

River, C., et al. (1970). *Economic Analysis of the Chromium Industry*. Report No. PB 196057, U.S. Department of Commerce, National Technical Information Service, Springfield, VA.

Sittig, M. (1978). *Inorganic Chemical Industry: Processes, Toxic Effluents and Pollution Control*. Noyes Data Corporation, Park Ridge, NJ, pp. 124–134.

Streicher, M.A. (1977). "Stainless steels: Past, present and future." In *Stainless Steel 1977*. Climax Molybdenum Inc., London.

Sully, A.H. and Brandes, E.A. (1967). *Chromium*, 2nd ed. Butterworths, London.

Udy, M.J. (1956). *Chromium*. Reinhold, New York, 2 vols.

Westbrook, J.H. (1979). "Chromium and chromium alloys." In *Encyclopedia of Chemical Technology*. John Wiley & Sons, New York, Vol. 6, pp. 54–82.

Wolfe, J.A. (1984). *Mineral Resources: A World Review*. Chapman & Hall, New York, pp. 104–108.

Yassi, A., and Nieboer, E. (1988). "Carcinogenicity of chromium compounds." In J.O. Nriagu and E. Nieboer, Eds., *Chromium in the Natural and Human Environment*. John Wiley & Sons, New York, Chapter 17.

Zapffe, C.A. (1949). *Stainless Steels*. American Society for Metals, Cleveland, Ohio.

4

ATMOSPHERIC EMISSIONS OF CHROMIUM FROM NATURAL AND ANTHROPOGENIC SOURCES

Jozef M. Pacyna

Norwegian Institute for Air Research
Lillestrøm, Norway

Jerome O. Nriagu

National Water Research Institute
Burlington, Ontario, Canada

1. INTRODUCTION

Chromium is an abundant element in the earth's crust and occurs in oxidation states ranging from Cr^{2+} to Cr^{6+}. Elemental chromium is never found in nature, and chromite, $FeOCr_2O_3$ is the only major commercial chromium mineral. Its deposits are mainly found in Rhodesia, the Soviet Union, the Republic of South Africa, New Caledonia, and the Philippines.

Widespread use of chromium in industrial and domestic products, as well as the toxicity of its hexavalent (both toxic and volatile) and trivalent forms (essential toxicity), have brought about a growing concern about chromium emissions into the atmosphere. In this chapter, the main emission sources of chromium are reviewed and some emission factors are presented.

2. NATURAL SOURCES OF CHROMIUM

Chemical composition measurements of aerosols in remote areas show the background chromium levels in the atmosphere to range from 0.01 ng/m^3 in Antarctica (Cunningham and Zoller, 1981), 0.09 ng/m^3 in Greenland (Flyger and Heidam, 1978), and 0.2 ng/m^3 over the East Atlantic Ocean (Duce et al., 1976) to 0.1–1.3 ng/m^3 in the Norwegian Arctic (Pacyna and Ottar, 1985). These concentrations were measured during the summer season, when no episodes of long-range transport from anthropogenic sources occurred. Thus, these chromium levels are the result of emissions from natural sources, particularly windblown dust and volcanoes.

The author's recent review of worldwide emissions of trace elements from natural sources (Pacyna, 1986a) estimates the chromium emissions from wind-blown dust and volcanoes at 50×10^3 and 3.9×10^3 t·$year^{-1}$, respectively. These calculated amounts assumed emission factors of 100 mg chromium kg^{-1} windblown dust and 390 mg chromium kg^{-1} volcanic particles (Pacyna, 1986b) and should be considered tentative, since there generally is a lack of hard, reliable information about the natural processes in the environment. For example, estimates of global production of windblown dust by various authors range from 6 to 1100×10^9 kg $year^{-1}$ (Nriagu, 1979). Zoller (1983), in his paper on perturbations of metal fluxes into the atmosphere, concluded that the emission rates of trace elements, among them chromium, from individual volcanoes can vary by more than two orders of magnitude.

The variable chromium concentrations in different soils and rocks also influence natural atmospheric emissions of the element. Balsberg-Påhlsson et al. al. (1982) reviewed bulk concentration ranges for Cr in various types of rocks and soil. The results are shown in Table 1, together with data from earlier work by Fleischer (1972). Although the chromium levels in soil are reportedly within the range 80–200 μg g^{-1} and an average value of 100 μg g^{-1} was reported by Balsberg-Påhlsson et al. (1982), the metal contents of some soils can be as high as

Table 1   Chromium Levels in Rocks and Soils

Rock Type	Cr Range μg g^{-1}	Rock Type	Cr Range μg g^{-1}
Basaltic igneous	40–600	Shale	30–600
Granitic igneous	2–100	Sandstone	35–90
Quartz	3–200	Phosphorite	300–3,000
Limestone	10–60	Sedimentary	
		iron ore	150–800
Dolomite	300–1,000	Soil	80–200

Source: After Balsberg-Påhlsson et al., 1982, and Fleischer, 1972.

1500 μg g^{-1} (in some American soils, as indicated by Lisk, 1972). Generally, the highest chromium levels are contained in soils derived from basalt and serpentine, where they range from 5 to 3,000 μg g^{-1} (Bowen, 1966).

The high, although variable levels of chromium in soils may cause windblown dust to be a significant source of the element in some areas. During recent years, several cases of a substantial transport of Saharan dust have been documented by direct measurements (SCOPE, 1979). Long-range transport of Asian dust has been observed also (Rahn et al., 1979; Darzi and Winchester, 1982). Thus, some influence of chromium release from this source can be expected on the total loading, including in Europe, but the extent is difficult to estimate.

Very scanty information exists on chromium behavior during volcanic eruptions. Phelan et al. (1982) report 65 ng m^{-3} as an average concentration in aerosol from the Mount St. Helens eruption on September 22, 1980. An average enrichment factor of chromium in the plume (with aluminum as the reference element and volcanic ash as reference material) places this element in a group of enriched trace elements.

Other natural sources, such as airborne sea salt particles and smoke from forest wildfires, as well as biogenic emissions from vegetation, do not seem to be important sources of chromium. However, Buat-Menard and Chesselet (1978) estimate the geometric mean crustal enrichment factor of chromium in North Atlantic marine aerosols to be more than 20. This may indicate an involvement of the element in the processes at the air–sea interface.

3. CHROMIUM EMISSIONS FROM THE METALLURGICAL INDUSTRY

The history of chromium usage in industry is given in Table 2 (NRCC, 1976). The metallurgical industry is a main user of chromite ore, and as a result the steel and alloy industry is a dominant source of anthropogenic chromium emissions into the atmosphere. Chromite ore, containing at least 50% chromic oxide, is

Table 2   History of Chromium Usage in Industry

Date	Event or Process	Comment
1798	Chromium discovery	Charcoal reduction of natural $PbCrO_4$
1800	Chromate manufacture	Chromite roasted with lime and soda ash
1816	Pigment manufacture	e.g., $PbCrO_4$
1820	Mordant dyeing	
1858	Chrome tanning	Commercialized in 1884
1879	Refractory brick	Declining production
1910	Metallurgy	Corrosion-resistant ferrous alloys
1926	Chrome plating	Increasing production

Source: After NRCC, 1976.

converted into one of various types of ferrochromium or chromium metal that are alloyed with iron or other elements, such as nickel and cobalt (NAS, 1974). Ferrochrome is then used to produce stainless steel and alloy steels. Other applications of ferrochrome include production of nonferrous steels and high-speed and high-temperature steels.

Chromium emissions from the metallurgical industry are almost exclusively in the form of particles and depend on the production technology employed. There are three main methods of steelmaking: the open hearth furnace, the basic oxygen furnace, and the electric-arc furnace. Most ferroalloys are produced in the electric-arc furnace. High-carbon ferrochrome of various grades and ferro-chrome-silicon are produced chiefly in electric-arc furnaces. Low-carbon ferrochrome is produced in another type of furnace, where lime melt reacts with ferrochrome-silicon. Using particulate emission factors and the chemical composition of the dust emitted from steel mills, Pacyna (1986b) estimated chromium emission factors for electric-arc furnaces equipped with different emission control devices (Table 3). They have been calculated for the oxygen lance electric-arc method of steelmaking in electric-arc furnaces. The emission factors for the nonoxygen lance electric-arc method are approximately 18% lower than the emission factors presented in Table 3 (Pacyna, 1986b). Control devices use for emission control in electric furnaces are mainly Venturi scrubbers (with an efficiency of 98%) and bag filters (with an efficiency of 99% or higher). Electrostatic precipitators (ESPs) are not as efficient because of the high electrical resistivity of the fume releases. The temperature of ferrochrome-silicon fumes is too low to be within an acceptable range to lower the electrical resistivity. The problem could be overcome by using a wet precipitator, but more water needs to be used than for a wet scrubber without recycle.

The metallic form of chromium is made commercially by two principal processes: electrolytic and aluminothermic reduction of chromic oxide with finely divided aluminum (IARC, 1973). These processes do not emit chromium into the atmosphere in amounts posing environmental problems.

Table 3   Chromium Emission Factors for
Electric-Arc Furnaces

Control device	Emission Factor (g t^{-1} Steel)
Uncontrolled	Up to 450
Venturi scrubber (>98% efficiency)	0.1–9.0
Bag filters (>99% efficiency)	0.1–4.0
Electrostatic precipitator	13.5–36.1

Source: Pacyna, 1986b.

4. CHROMIUM EMISSIONS FROM REFRACTORY MATERIAL AND THE CHEMICAL INDUSTRY

The production of refractory brick cast by electric furnace, and moreso for noncast refractory brick, emits chromium in such amounts that refractory processing is considered a main emission source of the element in an emission survey after ferrochrome production (GCA, 1973). In the beginning of the 1970s, chromium emissions from refractory processing contributed 10% to the total chromium emissions in the United States (GCA, 1973).

Chromate chemicals, mainly chromates and dichromates, are produced by smelting, roasting, and extraction processes. In common production technology, a mixture of chromite ore, limestone, and sode ash is calcined, and residues from this process contain certain amounts of various chromium compounds. Sodium chromate produced in the above process is then treated with sulfuric acid to obtain sodium dichromate. Both sodium chromite and sodium dichromate are the basic compounds for the production of all other chromium chemicals (U.S. EPA/ORNL, 1978). Manufacture of sodium dichromate is sometimes considered as a source of atmospheric emissions of chromium. However, the process is very well controlled (the efficiency of control devices is higher than 90%), and the contribution of emissions from sodium dichromate production in the United States to the total chromium releases into the atmosphere does not exceed 0.5% (GCA, 1973).

The uses of sodium dichromate (Fishbein, 1981) are many and include the production of chromium pigments, chromium trioxide, and chromium salt. These are only byproducts. Chromium pigments (e.g., lead chromate) are used in paints, inks, rubber, linoleum, and floor tile. Applications of chromium trioxide include chromium plating (e.g., in the production of automobiles) and secondary metal-finishing operations, such as aluminum anodizing, chemical conversion coatings, and the production of phosphate films on galvanized iron and steel. Chromium trioxide is used also as a corrosion inhibitor for ferrous alloys in recirculating water systems, in catalyst manufacture, and as an oxidant in organic synthesis. Chromium salt is used as a tanning agent for leather, a mordant in dyeing, a wood preservative, a fungicide, and an anticorrosion agent in cooling systems, boilers, and oil drilling muds (Flinn and Reimers, 1974).

The major sodium chromate applications include pigment production, manufacture of chromic acid, and textiles and dyes. Chromium emissions within the chemical industry may pose a serious problem only on a local scale. There is a lack of data in the literature on emission factors or emission rates of chromium from these sources, and the only indication of enhanced amounts of the element is increased chromium concentrations in the air. A daily maximum concentration of 370 ng m^{-3} was measured in the industrial zone of the Meuse valley, Belgium, where housing and steelworks are implicated indirectly (Rustagi, 1964). Chromium concentrations exceeding 150 ng m^{-3} for industrial regions in Belgium, Czechoslovakia, and the Soviet Union are reported by Rondia (1979) and for Poland by Tomza (1984). Chromium concentrations in the range 0.01–1.0 ng m^{-3} have been measured in remote areas (see Section 2) and below 10 ng m^{-3} in rural areas (Fishbein, 1981).

5. FUEL COMBUSTION AS A SOURCE OF ATMOSPHERIC CHROMIUM

The behavior of chromium during combustion of fuels is affected mainly by (1) chromium affinity for pure coal and mineral matter, (2) the chromium concentrations in coal and crude oil, and (3) chromium physical-chemical properties. Combustion conditions also play an important role. Kuhn et al. (1980), in a review of the occurrence of minor elements in the organic and mineral fractions of coal, assigned chromium to the group of elements displaying behavior intermediate between the volatile chalcophiled species and the nonenriched lithophiled elements in coal. They showed that information on chromium affinity for coal could be used to determine its chemical forms, to estimate the theoretical percentage of the element that can be removed by coal cleaning, and to predict material balances in the coal products and wastes. The chromium concentrations in coal vary from 1 to 100 μg g^{-1} and from 0.005 to 0.73 μg g^{-1} in crude oils (Pacyna, 1986c). Considering the large differences in metal concentrations from different production fields, it is difficult to generalize on chromium impurities in coal. Nevertheless, according to the literature (Gluskoter, et al. 1977; Pacyna, 1981), lignite and subbituminous coals are less contaminated by chromium than are medium-, low-, and high-volatility bituminous coals. For example, coal from the Western Basin in the United States contains less chromium than do coals from the Eastern and Illinois Basins. In Europe, German brown coals (lignite) are cleaner than bituminous and subbituminous coals from Czechoslovakia and Poland (Pacyna, 1980; Heinrichs, 1982). However, production of a specific amount of energy (i.e., electricity) requires a lignite charge almost twice as high as that required for a bituminous charge (Dvorak and Lewis, 1978), resulting in enhanced emissions of chromium and other trace elements.

During combustion, the volatile species in the coal evaporate in the furnace and subsequently condense as submicron aerosol particles or on the surfaces of ash particles as the flue gas cools in the convective sections. Natusch and Wallace (1974) used a scanning electron microscope to show that chromium is significantly more concentrated on the surfaces of fine particles than in their interior. $Cr(CO)_6$ is a volatile chromium compound that may be produced from chromium in the presence of carbon monoxide (Gerasimov and Sharifov, 1958). Chromium evaporation does not seem to be very effective because the enrichment of the element from coal through bottom- and fly ashes to stack dust is low (Gluskoter et al., 1977; Pacyna, 1981).

The temperature in the combustion chamber is one of the key parameters affecting chromium release into the atmosphere. Larger amounts of the element are emitted from conventionally fired boiler systems (e.g., stoker boiler or cyclone boiler), burning fuels at temperatures higher than 1,650 K, as compared to fluidized-bed systems with temperatures between 1,100 and 1,200 K. The chromium metal has an appreciable vapor pressure at elevated temperatures (1 torr at 1,889 K).

The type of fly ash emission control system and its efficiency also influence chromium emissions. ESPs and wet scrubbers are the most common systems used in electric power plants. Ondov et al. (1979) studied the penetration of several elements contained in particles from coal-fired power plants equipped with the two control systems. The chromium penetration through ESP varied between 1.2 and 12.1%, whereas that for Venturi scrubbers was between 0.6 and 36.0%. This efficiency information has been used to estimate chromium emission factors for coal combustion in electric utilities and industrial and commercial applications (Pacyna, 1982). The chromium emission factors for coal-fired power plants, for coal with 10% of ash and more than 99% control efficiency, are shown in Table 4. Similar data for chromium emissions from industrial, commercial, and residential furnaces are given in Table 5.

Chromium in crude oil concentrates in heavy distillate residuals and in liquid and solid waste streams during the refining process. The amounts of chromium, as

Table 4   Chromium Emission Factors for Coal-Fired Power Plants

Control system/boiler	ESP (μg MJ^{-1})[a]			Wet scrubber (μg MJ^{-1})[a]		
	Cyclone	Stoker	Pulverized	Cyclone	Stoker	Pulverized
Bituminous coal	120	200	85	123	206	87
Subbituminous coal	145	242	103	150	250	106
Lignite coal	202	337	143	208	347	147

Source: After Pacyna, 1982.

[a] 10% ash and more than 99% control efficiency.

Table 5   Chromium Emission Factors for Coal Combustion in Industrial, Commercial, and Residential Furnaces

Furnace	Emission Factor $(g\ t^{-1}\ Coal)$
Industrial	
Cyclone	1.7
Stoker	12.0
Pulverized	7.0
Commercial and residential	4.2

Source: After Pacyna, 1982.

Table 6   Chromium Factors for Oil Combustion in Electric Utility, Industrial, Commercial, and Residential Furnaces

Boiler	Unit	Emission Factor[a]
Electric utility	$\mu g\ MJ^{-1}$	43
	$g\ 10^{-3}\ L$	0.9
Industrial	$g\ 10^{-3}\ L$	2.2
Commercial and residential	$g\ 10^{-3}\ L$	1.1

Source: After Pacyna, 1982.
[a]Particle emission factor of 1.6 kg 10^{-3} L of oil.

well as other elements, discharged during combustion depend mainly on their concentration in crude oil, the efficiency of combustion, and the buildup of boiler deposits. Poor mixing of oil and air in a boiler, low flame temperatures, and short residence time in the combustion zone result in emission of larger particles with a higher content of combustible matter and in higher particle loadings containing chromium and other elements. Emission factors of chromium from electric utility, industrial, residential, and commercial boilers are shown in Table 6 (Pacyna, 1982), assuming a particle emission factor of 1.6 kg 10^{-3} L of residual oil.

6.   OTHER ANTHROPOGENIC SOURCES OF CHROMIUM

The cement industry is another potential source of atmospheric chromium. Manufacture of cement is a high-temperature process, processing more than 30 raw materials, with limestone as a major ingredient. The chromium concentration ranges in limestone are shown in Table 1. Operation of the kiln is the largest source of atmospheric emissions within cement plants. It may be considered to have three units: feed system, fuel-firing system, and clinker-cooling and handling systems. During the kiln operation, raw material is calcined, driving off CO_2 from $CaCO_3$, and heated to about 1,700 K to form clinker. Chromium from limestone is emitted during this process in varying amounts depending on the element concentration in limestone, production technology (wet or dry kiln process), and the efficiency of emission control equipment. The last is especially important because of the high electrical resistivity of cement dust (Pacyna, 1986d). Other emission sources of chromium-containing dust in cement production include quarry and raw plant, clinker coolers, and clinker mills. A total chromium emission factor of 1.6 g Cr t^{-1} cement produced in all the above operations was estimated by Pacyna (1982). This factor does not include chromium emitted in fugitive emissions during transport and handling of clinker and cement, since emission estimates for these are very imprecise, and emission factors change from one cement plant to another.

Only limited information is available on chromium emissions from refuse incineration, particularly from municipal incinerators. These emissions are influenced by the combustible/noncombustible portion of the refuse input, the chemical composition of the refuse input, the chamber design, and the efficiency of control devices. Chromium in incinerator effluents comes from noncombustible materials. Pacyna (1982) estimated chromium emission factors of 14.6 g Cr t^{-1} refuse burned in incinerators with uncontrolled emissions and 1.1 g Cr t^{-1} refuse burned in plants with 85% efficiency dust-removed installations.

The potential for chromium volatilization during sludge combustion is low, although the element concentrations in municipal sewage sludges vary from 22,000 to 30,000 mg Cr kg^{-1} dry sludge (U.S. EPA, 1976). A major part of chromium seems to go into the fly ash stream, but generally the particles are not enriched with the element over the amount in the sludge (U.S. EPA, 1972). Takeda and Hiraoka (1976) have suggested that chromium behaves similarly to Zn, in that 60–70% of the amount in the sludge was found in the ash. Data relating the chromium emissions to the sludge content show that <1% is emitted as fine particles or fume (Gerstle and Albrinck, 1982). Pacyna (1982) estimated an average chromium emission factor for sewage sludge incineration to be 9.7 μg chromium t^{-1} sludge incinerated.

Chromium base catalysts are an attractive alternative to the more expensive noble metal catalysts in the emission control from internal combustion engines. Catalysts, such as copper chromite, however, have been found in high concentrations in the ca. 0.01 μm diameter particles emitted at combustion temperatures ranging from 460 to 1,070 K (Balgord, 1973). Thus, the increased use of catalytic emission control devices is an additional potential source of chromium emissions to the urban atmosphere.

Production of phosphoric acid in a thermal process also involves emission of chromium into the atmosphere. Chromium is released when the chromium-containing phosphate rock is heated together with siliceous flux and coke in an electric furnace to produce elemental phosphorus, a byproduct in phosphoric acid manufacture.

Another source of chromium atmospheric emissions is the wearing away of asbestos-containing brake linings and production of asbestos textiles. According to Schroeder (1970), asbestos can contain as much as 1,500 μg Cr g^{-1}.

7. EMISSION SURVEYS OF CHROMIUM FROM ANTHROPOGENIC SOURCES

During the last decade, a growing interest in estimating chromium emissions has been documented in the literature. Chromium emission surveys on the continental scale have been performed for the United States (Lee and Duffield, 1977), Europe (Pacyna, 1982), and the Soviet Union (NILU, 1984; Pacyna, 1982). Total chromium emissions from anthropogenic sources in the United States were estimated to be 15 $\times$ 10^{3}t in 1978 (Lee and Duffield, 1977) and only about 5 $\times$ 10^3 t $\cdot$ year^{-1} in 1984 (Table 7). About half of the emission in the United States comes from the Great Lakes area, the southeast, and the east coast south of New York (U.S. EPA, 1984). Estimates for Europe (Pacyna, 1982) and the Soviet Union (Pacyna, 1982) in 1979–1980 are 11.8 $\times$ 10^3 t $\cdot$ year^{-1} and 7.1 $\times$ 10^3 t $\cdot$ year^{-1}, respectively. The Soviet Union results possibly are underestimates, since only the European part of the country was considered. Based on the NILU (1984) estimates, an additional amount of at least 10.0 $\times$ 10^3 t $\cdot$ year^{-1} should account for the other parts of the Soviet Union. The percentage contributions of the chromium emissions from metallurgical processing, fuel combustion, and other sources in the United States, Europe, and the Soviet Union are presented in Figure 1. It is likely that the contributions of chromium emissions from "other sources" in Europe and the Soviet Union can be higher because of possibly larger emissions of the element from the chemical process industry in these regions.

Table 7   Anthropogenic Emissions of Chromium to the Air in the United States

Source	Chromium Emissions, t/year
Chrome ore refining	3
Ferrochromium production	43
Chromium chemical production	450–900
Refractory production	90
Sewage sludge incineration	25
Steel production	2,870
Utility cooling towers	5
Cement production	16
Combustion of coal and oil	
Boilers	737
Process heaters	556
Total	4,825–5,275

Source: U.S. EPA, 1984.

Information about chromium emissions from the chemical industry in Europe and the Soviet Union is incomplete because of a general lack of data in the literature.

The chromium emissions for Europe were estimated for each country separately. The chromium releases into the atmosphere from the particular countries were calculated using: (1) chromium emission factors estimated separately for all the sources considered, and (2) statistical information on the consumption of ores, rocks, and fuel and the production of various types of industrial goods. The results are presented in Table 8. The largest chromium emissions in Europe appear to come from five regions: (1) the Be-Ne-Lux countries and western part of the Federal Republic of Germany, (2) the southern part of Poland and Czechoslovakia, (3) the central part of England, (4) the Donetsk area in the Soviet Union, and (5) the Urals and Moscow region in the Soviet Union.

There are also chromium emission surveys prepared on the national scale. For example, an annual emission of 160 t chromium has been estimated for Sweden (SNV, 1982), which is in a good agreement with the data in Table 8.

In order to compare global chromium emissions from anthropogenic sources with those from natural sources, the atmospheric chromium releases from Japan and China should be considered in addition to those from the United States, Europe, and the Soviet Union. The literature on chromium emissions in these countries is very scanty. Nevertheless, it can be assumed that metallurgical

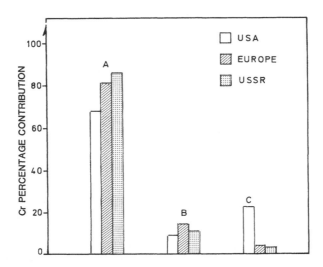

Figure 1. Percentage contribution of atmospheric emissions of Cr from metallurgical processes (A), fossil fuel combustion (B), and other anthropogenic sources (C) in the United States, Europe, and the Soviet Union.

Table 8   Chromium Emissions from Anthropogenic Sources in Europe in 1979/1980

Country	Metallurgical Processes $(t \cdot year^{-1})$	Fuel Combustion $(t \cdot year^{-1})$	Other Sources[a] $(t \cdot year^{-1})$	Total $(t \cdot year^{-1})$
Albania		3.6	1.4	5
Austria	175	15	9	199
Belgium	550	84	13	647
Bulgaria	100	73	8	181
Czechoslovakia	620	155	16	791
Denmark	28	15	7	50
Finland	94	17	3	114
France	950	97	50	1,097
German Democratic Republic	280	225	20	525
Federal Republic of Germany	1,860	217	76	2,153
Greece	40	19	18	77
Hungary	160	33	8	201
Iceland		0.1	0.2	0.3
Ireland	3	5	3	11
Italy	980	81	64	1,125
Luxemburg	194	1	0.2	195.2
Netherlands	220	22	9	251
Norway	33	4	4	41
Poland	750	374	35	1,159
Portugal	15	4	7	26
Romania	480	120	22	622
Spain	470	58	47	575
Sweden	180	15	4	199
Switzerland	32	2	6	40
Turkey[b]	66	56	25	147
USSR[b]	6,130	810	203	7,143
United Kingdom	870	222	41	1,133
Yugoslavia	140	51	14	205
Total	15,420	2,779	714	18,913

[a]Chromium emissions from chemical industry are not included.
[b]Only the European section of the country is considered.

processes are the main emission sources of the element in Japan and China, similar to the United States, Europe, and the Soviet Union (Fig. 1). In 1982, pig iron and ferroalloy production in Japan was double that of the United States, whereas the production data for China and the United States show similar figures (U.N., 1985). Thus, it is estimated that at least 20×10^3 t chromium year^{-1} are emitted in Japan and 10×10^3 t chromium year^{-1} in China. No possible differences in production technology, chromium concentrations in raw materials, and control device use in Japan, China, and the United States were considered, resulting in probably underestimated chromium emissions in Japan and China (Zonghua, 1982; Tsubaki, 1982). Considering these data for the United States, Europe, the Soviet Union, Japan, and China, the global anthropogenic emission of chromium amount to at least 74×10^3 t·year^{-1}, compared to 54×10^3 t·year^{-1} from natural sources (see Section 2). Thus, the anthropogenic flux for chromium slightly exceeds the flux from natural sources. An interesting method for assessing the importance of the anthropogenic flux for several metals has been suggested by Lantzy and Mackenzie (1979). They estimated the interference factor, and a value of 100% indicates that the anthropogenic flux equals the total natural flux. The value of 161% was estimated for chromium by Lantzy and Mackenzie (1979), whereas at least 137% can be deduced from this work. Considering the general lack of data on the chromium emissions from sources other than metallurgical processes for Japan and China, the releases of the metal from the chemical industry in the Soviet Union, Europe, Japan, and China, and anthropogenic emissions from the other parts of the world, the agreement between Lantzy and Mackenzie (1979) and this work is quite acceptable.

8. SOURCE APPORTIONMENT OF CHROMIUM IN THE ATMOSPHERE

The sources and rates of chromium emission described can be related to air quality using two basic models: (1) source dispersion model, which studies the behavior and fate of a parcel of pollutant, taking into account such factors as micrometeorology and macrometeorology, topography, spatial distribution of sources and receptors, chemical reactions, and disposition processes, and (2) receptor model, which basically is a chemical mass balancing of a given element in a monitored air sample by means of a variety of multivariate statistical techniques. The dispersion model is described in detail in most texts on air pollution and is not considered here.

The receptor model assumes that the mass of an element is conserved so that its concentration C_i is given by the relationship:

$$C_i = \sum_j f_{ij} \times a_{ij} \times m_j$$

where m_j is the mass contribution of material from source j to the total suspended particulates, a_{ij} is the fraction of element i in particulate matter from the source j, and f_{ij} is the coefficient of fractionation as the particles travel between the source and receptor. Several different approaches to receptor model analysis have been reported in the literature (Scheff et al., 1984; Hopke, 1986; Cass and McRae, 1986). Details of the model calibration and validation are not relevant to this report.

The results of source apportionment by chemical mass balance methods are summarized in Table 9 for a number of cities. As would be expected, the principal

Table 9   Chemical Mass Balance Model Results for Chromium in Air at Selected Urban Areas

Location	Measured Concentration (ng/m^3)	Percentage Emission							
		Soil	Iron and Steel	Mobile Sources	Coal	Oil	Refuse	Fugitive Sources	Mineral Industry
Los Angeles[a]	—	—	25	4.1	—	47	—	22	1.3
Washington, DC[b]	14	41	—	—	42	3.1	11	—	1.2
Bombay[c]	19	16	—	—	83	—	—	—	—
Chicago[d]	95	22	47	1.3	6.5	26	5.1	—	5.5
Houston[a]	—	—	45	—	—	46	—	—	—

Source: [a]Cass and McRae, 1986; [b]Kowalczyk et al., 1978; [c]Kamath and Kelkar, 1981; [d]Scheff et al., 1984.

sources of chromium aerosols are different depending on the principal industrial base for each city. In Los Angeles, about 70% of the airborne chromium originates from oil and gas combustion and from fugitive (automotive) sources. The dominant sources in Chicago are the massive iron and steel plants nearby, whereas in Houston, most of the chromium pollution comes from the oil and gas industries. In both Washington DC and Bombay, which have many small and medium-sized industries but no single overriding manufactury, the burning of coal is shown to be the principal source of airborne chromium (Table 9). These model results are consistent with, and indeed have been used to calibrate, the chromium emission inventories for these and other cities (Cass and McRae, 1986).

9. PARTICLE SIZE OF CHROMIUM EMISSIONS

The size of trace metal-containing particles is one of the most important factors affecting qualitative estimates of atmospheric releases because it determines the toxicity of atmospheric emissions, the migration of pollutants through individual environmental media, and the long-range transport of the pollutant. Chromium distribution in particles emitted from coal-fired power plants, electric-arc furnaces, plants producing chromate chemicals, and chromite refractories is shown in Figure 2 (Gafafer, 1953; Davison et al., 1974; Lee et al., 1975; Smith et al., 1979). A pronounced relationship is observed between the chromium concentration and particle size for the coal-fired power plants emissions, with chromium concentrated predominantly in particles below 1.0 μm in diameter. Chromium from electric-arc furnace steel plants is found mainly in particles

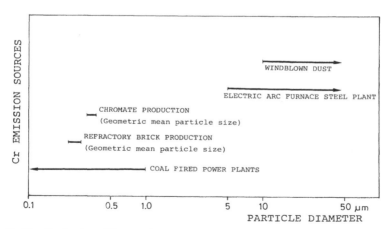

Figure 2. Size distribution of Cr-containing particles emitted from various sources. (After Gafafer, 1953; Davison et al., 1974; Lee et al., 1975; Smith et al., 1979.)

larger than 10 μm. The significant differences in chromium size distribution of particles from coal-fired power plants and electric-arc furnace steel plants must be considered by source apportionment analyses in studies of long-range transport of pollution (Pacyna, 1986e). Geometric mean particle sizes of 0.32–0.37 μm diameter for emissions from chromate production and 0.22–0.28 μm diameter for refractory brick production are reported by Gafafer (1953). The information about chromium emission from the two latter sources is especially important in the assessment of the total chromium exposure of workers in occupational environments and the general population living near chromate chemical and chromite refractory plants.

10. SUMMARY

A broad literature review shows that chromium belongs to the group of elements emitted from natural and anthropogenic sources in about comparable amounts. The largest single sources are windblown dust and metallurgical processes. These two sources release about 80% of the chromium measured in the atmosphere. Chromium emissions from the various sources have quite different particle size distributions. Combustion of coal produces chromium in <1.0 μm diameter particles that can be transported over long distances. Chromium from chromate and refractory brick production can also be emitted in <1.0 μm diameter particles. The relatively small stack height in these industries limits the possibility of long-range transport of emitted particles. Chromium from this source and from anthropogenic sources releasing the element in larger particles is deposited locally and can migrate through individual particular environmental media.

ACKNOWLEDGMENTS

We thank Dr. Val Vitols for helpful discussions.

REFERENCES

Balgord, W.D. (1973). "Fine particles produced from automotive emissions—Control catalysts." *Science* **180**, 1168–1169.

Balsberg-Påhlsson, A.-M., Lithner, G., and Tyler, G. (1982). *Krom i miljön*, Statens Naturvårdsverk Rapport SNV pm 1570, Solna, Sweden (in Swedish).

Bowen, H.J.M. (1966). *Trace Elements in Biochemistry*. Academic Press, London and New York.

Buat-Menard, P., and Chesselet, R. (1978). "Variable influence of the atmospheric flux on the trace metal chemistry of oceanic suspended matter." *Earth Planet. Sci. Lett.* **42**, 399–411.

Cass, G.R., and McRae, G.J. (1986). "Emissions and air quality relationships for atmospheric trace metals." *Adv. Environ. Sci. Technol.* **17**, 145–171.

Cunningham, W.C., and Zoller, W.H. (1981). "The chemical composition of remote area aerosols." *J. Aerosol Sci.* **12**, 367–384.

Darzi, J., and Winchester, J.W. (1982). "Aerosol characteristics at Mauna Loa Observatory, Hawaii, after East Asian dust storm episodes." *J. Geophys. Res.* **87**, 1251–1258.

Davison, R.L. Natusch, D.F.S., Wallace, J.R., and Evans, C.A. Jr. (1974). "Trace elements in fly ash. Dependence of concentration on particle size." *Environ. Sci. Technol.* **13**, 1107–1113.

Duce, R.A., Ray, B.J., Hoffmann, G.L., and Walsh, P.R. (1976). "Trace metal concentration as a function of particle size in marine aerosols from Bermuda." *Geophys. Res. Lett.* **3**, 339–342.

Dvorak, A.J., and Lewis, B.G. (1978). *Impacts of Coal-Fired Power Plants on Fish Wildlife and Their Habitats.* U.S. Dept. of Interior, Report FWS/OBS-78/29, Washington, DC.

Fishbein, L. (1981) "Sources, transport and alterations of metal compounds: An overview. I. Arsenic, beryllium, cadmium, chromium and nickel." *Environ. Health Prospect.* **40**, 43–64.

Fleischer, M. (1972). "An overview of distribution patterns of trace elements in rocks." *Ann. NY Acad. Sci.* **199**, 6–16.

Flinn, J.E., and Reimers, R.S. (1974). *Development of Predictions of Future Problems.* EPA Report No. 600/5-74-005, Environmental Protection Agency, Washington, DC.

Flyger, H., and Heidam, N.Z. (1978). "Ground level measurements of the summer tropospheric aerosol in Northern Greenland." *J. Aerosol Sci.* **9**, 157–168.

Gafafer, W.M. (1953). *Health of Workers in the Chromate Producing Industry.* U.S. Fed. Public Health Service Pub. 192.

GCA. (1973). *National Emissions Inventory of Sources and Emissions of Chromium.* U.S. National Technical Information Service, Springfield, VA (PB 230-034).

Gerasimov, Y.J., and Sharifov, K.A. (1958). "Equilibrium constants of reactions forming the hexacarbonyls of Cr, Mo and W." *Chem. Abst.* **52**, 7829.i.

Gerstle, R.W., and Albrinck, D.N. (1982). "Atmospheric emissions of metals from sewage sludge incineration." *J. Air Poll. Cont. Assoc.* **11**, 1119–1123.

Gluskoter, H.J., Ruch, R.R., Miller, W.G., et al. (1977). *Trace Elements in Coal: Occurrence and Distribution.* Circular 499, Illinois State Geological Survey, Urbana, IL.

Heinrichs, H. (1982). "Trace element discharge from a brown coal fired power plant." *Technol. Lett.* **3**, 127–136.

Hopke, P.K. (1986). "Quantitative source attribution of metals in the air using receptor models." *Adv. Environ. Sci. Technol.* **17**, 173–200.

IARC (1973). *Monographs on the Evaluation of Carcinogenic Risk of Chemicals to Man. Some Inorganic and Organometallic Compounds,* International Agency for Research on Cancer, Lyon, France, Vol. 2.

Kamath, R.R., and Kelkar, D.N. (1981). "Preliminary estimates of the primary source contributions to winter aerosols in Bombay, India." *Sci. Total Environ.* **20**, 195–201.

Kowalczyk, G.S., Choquette, C.E., and Gordon, G.E. (1978). "Chemical element balances and identification of air pollution sources in Washington, DC." *Atmos. Environ.* **12**, 1143–1153.

Kuhn, J.K., Fiene, F.L., Cahill, R.A., et al. (1980). *Abundance of Trace and Minor Elements in Organic and Mineral Fractions of Coal.* Illinois Institute of Natural Resources, Environ. Geology Notes 88, Urbana, IL.

Lantzy, R.J., and Mackenzie, F.T. (1979). "Atmospheric trace metals: Global cycles and assessment of man's impact." *Geochim. Cosmochim. Acta* **43**, 511.

Lee, R.E. Jr., and Duffield, F.V. (1977). "Sources of environmentally important metals in the atmosphere." *Adv. Chem. Ser.* **172**, 146.

Lee, R.E. Jr., Crist, H.L., Riley, A.E., and MacLeod, K.E. (1975). "Concentration and size of trace metal emissions from a power plant, a steel plant, and a cotton gin." *Environ. Sci. Technol.* **7**, 643–647.

Lisk, D.J. (1972). "Trace metals in soils, plants and animals." *Adv. Agron.* **24**, 267–325.

NAS. (1974). *Chromium.* National Academy of Sciences, Washington, DC.

Natusch, D.F-S., and Wallace, J.R. (1974). "Urban aerosol toxicity: The effect of particle size." *Science* **186**, 695–699.

NILU. (1984). *Emission Sources in the Soviet Union.* NILU Report 4/84. The Norwegian Institute for Air Research, Lillestrøm, Norway.

NRCC. (1976). *Effects of Chromium in the Canadian Environment.* NRCC Report No. 15017, National Council Canada, Ottawa, Canada.

Nriagu, J.O. (1979). "Global inventory of natural and anthropogenic emissions of trace metals to the atmosphere." *Nature* **279**, 409–411.

Ondov, J.M., Ragaini, R.C., and Bierman, A.H. (1979). "Elemental emissions from a coal-fired power plant. Comparison of a Venturi wet scrubber system with a cold-side electrostatic procipitator." *Environ. Sci. Technol.* **13**, 588–601.

Pacyna, J.M. (1980). *Coal-Fired Power Plants as a Source of Environmental Contamination by Trace Metals and Radionuclides.* Habilitation thesis, Technical University of Wroclaw, Wroclaw, Poland (in Polish).

Pacyna, J.M. (1981). *Emission Factors for Trace Metals from Coal-Fired Power Plants.* NILU Report 14/81. The Norwegian Institute for Air Research, Lillestrøm, Norway.

Pacyna, J.M. (1982). *Trace Element Emission from Anthropogenic Sources in Europe.* NILU Report 10/82. The Norwegian Institute for Air Research, Lillestrøm, Norway.

Pacyna, J.M. (1986a). "Atmospheric trace elements from natural and anthropogenic sources." In J.O. Nriagu and C. Davidson, Eds., *Toxic Metals in the Atmosphere.* John Wiley & Sons, New York, pp. 33–52.

Pacyna, J.M. (1986b). "Emission factors of atmospheric elements." In J.O. Nriagu and C. Davidson, Eds., *Toxic Metals in the Atmosphere.* John Wiley & Sons, New York, pp. 1–32.

Pacyna, J.M. (1986c). "Sources of air pollution." In E.E. Pickett, Ed., *Atmospheric Pollution.* Hemisphere Publishing Corp., Washington, DC.

Pacyna, J.M. (1986d). "Methods for air pollution abatement." In E.E. Pickett, Ed., *Atmospheric Pollution.* Hemisphere Publishing Corp., Washington, DC.

Pacyna, J.M. (1986e). "Source-receptor relationships for trace elements in Northern Europe and the Norwegian Arctic." *J. Water Air Soil Poll.* **30**, 825–835.

Pacyna, J.M., and Ottar, B. (1985). "Transport and chemical composition of the summer aerosol in the Norwegian Arctic." *Atmos. Environ.* **19** (12), 2109–2120.

Pacyna, J.M., Vitols, V., and Hanssen, J.E.H. (1984). "Size-differentiated composition of the Arctic aerosol at Ny-Ålesund, Spitsbergen." *Atmos. Environ.* **18**, 2447–2459.

Phelan, J.M., Finnegan, D.L., Ballantine, D.S., et al. (1982). "Airborne aerosol measurements in quiescent plume of Mount St. Helens: September, 1980." *J. Geophys. Res.* **9**, 1093–1096.

Rahn, K.A., Borys, R.D., Shaw, G.E., et al. (1979). "Long-range impact of desert aerosol on atmospheric chemistry: Two examples." In *Saharan Dust, SCOPE 14.* John Wiley & Sons, New York.

Rondia, D. (1979). "Sources, modes and levels of human exposure to chromium and nickel. In E. DiFerrante, Ed., *Trace Metals, Exposure and Health Effects.* Pergamon Press, Oxford.

Rustagi (1964). "Stochastic behaviour of trace substances." *Arch. Environ. Health* **8**, 76.

Scheff, P.A., Wadden, R.A., and Allen, R.J. (1984). "Development and validation of chemical element mass balance for Chicago." *Environ. Sci. Technol.* **18**, 923–931.

Schroeder, H.A. (1970). *Chromium Air Quality.* Monograph No. 7-15. American Petroleum Institute, Washington, DC.

SCOPE. (1979). *Saharan Dust. Scientific Committee on Problems of the Environment.* John Wiley & Sons, New York.

Smith, R.D., Campbell, J.A., and Nielsen, K.K. (1979). "Characterization and formation of submicron particles in coal-fired plants." *Atmos. Environ.* **13**, 607–617.

SNV. (1982). "Tungmetaller och organiska miljögifter i svensk natur. (Heavy metals and organic compounds in the Swedish environment)." *Monitor 1982* Statens naturvårdsverk, Solna, Sweden. (in Swedish).

Takeda, N., and Hiroaka, M. (1976). "Combined process of pyrolysis and combustion for sludge disposal." *Environ. Sci. Technol.* **10**, 1147.

Tomza, U. (1984). *Trace Elements in the Atmospheric Aerosol in Katowice, Poland.* Technical Report, Rijksuniversiteit Gent, Gent, Belgium.

Tsubaki, H. (1982). "Resource recycling in the Japanese steel industry." *U.N. Indus. Environ.* **5** (4), 18–23.

U.N. (1985). *Statistical Yearbook 1982.* United Nations, New York.

U.S. EPA. (1972). *Sewage Sludge Incineration.* U.S. Environmental Protection Agency, Sewage Sludge Incineration Task Force, National Technical Information Service, PB-211 323, Springfield, VA.

U.S. EPA. (1976). *Municipal Sludge Management, An Overview of the Sludge Management Situation.* EPA Report 430/9-76-009, U.S. Environmental Protection Agency, Washington, DC.

U.S. EPA. (1984). *Health Assessment Document for Chromium.* Report No. EPA-600/8-83-014F, U.S. Environmental Protection Agency, Environmental Criteria and Assessment Office, Research Triangle Park, NC.

U.S. EPA/ORNL. (1978). *Reviews of the Environmental Effects of Pollutants. III. Chromium.* Environmental Protection Agency, Oak Ridge National Laboratory, U.S. National Technical Information Service, Springfield, VA.

Zoller, W.H. (1983). "Anthropogenic perturbation of metal fluxes into the atmosphere." In J.O. Nriagu, Ed., *Changing Metal Cycles and Human Health.* Dahlem Konferenzen, Springer-Verlag, Berlin.

Zonghua, W. (1982). "Comprehensive utilization and environmental protection in the iron and steel industry of China." *U.N. Indust. Environ.* **5** (4), 2–4.

5

DISTRIBUTION AND CHARACTERISTIC FEATURES OF CHROMIUM IN THE ATMOSPHERE

Jerome O. Nriagu

National Water Research Institute
Burlington, Ontario, Canada

Jozef M. Pacyna

Norwegian Institute for Air Research
Lillestrøm, Norway

Jana B. Milford

Department of Engineering and Public Policy
Carnegie-Mellon University
Pittsburgh, Pennsylvania

Cliff I. Davidson

Departments of Civil Engineering and Engineering & Public Policy
Carnegie-Mellon University
Pittsburgh, Pennsylvania

1. INTRODUCTION

The atmosphere has become a major pathway for the long-range transfer of pollutant chromium to different ecosystems, and any significant increase in the input of the more bioavailable forms of chromium must be regarded as a potential hazard to the receiving fauna and flora. Because of the growing concern about the effects of elevated environmental levels of chromium, most multielement analyses of aerosols include measurements for chromium (Alian and Sansoni, 1985). A large database, scattered over a wide variety of literature sources, has thus become available on the atmospheric cycle of chromium. This chapter presents an overview of current research on the distribution, trends (temporal and spatial), physical and chemical properties, and the removal of chromium in the atmosphere. Data on the distribution and properties of airborne chromium provide insight into the extent of contamination of the ambient air and are a prerequisite in the establishment of guidelines and limits on tolerable atmospheric levels for this element.

2. SPATIAL DISTRIBUTION OF AIRBORNE CHROMIUM

Chromium in the atmosphere comes from a wide variety of natural and anthropogenic sources (see Chapter 3). At any particular time and location, the amount of chromium in the atmosphere depends on the intensity of entrainment processes or proximity to the source, the amount of chromium released, the degree of mixing determined by meteorologic factors, the aging history, and the rate and mechanism of removal of the suspended particulate matter. The spatial distribution is particularly influenced by the land use patterns, with urban centers generally being mild hot spots of chromium pollution. The large number of impacting variables often lead to large-scale variations in the geographic distribution of chromium in the atmosphere.

2.1 Remote Locations

Typical concentrations of chromium at remote areas are less than $1.0 \, \text{ng m}^{-3}$ and average about $0.3 \, \text{ng m}^{-3}$ (Table 1). As is to be expected, the lowest concentrations anywhere ($0.005\text{--}0.013 \, \text{ng m}^{-3}$) have been observed near the South Pole. On the other hand, the lowest levels reported in the Arctic region vary from 0.03 to $0.07 \, \text{ng m}^{-3}$. The 6-fold to 10-fold difference in atmospheric levels of chromium in the Arctic and Antarctica reflects primarily the emissions from industrial sources and, to a lesser extent, the difference in the hemispheric distribution of the land masses. It should be noted that the chromium levels in the Arctic region show a pronounced seasonality and can exceed $2.2 \, \text{ng m}^{-3}$, the incidence of highly elevated chromium concentrations being related to the Arctic haze phenomenon (Pacyna et al., 1985).

Chromium concentrations at remote continental locations commonly fall in the range of 0.2 to $0.6 \, \text{ng m}^{-3}$ (Table 1). The concentrations at the offshore marine locations vary from less than $0.1 \, \text{ng m}^{-3}$ (Eastern Equatorial Pacific) to over $1.0 \, \text{ng m}^{-3}$ (North Atlantic). The elevated tropospheric chromium level in the western North Atlantic can be attributed to the distant industrialized regions of eastern North America (Chester et al., 1974; Harriss et al., 1984). Because of the high crustal abundance of chromium, natural dust loading from major wind systems also can affect strongly the airborne chromium levels at remote locations. For example, dust storms of the Kalahari Desert have been implicated in the observed high chromium content of variable winds off the coast of South Africa (Chester et al., 1974). The episodic elevation of chromium levels in the subtropical North Atlantic likewise has been attributed to the injection of Sahara dust by the tradewinds (Schneider, 1985).

Table 1 Chromium Concentrations in Air at Remote Locations

Sample Location and Date	Cr Concentration (ng m^{-3})	Enrichment Factor	Reference
South Pole			
Summer 1970 and 1971	0.005	6.9	Zöller et al., 1974
Summer 1975 and 1976	0.011		Cunningham and Zöller, 1981
Winter 1975 and 1976	0.013		Cunningham and Zöller, 1981
Eastern Equatorial Pacific			
April–October 1979	<0.09		Maenhaut et al., 1983
Indian Ocean, S. Latitute	0.23	16	Egorov et al., 1970
South China Sea (sand-sized particles) 1971–1972	0.05	1.1	Chester et al., 1974
Java Sea (sand-sized particles) 1971–1972	0.05	0.4	Chester et al., 1974
Chacaltaya Mt., Bolivia 1972	0.68	2.4	Rancitelli et al., 1976
Chacaltaya Mt., Bolivia 1972	0.57	1.8	Adams et al., 1977; 1983
Enewetak April–May 1979	0.14	1.6	Duce et al., 1983
Enewetak June–August 1979	0.04	5.0	Duce et al., 1983
Bermuda 1973	0.28	1.8	Duce et al., 1976
Gulf of Guinea November 1970	0.98	6.8	Rahn, 1976
Jungfraujoch Mt., Switzerland	0.36	5.8	Dams and Jonge, 1976
Novaya Zemlya (Arctic Island), USSR	0.34		Egorov et al., 1970

Location			Reference
Great Wall (north of Beijing), China April 1980	0.43	25	Winchester et al., 1981
Nord, Greenland 1972	0.8	4.4	Rancitelli et al., 1976
Nord, Greenland 1979–1980	0.16	5.6	Heidam, 1986
Thule, Greenland Winter 1978	0.21		Heidam, 1981
Thule, Greenland 1979–1980	0.03	5.9	Heidam, 1986
Godhavn, Greenland 1979–1980	0.14	1.8	Heidam, 1986
Prins Christianssund, Greenland Winter 1978	0.07		Heidam, 1981
September 1979–January 1980	0.06		Heidam, 1984
Ny Ålesund, Spitsbergen March 1983	1.2	10	Pacyna et al., 1985
August–September 1983	0.8	10	Pacyna and Ottar, 1985
Twin Gorges, NWT, Canada July–August 1970	0.59	7.3	Rahn, 1976
Jaspar National Park, Alberta July–August 1970	0.7	3.9	Rahn, 1976
Seney National Wildlife Refuge, Michigan June 1979	0.5	2.0	Alkezweeney and Busness, 1984
Mauna Loa, Hawaii 1972	0.32		Rancitelli et al., 1976
Hawaii, coast	0.2		Gordon et al., 1978

The enrichment factor (EF) for airborne particulate chromium is often defined (Zöller et al., 1974) as:

$$EF = \frac{(Cr_{air}/Al_{air})}{(Cr_{crust}/Al_{crust})}$$

where Cr_{air} and Al_{air} refer to the airborne concentrations of chromium and aluminum and Cr_{crust} and Al_{crust} are the average concentrations of chromium and aluminum in the earth's crust, respectively. Values of EF close to unity suggest that the chromium is derived primarily from crustal weathering (natural sources), whereas much greater values of EF are generally believed to be indicative of anthropogenic origin.

The EFs for chromium in aerosols at remote locations generally fall between 1.0 and 6.0 and average about 3.8 (Table 1). For crust-dominated, sand-sized particles of the lower troposphere, the EFs fall in the range of only 0.4 to 3.4 and average about 1.3 (Chester et al., 1974). Heidam (1986) has suggested that the elevated EF value is the result of anthropogenic contribution of chromium to the eolian particles in the Arctic air. Volcanoes, however, also emit large amounts of chromium with highly variable EF values that most often exceed 1.0 (Table 2). The complete overlap of the EF for volcanic and anthropogenic particles make this particular feature unreliable in source apportionment for chromium.

2.2   Rural and Semirural Areas

The data compiled in Table 3 show that the chromium concentrations in air in rural and semirural areas generally fall in the range of 1 to 10 ng m^{-3} and average about 4 ng m^{-3}. The observed concentrations in the nonurban areas of the United States vary from below detection to over 75 ng m^{-3} and averaged about 3.0 ng m^{-3} between 1970 and 1974 (Akland, 1976). The average between 1972 and 1980 for rural areas of Britain was 4.0 ng m^{-3}, which is very close to the average figure for the United States.

The available data thus suggest that the average chromium concentration in rural air is about 10-fold higher than the mean level at remote locations. The difference can be attributed to (1) anthropogenic interference and (2) the location of sampling stations in rural areas on land (which can be influenced by wind entrainment of local dust particles), whereas the stations in remote areas are either offshore or on snowfields. The similarity of EF values at the remote and rural locations clearly stresses the possible importance of the sampling strategy in the observed difference.

2.3   Urban Areas

Representative chromium concentrations in the air in urban areas are summarized in Table 4. From these data, the mean atmospheric chromium levels in the United

Table 2   Chromium in Volcanic Emanations

Volcanic Activity	Cr Concentration ($ng\ m^{-3}$)	Enrichment Factor	Reference
Augustine, Alaska			Lepel et al., 1978
February 1, 1976			
0–8 km	<13	<66	
8–32 km	<31	<54	
Average (0–32)	<13	<66	
February 18, 1976			
0–8	35	3.5	
8–32	11	1.1	
February 20, 1976	8.2	0.5	
February 21, 1976	2.4	0.6	
Mount Etna, Sicily			Buat-Menard and
1976			Arnold, 1978
Main plume	67	2.8	
Hot vents	240	49	
Tolbachik, Kamchatka, USSR			Menyailov et al., 1982
July 1975–December 1976			
South vent, gases from lava	110	32	
North vent, fumaroles at Cone I	4500	32	
North vent, fumaroles at Cone II	180	32	

States, Canadian (mostly Ontario), and European cities are estimated to be 20, 13, and 30 $ng\ m^{-3}$, respectively. The concentrations in urban areas thus exceed the levels at remote locations by 40-fold to over 100-fold, on the average. Urban centers thus are weak hot spots of Cr pollution and potentially represent a source area of the metal to the adjoining rural and semirural environments.

In general, towns with major iron and steel industries show higher ambient chromium levels. In the industrial zone near the steel mills of northwest Indiana, the average concentration often exceeds 50 $ng\ m^{-3}$, and the ferroalloy industries account for the considerably higher values (121 $ng\ m^{-3}$) recorded in Steubenville, Ohio (EPA, 1984). Downwind of the steel mills in Hamilton, Ontario (which accounts for about 60% of Canadian steel), the chromium concentration was found to be 31 $ng\ m^{-3}$ compared to the provincewide average of only 13 $ng\ m^{-3}$ (Barton et al., 1975).

The EFs for urban aerosols vary from about 1.0 to over 25 and average about 10 (Table 4). Thus, the EF values in urban areas are not very different from those of rural and even remote locations. This similarity is consistent with the fact that aerosols from industrial sources do not show high chromium enrichment, the

Table 3 Chromium Concentration in Air at Rural and Semirural Locations

Sampling Location and Date	Cr Concentration (ng m^{-3})	Enrichment Factor	Reference
Chilton, Oxon			Cawse, 1974–1981,
1972	4.4	4.9	1985
1973	3.9	4.0	
1974	1.7	2.6	
1975	2.0	2.4	
1976	2.4	2.1	
1977	1.2	1.8	
1978	2.0	3.0	
1979	1.3	2.1	
Average, 1972–1981	2.4	2.9	
Leiston, Suffolk			Cawse, 1974–1981,
1972	7.5	9.2	1985
1973	3.8	4.7	
1974	1.7	2.8	
1975	3.4	4.9	
1976	2.9	2.78	
1977	1.6	2.4	
1978	1.7	2.7	
Average, 1972–1978	3.2	4.2	
Collafirth, Shetland Islands			Cawse, 1974–1981,
1972	1.1	5.4	1985
1973	0.71	5.6	
1974	0.94	4.9	
1975	0.60	3.4	
1976	0.39	2.6	
1977	<0.2	<2.0	
Average, 1972–1977	0.64	3.8	
Plynlimon, Powys			Cawse, 1974–1981,
1972	2.0	4.9	1985
1973	3.3	7.1	
1974	1.7	2.2	
1975	0.81	2.2	
1976	1.2	1.9	
1977	1.2	1.5	
Average, 1972–1977	1.7	3.3	
Styrrup, Notts			Cawse, 1974–1981,
1972	17	6.9	1985
1973	20	8.9	
1974	9.5	7.1	
1975	8.6	5.6	
1976	8.7	5.6	

Table 3 (*Continued*)

Sampling Location and Date	Cr Concentration (ng m^{-3})	Enrichment Factor	Reference
1977	7.3	4.6	
1978	7.8	7.2	
1979	7.2	5.4	
Average, 1972–1981	9.9	6.5	
Trebanos, W. Glamorgs			Cawse, 1974–1981,
1972	9.6	6.7	1985
1973	13	11	
1974	5.6	5.0	
1975	12	13	
1976	12	13	
1977	6.3	7.4	
1978	10	16	
1979	6.5	9.4	
Average, 1972–1981	8.6	9.3	
Wraymires, Cumbria			Cawse, 1974–1981,
1972	1.5	3.8	1985
1973	2.4	4.5	
1974	1.5	3.7	
1975	1.8	3.5	
1976	2.4	3.0	
1977	1.0	1.7	
1978	1.5	3.2	
1979	1.0	2.7	
Average, 1972–1981	1.8	3.4	
Selby, Yorkshire			Cawse, 1979
June–August 1976	8.5		
Central Swansea			Cawse, 1974
1972–1973	6.1		
Arran Bute, Scotland			Peirson et al., 1974
June 1972–May 1973	2.8		
Gas platform, North Sea			Peirson et al., 1974
June 1972–May 1973	3.7		
Salehard, USSR	1.6		Egorov et al., 1970
Sevastopol, USSR	6.7	4.2	Egorov et al., 1970
Petropavlovsk, Kamchatka,			Egorov et al., 1970
USSR	2.7	25	
Magada, USSR	6.0	57	Egorov et al., 1970
Spitsbergen, Norway			Heintzenberg et al.,
April–May 1979	2.6	4.2	1981
Sjoangen, Sweden	2.8		Lannefors et al., 1983

(*Continued*)

Table 3 (*Continued*)

Sampling Location and Date	Cr Concentration (ng m^{-3})	Enrichment Factor	Reference
Skoganvarre, Northern Norway			Rahn, 1976
November 1971–May 1972	0.62	12	
Birkenes, Norway			Semb, 1978
Spring, 1973	0.90	6.5	
Fall, 1973	1.2	12	
Birkenes, Norway			Pacyna et al., 1984
August 1978–June 1979	1.0		
Skoganvarre, Norway			Semb, 1978
Spring, 1973	0.35	4.2	
Dourbes, Belgium	0.78	8.6	Rahn, 1976
Botrange, Belgium			Kretzschmar et al.,
1972–1975	14		1980
Oostende, Belgium			Kretzschmar et al.,
1972–1975	19		1980
Mol, Belgium			Kretzschmar et al.,
1972–1975	13		1980
Louvain-la-Neuve, Belgium	4.6		Ronneau, 1985
Rena, Norway	0.96	4.1	Rahn, 1976
Tveiten, Norway	1.1	4.7	Rahn, 1976
Tange, Denmark	1.7	2.7	Rahn, 1976
East China Sea (sand-sized particles			Chester et al., 1974
1971–1972	0.69	0.4	
Inland Sea, Japan (sand-sized particles)			Chester et al., 1974
1971–1972	3.7	3.4	
South Japanese coast (sand-sized particles)			Chester et al., 1974
1971–1972	0.76	1.6	
Sea of Japan (sand-sized particles)			Chester et al., 1974
1971–1972	1.1	3.4	
Trombay, India			Kamath and Kelkar,
December 1978–			1981
January 1979	4.4		
Bagauda, northern Nigeria			Beavington and
January–December 1976	7.0		Cawse, 1978
Abidjan, Ivory Coast			Rahn, 1976
May 1971	8.4	24	
Niangbo, Ivory Coast			Rahn, 1976
May 1971	6.5	4.6	

Table 3 (*Continued*)

Sampling Location and Date	Cr Concentration (ng m^{-3})	Enrichment Factor	Reference
Singrobo, Ivory Coast			Rahn, 1976
May 1971	7.9	37	
Sudan, rural	5.3	2.6	Rahn, 1976
Pelindaba near Pretoria,			Beavington and
South Africa	10		Cawse, 1978
Bermuda, Bahamas			Sievering et al., 1982
1980	1.3	14	
Algonquin Provincial Park,			Rahn, 1976
Ontario			
July–August 1970	1.4	3.9	
Prince George's County			Rahn, 1976
Maryland	2.8	2.5	
Willmette Valley, Oregon			Rahn, 1976
September 1972	6.0	11	
Walker Branch, Oak Ridge,			Rahn, 1976
Tennessee	1.7	1.7	
Acadia National Park, Maine			EPA, 1984
1977	5.2		
Iberville Parish, Louisiana			EPA, 1984
1977	6.3		
1978	5.9		
1980	5.2		
Whiteface Mt., New York			Parekh and Husain,
Summer 1975	4.0	15	1981
Summer 1977	7.0	53	
Northeast Minnesota			Ritchie and Thingvold,
1976–1978	4.0		1985
Colstrip, Montana			Crecelius et al., 1980
May–September 1975	1.4	1.2	
Arizona, rural areas			Moyers et al., 1977
January–December 1974	3.1	2.1	
Grand Canyon National			EPA, 1984
Park, Arizona			
1977	5.8		
Black Hills National Forest,			EPA, 1984
South Dakota			
1978	9.0		
Hawaii County, Hawaii			EPA, 1984
1977	6.3		
Kinshasa, Zaire			Maenhaut and
1985	2.3	1.0	Akilimali, 1985

(*Continued*)

Table 4 Concentrations of Chromium in Air at Selected Urban Areas

Sampling Location and Date	Cr Concentration ($ng\ m^{-3}$)	Enrichment Factor	Reference
Antwerp, Belgium May 1972–April 1975	19		Kretzschmar et al., 1980
Brussels, Belgium May 1972–April 1975	20		Kretzschmar et al., 1980
Liege, Belgium May 1972–April 1975	32		Kretzschmar et al., 1980
Gent, Belgium 1971–1973			Heindryck, 1976
Urban	30	3.8	
Industrial	15	7.7	
Residential	10	7.1	
Rotterdam, Holland 1971–1972	18	8.6	Evendijk, 1974
Schiedam, Holland 1971–1972	71	26	Evendijk, 1974
Maasslius, Holland 1971–1972	20	11	Evendijk, 1974
Novosibirsk, USSR	115		Egorov et al., 1970
Tashkent, USSR	92		Egorov et al., 1970
Sevastopol, USSR	14		Egorov et al., 1970
Central Swansea 1972–1973	7.0		Pattenden, 1974
Fleet Street, London 1972	29		Cawse, 1974
Oslo, Norway 1962–1963	9.6		Cawse, 1974

Location			Reference
Glasgow (14 sites)	35	11	McDonald and Duncan, 1978
Heidelberg	4.6	2.0	Rahn, 1976
April–July 1971			
Munich (6 stations)	78		Schramel et al., 1974
1971			
Paris	15	14	Rahn, 1976
1970			
Beijing, China			Winchester and Bi, 1984
July–December 1980			
Fine particles	91		
Coarse particles	10		
Trombay, India	35		Sadasivan, 1981
November–December 1976			
Ankara, Turkey	21	5	Olmez and Aras, 1977
Bagdad, Iraq	100	2.1	Kanbour et al., 1985
1983–1984			
Residential (Jadiria)	95		
Industrial (Wazieria)	52		
Business (Allawie)	134		
Business (Tal-Muhamed)	213		
Ontario (Canada) cities, 32 sites			Barton et al., 1975
1971	13		
New York City			Kleinman et al., 1980;
1969	33		Lioy et al., 1978
1972	12		
1973	9		
1974	10		
1975	4.1		
Average, 1972–1975	8.5		

(*Continued*)

Table 4 (*Continued*)

Sampling Location and Date	Cr Concentration ($ng\ m^{-3}$)	Enrichment Factor	Reference
Niagara Falls, New York			EPA, 1984
1979	39		
1980	14		
Trenton, New Jersey			Lioy et al., 1978
1972–1974	5.3		
Newark, New Jersey			Lioy et al., 1978
1972–1974	15		
Newark, New Jersey			EPA, 1984
1978	18		
1979	13		
1980	9		
Elizabeth, New Jersey			Lioy et al., 1978
1972–1974	6.3		
Waterbury, Connecticut			EPA, 1984
1978	34		
1979	33		
Bridgeport, Connecticut			Lioy et al., 1978
1972–1974	6.0		
Boston, Massachusetts	6.8	4.7	Gordon et al., 1973
Worcester, Massachusetts			EPA, 1984
1977	6.3		
1978	9.9		
1979	6.7		
Baltimore, Maryland			EPA, 1984
1977	157		
1978	94		

Location	Value		Reference
Washington, DC			Saltzman et al., 1985
1968–1969			
Summer, 1974	10		
Norfolk, Virginia	11		Kowalczyk et al., 1978
			EPA, 1984
1977	6.7		
1978	6.9		
1979	8.3		
1980	12		
Atlanta, Georgia			EPA, 1984
1977	8.9		
1978	6.2		
Chattanooga, Tennessee			EPA, 1984
1977	12		
1978	14		
1979	11		
1980	15		
St. Louis, Missouri			Tanner et al., 1974
July 1973			
Downtown	20	1.1	
32 km downwind	8	0.6	
64 km downwind	4	0.6	
96 km downwind	10	0.5	
Houston, Texas	9.0		Saltzman et al., 1985
March 1970–February 1971			
Cleveland, Ohio	19		King et al., 1976
1971			
Akron, Ohio			EPA, 1984
1977	13		
1978	19		
1979	17		
1980	20		

(*Continued*)

Table 4 (*Continued*)

Sampling Location and Date	Cr Concentration ($ng\ m^{-3}$)	Enrichment Factor	Reference
Cincinnati, Ohio			EPA, 1984
1977	8.3		
1978	12		
1980	15		
1970/1971	12		Saltzman et al., 1985
Steubenville, Ohio			EPA, 1984
1978	52		
1979	121		
Northwest Indiana			Rahn, 1976
June 1969			
Industrial	54	24	
Suburban and semirural	11	4.5	
Chicago			
April 4, 1968	17	9.9	Rahn, 1976
July–November 1970	30	6.1	Henry and Blosser, 1971
March 1970–February 1971	12		Saltzman et al., 1985
July 1981–January 1982	95		Scheff et al., 1984
Kansas City, Missouri			EPA, 1984
1977	17		
1978	28		
1979	5.2		
San Francisco Bay area, July 1970	8.2	7.7	John et al., 1973
Sacramento, California January 1974	11		Flocchini et al., 1976

140

Location			Reference
Los Angeles 1977	19		EPA, 1984
Osaka, Japan Residential	26	4.5	Rahn, 1976
Industrial	81	14	Rahn, 1976
Sydney, Australia June–September 1974	40		Lee et al., 1977
Melbourne, Australia June–September 1974	50		Lee et al., 1977
Brisbane, Australia June–September 1974	115		Lee et al., 1977
Rio de Janeiro, Brazil	22		Trindade et al., 1981
Caracas, Venezuela	5		Escalona and Sanhueza, 1981
Santiago, Chile 1972	8.3	0.6	Ranticelli et al., 1976
Lima, Peru 1972	4.2	1.6	Rancitelli et al., 1976
San Juan, Puerto Rico 1972	1.3	2.6	Rancitelli et al., 1976
Toronto, Ontario 1982–1984	80		Brzezinska-Paudyn et al., 1986
Katowice, Poland 1979 Summer	46		Tomza et al., 1982
1979 Winter	104		

reported EF for anthropogenic particles typically being in the range of 5 to 30 (Block and Dams, 1976; Hansen and Fisher, 1980; Cox et al., 1985).

2.4 Point Sources

Any factory involved in the production or use of chemical compounds and alloys that contain chromium is a potential emitter (point source) of this metal to the environment (Table 5). Factories of particular note include chromite and ferro-chrome smelting plants, welding shops, ferrochrome and chromeplating plants, chromium pigment, tanning, and leatherworking shops (Stern, 1982). Until recently, when concerted efforts have been made to clean up the chromium shops, it was not unusual to find chromium concentrations in excess of $1,000 \, \mu g \, m^{-3}$ in the immediate vicinity of the plants. Bourne and Rashin (1950) observed the following concentrations of chromium downwind from an old chromate plant between June and December 1949:

Distance from Plant (m)	Cr Concentration ($\mu g \, m^{-3}$)
100	120
500	100
1,100	80
2,000	60
10,000	1.0

Kuperman (1964) found chromium concentrations of 15–450 $\mu g \, m^{-3}$ within a radius of 0.5 km from two chromate-producing plants. The plume did not decay to the background levels (2 ng m^{-3}) until a distance of 2.0 km from the plants. From such observations, the sanitary clearance zone for chromium-producing plants is generally set at 1,000 m for plants that discharge 200 kg/day and 2,000 m for plants that discharge up to 1,000 kg/day (NAS, 1974).

Table 5 Chromium Concentrations in Air Around Point Sources

Source	Cr Concentration ($\mu g \, m^{-3}$)	Reference
Iron foundry workshop		
Indoor	0.96	Zhang et al., 1985
Outdoor	0.005	Zhang et al., 1985
Powerplant		
Inlet	300	Lee et al., 1975
Outlet	0.7	Lee et al., 1975
Steel mill		
Inlet	1,200	Lee et al., 1975
Outlet	1.2	Lee et al., 1975
Cotton gin		
Inlet	3.0	Lee et al., 1975
Outlet	<0.1	Lee et al., 1975

Table 5   (*Continued*)

Source	Cr Concentration (μg m^{-3})	Reference
Steel foundry		
Chromeplating shop	10	Rondia and Closset, 1985
Electroplating shop	7.7	Rondia and Closset, 1985
Stainless steel welding shops	40–400	Stern, 1982
Chromate production plants	100–500	Stern, 1982
Chromeplating		
Old plants	50–1,000	Stern, 1982
New plants	2–25	Stern, 1982
Ferrochrome plants	10–190	Stern, 1982
Chromium pigment shops	60–600	Stern, 1982
Tanning [Cr(III)] shops	10–50	Stern, 1982

The burning of fossil fuels, especially by power plants, is an industrial activity that represents a significant point source of chromium (see Chapter 3). Industrial cooling waters that contain chromate salts as a corrosion inhibitor can also add to atmospheric chromium pollution. Alkezweeny et al. (1975) reported chromium concentrations of 50 ng m^{-3} at distances of up to 200 m from the Oak Ridge (Tennessee) Gaseous Diffusion Plant using chromated cooling waters. The chromium deposition rate associated with this source was found to be about 1.0 mg m^{-2}hour^{-1} at 30 m and only 0.01 mg m^{-2}hour^{-1} at a distance of 200 m.

2.5   Effect of Elevation

The change of chromium concentration as a function of altitude is an important question that has yet to be studied properly. Rancitelli and Perkins (1970) found no systematic variation with elevations up to 12 km, above which a sharp decrease was observed (Table 6). The chromium profile may differ from those of other trace metals and suspended particles that generally decrease with elevation in the stratosphere (Nriagu and Davidson, 1980; Harriss et al., 1984), although conclusive evidence is not available.

3.   CHEMICAL AND PHYSICAL PROPERTIES

The growing concern about the occupational health effects of chromium-rich fumes in the workplace has led to several studies on the specific chemical forms of chromium in a number of industrial emissions. Although the water-soluble component of welding fumes can reach 5%, the most typical range is 0.01–1.5% (Stern, 1982). The solubility is determined primarily by Cr(VI) content, whereas the insoluble fraction includes Cr(VI), Cr(III), and Cr(O). The fumes from metal inert gas welding shops consist mostly of Fe_3O_4 spinel, with the chromium uniformly substituted throughout the matrix. In fumes generated from manual

Table 6 Variation of Chromium Concentration with Elevation Above Ground Level

Altitude (km)	Sampling Date	Cr Concentration (ng m^{-3})
3.1	August 21, 1967	8.0
3.1	September 5, 1967	4.0
3.1	October 23, 1967	5.0
4.6	September 5, 1967	4.0
6.1	August 21, 1967	7.0
6.1	September 5, 1967	4.0
8.8	October 28, 1967	2.7
11.4	February 24, 1968	4.0
15.2	April 17, 1968	0.4

Source: Rancitelli and Perkins, 1970.

metal arc welding, the chromium often is in association with potassium and only occasionally with iron, and is localized in just a small number of the particles (Stern, 1982). The difference in the form and distribution of chromium is believed to have a profound effect on the toxicity of the welding fumes.

The particles from chromeplating operations consist mostly of CrO_3 droplets of 100 μm diameter (Stern, 1982). Particulate emissions from tanning and leatherworks typically contain 2–6% of organically bound Cr(III) (Stern, 1982). About 40% of the acid-extractable chromium in dusts from primary ferrochrome smelters consist of Cr(VI) in the form of $Cr_2O_7^{2-}$ and CrO_4^{2-}. This Cr(VI) predominates in submicron particles, whereas Cr_2O_3, which accounts for most of the remaining forms of chromium, is located in large particles similar to the original chromite ore (Cox et al., 1985). These studies on particulates from specific point sources suggest that the composition of chromium in ambient air is complex and can vary from pure inorganic compounds and minor polydispersed phases to organically bound chromium species.

In general, the solubility of chromium in particulate emissions from different sources is low in water and dilute acids. Dreesen et al. (1977) found that less than 1.0% of the chromium in fly ash from a coal-fired power plant was leachable by water, dilute (less than 0.01 M) acid, and ammonia solutions. The solubility, however, was found to increase with the acidity of the extract, reaching 12–15% in 0.1 M citric acid and 1.0 M nitric and hydrochloric acids. Hansen and Fisher (1980) observed much higher chromium solubility in coal fly ash, which increased from 39% in 0.029 M HCl to about 65% in 0.57 M HCl. Concentrated boric acid could remove only 3.6% of chromium in the fly ash, whereas the solubility in hydrofluoric acid increased from 6.1% in 0.029 M solution to 84% in 0.57 M solution (Hansen and Fisher, 1980). Cox et al. (1985) also found that about 50% of the chromium in dusts from primary ferrochrome smelter was extractable by mild acid and base treatments.

The solubility of chromium apparently is strongly dependent on chromium dispersion in the particles. In coal fly ash, chromium has been found to be evenly distributed between the aluminosilicate matrix and the nonmatrix material (Hansen and Fisher, 1980). Elements, such as lead, cadmium, zinc, and copper, that predominate in the nonmatrix component usually are accumulated preferentially on the particle surface and hence are extractable readily by mild reagents. In contrast, elements, such as Na, K, Ca, and Mg, held in the matrix exhibit low extractability by weak acid or base treatment. Chromium in fly ash is only moderately extractable, which is consistent with the observed even distribution between matrix and nonmatrix material. It should be noted that the ratio of chromium concentration in surface (300 Å) microlayer to the bulk concentration in coal fly ash particles was 1400:400, or 7:2, compared to 12:1 for zinc and 29:1 for cadmium (Natusch, 1978).

Few solubility measurements have been made using the particles in ambient air. About 4% of the chromium in bulk precipitation samples collected in northern Ontario was in solution; the solubility conceivably was enhanced strongly by the acidity of the rainfall in the region (Kramer, 1976). In general, only 1–2% of the chromium in bulk precipitation is water soluble. In Bagauda, northern Nigeria, soluble chromium accounted for 1.7% of the total chromium in the bulk precipitation samples (Beavington and Cawse, 1979). The study by Thornton et al. (1980), however, points to the fact that up to 20% of the chromium in bulk precipitation is leachable by dilute acids.

The use of wet chemical extraction techniques has provided some insights into the partitioning of chromate in atmospheric aerosols. Chester et al. (1986) found that over 80% of the total chromium content of crust-dominated, soil-sized particles from the Atlantic northeast tradewinds was held in the refractory or residual fractions. About 6% of the chromium was found to be associated with oxides and 7% was affiliated with organic matter. In contrast, only about 18% of the chromium in particles from the ship's stack was in refractory fraction. Roughly 44% and 24% of the chromium had oxide and organic association, respectively, whereas 14% of the ship's stack chromium emissions was held loosely or associated with carbonate and surface oxides (Chester et al., 1986). An analysis of chromium in particulate matter from St. Louis, Missouri (used as NBS reference material), found that 51% was refractory, 25% was hosted by oxides, and 16% by organics (Lum et al., 1982). These studies suggest that urban-derived anthropogenic particles contain significantly less refractory chromium that do aerosols from natural sources.

3.1 Size Distributions of Airborne Chromium

The way in which airborne particles interact with the environment is greatly affected by particle size. For example, deposition from the atmosphere onto soil, vegetation, and other surfaces is a function of particle size. The extent to which airborne particles penetrate the human respiratory system is determined mainly by size, with possible health effects resulting from the presence of toxic substances.

This section discusses the sizes of airborne particles containing chromium as reported in the literature. First, measurement methods and errors associated with the measurements are discussed. Several distributions of airborne chromium are then presented, and as examples of the use of the data, the distributions are used to estimate deposition efficiencies in the human respiratory tract and dry deposition onto natural grassland. Much of the information in this section has been taken from the literature review of Milford and Davidson (1985).

3.1.1 Measurement Methods and Experimental Errors

Nearly all of the size distribution data reported in the literature have been obtained with cascade impactors. These devices collect atmospheric aerosols by inertial impaction, depositing particles of decreasing size on consecutively cascaded stages. Usually, a final backup filter is used to capture particles that are too small to be collected by inertial deposition. Although they provide a convenient method of fractionating aerosols for subsequent elemental analysis, impactors are highly imperfect.

Obtaining a representative sample requires that the impactor inlet be faced directly into the wind and that the inlet air speed be identical to the ambient wind speed (isokinetic sampling). Several investigators have shown that size distributions measured with commercial impactors tend to underestimate particle concentrations in the largest size ranges due to nonisokinetic sampling (Davidson, 1977; Wedding et al., 1977; Liu and Pui, 1981). Supermicron particles are more likely than smaller particles to deposit on walls of the inlet and on internal walls of the impactor (McFarland et al., 1977; Cushing et al., 1979; Chan and Lawson, 1981).

Particles may bounce off impaction surfaces, becoming reentrained in the flow and depositing on lower stages. This problem is usually minimized by applying adhesive coatings, although some particle bounce may still occur, and the coatings may interfere with subsequent chemical analysis (Dzubay et al., 1976; Lawson, 1980). Recent studies have shown that the severity of bounceoff depends on particle characteristics, relative humidity, and other factors; for example, hygroscopic seaspray aerosols are retained on dry surfaces better than are soil-derived elements (Walsh et al., 1978; Wesolowski et al., 1980). Adhesive-coated surfaces have been used in most of the studies reported here, although none of the investigators have attempted to quantify the effect of bounceoff on their data.

Aside from these difficulties, the fractionation of particles into desired size ranges is imperfect, since small particles may be collected to a certain extent on the upper stages, and some of the large particles may not be impacted until reaching the lower stages. Such problems are the result of nonideal air flow through the impactor, particularly variations in the air velocity across each jet (Marple and Liu, 1974; Marple et al., 1974). To define their size ranges, most researchers have merely adopted the manufacturers' suggested cutoff diameters, which are often based on ideal air flow.

Another potential problem with the data reported here is the error associated with analytic results. Of primary importance is possible contamination during sampling, sample handling, and analysis. The potential for contamination is particularly high in impactor sampling, since the small amount of total airborne material is fractionated by size into even smaller samples for analysis. Analytic results are highly uncertain also when the mass of material approaches the limit of detection, a situation often encountered during analysis of size-fractionated aerosol samples. Other analytic problems include difficulties specific to the method of analysis, for example, atomic absorption analysis may be complicated by interference from other metals or from organic material in the samples (Thompson et al., 1979; Waughman and Brett, 1980). Neutron activation suffers from uncertainties in the position of the samples with respect to the detector and from counting statistics limitations (Zöller and Gordon, 1970). Both x-ray fluorescence and proton-induced x-ray emission analyses are prone to errors if the sample is too thick or not uniformly deposited on the substrate (Dzubay and Stevens, 1973; Johansson et al., 1975). All four of these analytic methods have been used to obtain the chromium size distribution data reported in this section.

Finally, the small number of experiments conducted in these studies suggests that the representativeness of the data must be questioned. All of the size distribution data summarized here involved a limited number of samples. These problems suggest that the chromium size distributions reported in the literature are subject to considerable uncertainty. Nevertheless, the data show consistencies in the shapes of the distributions.

3.1.2 Presentation of the Data

A recent literature review has indicated that at least 10 separate airborne trace element sampling programs have included analysis for chromium (Milford and Davidson, 1985). Two of these data sets involved incomplete distributions. The remaining eight sets of data are summarized in Figure 1. These distributions each represent the average of two or more individual measurements. The ordinate is the normalized mass distribution function $(\Delta C/C_T)/\Delta \log d_p$, where ΔC is the airborne mass concentration of chromium in a given size range (which extends from $d_{p\,min}$ to $d_{p\,max}$), C_T is the total concentration of chromium in all size ranges, and $\Delta \log d_p$ is the difference $\log d_{p\,mas} - \log d_{p\,min}$. The aerodynamic diameter of a particle d_p represents the size of a unit density sphere with aerodynamic transport characteristics that are identical to those of the original particle. Note that the area under the curve between any two particle diameters is proportional to the fraction of airborne mass in that size interval.

For consistency, overall minimum and maximum aerodynamic diameters of 0.05 and 40 μm, respectively, have been assumed (Hidy, 1974; Davidson, 1977). Airborne particles as large as 40 μm have been documented in the

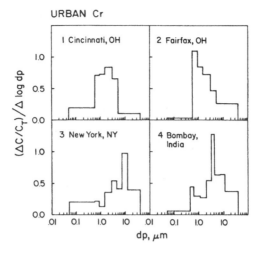

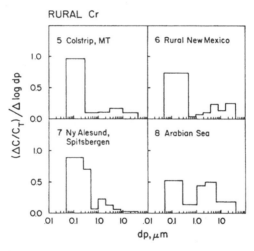

Figure 1. Histogram of sizes of airborne chromium particles.

literature (Johnson, 1976; Davidson, 1977), although such large particles were probably sampled with low efficiency in the studies reported here. Much of the mass reported for the largest size ranges in Figure 1 may represent small particles collected by the nonideal upper impactor stages.

All distributions have been plotted as though the particles are uniformly distributed in log d_p in each size range. The resulting histograms are thus only approximations to the actual smooth-curve distributions. Although techniques for estimating the true size spectra have been reported in the literature (Natusch and Wallace, 1976; Esmen, 1977; Raabe, 1978), this style of graphing has been chosen to allow direct observation of those size ranges containing the bulk of the airborne mass and to emphasize differences in the shapes of the distributions. Data to construct these graphs were obtained from differential or cumulative distribution plots or from tables in the original references. Personal communication with the authors was necessary in some instances to clarify or update the data.

Pertinent information associated with each distribution is given in Table 7. Note that the aerodynamic mass median diameters (MMD) for the urban distributions are considerably greater than those of the rural areas, a fact that probably reflects proximity to sources in the former category. The Arabian Sea distribution is an exception, with a relatively large MMD, possibly related to the sizes of seaspray aerosol.

3.1.3   Use of the Data to Calculate Deposition

The size data have been used with information on deposition in the human respiratory tract and on dry deposition from the atmosphere onto surfaces to estimate chromium transport, following procedures described earlier (Milford and Davidson, 1985). Minimum and maximum estimates are given for fractional deposition in the pulmonary region of the respiratory system (lower lung). Estimates are given also for the tracheobronchial and nasopharyngeal regions of the upper respiratory system. The lung deposition data summarized by the United States Environmental Protection Agency (EPA) (1982) have been used for the calculations. Dry deposition data obtained in wind tunnel experiments (Chamberlain, 1967; Sehmel and Hodgson, 1978), as well as a deposition model based on ambient wind data over a field of wild grass (Davidson et al., 1982), have been used for the dry deposition estimates. Values of friction velocity U_* and roughness height z_o are comparable for all three sets of dry deposition calculations. Estimates have been obtained for each of the four urban distributions, and for the first three rural distributions in Figure 1. Results are shown in Table 8.

Overall, the calculated lung deposition fractions show greater values for the urban size data. The difference between urban and rural deposition fractions is greater in the tracheobronchial and nasopharyngeal regions, since deposition in these parts of the lung increases with particle size. The pulmonary region fractions are similar for urban and rural distributions. Deposition in the lower lung is greatest for midsized particles in the range 1–5 μm (U.S. EPA, 1982). The urban and rural size data show similar fractions of airborne chromium mass in this size range.

The dry deposition velocities based on the urban distributions are considerably greater than those of the rural distributions. As with the upper lung data, dry deposition velocities increase with particle size for supermicron particles. The larger particle sizes in urban areas are responsible.

Table 7 Descriptions of Chromium Airborne Mass Size Distributions

Date of Sampling	Location of Sampling	Type of Sampler	C_T (ng m^{-3})	Approximate MMD μm	Reference
September 1966 Average of 24-hour samples taken continuously for 2 weeks	Cincinnati, Ohio 1.5 m above ground	Andersen cascade impactor	310	1.5	Lee et al., 1968
February 1967 Average of three 4-day samples	Fairfax, Ohio 1.5 m above ground	Andersen cascade impactor	280	1.8	Lee et al., 1968
August 1976 Average of two 5-day samples	New York, New York Roof of a 15-story building	Andersen cascade impactor	28	4.7	Bernstein and Rahn, 1979
November–December 1976 Average of seven to 2 5-day samples, except the backup filter: average of two samples)	Trombay, Bombay, India 15 m above ground	Andersen cascade impactor	35	4.2	Sadasivan, 1981
May–September 1975 Average of four samples	Near Colstrip, Montana 7 m above ground	Lundgren cascade impactor	2.2	0.17	Crecelius et al., 1980
June 1977–January 1978 Average of seven background samples	Northwestern New Mexico, 8 km west of a coal-fired power plant, 7 m above ground	Cascade impactor	18	0.24	Wangen, 1981
March–April 1983 Average of seven 1 to 3-day samples	Ny-Ålesund, Spitsbergen	Battelle-style cascade impactor	2.1	0.18	Pacyna et al., 1984
May–June 1973 Average of three 12 to 16.5-hour samples	Arabian Sea cruises along the line 20 N latitude 8–10 m above water	Casella Mark II cascade impactor	8.1	1.2	Sadasivan, 1978

Table 8 Calculated Fractions of Deposition in the Human Respiratory Tract and Calculated Dry Deposition Velocities, Based on the Distributions of Figure 1

Fraction Deposited of Total Mass Inhaled	Urban Distributions[a]	Rural Distributions[b]
Pulmonary region (min.)	0.12 ± 0.03[c]	0.08 ± 0.01
Pulmonary region (max.)	0.30 ± 0.05	0.22 ± 0.02
Tracheobronchial region	0.14 ± 0.07	0.05 ± 0.01
Nasopharyngeal region	0.27 ± 0.09	0.08 ± 0.06
Dry Deposition Velocity, cm/s		
Model of Chamberlain (1967) $U_* = 70$ cm/s, $z_o = 6$ cm	0.75 ± 0.38	0.30 ± 0.21
Model of Sehmel and Hodgson (1978) $U_* = 30$ cm/s, $z_o = 3$ cm	0.65 ± 0.40	0.21 ± 0.12
Model of Davidson et al. (1982) $U_* = 64$ cm/s, $z_o = 5$ cm Measured wind profile over a field of *Dactylus glomerata*	1.5 ± 0.74	0.59 ± 0.44

Source: U.S. EPA, 1982.

[a]Based on distributions 1–4 in Figure 1.
[b]Based on distributions 5–7 in Figure 1.
[c]Average $\pm$ SD.

4. TIME-DEPENDENT VARIABILITY

Retrospective studies using sediment cores, peat, and bog profiles point to a recent general increase in rate of atmospheric fallout of anthropogenic chromium. For example, the records preserved in some Danish bogs suggest that the chromium content of the sphagnum species has been increasing at the rate of about 100 ng/decade since ca. 1,000 AD (Aaby and Jacobsen, 1978). Such an increase can only be attributed to changes in local and regional release of chromium to the atmosphere. Similar enrichments of chromium in superficial bog and peat layers have been reported in many other parts of Europe and North America (Glooschenko, 1986).

Supporting evidence on the recent increase of the atmospheric chromium burden comes from lake sediments. The present-day rate of chromium deposition in sediments of remote Woodhull Lake (Adirondack Mountains, New York) of 1.2 mg cm^{-2}year^{-1} is about 10-fold higher than the rate in prehistoric times (Galloway and Likens, 1979). The current rates of chromium deposition into sediments of the lower Great Lakes (Ontario and Erie) are generally twofold to fivefold higher than rates in precolonial times (Kemp et al., 1977). However, in southern Lake Huron, which is much less polluted, the present rate of chromium accumulation in sediments is only 20% higher than the prehistoric rate (Robbins,

1980). It should be noted that unlike Woodhull Lake, which imports most its chromium through the atmosphere, only 20–40% of the chromium input into the Great Lakes comes via this route (Allen and Halley, 1980). Pollution-derived enrichments of chromium in the most recent lake sediments have been documented in many other parts of the world (MARC, 1985).

The available data suggest furthermore that the ambient concentrations and fallout of chromium have been decreasing since the mid-1970s in response to controls imposed by various countries on industrial emissions of particulate matter. In rural areas of the United Kingdom, the airborne chromium levels declined at a rate of about 7–12%/year between 1972 and 1981 (Cawse, 1985). Between 1965 and 1974, atmospheric chromium levels in urban areas of the United States showed no change in the median value but a 17% decrease in the 90th percentile (Faoro and McMullen, 1977). A more recent study found declines of up to 30% in cities in the United States between 1968–1971 and 1983 (Saltzman et al., 1985). The concentrations of chromium in forest moss (which derives its mineral requirements primarily from the atmosphere) in southern Sweden show a drastic decline between 1968–1969 and 1980 (Fig. 2). Correspondingly, atmospheric emissions of chromium in Sweden were reduced from 789 to 140 t/year during this period (Ruhling and Tyler, 1984). Other available survey data on long-term air quality in Canada, Denmark, Belgium, and elsewhere likewise show significant downward trends during the last decade (Kretzschmar et al., 1980; Cawse, 1982).

Usually superimposed on the long-term trends are the year-to-year variations. Kretzschmar et al. (1980), for instance, found that the arithmetic means of the annual concentrations of the elements in air in Belgium vary by up to twofold, and the 98th percentiles vary by as much as a factor of 3.0. Similar ranges in annual averages have been reported in rural areas of Britain between 1972 and 1981 (Table 3). These variations presumably reflect fluctuations in emission rates that

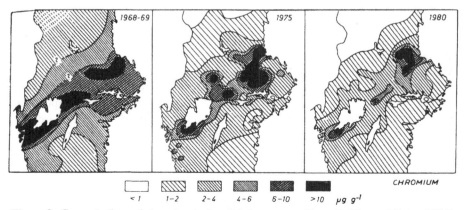

Figure 2. Concentrations of chromium in Swedish forest moss (From Ruhling and Tyler, 1984).

may be related to annual cycles in economic and industrial activities in the various countries.

Weak seasonality generally has been observed in atmospheric levels and fallout of chromium at many locations. From 1972 to 1981, the average winter/summer ratios in total deposition of chromium at Chilton, Oxon, and Styrrup, Notts, were 5:6 and 5:9, respectively. In contrast, the chromium concentrations were higher in summer in Wraymires, Cumbria, the average winter/summer ratio being only 5:4 (Cawse, 1985). In most cases, the seasonal variation in aerosol pollution pattern reflects changes in fuel composition and rate of fuel consumption associated with space heating, especially in temperate countries. The observed weakness of the seasonal effect stems from the fact that fuel consumption is not usually the major source of airborne chromium (see Chapter 3). The winter maximum in chromium concentrations in Arctic air, however, is the result of nonremoval of particulates, which increases the normal residence in the atmosphere by about one order of magnitude (Heidam, 1986).

Short-term (diurnal and day-to-day) variations in chromium levels are usually due to shifts in prevailing winds, especially where the sources of chromium are unevenly distributed around the sampling station. Numerous examples of pollution episodes associated with high chromium concentrations abound in the literature (Demuynck et al., 1976; Pacyna and Ottar, 1985; Lioy et al., 1985). A good example of diurnal variability comes from Beersheba and Shivta, Israel, where the daytime concentration of chromium was more than three times higher than the levels at night (Shani and Haccoun, 1976). It was noted that the small industries that emit most of the chromium in the two cities did not operate at night.

The peak-to-mean ratios for total chromium concentrations at specified time intervals can provide a general insight into the nature of chromium emissions. Relatively constant chromium levels or small peak-to-mean ratios reflect fairly constant and homogeneous source area, whereas large (over 3.0) peak-to-mean ratios or highly variable chromium concentrations typify contributions from either intermittent or heterogeneous sources. The peak-to-mean ratios for a number of locations in California are shown in Table 9. About 50% of the stations had ratios less than 3.0, that is, about half of the sites receive a steady supply of chromium, whereas the other sites are impacted by spatially nonuniform or intermittent sources. The fact that the sites with the highest ratios (San Bernardino–West 3rd, and Torrance–Carson) also show the highest peak concentrations and highest mean concentrations can be related to the short residence time of airborne chromium, so that any cutback in rate of emission results in sharp declines in the ambient chromium levels.

5. DEPOSITION OF AIRBORNE CHROMIUM

Atmospheric fallout has become a major contributor to the chromium budget of many ecosystems. The mechanisms and rates of depuration of airborne chromium by processes of dry deposition (diffusion, sedimentation, and impaction) and wet

Table 9 Concentrations and Peak-to-Mean Ratios for Chromium at Selected Locations in California

Sampling Location	EPA Site Number	Peak Chromium ($\mu g\ m^{-3}$)	Mean Chromium ($\mu g\ m^{-3}$)	Peak: Mean Ratio	n^a
Anaheim Harbor	0230001	28.3	7.1	3.99	106
Bakersfield	0520003	22.5	9.4	2.39	28
Berkeley–Berkeley Way	0740001	72.9	7.5	9.72	61
Burbank–West Palm	0900002	23.6	10.5	2.25	30
Fresno–South Cedar	2800002	32.4	9.5	3.41	58
Long Beach–Pine	4100001	41.0	17.9	2.29	27
Los Angeles–South San Pedro	4180001	66.6	15.5	4.30	93
Los Angeles–Downtown	4180103	35.2	17.3	2.03	25
Oakland–Fifth Street	5300001	60.3	19.0	3.17	29
Ontario–Airport	5380001	50.5	15.6	3.24	137
Oxnard	5560001	15.0	6.0	2.50	26
Pasadena	5760002	32.8	13.5	2.43	29
San Bernardino–West 3rd	6680001	280.8	23.6	11.90	99
San Diego–Island	6800004	40.8	9.1	4.48	94
San Francisco–Grove	6860001	14.8	5.8	2.55	30
San Francisco	6860004	21.8	9.3	2.34	23
San Jose–North 4th	6980004	36.3	12.2	2.98	99
Santa Ana–Ross	7180001	24.4	9.1	2.68	64
Torrance–Carson	8260001	315.3	30.5	10.34	29

Source: CARB, 1986.

[a]Number of samples analyzed.

deposition (rainout and washout during rain, snow, ice, and sleet), however, remain a critical area for further research, Hoff and Barrie (1986).

5.1 Dry Deposition

Since most of the chromium is associated with particulate matter, dry deposition is an important mechanism for removal of this metal from the atmosphere. The rates of dry deposition of chromium at remote locations typically are 0.001–0.03 mg m^{-2} year^{-1}, whereas the rates at rural and urban areas generally are 0.2–1.5 and 20–60 mg m^{-2} year^{-1} (Table 10). Thus, the fallout rates in urban areas exceed the rates at rural locations by more than 10-fold and the rates at remote locations by 100-fold to 2000-fold. It is obvious why urban areas are rapidly becoming hot spots for chromium. The dry deposition rate in the Arctic is about 10-fold higher than that of the South Pole; this difference can be related to much higher anthropogenic emissions and the presence of more land mass in the Northern Hemisphere compared to the Southern Hemisphere.

The principal processes that affect dry deposition include diffusion, interception, sedimentation, and inertial impaction. Model descriptions of the role of these processes are described in detail by Davidson et al. (1982) and Sehmel and Hodgson (1978). The dry deposition velocities at remote and rural/urban areas are described and discussed in Section 3.1.3.

5.2 Wet Deposition

The available literature shows a wide and variable range of concentrations of chromium in rainwater (Table 11). Rainfall intensity, scavenging processes, wind direction, origin of the air masses, event type (rain or snow), and duration all contribute to the observed variability in chromium concentration in rainwater. There is also the analytic problem in the determination of insoluble particulate material in the rainfall. Apparently no standardized technique currently exists for measuring the particulate phases that account for most of chromium in rainwater samples. Schutyser et al. (1978) found that about 72% of the chromium in rainwater samples collected at Ghent was bound to insoluble particles. Beavington and Cawse (1979), however, found that a lower fraction (45%) of the chromium in bulk precipitation at Bagauda, Nigeria, was insoluble. Because of the difficulty and differences in methods used in quantifying the insoluble component, the data on wet deposition of chromium shown in Table 11 must be used in a circumspect manner.

5.3 Bulk Deposition

Several methods have been used for determining the bulk atmospheric deposition of chromium at any given location. Most of the available data, however, are based on the use of bulk sample collectors or biomonitoring techniques. The two

Table 10 Rates of Dry Deposition of Chromium from the Atmosphere

Sampling Location and Date	Dry Fallout Rate $(mg\ m^{-2}\ year^{-1})$	Reference
Remote		
South Pole, 1975–1976	0.0012	Estimated[a]
Nord, Greenland, 1979–1980	0.015	Estimated[a]
Thule, Greenland, 1978	0.020	Estimated[a]
Mauna Loa, Hawaii, 1972	0.031	Estimated[a]
Southern Indian Ocean	0.022	Estimated[a]
Rural		
Chilton, Berks, 1972–1979	0.68	Cawse, 1974–1980
Leiston, Suffolk, 1972–1978	0.43	Cawse, 1974–1980
Collafirth, Shetland Islands, 1972–1976	0.17	Cawse, 1974–1978
Plynlimon, Montgoms., 1972–1976	0.53	Cawse, 1974–1978
Styrrup, Notts., 1972–1979	4.1	Cawse, 1974–1980
Trebanos, Glamorgs., 1972–1976	1.7	Cawse, 1974–1978
Wraymires, Lancs., 1972–1978	0.44	Cawse, 1974–1980
Perwez, Belgium, 1982	0.93	Ronneau and Cara, 1984
Cherleroi, Belgium, 1980–1981	1.3	Hallett et al., 1982
Walker Branch Watershed, Tennessee, 1974	0.05	Lindberg et al., 1975
Northeast Minnesota (9 stations), 1976–1978	0.57	Ritchie and Thingvold, 1985
Lake Michigan	5.4	Allen and Halley, 1980
USA, several stations	<2.0	DOE, 1979
Bagauda, northern Nigeria, 1976	1.4	Beavington and Cawse, 1979
Urban		
New York City, 1972–1974	48	Kleinman et al., 1975
UK (unspecified), 1980–1981	26	Pattenden et al., 1982
USA, average	9.6	Estimated[b]
Canada, average	6.2	Estimated[b]
Europe, average	14	Estimated[b]

[a]Estimated using dry deposition velocity of 0.3 cm/s and the airborne concentrations of chromium in Table 4.

[b]Estimated using the dry deposition velocity of 0.75 cm/s and the averages of the data listed in Table 4.

Table 11 Representative Concentrations of Chromium in Atmospheric Precipitation

Sampling Location and Date	Concentration (μg/L)	Reference
Canada, 29 stations, 1980	0.97 (rain)	Hamilton and Chatt, 1982
Chibougamu, Quebec, 1980–1981	0.57 (snowpack)	Barrie and Vet, 1984
Quebec City, Quebec, 1980–1981	0.34 (snowpack)	Barrie and Vet, 1984
Dorset, Ontario, 1980–1981	0.21 (snowpack)	Barrie and Vet, 1984
Chalk River, Ontario, 1980–1981	0.42 (snowpack)	Barrie and Vet, 1984
New York City	2.8 (rain)	Galloway et al., 1982
Walker Branch Watershed, Tennessee, 1973–1974	1.9 (rain)	Andren and Lindberg, 1977
Quillayute, Washington	<0.1 (rain)	Tanner et al., 1972
Minneapolis, Minnesota	0.51 (rain/snow)	Thornton et al., 1980
North Dakota	0.44 (rain/snow)	Thornton et al., 1980
Northern Minnesota, 1978–1979	0.32 (rain/snow)	Thornton et al., 1980
Northeastern Minnesota, 1978–1979	0.24 (rain/snow)	Thornton et al., 1980
Lake Superior basin	1.5 (rain)	Eisenreich et al., 1978
Lake Erie basin	7.0 (bulk)	Konasewich et al., 1978
Lake Huron basin	<2.0 (bulk)	Konasewich et al., 1978
Penage, Ontario, 1970–1975	0.4 (bulk)	Kramer, 1976
Skead, Ontario, 1970–1975	1.5 (bulk)	Kramer, 1976
Killarney, Ontario, 1970–1975	0.5 (bulk)	Kramer, 1976
Gore Bay, Ontario, 1970–1975	2.3 (bulk)	Kramer, 1976
Jamot, Ontario, 1970–1975	1.5 (bulk)	Kramer, 1976
Windy Lake, Ontario, 1970–1975	1.1 (bulk)	Kramer, 1976
Gogama, Ontario, 1970–1975	1.3 (bulk)	Kramer, 1976
Temagami, Ontario, 1970–1975	1.6 (bulk)	Kramer, 1976
Espanola, Ontario, 1970–1975	1.0 (bulk)	Kramer, 1976
Sudbury (South), Ontario, 1970–1975	1.2	Kramer, 1976
Moscow, USSR, 1975–1976	2.3 (bulk)	Miklishanskiy et al., 1977
Gottingen, 1971	>14 (particulate)	Ruppert, 1975
Ghent, Belgium, 1976		Schutyser et al., 1978
Soluble fraction	1.1	
Insoluble fraction	2.9	
Total	4.0	
Heidelberg, FRG	3.6 (rain)	Bogen, 1974
Rural Sweden	0.2 (rain/snow)	Soderlund, 1975
Rural UK, 1972–1973	6.0 (bulk)	Cawse, 1974
Swansea, UK	8.3 (bulk)	Pattenden, 1974
Bagauda, northern Nigeria, 1976	2.9 (bulk)	Beavington and Cawse, 1979
	2.4 (particulate)	Beavington and Cawse, 1979
Toronto area, Ontario, 1983–1984	5.0 (snow)	Brzezinska-Paudyn et al., 1986
Oakville, Ontario, 1984	3.0 (snow)	Brzezinska-Paudyn et al., 1986
Kingston, Ontario, 1984	3.0 (snow)	Brzezinska-Paudyn et al., 1986

monitoring strategies often yield data that may not be entirely comparable (Glooschenko, 1986).

The available data (Table 12) show that the bulk deposition of chromium at remote locations is generally less than $0.2\ mg\ m^{-2}\ year^{-1}$. As to be expected from the locations of industrial sources, the fallout in the Arctic (about $0.04\ mg\ m^{-2}\ year^{-1}$) is about 20-fold higher than the rate at the South Pole ($0.002\ mg\ m^{-2}\ year^{-1}$). The wind circulation patterns also affect the bulk deposition of chromium at remote locations. Thus, the bulk deposition rate in the tropical North Atlantic ($0.14\ mg\ m^{-2}\ year^{-1}$) is much higher than the rate in the tropical North Pacific ($0.06\ mg\ m^{-2}\ year^{-1}$).

Bulk deposition of chromium in rural and semirural areas typically falls in the range of 0.5–$5\ mg\ m^{-2}\ year^{-1}$ (Table 12). Representative fallout rates include $1.3\ mg\ m^{-2}\ year^{-1}$ in northeastern United States (Groet, 1976), $2.2\ mg\ m^{-2}\ year^{-1}$ in rural Belgium (Ronneau, 1985), $0.98\ mg\ m^{-2}\ year^{-1}$ in northeastern Minnesota (Thornton et al., 1980), $2.0\ mg\ m^{-2}\ year^{-1}$ in the Great Lakes basin (Allen and Halley, 1980), $1.5\ mg\ m^{-2}\ year^{-1}$ in rural areas of Switzerland (Imboden et al., 1975), and 2–$6\ mg\ m^{-2}\ year^{-1}$ in rural areas of Britain (Cawse, 1985).

In general, the bulk deposition of chromium in urban areas is over $10\ mg\ m^{-2}\ year^{-1}$ (Table 12). Some examples include $16\ mg\ m^{-2}\ year^{-1}$ in the inner city of Copenhagen (Andersen et al., 1978), $57\ mg\ m^{-2}\ year^{-1}$ in Tulsa, Oklahoma (Tate and Bates, 1984), $32\ mg\ m^{-2}\ year^{-1}$ in New York City (Kleinman et al., 1975), and $34\ mg\ m^{-2}\ year^{-1}$ in urban areas of Britain (Pattenden et al., 1982). The effects of the elevated flux of chromium into urban ecosystems have yet to be investigated.

5.4   Roadside Deposition of Chromium

The automobile industry is a major consumer of chromium, primarily (1) as an additive in unleaded gasoline, (2) in recirculating cooling waters (antifreeze), (3) as paint on automobile bodies, and (4) as trims and alloys on automotive parts (see Chapter 2). Several studies have now demonstrated that the wastage of automotive chromium often results in a buildup of chromium in roadside ecosystems and in street runoff. Fergusson and Ryan (1984) showed that the chromium content of street dust is much higher in large urban centers (average $121 \pm 52\ \mu g\ g^{-1}$ for London and New York) than in small urban or rural areas (average $51 \pm 21\ \mu g\ g^{-1}$ for Halifax, Nova Scotia, Kingston, Ontario, and Christchurch, New Zealand). In a well-known case study, Ward et al. (1977) found a strong correlation between traffic density and the concentrations of chromium in roadside soils and vegetation of Auckland, New Zealand. A large fraction of the chromium burden of the vegetation was removable by washing, and the decrease in chromium concentration with soil depth further attests to the fact that most of the chromium was derived from automobiles.

High rates of chromium fallout have been reported especially near major highways. Dry and wet depositions of chromium in a viaduct in Pullman,

Table 12 Bulk Deposition of Chromium from the Atmosphere

Sampling Location and Date	Fallout Rate (mg m^{-2} year^{-1})	Reference
Southern New England coast	1.8 (moss samples)	Groet, 1976
Inland Massachusetts	1.4 (moss samples)	Groet, 1976
Southern New Hampshire	1.7 (moss samples)	Groet, 1976
Inland Connecticut and Rhode Island	1.0 (moss samples)	Groet, 1976
Central and southern New York State	1.3 (moss samples)	Groet, 1976
Northern New Hampshire	0.9 (moss samples)	Groet, 1976
Cape Cod, Massachusetts	10.6 (moss samples)	Groet, 1976
Eastern coastal Maine	0.7 (moss samples)	Groet, 1976
Northeastern USA (average)	1.3 (moss samples)	Groet, 1976
Copenhagen, Norway	(moss samples)	Andersen et al., 1978
City center	16	
Suburban areas	6.3	
Rural areas	5.6	
Switzerland (several stations), 1973–1974	1.5	Imboden et al., 1975
Gottingen, 1973	9.2	Ruppert, 1975
Perwez, Belgium, 1982	3.5	Ronneau and Cara, 1984
Charleroi, Belgium, 1980–1981	28	Hallet et al., 1982
Rural Belgium	2.2	Ronneau, 1985
Urban UK, 1980–1981	34	Pattenden et al., 1982
Chilton, Oxon, 1972–1981	1.9	Cawse, 1985
Leiston, Suffolk, 1972–1975	2.1	Cawse, 1974–1977
Collafirth, Shetland Islands, 1972–1975	2.0	Cawse, 1974–1977
Plynlimon, Montgoms, 1972–1975	4.0	Cawse, 1974–1977
Styrrup, Notts, 1972–1981	6.1	Cawse, 1985
Trebanos, W. Glamorgs., 1972–1975	<0.5	Cawse, 1974–1977
Wraymires, Cumbria, 1972–1981		Cawse, 1985
Walsall, W. Midlands, 1979	1.2	Cawse, 1985
Central Swansea, 1972–1973	5.3	Pattenden, 1974
Sutton Bonington, Leics., 1963	13	Hallsworth and Adams, 1973
New York City, 1972–1974	32	Kleinman et al., 1975
Tulsa, Oklahoma	2.6	Tate and Bates, 1984
Walker Branch Watershed	0.3	Lindberg et al., 1975
Northwestern Minnesota, 1976–1978	0.6	Ritchie and Thingvold, 1978
Lake Superior	1.0	Eisenreich et al., 1978
1978–1979	1.7–3.1	Allen and Halley, 1980
Lake Michigan	0.4	Winchester and Nifong, 1971
	0.25	Gatz, 1975
	1.2	Klein, 1975
1979–1980	7.2–8.1	Allen and Halley, 1980

(Continued)

Table 12 *(Continued)*

Sampling Location and Date	Fallout Rate (mg m^{-2} year^{-1})	Reference
Lake Erie, 1979–1980	2.8–3.7	Allen and Halley, 1980
Lake Ontario	0.3	Kramer, 1976
Crane Creek, Ohio, 1978–1979	2.8	Allen and Halley, 1980
Conneaut, Ohio, 1978–1979	2.5	Allen and Halley, 1980
Rainesville, Ohio, 1978–1979	1.5	Allen and Halley, 1980
Vermillion, Ohio, 1978–1979	2.2	Allen and Halley, 1980
Paradise, Michigan, 1978–1979	1.8	Allen and Halley, 1980
Grand Marais, Michigan, 1978–1979	1.1	Allen and Halley, 1980
Baraga, Michigan, 1978–1979	1.8	Allen and Halley, 1980
Houghton, Michigan, 1978–1979	1.9	Allen and Halley, 1980
Copper Harbor, Michigan, 1978–1979	1.7	Allen and Halley, 1980
Ontonagun, Michigan, 1978–1979	2.2	Allen and Halley, 1980
Beaver Island, Michigan, 1978–1979	1.7	Allen and Halley, 1980
St. Joseph, Michigan, 1978–1979	1.9	Allen and Halley, 1980
Chicago, Illinois, 1978–1979	17	Allen and Halley, 1980
Wilmette, Illinois, 1978–1979	1.9	Allen and Halley, 1980
Zion, Illinois, 1978–1979	4.4	Allen and Halley, 1980
Cudahy, Wisconsin, 1978–1979	3.8	Allen and Halley, 1980
Two Rivers, Wisconsin, 1978–1979	1.5	Allen and Halley, 1980
Bayfield, Wisconsin, 1978–1979	1.6	Allen and Halley, 1980
Superior, Wisconsin, 1978–1979	2.6	Allen and Halley, 1980
Grand Marais, Minnesota, 1978–1979	1.5	Allen and Halley, 1980
Silver Bay, Minnesota, 1978–1979	3.1	Allen and Halley, 1980
Two Harbors, Minnesota, 1978–1979	3.1	Allen and Halley, 1980
Penage, Ontario, 1970–1975	0.07	Kramer, 1976
Skead, Ontario, 1970–1975	7.4	Kramer, 1976
Killarney, Ontario, 1970–1975	9.2	Kramer, 1976
Gore Bay, Ontario, 1970–1975	22	Kramer, 1976
Jamot, Ontario, 1970–1975	7.8	Kramer, 1976
Gogama, Ontario, 1970–1975	0.33	Kramer, 1976
Sudbury (South), Ontario, 1970–1975	8.6	Kramer, 1976
Tropical North Pacific	0.06	Buat-Menard, 1984
Tropical North Atlantic	0.14	Buat-Menard, 1984
Bermuda	0.09	Buat-Menard, 1984
Western Mediterranean	0.49	Buat-Menard, 1984
North Sea	2.1	Buat-Menard, 1984
Baguada, northern Nigeria, 1976	3.3	Beavington and Cawse, 1979
Campo de Gibralter, Spain, 1982–1983	40	Usero and Gracia, 1986
Seville, Spain	11	Usero and Gracia, 1986
Almeria, Spain	13	Usero and Gracia, 1986

Washington, were found to be 190 and 2,400 mg m^{-2} year^{-1}, respectively (Bourcier and Hindin, 1979). These rates are several-fold higher than the average values in urban areas. The high chromium flux is often reflected in the highly elevated chromium concentrations in drainage waters. The reported concentrations in street runoff waters include 140 μg L^{-1} in Pullman, Washington, 230 μg L^{-1} in Durham, North Carolina, and 30 μg L^{-1} in Lodi, New Jersey (Bourcier and Hindin, 1979). The average for street drainage in the United States of about 800 μg L^{-1} (Pitt and Amy, 1973) is at least an order of magnitude higher than the levels in even polluted surface waters (see Chapter 6). Studies by Bourcier and Hindin (1979) show that most of the chromium in street runoff is in particulate form, with less than 5% present as dissolved or colloidal fraction.

Even the weathering of highway bridges can result in the contamination of runoff waters with chromium. For example, Yousef et al. (1984) found the chromium concentrations in runoff samples from bridges over Lake Ivanhoe and on Maitland Interchange (both in Orlando, Florida) to be 13 μg L^{-1}, with about 75–85% of the chromium in particulate form.

6. SOURCE–RECEPTOR RELATIONS FOR ATMOSPHERIC CHROMIUM

An assessment of the contribution of source emissions to the ambient levels of air pollutants is a key requirement in air pollution studies. To obtain this information, data from emission surveys of air pollutants are related to the measured concentrations, usually at remote areas, by applying various long-range transport models.

In this section, the emission data of chromium from various anthropogenic sources in Europe (see Chapter 3) are compared with measured air concentrations in Northern Europe and the Norwegian Arctic using trajectory model calculations and trajectory sector statistics.

6.1 Source–Receptor Relations for Episodes of Long-Range Transport of Atmospheric Chromium

The origin of aerosols, measured during the episodes of long-range transport of air pollutants, can be analyzed using a simple trajectory model. The model is based on the mass-balance equation:

$$\frac{dq}{dt} = (1 - \alpha) \times \frac{Q}{h} - k \times q$$

where:

q = trace element concentration in air, ng^{-3}
t = time in s

Q = trace element emission per unit area and time, ng M^{-2} s^{-1}

h = height for mixing layer in m

k = decay rate for the trace element considered (wet and dry deposition), s^{-1}

α = fraction of trace element emission deposited in the same grid element as it is emitted. (This local deposition supplements the deposition included in the decay rate, k).

A part of the emitted trace elements is deposited in the same grid element as it is released. Dry deposition studies in the surroundings of coal-fired power plants (Pacyna, 1980) in Poland were used to assess the local deposition of the pollutants (coefficient α in equation). It was found that ca. 15% of the chromium releases were deposited in the emission area. As mentioned in Chapter 3, most of the chromium from coal-fired power plants is emitted on particles with diameter <1 μm. These particles are a subject of long-range transport. The most important sources of the chromium emission, that is, metallurgical processes, produce larger particles, which are deposited mostly in the emission area. Thus, the local deposition of chromium around steel plants is larger than the deposition around coal-fired plants.

Model concentrations are inversely proportional to the mixing height, h. From radiosonde data for Europe (Eliassen and Saltbones, 1982), a constant mixing height of 1,000 m can be assumed as a preliminary estimate.

Decay rates were estimated from deposition velocities for constant mixing height. Wet deposition was not considered in calculating the decay rates. Thus, only episodes with no or very little precipitation between emission and study area were selected, possibly resulting in a slight overestimation of concentrations.

This model, described in detail by Pacyna et al. (1984), was used to calculate the chromium concentrations at Ny Ålesund, Spitsbergen, in March 1983. Eight cases were selected for the model calculations. The measured and predicted concentrations of chromium are shown in Figure 3. A good agreement, within a factor of 2, shows that the chromium concentrations from long-range transport of air pollutants measured at Ny Ålesund, can be related to the estimated anthropogenic emissions of the element.

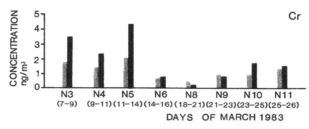

Figure 3. Measured (stippled lines) and predicted (solid lines) concentrations of Cr at Ny Ålesund, Spitsbergen.

6.2 Application of Chromium Data in Sector Statistics to Assess the Origin of Aerosols

Chromium, together with other elements, can be used to assess the origin of remote aerosols not only from the measurements during episodes of long-range transport but also from sector analysis of daily mean concentrations over a sampling period. An example of such study was described by Pacyna (1986) for the chromium and other element transport to Scandinavia. The average sectoral concentrations of trace elements in the fine fraction of particles ($<2.0 \mu$m diameter) were used to draw elemental diagrams for various wind sectors at three stations: Birkenes (Southern Norway), Rörvik (Sweden), and Vivolahti (Southern Finland). The elements that showed the largest differences between sectional concentrations were selected to serve as elemental tracers for the European aerosol measured in Scandinavia. These elements included chromium, together with Cu, As, Se, Mo, and Sn, with V as a reference element. To distinguish between the east and west European aerosols affecting receptors in Scandinavia, the concentrations of the above tracers were calculated separately for days, with air pollution transport from Western Europe, Eastern Europe, and the Moscow and Urals areas to Scandinavia deduced from 850 mb trajectories (Table 13).

It should be noted that similar estimates of elemental tracers for total concentrations in this work have resulted in smaller differences between the east and west aerosols. Thus, total aerosols seem to be less suitable tracers than the fine aerosol fraction.

Table 13   Elemental Ratios for the East and West European Aerosol at Birkenes and Rørvik

Ratio[a]	West European[b] 10 Samples	East European[c] 11 Samples	Moscow and Urals 11 Samples
Cr/V	0.23–0.37 0.30	0.12–0.17 0.14	0.5–0.8 0.70
Cu/V	0.70–1.10 0.92	0.07–0.20 0.13	1.2–1.4 1.1
As/V	0.80–1.10 0.91	0.15–0.37 0.28	3.4–4.1 3.8
Se/V	0.13–0.28 0.15	0.02–0.07 0.04	0.11–0.16 0.14
Mo/V	0.12–0.13 0.12	0.03–0.07 0.05	0.10–0.15 0.14
Sn/V	0.54–0.70 0.62	0.24–0.27 0.25	0.80–1.20 1.00

[a] Based on concentrations in particles $<2 \mu$m EAD.

[b] Including the UK, France, the Netherlands, Belgium, Luxemburg, and Federal Republic of Germany.

[c] Including German Democratic Republic, Poland, Czechoslovakia, Hungary, and the following republics of the USSR: Lithuania, Latvia, Ukraine, and White Russia.

REFERENCES

Aaby, B., and Jacobsen, J. (1978). "Changes in biotic conditions and metal deposition in the last millennium as reflected in obrotrophic peat in Draved Mose, Denmark." *Danm. Geol. Unders. Arbog* pp. 5–43.

Adams, F., van Espen, P., and Maenhaut, W. (1983). "Aerosol composition at Chacaltaya, Bolivia, as determined by size-fractionated sampling." *Atmos. Environ.* **17**, 1521–1536.

Adams, F., Dams, R., Guzman, L. and Winchester, J.W. (1977). "Background aerosol composition on Chacaltaya Mountain, Bolivia." *Atmos. Environ.* **11**, 629–634.

Akland, G.G. (1976). *Air Quality Data for Metals, 1970 Through 1974, from the National Air Surveillance Networks.* Report No. EPA-600/4-76-041, U.S. Environmental Protection Agency, Office of Research and Development, Research Triangle Park, NC.

Alkezweeny, A.J., and Busness, K.M. (1984). "Observations of aerosol chemical composition and acidity in northwest and southeast regions of the United States." *Sci. Total Environ.* **39**, 125–133.

Alkezweeny, A.J., Glover, D.W., Lee, R.N., et al. (1975). "Measured chromium distributions resulting from cooling-tower drift." In: S.R. Hannah and J. Pell, Eds., *Cooling Tower Environment.* U.S. Energy Research and Development Administration, Washington, DC, pp. 558–572.

Alian, A., and Sansoni, B. (1985). "A review on activation analysis of air particulate matter." *J. Radioanal. Nuclear Chem.* **89**, 191–275.

Allen, H.E., and Halley, M.A. (1980). "Assessment of airborne inorganic contaminants in the Great Lakes." In *1980 Annual Report, Appendix B.* Great Lakes Science Advisory Board, International Joint Commission on the Great Lakes, Windsor, Ontario.

Andersen, A., Hovmand, M.F., and Johnsen, I. (1978). "Atmospheric heavy metal deposition in Copenhagen area." *Environ. Pollut.* **17**, 133–149.

Andren, A.W., and Lindberg, S.E. (1977). "Atmospheric input and origin of selected elements in Walker Branch Watershed, Oak Ridge, Tennessee." *Water Air Soil Pollut.* **8**, 199–215.

Barrie, L.A., and Vet, R.J. (1984). "The concentration and deposition of acidity, major ions and trace metals in the snowpack of the eastern Canadian Shield during the winter of 1980–81." *Atmos. Environ.* **18**, 1459–1469.

Barton, S.C., Shenfeld, L., and Thomas, B.A. (1975). "A review of heavy metal measurements in Ontario." In *Abstracts Volume, International Conference on Heavy Metals in the Environment.* Toronto, Ontario, pp. C91–C93.

Beavington, F., and Cawse, P.A. (1978). "Comparative studies of trace elements in air particulate in northern Nigeria." *Sci. Total Environ.* **10**, 239–244.

Beavington, F., and Cawse, P.A. (1979). "The deposition of trace elements and major nutrients in dust and rainwater in northern Nigeria." *Sci. Total Environ.* **13**, 263–274.

Bernstein, D.M., and Rahn, K.A. (1979). "NY summer aerosol study: Trace element concentration as a function of particle size." *Ann. NY Acad. Sci.* **322**, 87.

Block, C., and Dams, R. (1976). "Study of fly ash emission during combustion of coal." *Environ. Sci. Technol.* **10**, 1011–1017.

Bogen, J. (1974). "Trace elements in precipitation and cloud water in the area of Heidelberg measured by instrumental neutron activation analysis." *Atmos. Environ.* **8**, 835–844.

Bourcier, D.R., and Hindin, E. (1979). "Lead, iron, chromium, and zinc in road runoff at Pullman, Washington." *Sci. Total Environ.* **12**, 205–215.

Bourne, H.G., and Rushin, W.R. (1950). "Atmospheric pollution in the vicinity of a chromate plant." *Indust. Med.* **19**, 568–569.

Brzezinska-Paudyn, A., van Loon, J., and Balicki, M.R. (1986). "Multielement analysis and mercury speciation in atmospheric samples from the Toronto area." *Water Air Soil Pollut.* **27**, 45–56.

Buat-Menard, P.E. (1984). "Fluxes of metals through the atmosphere and oceans." In: J.O. Nriagu, Ed., *Changing Metal Cycles and Human Health.* Springer-Verlag, Berlin, pp. 43–69.

Buat-Menard, P.E., and Arnold, M. (1978). "The heavy metal chemistry of atmospheric particulate matter emitted by Mount Etna volcano." *Geophys. Res. Lett.* **5**, 245–248.

CARB (1986). *Report of the Joint Meeting of the Air Resources Board and Scientific Review Panel on Toxic Air Contaminants.* California Air Resources Board, Sacramento, CA.

Cawse, P.A. (1974–1981). *A Survey of Atmospheric Trace Elements in the U.K., 1972–1981.* Environmental and Medical Sciences Division, A.E.R.E. Reports, Her Majesty's Stationery Office, London.

Cawse, P.A. (1982). "Inorganic particulate matter in the atmosphere." In *Environmental Chemistry.* Royal Society of Chemistry, London, Vol. 2, pp. 1–69.

Cawse, P.A. (1985). *Trace and Major Elements in the Atmosphere at Rural Locations in Great Britain, 1972–1981. Proc. of Meeting at Bristol on Pollutant Transport and Fate in Ecosystems.* Special Publications Series, British Ecological Society.

Chamberlain, A.C. (1967). "Transport of *Lycopodium* spores and other small particles to rough surfaces." *Proc. Royal Soc. London* **296**, 45.

Chan, T.L., and Lawson, D.R. (1981). "Characteristics of cascade impactors in size determination of diesel particles." *Atmos. Environ.* **15**, 1237.

Chester, R., Aston, S.R., Stoner, J.H., and Bruty, D. (1974). "Trace metals in soil-sized particles from the lower troposphere over the world ocean." *J. Recherches Atmos.* pp. 777–789.

Chester, R., Murphy, K.J.T., Towner, J., and Thomas, A. (1986). "The partitioning of elements in crust-dominated marine aerosols." *Chem. Geol.* **54**, 1–15.

Cox, X.B., Linton, R.W., and Butler, F.E. (1985). "Determination of chromium speciation in environmental particles. Multitechnique study of ferrochrome smelter dust." *Environ. Sci. Technol.* **19**, 345–352.

Crescelius, E.A., Lepel, E.A., Laul, J.C., et al. (1980). "Background air particulate chemistry near Colstrip, Montana." *Environ. Sci. Technol.* **14**, 422.

Cunningham, W.C., and Zöller, W.H. (1981). "The chemical composition of remote area aerosols." *J. Aerosol Sci.* **12**, 367–384.

Cushing, K.M., McCain, J.D., and Smith, W.B. (1979). "Experimental determination of sizing parameters and wall losses of five source-test cascade impactors." *Environ. Sci. Technol.* **13**, 726.

Dams, R., and de Jonge, J. (1976). "Chemical composition of Swiss aerosols from the Jungfraujoch." *Atmos. Environ.* **10**, 1079–1084.

Davidson, C.I. (1977). "The deposition of trace metal-containing particles in the Los Angeles area." *Powder Technol.* **18**, 117.

Davidson, C.I., Miller, J.M., and Pleskow, M.A. (1982). "The influence of surface structure on predicted particle dry deposition to natural grass canopies." *Water Air Soil Pollut.* **18**, 25–43.

Demuynck, M., Rahn, K.A., Janssens, M., and Dams, R. (1976). "Chemical analysis of airborne particulate matter during a period of unusually high pollution." *Atmos. Environ.* **10**, 21–26.

DOE (1979). *The Chemical Composition of Atmospheric Deposition.* Environ. Measurements Lab. Quarterly EML-363, U.S. Department of Energy, Washington, DC.

Dreesen, D.R., Gladney, E.S., Owens, J.W., et al. (1977). "Comparison of levels of trace elements extracted from fly ash and levels found in effluent waters from a coal-fired power plant." *Environ. Sci. Technol.* **10**, 1017–1019.

Duce, R.A., Ray, R., Unni, C.K., and Harder, P.J. (1983). "Atmospheric elements at Enewetak Atoll. 1. Concentrations, sources and temporal variability." *J. Geophys. Res.* **88**, 5321–5342.

Duce, R.A., Hoffman, G.L., Ray, B.J., et al. (1976). "Trace metals in marine atmosphere: Sources and fluxes." In H. Windom and R.A. Duce, Eds., *Marine Pollutant Transfer.* D.C. Heath, Lexington, MA, pp. 77–120.

Dzubay, T.G., and Stevens, R.K. (1973). *Applications of X-ray Fluorescence to Particulate Measurements.* Second Joint Conference on Sensing of Environmental Pollutants, Instrumental Society of America, Washington, DC.

Dzubay, T.G., Hines, L.E., and Stevens, R.K. (1976). "Particle bounce errors in cascade impactors." *Atmos. Environ.* **10**, 229.

Egorov, V.V., Zhigalovskaya, T.N., and Malakhov, G.M. (1970). "Microelement content of surface air above the continent and the ocean." *J. Geophys. Res.* **75**, 3650–3656.

Eisenreich, S.J., Hollod, G.J., and Langevin, S. (1978). *Precipitation Chemistry and Atmospheric Deposition of Trace Elements in Northeastern Minnesota.* Report to Minnesota Environmental Quality Council, Minneapolis.

Eliassen, A., and Saltbones, J. (1979). *Modelling of Long-Range Transport of Sulphur Over Europe: A Two-Year Model Run and Some Model Experiments.* EMEP/MSC-W Report 1/82. The Norwegian Meteorological Institute, Oslo, Norway.

EPA (1984). *Health Assessment Document for Chromium.* Report No. EPA-600/8-83-014F, U.S. Environmental Protection Agency, Environmental Criteria and Assessment Office, Research Triangle Park, NC.

Escalona, L., and Sanhueza, E. (1981). "Elemental analysis of the total suspended matter in the air in downtown Caracas." *Atmos. Environ.* **15**, 61–64.

Esmen, N.A. (1977). *An Iterative Impactor Data Analysis Method.* 51st Colloid and Interface Science Symposium, New York.

Faoro, R.B., and McMullen, T.B. (1977). *National Trends in Trace Metals in Ambient Air, 1965–1974.* Report No. EPA-450/1-77-003, U.S. Environmental Protection Agency, Office of Air and Waste Management, Research Triangle Park, NC.

Fergusson, J.E., and Ryan, D.E. (1984). "The elemental composition of street dust from large and small urban areas related to city type, source and particle size." *Sci. Total Environ.* **34**, 101–116.

Flocchini, R.G., Cahill, T.A., Shadoan, D.J., et al. (1976). "Monitoring California's aerosols by size and elemental composition." *Environ. Sci. Technol.* **10**, 76–82.

Galloway, J.N., and Likens, G.E. (1979). "Atmospheric enhancement of metal deposition in Adirondack lake sediments." *Limnol. Oceanogr.* **24**, 427–433.

Galloway, J.N., Thornton, J.D., Norton, S.A., et al. (1982). "Trace metals in atmospheric deposition: A review and assessment." *Atmos. Environ.* **16**, 1677–1700.

Gatz, D.F. (1975). "Pollutant aerosol deposition into southern Lake Michigan." *Water Air Soil Pollut.* **5**, 239–251.

Glooschenko, W.A. (1986). "Monitoring the atmospheric deposition of metals by the use of bog vegetation and peat profiles." *Adv. Environ. Sci. Technol.* **17**, 507–533.

Gordon, G.E., Zöller, W.H., and Gladney, E.S. (1973). "Abnormally enriched trace elements in the atmosphere." *Trace Substances Environ. Health* **7**, 167–173.

Gordon, G.E., Moyers, J.L., Rahn, K.A., et al. (1978). *Atmospheric Trace Elements: Cycles and Measurements.* Report to the National Science Foundation Atmospheric Chemistry Workshop, National Center for Atmospheric Research, Boulder, CO.

Groet, S.S. (1976). "Regional and local variations in heavy metal concentrations of bryophytes in the northeastern United States." *OIKOS* **27**, 445–456.

Hallet, J.Ph., Lardinois, P., Ronneau, C., and Cara, J. (1982). "Elemental deposition as a function of distance from an industrial zone." *Sci. Total Environ.* **25**, 99–109.

Hallsworth, E.G., and Adams, W.A. (1973). "The heavy metal content of rainfall in the east Midlands." *Environ. Pollut.* **4**, 231–235.

Hamilton, E.P., and Chatt, A. (1982). "Determination of trace elements in atmospheric wet precipitation by instrumental neutron activation analysis." *J. Radioanal. Chem.* **71**, 29–45.

Hansen, L.D., and Fisher, G.L. (1980). "Elemental distribution in coal fly ash particles." *Environ. Sci. Technol.* **14**, 1111–1117.

Harriss, R.C., Browell, E.V., Sebacher, D.I., et al. (1984). "Atmospheric transport of pollutants from North America to the North Atlantic Ocean." *Nature* **308**, 722–724.

Heidam, N.Z. (1981). "On the origin of the Arctic aerosol: A statistical approach." *Atmos. Environ.* **15**, 1421–1427.

Heidam, N.Z. (1984). "The composition of the Arctic aerosol." *Atmos. Environ.* **18**, 329–343.

Heidam, N.Z. (1986). "Trace metals in the Arctic aerosol." *Adv. Environ. Sci. Technol.* **17**, 267–294.

Heindryckx, R. (1976). "Comparison of the mass-size functions of the elements in the aerosols of the Ghent industrial district with data from other areas. Some physico-chemical implications." *Atmos. Environ.* **10**, 65–71.

Heintzenberg, J., Hansson, H.C., and Lannefors, H. (1981). "The chemical composition of Arctic haze at Ny Ålesund, Spitsbergen." *Tellus* **33**, 162–171.

Henry, W.M., and Blosser, E.R. (1971). *Identification and Estimation of Ions, Molecules and Compounds in Particulate Matter Collected from Ambient Air.* Technical Report CPA-70-159, Battelle Columbus Laboratories, Columbus, OH.

Hidy, G.M. (1974). *Characterization of Aerosols in California.* Final Report, Air Resources Board Contract No. 358, Science Center, Rockwell International.

Hoff, R.M., and Barrie, L.A. (1986). "Air chemistry observations in the Canadian Arctic." *Water Sci. Technol.* **18**, 97–107.

Imboden, D.M., Hegi, H., and Zobrist, J. (1975). "Atmospheric loading of metals in Switzerland." *EAWAG News* (Dubendorf, Switzerland) **4**, 5–7.

Johansson, T.B., Van Grieken, R.E., Nelson, J.W., and Winchester, J.W. (1975). "Elemental trace analysis of small samples by proton induced x-ray emission." *Anal. Chem.* **47**, 855.

John, W., Kaifer, R., Rahn, K. and Wesolowski, J.J. (1973). "Trace element concentrations in aerosols from the San Francisco Bay area." *Atmos. Environ.* **7**, 107–118.

Johnson, D.B. (1976). "Ultragiant urban aerosol particles." *Science* **194**, 941.

Kamath, R.R., and Kelkar, D.N. (1981). "Preliminary estimates of the primary source contributions to winter aerosols in Bombay, India." *Sci. Total Environ.* **20**, 195–201.

Kanbour, F.I., Kitto, A.M.N., Yassein, S., et al. (1985). "Elemental analysis of total suspended particulate matter in the ambient air at Baghdad." *Environ. Int.* **11**, 459–463.

Kemp, A.L.W., Thomas, R.L., and Williams, J.D.H. (1977). *Major Elements, Trace Elements, Sediment Particle Size, Water Content, Eh, and pH in 26 Cores from Lakes Superior, Huron, Erie and Ontario.* Unpubl. report, Canada Centre for Inland Waters, Burlington, Ontario.

King, R.B., Fordyce, J.S., Antoine, A.C., et al. (1976). "Elemental composition of airborne particulates and source identification: An extensive one year survey." *J. Air Pollut. Control Assoc.*, **26**, 1073–1078.

Klein, D.H. (1975). "Fluxes, residence times, and sources of some elements to Lake Michigan." *Water Air Soil Pollut.* **4**, 3–8.

Kleinman, M.T., Kneip, T.J., Bernstein, D.M., and Eisenbud, M. (1975). "Fallout of toxic trace metals in New York City." In *Proc. First International Conference on Heavy Metals in the Environment.* Toronto, Ontario, pp. 144–152.

Kleinman, M.T., Pasternack, B.S., Eisenbud, M., and Kneip, T.J. (1980). "Identifying and estimating the relative importance of sources of airborne particulates." *Environ. Sci. Technol.* **14**, 62–65.

Konasewich, D., Traversy, W., and Zar, H. (1978). *Organic and Heavy Metal Contamination in*

Lakes Erie, Michigan, Huron and Superior Basins. Status Report, Great Lakes Water Quality Board, International Joint Commission on the Great Lakes, Windsor, Ontario.

Kowalczyk, G.S., Choquette, C.E., and Gordon, G.E. (1978). "Chemical element balances and identification of air pollution sources in Washington, DC." *Atmos. Environ.* **12**, 1143–1153.

Kramer, J.R. (1976). *Fate of Atmospheric Sulfur Dioxide and Related Substances as Indicated by Chemistry of Precipitation.* Department of Geology Report, McMaster University, Hamilton, Ontario.

Kretzschmar, J.G., Delespaul, I., and de Rijck, Th. (1980). "Heavy metal levels in Belgium: A five-year survey." *Sci. Total Environ.* **14**, 85–97.

Kuperman, E.F. (1964). "Maximal allowable hexavalent chromium concentration in atmospheric air." In V.A. Ryazanov and M.S. Gol'dberg, Eds., *Maximum Permissible Concentrations of Atmospheric Pollutants.* Meditsina Press, Moscow, Vol. 8.

Lannefors, H., Hansson, H.C., and Granat, L. (1983). "Background aerosol composition in southern Sweden." *Atmos. Environ.* **17**, 87–101.

Lawson, D.R. (1980). "Impaction surface coatings: Intercomparison and measurements with cascade impactors." *Atmos. Environ.* **14**, 195.

Lee, M.M., Chaudhri, M.A., Rouse, J.L., and Spicer, B.M. (1977). "Environmental and ecological studies using proton activation analysis." *J. Radioanal. Chem.* **37**, 889–896.

Lee, R.E., Patterson, R.K., and Wagman, J. (1968). "Particle size distribution of metal components in air." *Environ. Sci. Technol.* **2**, 288.

Lee, R.E., Crist, H.L., Riley, A.E., and MacLeod, K.E. (1975). "Concentration and size of trace metal emissions from a power plant, a steel plant and a cotton gin." *Environ. Sci. Technol.* **9**, 643–647.

Lepel, E.A., Stefansson, K.M., and Zöller, W.H. (1978). "The enrichment of volatile elements in the atmosphere by volcanic activity: Augustine Volcano 1976." *J. Geophys. Res.* **83**, 6213–6220.

Lindberg, S.E., Andren, A.W., Raridon, R.J., and Fulkerson, W. (1975). "Mass balance of trace elements in Walker Branch Watershed: Relation to coal-fired steam plants." *Environ. Health Perspect.* **12**, 9–18.

Lioy, P.J., Wolff, G.T., and Kneip, T.J. (1978). "Toxic airborne elements in the New York area." *J. Air Pollut. Control Assoc.* **28**, 510–512.

Lioy, P.J., Daisey, J.M., Greenberg, A., and Harkov, R. (1985). "A major wintertime (1983) pollution episode in northern New Jersey." *Atmos. Environ.* **19**, 429–436.

Liu, B.Y.H., and Pui, D.Y.H. (1981). "Aerosol sampling inlets and inhalable particles." *Atmos. Environ.* **15**, 589.

Lum, K.R., Betteridge, J.S., and Macdonald, R.R. (1982). "The potential availability of P, Al, Cd, Co, Cr, Fe, Mn, Ni, Pb, and Zn in urban particulate matter." *Environ. Technol. Lett.* **3**, 57–62.

Maenhaut, W., and Akilimali, K. (1985). "Trace elements in air particulates in Kinshasa (Zaise) and Butare (Rwanda). *Proc. Belgian SCOPE Meeting on Belgian Research on Metals Cycling.* Brussels, October 11–12.

Maenhaut, W., Raemdonck, H., Selen, A., et al. (1983). "Characterization of the atmospheric aerosol over the Eastern Equatorial Pacific." *J. Geophys. Res.* **88** (No. C9), 5353–5364.

MARC (1985). *Historical Monitoring.* Technical Report 31, Monitoring and Research Assessment Center, University of London.

Marple, V.A., and Liu, B.Y.H. (1974). "Characteristics of laminar jet impactors." *Environ. Sci. Technol.* **8**, 648.

Marple, V.A., Liu, B.Y.H., and Whitby, K.T. (1974). "Fluid mechanics of the laminar flow aerosol impactor." *J. Aerosol Sci.* **5**, 1.

McDonald, C., and Duncan, H.J. (1978). "Variability of atmospheric levels of metals in an industrial environment." *J. Environ. Sci. Health* **A13**, 687–695.

McFarland, A.R., Wedding, J.T., and Cermak, J.E. (1977). "Wind tunnel evaluation of a modified Andersen impactor and an all weather sampler inlet." *Atmos. Environ.* **11**, 535.

Menyailov, I.A., Nikitina, L.P., Shapar, V.N., and Miklishanskiy, A.Z. (1982). "The role of active volcanism in enrichment of the atmosphere in chalcophile elements." *J. Geophys. Res.* **87**, 11,113–11,118.

Miklishanskiy, A.Z., Pavlotskaya, F.I., Savel'yev, B.V., and Yakovlev, Yu. V. (1977). "Content and mode of occurrence of trace elements in the near-surface air layer and in atmospheric precipitation." *Geochem. Int.* **14**, 54–62.

Milford, J.B., and Davidson, C.I. (1985). "The sizes of particulate trace elements in the atmosphere: A review." *J. Air Pollut. Control Assoc.* **35**, 1249–1260.

Moyers, J.L., Ranweiler, L.E., Hopf, S.B., and Korte, N.E. (1977). "Evaluation of particulate trace species in southwest desert atmosphere." *Environ. Sci. Technol.* **11**, 789–795.

NAS (1974). *Chromium.* U.S. National Academy of Sciences, Washington, D.C.

Natusch, D.F.S. (1978). "Potentially carcinogenic species emitted to the atmosphere by fossil-fueled power plants." *Environ. Health Perspect.* **22**, 79–90.

Natusch, D.F.S., and Wallace, J.R. (1976). "Determination of airborne particle size distributions: Calculation of cross-sensitivity and discreteness effects in cascade impaction." *Atmos. Environ.* **10**, 315.

Nriagu, J.O., and Davidson, C.I. (1980). "Zinc in the atmosphere." In *Zinc in the Environment.* John Wiley & Sons, New York, Vol. 1, pp. 113–159.

Olmez, I., and Aras, N.K. (1977). "Trace elements in the atmosphere determined by nuclear activation analysis and their interpretation." *J. Radioanal. Chem.* **37**, 671–677.

Pacyna, J.M. (1986). "Source-receptor relationships for trace elements in Northern Europe." *Water Air Soil Pollut.* (in press).

Pacyna, J.M. (1980). *Coal-Fired Power Plants as a Source of Environmental Contamination by Trace Metals and Radionuclides.* Wroclaw Tech. Univ. Ed., Wroclaw, Poland, 1–170.

Pacyna, J.M., and Ottar, B. (1985). "Transport and chemical composition of the summer aerosols in the Norwegian Arctic." *Atmos. Environ.* **19**, 2109–2120.

Pacyna, J.M., Semb, A., and Hanssen, J.E. (1984). "Emission and long-range transport of trace elements in Europe." *Tellus* **36B**, 163–178.

Pacyna, J.M., Ottar, B., Tomza, U., and Maenhaut, W. (1985). "Long-range transport of trace elements to Ny Ålesund, Spitsbergen." *Atmos. Environ.* **19**, 857–865.

Parekh, P.P., and Hussain, L. (1981). "Trace element concentrations in summer aerosols at rural sites in New York State and their possible sources." *Atmos. Environ.* **15**, 1717–1725.

Pattenden, N.J. (1974). *Atmospheric Concentrations and Deposition Rates of Some Trace Elements Measured in the Swansea/Neath/Port Talbot Area.* Environmental and Medical Sciences Division, A.E.R.E. Report R-7729, Her Majesty's Stationery Office, London.

Pattenden, N.J., Branson, J.R., and Fisher, E.M.R. (1982). "Trace element measurement in wet and dry deposition and airborne particulate at an urban site." In H.W. Georgii and J. Pankrath, Eds., *Deposition of Atmospheric Pollutants.* D. Reidel, Dordrecht, Holland, pp. 173–184.

Peirson, D.H., Cawse, P.A., and Cambray, R.S. (1974). "Chemical uniformity of airborne particulate material and a maritime effect." *Nature* **251**, 675–679.

Pitt, R.E., and Amy, G. (1973). *Toxic Materials Analysis of Street Surface Contaminants.* Report No. EPA-R2-73-283, U.S. Environmental Protection Agency, Office of Research and Monitoring, Washington, DC.

Raabe, O.G. (1978). "A general method for fitting size distributions to multicomponent aerosol data using weighted least-squares." *Environ. Sci. Technol.* **12**, 1162.

Rahn, K.A. (1976). *The Chemical Composition of Atmospheric Aerosol.* Technical Report, Graduate School of Oceanography, University of Rhode Island, Kingston, RI.

Rancitelli, L.A., and Perkins, R.W. (1970). "Trace element concentrations in the troposphere and lower stratosphere." *J. Geophys. Res.* **75**, 3055–3064.

Rancitelli, L.A., Cooper, J.A., and Perkins, R.W. (1976). "Multielement characterization of atmospheric aerosols by instrumental neutron activation analysis and x-ray fluorescence analysis." In F. Coulston and F. Korte, Eds., *Environmental Quality and Safety.* Academic Press, New York, Vol. 5, pp. 152–166.

Ritchie, I.M., and Thingvold, D.A. (1985). "Assessment of atmospheric impacts of large-scale copper-nickel development in northeastern Minnesota." *Water Air Soil Pollut.* **25**, 145–160.

Robberecht, H., Deelstra, H., Vanden Berghe, D., and van Grieken, R. (1983). "Metal pollution and selenium distributions in soils and grass near a non-ferrous plant." *Sci. Total Environ.* **29**, 229–241.

Robbins, J.A. (1980). *Sediments of Southern Lake Huron.* Report No. EPA-600/3-80-080, U.S. Environmental Protection Agency, Environmental Research lab., Duluth, Minnesota.

Rondia, D., and Closset, J. (1985). "Aerosol versus solution composition in occupational exposure." *Sci. Total Environ.* **46**, 107–112.

Ronneau, C. (1985). "Transfer of anthropogenic airborne elements towards the rural environment." *Agric. Ecosystems Environ.* **13**, 191–203.

Ronneau, C., and Cara, J. (1984). "Correlations of element deposition on pastures with analysis of cow's hair." *Sci. Total Environ.* **39**, 135–142.

Ruhling, A., and Tyler, G. (1984). "Recent changes in the deposition of heavy metals in northern Europe." *Water Air Soil Pollut.* **22**, 173–180.

Ruppert, H. (1975). "Geochemical investigations on atmospheric precipitation in a medium-sized city (Gottingen, F.R.G.)." *Water Air Soil Pollut.* **4**, 447–460.

Sadasivan, S. (1981). "Trace elements in size separated atmospheric particulates at Trombay, Bombay, India." *Sci. Total Environ.* **20**, 109–115.

Saltzman, B.E., Cholak, J., Schafer, L.J., et al. (1985). "Concentrations of six metals in the air of eight cities." *Environ. Sci. Technol.* **19**, 328–333.

Scheff, P.A., Wadden, R.A., and Allen, R.J. (1984). "Development and validation of chemical element mass balance for Chicago."*Environ. Sci. Technol.* **18**, 923–931.

Schneider, B. (1985). "Sources of atmospheric trace metals over the Subtropical North Atlantic." *J. Geophys. Res.* **90**, 10,744–10,746.

Schramel, P., Samsahl, K., and Pavlu, J. (1974). "Determination of 12 selected microelements in air particles by neutron activation analysis." *J. Radioanal. Chem.* **19**, 329–337.

Schutyser, P., Maenhaut, W., and Dams, R. (1978). "Instrumental neutron activation analysis of dry atmospheric fall-out and rain-water." *Anal. Chim. Acta* **100**, 75–85.

Sehmel, G.A., and Hodgson, W.H. (1978). *A Model for Predicting Dry Deposition of Particles and Gases to Environmental Surfaces.* Battelle Pacific Northwest Labs, Report PNL-SA-6721.

Semb, A. (1978). *Deposition of Trace Elements from the Atmosphere in Norway.* SNSF Project Report FR 13/78, Norwegian Institute for Air Pollution Research, Lillestrøm, Norway.

Shani, G., and Haccoun, A. (1976). "Nuclear methods used to compare air pollution in a city and a pollution-free area."

Sievering, H., Dave, M., Dolske, D., and McCoy, P. (1981). "Transport and dry deposition of trace metals over southern Lake Michigan." In S.J. Eisenreich, Ed., *Atmospheric Pollutants in Natural Waters.* Ann Arbor Sci., Ann Arbor, Michigan, pp. 285–325.

Soderlund, R. (1975). *Some Preliminary Views of Atmospheric Transport of Matter to the Baltic Sea.* Report No. AC-31 MISU-IMI, Arrhenius Laboratory, Stockholm, Sweden.

Stern, R.M. (1982). "Chromium compounds: Production and occupational exposure." In S. Langård, Ed., *Biological and Environmental Aspects of Chromium.* Elsevier Biomedical Press, Amsterdam, pp. 16–47.

Tanner, T.M., Rancitelli, L.A., and Haller, W.A. (1972). "Multielement analysis of natural waters by neutron activation, group chemical separations, and gamma ray spectrometric techniques." *Water Air Soil Pollut.* 1, 132–143.

Tanner, T.M., Young, J.A., and Cooper, J.A. (1974). "Multielement analysis of St. Louis aerosols by nondestructive techniques." *Chemosphere* 5, 211–220.

Tate, M.B., and Bates, M.H. (1984). "Bulk deposition of metals in Tulsa, Oklahoma." *Water Air Soil Pollut.* 22, 15–26.

Thompson, M., Walton, S.J., and Wood, S.J. (1979). "Statistical appraisal of interference effects in the determination of trace elements by atomic absorption spectrophotometry in applied geochemistry." *Analyst* 104, 299.

Thornton, J.D., Eisenreich, S.J., Munger, J.W., and Gorham, E. (1980). "Trace metal and strong acid composition of rain and snow in northern Minnesota. In S.J. Eisenreich, Ed., *Atmospheric Pollutants in Natural Waters.* Ann Arbor Science, Ann Arbor, Michigan, pp. 261–283.

Tomza, U., Maenhaut, W., and Cafmeyer, J. (1982). "Trace elements in atmospheric aerosols at Katowice, Poland." In D.D. Hemphill, Ed., *Trace Substances in Environmental Health.* University of Missouri, Columbia, Vol. 16, pp. 105–115.

Trindade, H.A., Pfeiffer, W.C., Londres, H., and Costa-Ribeiro, C.L. (1981). "Atmospheric concentration of metals and total suspended particulates in Rio de Janeiro." *Environ. Sci. Technol.* 15, 84–89.

U.S. EPA (1982). *Air Quality Criteria for Particulate Matter and Sulfur Oxides.* EPA Report No. 600/8-82-029a, Vol. 3.

Usero, J., and Gracia, I. (1986). "Trace and major elements in atmospheric deposition in the 'Campo de Gibralter' region." *Atmos. Environ.* 20, 1639–1646.

Walsh, P.R., Rahn, K.A., and Duce, R.A. (1978). "Erroneous elemental mass-size functions from a high-volume cascade impactor." *Atmos. Environ.* 12, 1793.

Wangen, L.E. (1981). "Elemental composition of size-fractionated aerosols associated with a coal-fired power plant plume and background." *Environ. Sci. Technol.* 15, 1080.

Ward, N.I., Brooks, R.R., Roberts, E., and Boswell, C.R. (1977). "Heavy-metal pollution from automotive emissions and its effect on roadside soils and pasture species in New Zealand." *Environ. Sci. Technol.* 9, 917–920.

Waughman, G.J., and Brett, T. (1980). "Interference due to major elements during the estimation of trace heavy metals in natural materials by atomic absorption spectrophotometry." *Environ. Res.* 21, 385.

Wedding, J.B., McFarland, A.R., and Cermak, J.E. (1977). "Large particle collection characteristics of ambient aerosol samplers." *Environ. Sci. Technol.* 11, 387.

Wesolowski, J.J., Alcocer, A.E., and Appel, B.R. (1980). "The validation of the Lundgren impactor." *Adv. Environ. Sci. Technol.* 9, 125–145.

Winchester, J.W., and Bi, M-T. (1984). "Fine and coarse aerosol composition in an urban setting: A case study in Beijing, China." *Atmos. Environ.* 18, 1399–1409.

Winchester, J.W., and Nifong, G.P. (1971). "Water pollution in Lake Michigan by trace elements from pollution aerosol fallout." *Water Air Soil Pollut.* 1, 50–64.

Winchester, J.W., Ferek, R.J., Lawson, D.R., et al. (1981). "Comparison of aerosol sulfur and crustal element concentrations in particle size fractions from continental U.S. locations." *Water Air Soil Pollut.* 12, 431–440.

Yousef, Y.A., Wanielista, M.P., Hvitveo-Jacobsen, T., and Harper, H.H. (1984). "Fate of heavy metals in stormwater runoff from highway bridges." *Sci. Total Environ.* **33**, 233–244.

Zhang, J., Billiet, J., and Dams, R. (1985). "Elemental composition and source investigation of particulates suspended in the air of an iron foundry." *Sci. Total Environ.* **41**, 13–28.

Zöller, W.H., and Gordon, G.E. (1970). "Instrumental neutron activation analysis of atmospheric pollutants utilizing Ge(le) y-ray detectors." Anal. Chem. **42**, 257.

Zöller, W.H., Gladney, E.S., and Duce, R.A. (1974). "Atmospheric concentrations and sources of trace metals at the South Pole." *Science* **183**, 198–200.

6

GEOCHEMISTRY OF CHROMIUM IN THE OCEANS

Lawrence M. Mayer

Program in Oceanography
Center for Marine Studies
Ira C. Darling Center
University of Maine
Walpole, Maine

1. INTRODUCTION

The chemical cycling of chromium in the oceans has received less attention than that of many of the first-row transition metals. This lack of attention probably results from a combination of analytic difficulties and the lack of pressing ecologic or environmental implications of chromium in the sea. Nevertheless, the few data sets available indicate a cycling rich in a variety of geochemical transformations, ranging from oxidation–reduction processes to particulate scavenging. This review assesses recent developments in the state of our knowledge of this cycling and points out the areas where deficiencies remain.

The geochemical aspects of the ocean of major pertinence to chromium cycling are that the sea is largely an oxidizing environment of low suspended particulate content. The exceptions to this generalization are the rare zones of anoxia found in areas of poor circulation and the sedimentary environments underlying the water column. Along with estuaries, which are typically oxidizing but often have substantial suspended particulate concentrations, these special environments are considered separately.

Several features of chromium chemistry cause it to respond in a unique fashion to this geochemical milieu. First, the element is sufficiently close to thermo-dynamic stability for both its trivalent and hexavalent forms to be found in considerable amounts. The unstable trivalent form can be maintained for extended periods of time due to either (1) slow kinetics of oxidation in the absence of suitable catalysts or (2) protection from oxidation by complexing with organic ligands. Second, the two forms differ enormously in their reactivity toward association with various organic and particulate substrates. The trivalent, cationic form is very reactive with respect to these reactions due in large part to its high charge density. The hexavalent, anionic form is quite unreactive and likely participates in such associations only after an initial reduction to the trivalent form. The hexavalent form is reduced to the trivalent form if a sufficiently reactive reductant is present but is evidently able to coexist for extended periods of time in the presence of dissolved organic matter in the oceans. Third, chromium is of little use in the biochemistry of most organisms in the sea, with a consequently low intensity of incorporation into the cycle of biologic uptake and remineralization that is typical of many transition metals.

2. CHROMIUM IN THE OCEANIC WATER COLUMN

2.1 Analytic Considerations

Sampling for chromium in the water column has been carried out largely with conventional sampling gear. Niskin or Go-Flo bottles on steel wire have been used for subsurface water samples, and surface samples often have been collected with handheld polyethylene bottles. No significant contamination with dissolved chromium has been found using steel cable (Jeandel and Minster, 1984), although contamination with particulate chromium has been noted (Martin and Knauer, 1973).

The analytic schemes found in the literature can be divided into three steps: (1) an optional filtration of the water sample, (2) a preconcentration step, and (3) detection and quantitation of the concentrated chromium.

Filtration has been carried out by most workers, but samples from the open ocean usually have shown no difference in concentration between filtered and unfiltered samples. It is perhaps advisable for such samples that this step be eliminated because of the potential for either loss of chromium to adsorption or contamination of the sample by the filtration assembly.

Analysis for dissolved chromium rarely has been achieved without pre-concentration steps. A variety of preconcentration methods have been used that generally rely on either a coprecipitation–adsorption with an oxide or a complexing with a chelating agent, which then is removed by solvent extraction. The most commonly used of these methods is coprecipitation with iron oxyhydroxides (Chuecas and Riley, 1966; Fukai and Vas, 1967; Grimaud and Michard, 1974; Shigematsu et al., 1977; Cranston and Murray, 1978; Cranston, 1980, 1983; Nakayama et al., 1981a; Murray et al., 1983). The most common variant of this technique uses two separate precipitation steps. A ferric salt can be added to the seawater sample to form ferric oxyhydroxides, which quantitatively scavenge dissolved, free (i.e., not complexed by organic ligands) trivalent chromium. A ferrous salt can be added that will scavenge both trivalent and hexavalent chromium, the latter by first reducing chromate to the trivalent form. Both stable and radioactive chromium spikes have been used to demonstrate that this method does indeed collect the species intended, under conditions of normal seawater, with excellent collection efficiencies and specificities. Coprecipitation with aluminum hydroxides (Shigematsu et al., 1977) has been used to concentrate trivalent chromium; hexavalent chromium similarly can be collected after an initial reduction step. Nakayama et al. (1981a) used hydrated bismuth oxide to collect hexavalent chromium, with free and organically complexed forms then collected with the same oxide after an exhaustive oxidation of the sample. Trivalent and hexavalent chromium has been collected using an initial acidification and reduction step followed by chelation of the resulting trivalent form with trifluoroacetylacetone and extraction of the chelate into hexane (Siu et al., 1983). DeJong and Brinkmann (1978) collected trivalent and hexavalent chromium separately using methyltri-n-alkylammonium salts under different solution conditions. Campbell and Yeats (1981, 1984) first oxidized the sample and then extracted the resulting hexavalent chromium with similar tri-n-alkylamines.

2.2 Open Ocean

2.2.1 Observed Distributions and Speciation

Use of the methods summarized has yielded a picture of chromium distribution in the open ocean that is only partially clear or consistent. In my opinion, the following conclusions appear warranted at this time. Total dissolved chromium concentrations generally fall in the range of 2 to 10 nM, with exceptions found at boundaries of the open ocean. Some profiles are shown in Figure 1. There is usually a depletion of up to a few tens of percent of total dissolved chromium in surface waters relative to the deep sea. This depletion is typical of elements incorporated into the biologic cycle but is of lower intensity for chromium than is found for many transition metals (Boyle et al., 1976, 1977; Bruland and Franks, 1983). An interocean concentration difference often is found for elements taken into the biologic cycle (Bruland and Franks, 1983), with higher concentrations in the deep waters of the Pacific relative to those of the Atlantic, but the data to date do not show this trend for chromium. There is commonly an increase in dissolved chromium concentrations just above the sediment–water interface, indicating regeneration of chromium from the sediments into the water column.

The speciation of chromium in the open ocean is less clear at this time. Data from the past decade appear consistent in showing that dissolved, free, trivalent chromium is in considerably lower concentration than hexavalent chromium (Fig. 1). However, Nakayama et al. (1981a) suggest that many of the data published up to the time of their paper did not quantitate the chromium present as organically complexed forms. Their technique included a persulfate oxidation in order to convert all of the chromium to a form amenable to their coprecipitation technique, resulting in total dissolved chromium values considerably higher than are found in most of the rest of the literature. The difference between the free trivalent plus the hexavalent chromium and the total, oxidized chromium was considered in their work to be organically bound chromium and constituted the major chromium fraction in the samples they studied. The only other study to use a strong oxidation, which would certainly recover all forms of chromium, was that of Campbell and Yeats (1981). The total dissolved chromium values of generally less than 5 nM found by Campbell and Yeats, however, are comparable with levels found by most other authors.

The question of the existence and importance of organically bound chromium is, therefore, an important one. This question cannot be answered definitively with the literature now available but does bear some discussion. Studies that have specifically searched for organically bound chromium in the open ocean water column are almost nonexistent. Mackey (1984) examined a number of water samples from the Australian shelf using SEP-PAK cartridges to remove hydrophobic organic compounds from seawater. He found values for organically bound chromium ranging from 0 to 160 pM, thereby at least indicating the existence of such compounds in these coastal waters. It should be noted, however, that SEP-PAK separations do not remove all, or even the major fraction, of the organic compounds from seawater (Fu and Pocklington, 1983).

It is worth examining the probability that the various preconcentration methods in use will extract organically bound chromium from seawater. The various iron oxyhydroxide coprecipitation methods should scavenge some organic material, since seawater organics have repeatedly been shown to adsorb

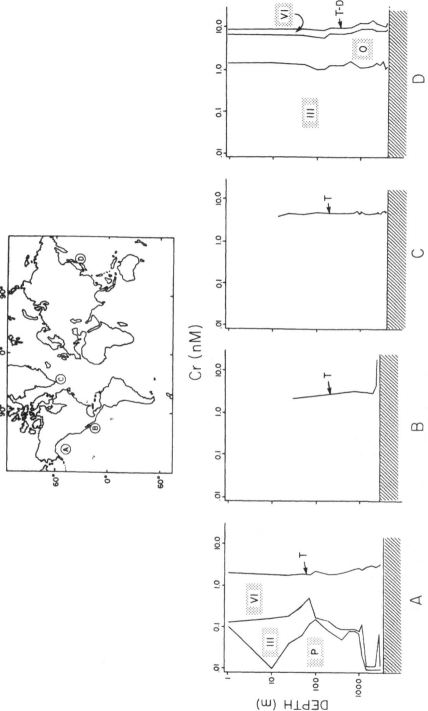

Figure 1. Chromium concentrations versus depth for four locations in the deep sea. Lines represent cumulative concentrations in plots A and D. P, particulate; III, dissolved, free, trivalent; VI, dissolved hexavalent; O, organic; T, total; T-D, total dissolved. These definitions are the operational descriptors used by the original authors. (*Sources:* A, Cranston, 1983; B, Jeandel and Minster, 1984; C, Campbell and Yeats, 1981; D, Nakayama et al., 1981a.)

177

readily onto these oxides (Jeffrey and Hood, 1958; Bader et al., 1960). However, organic chromium compounds are unlikely to be quantitatively scavenged by the recipes used for chromium extraction, for two reasons. First, the amount of iron used in most of these recipes, typically a few milligrams of iron per liter of seawater, is not sufficient to remove completely the dissolved organic matter from seawater. Second, Nakayama et al. (1981d) showed that organic complexes of chromium, including some with humic acids, can escape complete coprecipitation with iron oxyhydroxides. It, therefore, seems likely that organically bound chromium is partially, but probably not completely, scavenged by iron oxyhydroxides. Evidence for the incorporation of organically bound chromium into iron oxyhydroxides comes from the study of Emerson et al. (1979), in which the ferrous hydroxide coprecipitation method collected a significant amount of chromium inferred to be hexavalent. This hexavalent fraction evidently coexisted with hydrogen sulfide in the water column. Given the demonstrated ability of hydrogen sulfide to reduce hexavalent chromium (Smillie et al., 1981), it seems more likely that this fraction was organically bound. Cranston and Murray (1980) cited some unpublished data of Gohda et al. that indicate that iron hydroxides are partial scavengers of organic and colloidal fractions of chromium. Other methods of preconcentrating chromium that involve a preacidification step before removal of trivalent chromium (Jeandel and Minster, 1984; Siu et al., 1983), usually done in order to reduce the hexavalent to the trivalent form or to inhibit adsorption of chromium onto container walls, may release some fraction of the chromium to the free, aquated form. This previously complexed chromium would then be considered free, trivalent chromium.

The reader of this literature on open-ocean concentrations and speciation is thus left with questions. On the one hand, most of the literature indicates total dissolved chromium of rarely no greater than 6 nM and with no indication of organically bound chromium. On the other hand, the two studies with methodologies that truly should have measured total dissolved chromium (Nakayama et al., 1981a; Campbell and Yeats, 1981) yield values that may or may not be consistent with one another, depending on whether the Pacific Ocean has markedly higher levels of chromium than the Atlantic Ocean. Potentially missing is a major fraction of dissolved chromium, which affects conclusions about speciation and cycling of chromium in the deep sea. It also leaves unresolved the old controversy of whether or not the bulk of oceanic chromium is in the thermodynamically favored (Krauskopf, 1956; Elderfield, 1970) hexavalent state. This state of affairs is clearly unsatisfactory and requires concerted study, a point emphasized by Cranston (1983).

2.2.2 Cycling

A schematic diagram of chromium fluxes in the ocean is shown in Figure 2.

Delivery of chromium to the ocean appears to be dominated by riverine pathways, with atmospheric delivery and hydrothermal vent inputs probably

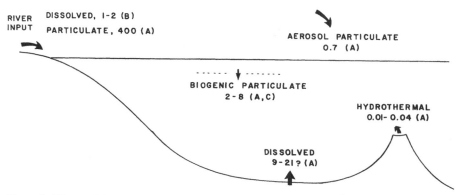

RIVER INPUT   DISSOLVED, 1-2 (B)
PARTICULATE, 400 (A)

AEROSOL PARTICULATE
0.7 (A)

BIOGENIC PARTICULATE
2-8 (A,C)

HYDROTHERMAL
0.01- 0.04 (A)

DISSOLVED
9-21? (A)

Figure 2. Fluxes of chromium in the ocean. All numbers are in units of 10^8 mol/year. (*Sources:* A, Jeandel and Minster, 1984; B, Campbell and Yeats, 1984; C, this work.

accounting for no more than 0.2% of the total (Jeandel and Minster, 1984). Most of the chromium delivered to the ocean by rivers is in the form of particulate material, with delivery of about 4×10^{10} mol year^{-1}. The delivery of dissolved chromium from rivers is uncertain, with estimates ranging from 1 to 8×10^8 mol year^{-1} (Campbell and Yeats, 1984; Martin and Meybeck, 1979; Jeandel and Minster, 1984). The lower end of this range seems more reasonable. Atmospheric delivery appears to be dominated by terrigenous particulate material with chromium concentrations of 2 to 5 times average crustal concentrations (Buat-Menard and Chesselet, 1979; Duce et al., 1983).

As noted previously, chromium participates in the oceanic biologic cycle by incorporation into planktonic material in surface waters and regeneration at depth. Chromium has no major biochemical function, and it is, therefore, not surprising that plankton do not concentrate it to any great extent. The organic fraction of marine plankton apparently has chromium concentrations of less than 1 ppm (Martin and Knauer 1973), which is low for transition metals. The work of several groups (Campbell and Yeats, 1981; Cranston, 1983; Murray et al., 1983) indicates that chromium is better correlated with silica than any of the other major nutrient elements, indicating its probable incorporation into diatom frustules. Consistent with this correlation, Kastner (1981) found the chromium concentration of siliceous tests to be about 8 ppm. This host phase implies a relatively deep water regeneration for chromium, as compared with a metal such as cadmium that is more closely related to tissue nutrients (e.g., phosphate) that have a shallow water regeneration cycle. There is evidently some incorporation into tissue material, however. Cranston (1983) found that although chromium was best correlated with silica, inclusion of phosphate into a multiple regression significantly improved the correlation. It is possible also that carbonate skeletal material is an important transport phase. Turekian et al. (1973) found chromium concentrations of 0.4–5 ppm in pteropod tests, and the vertical flux of calcium carbonate is generally higher than that of silica (Honjo et al., 1982).

The overall flux of biogenic particulate chromium across the thermocline can be estimated in two ways. First, the simple box model approach of Broecker (1974), assuming a concentration difference of 0.5 nM between surface and deep water, yields an estimate of $5–8 \times 10^8$ mol year^{-1}. Second, coupling of the biogenic particulate fluxes presented in Honjo et al. (1982) with the probable chromium concentrations in calcareous, siliceous, and organic particulates implies flux of about $2–4 \times 10^8$ mol year^{-1}. The agreement between these two estimates is reasonable given the simplicity of Broecker's model and the paucity of data with which to calculate actual fluxes.

The removal of chromium from the water column to the sediments occurs through a combination of sedimentation of biologic material and adsorptive scavenging of dissolved chromium onto sedimenting particulates. Given the relative particle reactivities of trivalent and hexavalent chromium, it is not surprising that the trivalent form has been found to exhibit much lower residence times in the water column (Amdurer et al., 1983). The removal of even hexavalent chromium from the water column likely involves a prior reduction to the trivalent form (Mayer and Schick, 1981; van der Weijden and Reith, 1982). Evidence for this pathway in open ocean systems is seen in the minimum in hexavalent chromium concentrations and a corresponding maximum in particulate chromium in the oxygen minimum zone of the eastern Pacific (Murray et al., 1983). Jeandel and Minster (1984) suggested that chromium scavenging may be particularly enhanced onto iron oxides precipitated near submarine hydrothermal vents.

Regeneration of chromium from the sediments to the water column has been inferred from elevated levels of dissolved chromium near the sediment–water interface. Cranston (1983) calculated that the bottom water maximum in the Cascadia Basin could be completely accounted for by sediment regeneration of planktonic debris. Similar maxima observed by Nakayama et al. (1981c) and Jeandel and Minster (1984) were hypothesized to result from an oxidation of trivalent to hexavalent chromium catalyzed by the manganese dioxide present in large concentrations in surficial sediments. The dominance of particulate inputs of chromium to the oceans suggests that sedimentary regeneration of a small fraction of this chromium may be more important than dissolved inputs.

Campbell and Yeats (1984) calculated a residence time of 25,000 to 40,000 years for dissolved chromium in the ocean. This figure was derived by dividing the measured dissolved chromium concentration in the ocean by the riverine delivery of dissolved chromium. This residence time is longer than that for most transition metals, largely due to the lack of biologic incorporation of chromium and the low adsorptivity of the hexavalent form, which comprises a large fraction of the dissolved metal. However, if sedimentary remobilization acts on chromium delivered in particulate form, the actual residence time would be somewhat lower.

2.3 Anoxic Basins

Basins with anoxic deep waters provide particularly useful laboratories in which to study the oxidation–reduction processes affecting chromium. A study in Saanich Inlet, Puget Sound, was used to examine the transformations between the trivalent and hexavalent forms (Cranston and Murray, 1978; Emerson et al., 1979; Cranston, 1980). Using the ferrous and ferric oxide coprecipitation methods described previously, the predicted dominance of the trivalent form in the deep anoxic waters and the hexavalent form in the oxidized surface waters was found. Increasing concentrations of the apparent hexavalent form with depth correlate with increasng sulfide, however, and suggest that a nonhexavalent form was collected by this technique. Decreases in the trivalent form above the oxic–anoxic boundary depth were modeled in order to determine the residence time of trivalent chromium diffusing above this interface. Residence times of trivalent chromium in individual water layers decreased with increasing oxygen concentration and height above the interface (Cranston, 1980), with overall residence times in the oxygenated layer of 6–20 days (Emerson et al., 1979). This range encompasses somewhat lower values than have been found for a variety of other environments (reviewed in Murray et al., 1983), which seems likely due to the presence of considerable concentrations of catalytically active manganese dioxide particulates at the redox interface in anoxic basins (Emerson et al., 1979). Some portion of the trivalent chromium removal may be due also to adsorptive scavenging by the metal oxides in this zone.

2.4 Estuaries

The only recent studies of dissolved chromium in estuarine water columns are those of Cranston and Murray (1980) in the Columbia River estuary and Campbell and Yeats (1984) in the St. Lawrence estuary. The former study, using the ferrous and ferric hydroxide coprecipitation method of Cranston and Murray (1978), found roughly conservative mixing of chromium during mixing of river water and seawater, with the potential for about a fourth of the chromium to be converted to particulate form by either flocculation processes or adsorptive scavenging. The latter study measured only total dissolved chromium, using the method ascribed to them, and found a rapid loss of a major portion of the chromium from the river water during early mixing with seawater (Fig. 3).

A variety of mechanisms can be invoked to explain the loss of chromium from the water column during estuarine mixing. Conversion of riverine, dissolved trivalent chromium to a particulate form during the early stages of mixing with seawater can result from its preferential association in river water with higher molecular weight organic and colloidal material (Mayer et al., 1984; Steinberg, 1980). This material undergoes flocculation quickly upon mixing. Adsorption of dissolved chromium onto suspended particulate matter would have been enhanced by both the higher concentrations of particulates at the lower salinities and the

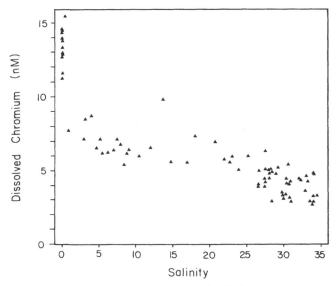

Figure 3. Dissolved total chromium concentrations in the St. Lawrence estuary as a function of salinity.

tendency for both trivalent and hexavalent chromium to adsorb at lower salinities (Mayer and Schick, 1981, and unpublished data). In particular, the tendency for hexavalent chromium to be reduced to the more particle-reactive trivalent form is enhanced at low estuarine salinities relative to both the river water endmember and higher estuarine salinities (Mayer and Schick, 1981).

On the basis of the two reported field studies, no clear picture of chromium passage through the estuarine environment can be drawn.

3. CHROMIUM IN SEDIMENTS

The concentrations of chromium in sediments, typically 60–100 ppm, are similar to those of crustal rock—100 ppm (Wedepohl, 1960), which reflects the lithogenic character of this element. Volcanic ashes can be particularly rich in chromium (Rehm, 1981, cited in Marchig et al., 1982), with concentrations of up to 500 ppm. This lithogenic character is confirmed by the few partitioning studies of the metal (Loring, 1979; Mayer and Fink, 1980), which have tended to show the dominance of aluminosilicate lattices as the host for the bulk of sedimentary chromium. The metal is not known to form carbonate, sulfide, or phosphate minerals. Concentrations in carbonate oozes are, therefore, especially low (Chester and Aston, 1976). Apparently, it is excluded from manganese dioxide phases in the deep sea to a degree anomalous for transition metals (Li, 1982) and thus tends to be in relatively low concentration in metalliferous sediments (Marchig et al., 1982). Feely et al. (1983) found chromium to be enriched in suspended matter also highly enriched in manganese oxides, but it is not clear from their data that the manganese-rich phase was in fact the host for the chromium.

There is very little published work on the sedimentary diagenesis of chromium. Downcore profiles of the metal concentration in the solid phase generally show little evidence of major redistribution due to diagenesis (Kato et al., 1983; Sutherland et al., 1984). Pore water concentrations are usually higher than are found in the water column, with levels reaching tens of nM (Brumsack and Gieskes, 1983; Kahn et al., 1983; Douglas et al., 1986; Hurd et al., personal communication). Complexing of chromium with dissolved organic matter in pore waters may be critical for its remobilizationin at least nearshore systems, as evidenced by the strong correlation between dissolved chromium and gelbstoff (Brumsack and Gieskes, 1983) and its adsorption onto XAD-2 resins (Douglas et al., 1986).

Remobilization of chromium from the sediments to the overlying water column is a process frequently discussed in the literature but not yet actually observed. This remobilization has been suggested to result, in the deep sea, from two processes that are perhaps related. First, siliceous tests with their 8 ppm chromium may simply dissolve at the sediment–water interface. Cranston (1983) calculated that the primary production in the Cascadia Basin could support enough diatom production to deliver the requisite amount of chromium to account for the bottom water enrichments he observed. A similar argument could be made for other biogenic material, such as calcareous skeletons. Second, trivalent chromium in the sediments may be catalytically oxidized by manganese dioxide present at the sediment–water interface, with resultant remobilization of the highly soluble hexavalent chromium. This reaction has been shown in laboratory experiments by Nakayama et al. (1981b), van der Weijden and Reith (1982), and Schroeder and Lee (1975), although to date it has been demonstrated only with dissolved trivalent chromium and particulate manganese dioxide. It would be useful to demonstrate the reaction using solid forms of trivalent chromium, such as those adsorbed by sedimentary iron oxides or organic material. Remobilization of chromium from nearshore sediments may be more intensive than in the deep sea, given the former's higher pore water concentrations (Douglas et al., 1986; Kahn et al., 1983) and irrigation rates.

4. SUMMARY AND CONCLUSIONS

Chromium is present in nanomolar concentrations throughout most of the oceanic water column. Most recent reports indicate a dominance by the hexavalent form of the element, although there are indications that a substantial amount of trivalent chromium may be stabilized by complexing with organic materials. Resolution of this question will require concerted effort to develop and test analytic procedures capable of unambiguously discerning the various inorganic and organic forms of chromium, and applying these procedures in a consistent fashion to different provinces of the ocean.

Chromium enters the ocean primarily in particulate form from rivers, which is rapidly sedimented and lost from the system. Dissolved chromium in rivers in partially scavenged in its passage through turbid coastal environments. Chromium is lost from the oceanic water column through some combination of incorporation into biologic material, primarily siliceous skeletons, and adsorption onto sedimenting particulates. Dissolution of this incorporated chromium occurs both in the water column and at the sediment–water interface, leading to deep and bottom water enrichments. Reduction of hexavalent chromium occurs in areas of high reductant concentration, such as anoxic basins and the oxygen minimum zone, and may lead to a major removal pathway from the ocean. Another potentially important removal mechanism is adsorption onto freshly formed iron oxides near hydrothermal areas.

Sedimentary chromium is dominated by its relatively inert presence in detrital mineral lattices. This dominance makes it difficult to discern authigenic and diagenetic reactions involving the element. Evidence points to some remobilization by (1) oxidative reactions that solubilize chromium by conversion to the hexavalent form, if catalyzed by manganese dioxide, and (2) organic complexation reactions that can solubilize trivalent chromium.

ACKNOWLEDGMENTS

I thank M. Wells and L. Schick, who assisted in the preparation of this manuscript. Partial support for this work came from the Land Water Resources Center, University of Maine, and the U.S. Department of the Interior. Contribution No. 191 from the Ira C. Darling Center.

REFERENCES

Amdurer, M., Adler, D., and Santschi, P.H. (1983). "Studies of the chemical forms of trace elements in sea water using radiotracers." In C.S. Wong, E. Boyle, K. Bruland, et al., Eds., *Trace Metals in Sea Water*. Plenum Press, New York, pp. 537–562.

Bader, R.G., Hood, D.W., and Smith, J.B. (1960). "Recovery of dissolved organic matter in seawater and organic sorption by particulate material." *Geochim. Cosmochim. Acta* 19, 236–243.

Boyle, E.A., Sclater, F., and Edmond, J.M. (1976). "On the marine geochemistry of cadmium." *Nature* 263, 42–44.

Boyle, E.A., Sclater, F.R., and Edmond, J.M. (1977). "The distribution of dissolved copper in the Pacific." *Earth Planet. Sci. Lett.* 37, 38–54.

Broecker, W.S. (1974). *Chemical Oceanography*. Harcourt, Brace, Jovanovich, New York, 214 pp.

Bruland, K., and Franks, R.P. (1983). "Mn, Ni, Cu, Zn and Cd in the western north Atlantic." In

C.S. Wong, E. Boyle, K. Bruland, et al., Eds., *Trace Metals in Sea Water*. Plenum Press, New York, pp. 395–414.

Brumsack, H.J., and Gieskes, J.M. (1983). "Interstitial water trace-metal chemistry of laminated sediments from the Gulf of California, Mexico." *Mar. Chem.* **14**, 89–106.

Buat-Menard, P., and Chesselet, R. (1979). "Variable influence of the atmospheric flux on the trace metal chemistry of oceanic suspended matter." *Earth Planet. Sci. Lett.* **42**, 399–411.

Campbell, J.A., and Yeats, P.A. (1981). "Dissolved chromium in the northwest Atlantic Ocean." *Earth Planet. Sci. Lett.* **53**, 427–433.

Campbell, J.A., and Yeats, P.A. (1984). "Dissolved chromium in the St. Lawrence estuary." *Est. Coast. Shelf Sci.* **19**, 513–522.

Chester, R., and Aston, S.R. (1976). "The geochemistry of deep-sea sediments." In J.P. Riley and R. Chester, Eds., *Chemical Oceanography*. Academic Press, London, pp. 281–390.

Chuecas, L., and Riley, J.P. (1966). "The spectrophotometric determination of chromium in sea water." *Anal. Chim. Acta* **35**, 240–246.

Cranston, R.E. (1980). "Cr species in Saanich and Jervis Inlets." In Freeland, H.J., D.M. Farmer, and C.D. Levings, Ed., *Fjord Oceanography*. Plenum Press, New York, pp. 689–692.

Cranston, R.E. (1983). "Chromium in Cascadia Basin, northeast Pacific Ocean." *Mar. Chem.* **13**, 109–125.

Cranston, R.E., and Murray, J.W. (1978). "The determination of chromium species in natural waters." *Anal. Chim. Acta* **99**, 275–282.

Cranston, R.E., and Murray, J.W. (1980). "Chromium species in the Columbia River and estuary." *Limnol. Oceanogr.* **25**, 1104–1112.

Douglas, G.S., Mills, G.L., and Quinn, J.G. (1986). "Organic copper and chromium complexes in the interstitial waters of Narragansett Bay sediments." *Mar. Chem.* **19**, 161–174.

Duce, R.A., Arimoto, R., Ray, B.J., et al. (1983). "Atmospheric trace elements at Enewetak Atoll. 1. concentrations, sources, and temporal variability." *J. Geophys. Res.* **88**, 5321–5342.

Elderfield, H. (1970). "Chromium speciation in sea water." *Earth Planet. Sci. Lett.* **9**, 10–16.

Emerson, S., Cranston, R.E., and Liss, P.S. (1979). "Redox species in a reducing fjord: Equilibrium and kinetic considerations." *Deep-Sea Res.* **26A**, 859–878.

Feely, R.A., Massoth, G.J., Paulson, A.J., and Gendron, J.F. (1983). "Possible evidence for enrichment of trace elements in the hydrous manganese oxide phases of suspended matter from an urbanized embayment." *Est. Coast. Shelf Sci.* **17**, 693–708.

Fu, T., and Pocklington, R. (1983). "Quantitative adsorption of organic matter from seawater on solid matrices." *Mar. Chem.* **13**, 255–264.

Fukai, R., and Vas, D. (1967). "A differential method of analysis for trivalent and hexavalent chromium in sea water." *J. Oceanogr. Soc. Japan* **23**, 298.

Grimaud, D., and Michard, G. (1974). "Concentration du chrome dans deux profils de l'Ocean Pacifique." *Mar. Chem.* **2**, 229–237.

Honjo, S., Manganini, S.J., and Cole, J.J. (1982). "Sedimentation of biogenic matter in the deep ocean." *Deep-Sea Res.* **29**, 609–625.

Jeandel, C., and Minster, J.-F. (1984). "Isotope dilution measurement of inorganic chromium(III) and total chromium in seawater." *Mar. Chem.* **14**, 347–364.

Jeffrey, L.M., and Hood, D.W. (1958). "Organic matter in sea water: An evaluation of various methods for isolation.." *J. Mar. Res.* **17**, 247–271.

Jong, G.J. de, and Brinkman, U.A.Th. (1978). "Determination of chromium(III) and chromium(VI) in sea water by atomic absorption spectrometry." *Anal. Chim. Acta* **98**, 243–250.

Kahn, D., Heggie, D., and Bender, M. (1983). "Trace metals (Mn, Cu, Ni, Cd, V, Cr) in surficial pore waters (from MANOP site M of the eastern Pacific)." *EOS* **64**, 243.

Kastner, M. (1981). "Authigenic silicates in deep-sea sediments: Formation and diagenesis." In C. Emiliani, Ed., *The Sea*. John Wiley & Sons, New York, Vol. 7, 939 pp.

Kato, T., Endo, M., and Kato, M. (1983). "Vertical distribution of various elements in sediment cores from the Japan Sea." *Mar. Geol.* 53, 277–290.

Krauskopf, K.B. (1956) "Factors controlling the concentration of thirteen rare metals in sea water." *Geochim. Cosmochim. Acta* 9, 1–32.

Li, Y.-H. (1982). "Interelement relationship in abyssal Pacific ferromanganese nodules and associated pelagic sediments." *Geochim. Cosmochim. Acta* 46, 1053–1060.

Loring, D.H. (1979). "Geochemistry of cobalt, nickel, chromium, and vanadium in the sediments of the estuary and open Gulf of St. Lawrence." *Can. J. Earth Sci.* 16, 1196–1209.

Mackey, D. (1984). "Trace metals and the productivity of shelf waters off North West Australia." *Aust. J. Mar. Freshw. Res.* 35, 505–516.

Marchig, V., Gundlach, H., Möller, P., and Schley, F. (1982). "Some geochemical indicators for discrimination between diagenetic and hydrothermal metalliferous sediments." *Mar. Geol.* 50, 241–256.

Martin, J.H., and Knauer, G.A. (1973). "The elemental composition of plankton." *Geochim. Cosmochim. Acta* 37, 1639–1653.

Martin, J.M., and Meybeck, M. (1979). "Elemental mass-balance of material carried by major world rivers." *Mar. Chem.* 7, 173–206.

Mayer, L.M., and Fink, L.K. Jr. (1980). "Granulometric dependence of chromium accumulation in estuarine sediments in Maine." *Est. Coast. Mar. Sci.* 11, 491–503.

Mayer, L.M., and Schick, L.L. (1981). "Removal of hexavalent chromium from estuarine waters by model substrates and natural sediments." *Environ. Sci. Technol.* 15, 1482–1484.

Mayer, L.M., Schick, L.L., and Chang, C.A. (1984). "Incorporation of trivalent chromium into riverine and estuarine colloidal material." *Geochim. Cosmochim. Acta* 48, 1717–1722.

Murray, J.W., Spell, B., and Paul, B. (1983). "The contrasting geochemistry of manganese and chromium in the Eastern Tropical Pacific Ocean." In C.S. Wong, E. Boyle, K. Bruland, J.D. et al., Eds., *Trace Metals in Seawater NATO Conference Series, IV*. Plenum Press, New York, Vol. 9, pp. 643–670.

Nakayama, E., Kuwamoto, T., Tokoro, H., and Fujinaga, T. (1981a). "Chemical speciation of chromium in sea water. Part 3. The determination of chromium species." *Anal. Chim. Acta* 131, 247–254.

Nakayama, E., Kuwamoto, T., Tsurubo, S., and Fujinaga, T. (1981b). "Chemical speciation of chromium in sea water. Part 2. Effects of manganese oxides and reducible organic materials on the redox processes of chromium." *Anal. Chim. Acta* 130, 401–404.

Nakayama, E., Tokoro, H., Kuwamoto, T., and Fujinaga, T. (1981c). "Dissolved state of chromium in seawater." *Nature* 290, 768–770.

Nakayama, E., Kuwamoto, T., Tsurubo, S., et al. (1981d). "Chemical speciation of chromium in sea water. Part 1. Effect of naturally occurring organic materials on the complex formation of chromium(III)." *Anal. Chim. Acta* 130, 289–294.

Schroeder, D.C., and Lee, G.F. (1975). "Potential transformations of chromium in natural waters." *Water Air Soil Pollut.* 4, 355–365.

Shigematsu, T., Gohda, S., Yamazaki, H., and Nishikawa, Y. (1977). "Spectrophotometric determination of chromium(III) and chromium(VI) in sea water." *Bull. Inst. Chem. Res. Kyoto Univ.* 55, 429–440.

Siu, K.W.M., Bednas, M.E., and Berman, S.S. (1983). "Determination of chromium in seawater by isotope dilution gas chromatography/mass spectrometry." *Anal. Chem.* 55, 473–476.

Smillie, R.H., Hunter, K., and Loutit, M. (1981). "Reduction of chromium(VI) by bacterially produced hydrogen sulphide in a marine environment." *Water Res.* 15, 1351–1354.

Steinberg, C. (1980). "Species of dissolved metals derived from oligotrophic hard water." *Water Res.* **14**, 1239–1250.

Sutherland, H.E., Calvert, S.E., and Morris, R.J. (1984). "Geochemical studies of the recent sapropel and associated sediment from the Hellenic Outer Ridge, eastern Mediterranean Sea. I. Mineralogy and chemical composition." *Mar. Geol.* **56**, 79–92.

Turekian, K.K., Katz, A., and Chan, L. (1973). "Trace element trapping in pteropod tests." *Limnol. Oceanogr.* **18**, 240–249.

Wedepohl, K.H. (1960). "Spurenanalytische Untersuchungen an Tiefseetonen aus dem Atlantik." *Geochim. Cosmochim. Acta* **18**, 200–231.

Weijden, C.H. van der, and Reith, M. (1982). "Chromium(III)—Chromium(VI) interconversions in seawater." *Mar. Chem.* **11**, 565–572.

7

OCCURRENCE AND DISTRIBUTION OF CHROMIUM IN NATURAL WATERS OF INDIA

B.K. Handa

Water Research Institute
Chandigarh, India

Editors' Note: The data in this chapter are generally higher than the concentrations that are now being reported by laboratories that have adopted stringent contamination control procedures in their trace metal analysis (see, e.g. Chapter 6, this volume). This chapter emphasizes the differences in the quality of the available databases that have shaped our understanding of the biogeochemistry of chromium in the environment.

1. INTRODUCTION

Despite the great importance of chromium in human and animal metabolism, not much information is available on the chromium content of natural waters in India. However, chromium salts are being used extensively by industry, and waste effluents are being released into the Indian environment. Handa (1978a,b) first described the pollution of groundwaters by chromium in Ludhiana (Punjab). Kakar and Bhatnagar (1981) have confirmed these findings, and more recently, Handa (1983a,b; 1984) and Handa et al. (1983a,b) have identified other areas in India where groundwater pollution by chromium has been detected. This paper discusses the occurrence and distribution of chromium in rainwater, surface waters, and groundwaters in some parts of India, with particular reference to areas where groundwater pollution by chromium has been detected.

2. EXPERIMENTAL DETAILS AND PRESENTATION OF DATA

Rainwater samples were collected in large polyethylene bottles fitted with large-diameter polyethylene funnels. These bottles were placed on a raised platform (about 2 m above ground level) to prevent contamination of the collected rainwater samples by splashes from the ground. The samples were filtered through Whatman No. 42 filter paper and acidified with reagent grade HNO_3 to pH 1.5.

Surface water samples were collected in clean 2.5-L polyethylene bottles and acidified with reagent grade HNO_3 to pH 1.5, usually without filtration, in order to determine total chromium in surface waters. Filtration through Whatman filter No. 42 paper before acidification was done wherever visible suspended matter was observed. Samples were collected from midstream with the samplers in boats.

Groundwater samples from dug wells, handpumps, and tubewells were collected in 2.5-L polyethylene bottles and acidified with reagent grade HNO_3 to pH 1.5 without prior filtration. Both the handpump and tubewell water samples were collected after the pump had been operated for 3–5 minutes and the water allowed to run to waste.

Most of the industrial waste effluents were found to contain considerable suspended matter. All these samples were filtered through Whatman No. 42 filter paper and then acidified to pH 1.5 with reagent grade HNO_3. Because of certain constraints, the sampling had to be done outside the factory premises at some convenient outlet point or drain.

An aliquot of the sample was concentrated at low heat, cooled, and made up to specific volume with distilled and deionized water. The chromium concentration was determined with an atomic absorption spectrophotometer (Perkin Elmer model 306) using the 357.9 nm resonance line and air-acetylene (reducing) flame.

Because there were many samples analyzed, frequency diagrams were prepared, assuming normal distribution of chromium in natural waters. Averages and standard deviations were also computed.

3. CHROMIUM IN RAINWATER

3.1 Sources of Chromium in the Indian Atmosphere

Little information is available on the release of chromium from natural and anthropogenic sources or on the dispersion of the chromium in the Indian subcontinent. A very rough calculation shows that the environmental release of chromium to the atmosphere is quite low, except locally, where specific emitting sources exist. For example, with consumption of 130 million t of coal annually and assuming that the average chromium content in coal is 10 mg/kg (Bertine and Goldberg, 1971; Taylor et al., 1979) and 10% is lost to the atmosphere, the total release to the atmosphere will be only 130 t, most of which will be present in the fly ash (Furr, et al., 1977; Natusch et al., 1975). Similarly, burning of oil (average chromium content 0.3 mg/kg; 10% loss to the atmosphere), which on the Indian subcontinent totals 30 million t annually, would release only 1 t of chromium to the environment.

With 20 million t of cement production and 10 million t of refractories, the loss to the atmosphere is around 100 t. In the production of ferrochromes (annual production 30,961 t), the efficiency of dust filters determines the mineral losses, which can be as high as 10–15% for low-efficiency filters. Mining activities also can release some chromium dust into the environment. However, most of these losses are in particulate aerosol forms and, as such, are located or concentrated near the places of emission in Orissa and Andhra. Their dispersal to distant places depends considerably on weather conditions, topography, and so on.

Forest fires (70,000 ha annually) and burning of municipal solid waste contribute small amounts of chromium to the atmosphere, although no specific data about India are available. Freedman and Hutchinson (1981) have estimated that incineration of refuse contributes 0.29 kg chromium per day to the atmosphere.

A study in Lucknow (U.P.) and New Delhi has shown that the chromium concentration in rainwater is quite small. Nearly 80% of the rainwater samples analyzed had less than 1 μg chromium L^{-1}. In fact, of 94 rainwater samples analyzed, chromium was not detected in 75% (i.e., <0.1 μg chromium L^{-1}). Further, the chromium content in rainwater samples from Lucknow averaged 1.0 μg chromium in L^{-1}, contrast to that from Delhi, which averaged 5 μg chromium L^{-1}. These values can be compared with the values quoted by Taylor et al. (1979) of 2–4 μg chromium L^{-1} on a global basis. The high values in Delhi could be attributed to anthropogenic sources (industrial, including a thermal power plant). The annual wet deposition of chromium in Lucknow (100 cm average rainfall; chromium content 1 μg L^{-1}) is 10 g/ha year^{-1}, whereas for Delhi the figure is 30 g/ha year^{-1} (average rainfall 60 cm; average chromium content 5 μg chromium L^{-1}. These values are comparable with the value of 12 g/ha year^{-1} computed by Navaree et al. (1980) for Belgium and 12–22 g/ha year^{-1} given by Scheffler and Schachtschabel (1982) for the Selling Forest area.

4.   CHROMIUM IN GROUNDWATER

4.1   Natural Sources of Chromium

Apart from wet precipitation and dry fallout, weathering of rocks constitutes the most important natural source of chromium in groundwater. Chromium is widely distributed in nature, with an average concentration in the continental crust of 125 mg kg^{-1} (Sittig, 1980). Hem (1970), quoting Horn and Adams (1966), gives 198 mg kg^{-1} chromium in igneous rocks, 120 mg kg^{-1} chromium in sandstone, 423 mg kg^{-1} chromium in shale, and 7.1 mg kg^{-1} chromium in carbonate rocks. According to Bertine and Goldberg (1971), 200,000 t of chromium are released to the environment by weathering processes. If it is assumed that 5% of this figure relates to the Indian subcontinent, with 178 million ha of annual renewable water resources, the average chromium content of natural waters (assuming no loss) is 6 μg chromium L^{-1}. Of course, the actual value is likely to be much lower because of adsorption and the fact that thermodynamic considerations show that at pH 8 and at Eh slightly lower than that imparted by saturation with oxygen, the chromium content should be less than 0.5 μg L^{-1} (Hem, 1970). However, Cr(VI) can occur at much higher concentrations depending on Eh and pH of water (Fig. 1).

4.2 Occurrence of Chromium in Groundwater in India

The studies on groundwater occurrence of chromium were confined to Uttar Pradesh, Rajasthan, Jammu, and Kashmir, Punjab, Haryana, Delhi, coastal areas of Saurashtra in Gujarat, and a few samples from Maharashtra and Kerala (Fig. 2). The data given in Figure 3 show that of 1,473 dug well waters analyzed, nearly 50% had less than 2 μg chromium L^{-1}. The 50% cumulative frequency value (Fig. 3) is higher at 4 μg chromium L^{-1} because it takes into account also the effect of higher concentrations of chromium. For tubewell waters (including both shallow and deep tubewells), the 50% cumulative frequency value is around 6 μg chromium L^{-1}. These data are comparable with those of Matzat and Shiraki (1978) from Japan, where the average chromium content is reported to be 4.3 ppb (or 4.3 μg chromium L^{-1}), the maximum value being 13.8 μg chromium L^{-1}. Matzat and Shiraki (1978) noted that in North American oil field waters, the chromium content was generally between 1 and 10 μg L^{-1}.

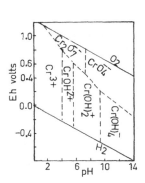

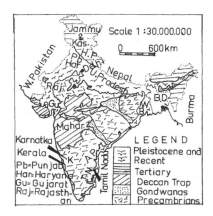

Figure 1. Chromium species in water at 25°C and 1 atm. $\Sigma Cr = 10^{-7}M$, $\Sigma S = 10^{-4}M$, $\Sigma C = 10^{-3}M$. (From Hem, 1977.)

Figure 2. Key map of India showing the main geologic formations.

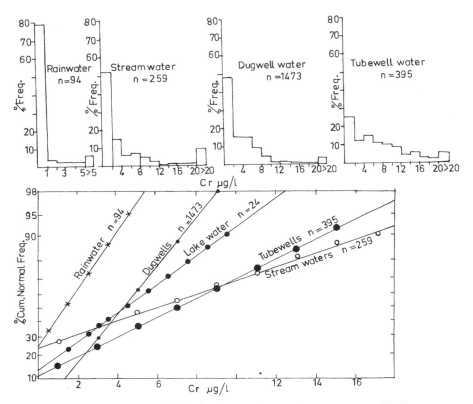

Figure 3. Chromium distribution in natural waters from some parts of India.

4.2.1   Uttar Pradesh

Most of the groundwater samples analyzed were from the Ganga plain, although some samples from the sub-Himalayan zone to the north as well as from the hard rock terrain (Bundelkhand granite and Vindhyan formations) in the south also were analyzed. The data show clearly that in most of the area, the chromium content is less than 5 μg chromium L^{-1} (Fig. 4). In fact the percent cumulative curve (Fig. 5) shows that 56% of the water samples are likely to have less than 3 μg chromium L^{-1}. Similar conclusions can be drawn if one examines the percent cumulative curve for chromium content in tubewell water samples from Uttar Pradesh. Of course both Figures 4 and 5 clearly show that there are some areas with anomolous concentrations of chromium in groundwater, as will be discussed.

4.2.2   Rajasthan

The area may be divided broadly into three major parts: the western desert (Thar), the stony and hilly region (mainly metamorphic series of Archaen rocks),

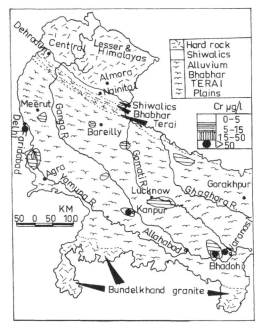

Figure 4. Chromium occurrences in shallow groundwaters from Uttar Pradesh in 1982.

and the eastern plains. The average chromium content was found to be 5.3 μg chromium L^{-1} (SD 5.5, $n = 71$).

4.2.3 Jammu and Kashmir

Lying between the Great Himalayan range in the north and the Pir Panjal in the south, the vast expanse of Kashmir valley is formed by the Karewas formation. The Karewas are a thick pile of alternating thin bands of sand, silt, and clay interspersed at two to three levels by glacial boulder beds. This formation of Plio-Pleistocene age lies disconformably over the older rocks ranging in age from Cambrian to Triassic. The Panjal volcanic formation [agglomerate slates, grits and effusive rocks, shales, slates with quartzites and limestone (Permo-carboniferous)] forms the hilly portion to the south. Groundwater occurs mainly in the form of springs. The analysis of these springwaters showed low chromium content (average 3.3 μg L^{-1}; SD 5.3, $n = 25$).

4.2.4 Coastal Saurashtra (Gujarat)

The area investigated is occupied mainly by Milolitic limestone of Pliocene to Recent age. The thickness of this formation ranges from 5 to 50 m. These are underlain by the Gaj formations (limestones, clays, and calcareous sandstones of Miocene age). The average chromium content in the premonsoon season was

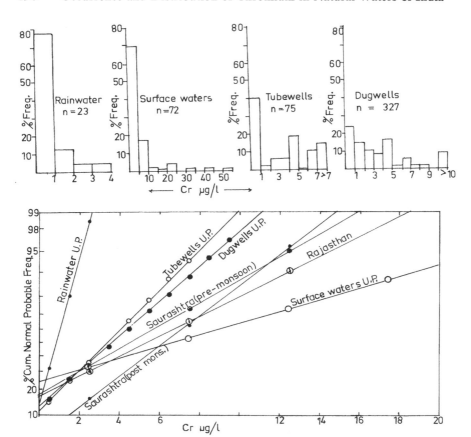

Figure 5. Chromium in Uttar Pradesh waters (1984). Dug well data for Rajasthan and Saurashtra are given also.

found to be 4.4 μg chromium L^{-1} (SD 4.8, $n = 100$), whereas in the postmonsoon period, the average was 6.3 μg chromium L^{-1} (SD 3.5, $n = 100$).

4.2.5 Punjab

The state of Punjab, lies largely in the Indus basin, with Siwalik Hills bordering the northeastern fringe of Gurdaspur and eastern fringes of the Hoshiarpur and Ropar districts. The alluvial deposits of Recent to sub-Recent age consist of clays, silts, sands, and gravels brought down by the various streams emanating from the Himalayas (Figs. 1 and 6). Groundwater occurs under unconfined, semiconfined, and confined conditions at various depths in the alluvium.

Most of the samples analyzed had less than 3 μg chromium L^{-1} (average 2.8 μg L^{-1}; SD 2.2), although a few samples from Ludhiana had very high chromium content.

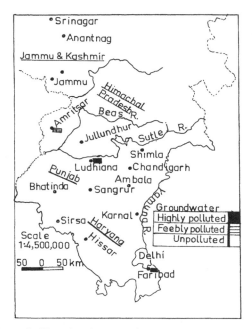

Figure 6. Chromium in parts of north and northwest India.

4.2.6 Delhi–Faridabad Area

The area is occupied by Quaternary alluvium and Pre-Cambrian metasediments of the Delhi System (Alwar Quartzites). The alluvial deposits (Pleistocene to Recent age), which are the main repository for groundwater, were laid down by the Yamuna river. These consist of interbedded, lenticular, and interfingering deposits of clay, silt, and sand with occasional gravel. Kankar (impure calcareous nodules) occur generally mixed with all these deposits and sometimes as pans. Groundwater occurs under both unconfined and confined conditions.

The Delhi groundwater had on the average 7.1 μg chromium L^{-1} (SD 3.5), whereas in Faridabad, the average content was 7.5 μg chromium L^{-1} (SD 5.2).

4.2.7 Coastal Kerala Tract

Sedimentary formations equivalent to the Cuddalore and Rajahmundry sandstones of east coast India occur as a narrow strip along the Kerala coast (Fig. 7) and are known as Warkalai beds. These consist of thin lenticular beds of clays and sands with thin lenticular beds of lignite. These are underlain by fossiliferous calcareous formations, or Quilon beds. These are overlain by laterites formed as residual deposits due to weathering of crystalline or sedimentary rocks. Recent formations include coastal sands, alluvium, and flood plain deposits and occur at various places. Although the laterites are likely to contain more chromium, the very heavy annual rainfall (ca 2,500 mm) precludes any accumulation of

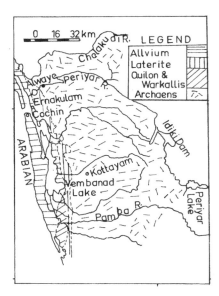

Figure 7. Chromium in Kerala waters.

chromium. Further, the laterites are generally quite porous, so that the residence time or contact time of water with lateritic rocks is short.

Most of the samples analyzed from this area had chromium levels below the detection limit ($<0.1 \mu g \, L^{-1}$), but a few samples had as much as 5 μg chromium L^{-1} (average 1.0 $\mu g \, L^{-1}$; SD 1.2, $n = 20$).

5. POLLUTION OF GROUNDWATER BY CHROMIUM IN INDIA

Examination of the data given in Tables 1, 2, and 3 shows clearly that in some areas the groundwater has exceptionally high chromium content, against the normal average background value of 4–7 μg chromium L^{-1} found in different parts of India. The chromium content in groundwater in such areas does not remain constant but shows considerable variations, apparently depending on the chromium content in the polluting plume and the local hydrogeologic features. In Kanpur (Fig. 8), the area near Anwarganj seems to be grossly polluted, whereas in Varanasi (Fig. 9) the area under the influence of effluent discharged by Diesel Locomotive Works Varansi is affected. In Ludhiana (Figs. 6 and 10) and in Faridabad (Fig. 6, Table 3), a high chromium content was observed. High chromium content in groundwaters has been reported elsewhere also (see Chapter 8, this volume). Davids and Lieber (1951) reported 450 μg chromium L^{-1} in groundwater, whereas Förstner and Wittman (1983) report that drinking water in Tokyo obtained from groundwater near Cr(VI)-containing soil heaps had over 10,000 μg chromium L^{-1}.

Table 1 Groundwater Samples from Varanasi and Bhadohi Areas with Anomolous Chromium Concentrations

Sample No.	Location	Type	Depth (m)	Sampling Date	Fe	Mn	Sr	Li	Rb	Cs	Cr
							(μg L^{-1})				
1	Diesel Locomotive Works, Varanasi	TW	82.0	6/6/84	120	8	320	9	3	6	133
2	Sunderpur, Varanasi	DW-1	12.5	2/26/84	109	8	340	21	4	7	108
3	Sunderpur, near Temple	DW-2	11.6	6/6/84	55	5	315	11	3	6	45
4	Sunderpur, Naria	DW-3	13.4	6/11/84	25	nd	418	15	4	4	110
5	Sunderpur	DW-4	12.6	11/26/83	5	10	760	18	25	7	70
6	Sunderpur	DW-5	17.5	10/17/81	120	16	300	20	nd	5	2,600
7	Cantt. Varanasi	DW-6	11.6	6/6/84	140	11	92	12	10	4	152
8	Saroi, Bhadohi	HP	13.5	6/7/84	1,930	56	435	14	220	12	200
9	Near Rly. Station, Bhadohi	DW-1	11.4	6/7/84	480	92	845	5	10	4	418
10	Manikpur, Bhadohi	DW-2	10.2	6/7/84	8	8	380	8	5	8	31
11	Near Bhadohi Yarn Dyers	HP	15.6	7/24/82	5,503	8,100	26	147	255	30	147

HP, handpump; DW, dug well; TW, tubewell.

Table 2 Anomolous Content of Chromium in Groundwater Samples from Kanpur, U.P., India

Sample No	Location	Type	Depth (m)	Date	Fe	Mn	Sr	Li	Rb	Cs	Cr
								(mgL^{-1})			
1	Kalpi Road, Fazalganj	HP	18	2/6/83	35	6	600	31	13	12	3,575
2		HP	18	3/16/83	64	330	560	33	20	17	3,900
3		HP	18	6/7/83	1,910	95	510	35	13	6	3,950
4		HP	18	2/6/84	97	53	280	30	7	8	4,300
5		HP	18	3/21/84	2,288	390	625	30	38	1	4,860
6		HP	18	7/1/84	530	59	285	31	13	13	5,650
7		HP	18	7/7/84	1,230	93	1,650	50	13	16	3,800
8		HP	18	9/4/84	3,550	97	615	30	44	2	4,850
9		HP	18	11/20/84	258	495	230	31	6	6	4,500
10	Kalpi Road, near Sardarji Hotel	HP	19	1/27/83	15	5	230	44	18	16	10,400
11		HP	19	3/16/83	31	580	480	47	27	31	19,000
12		HP	19	6/7/83	65	69	270	45	6	8	11,250
13		HP	19	2/6/84	35	55	400	39	80	14	7,800
14		HP	19	3/21/84	430	81	400	35	33	6	6,640
15		HP	19	7/1/84	650	79	770	39	15	19	7,200
16		HP	19	9/4/84	1,030	150	800	34	31	10	3,710
17		HP	19	11/20/84	80	61	370	37	8	17	7,500
18		DW	14	3/16/83	110	62	1,010	56	20	18	64
19		DW	14	7/7/84	590	6	1,660	26	10	8	50

20	Kalpi Road, Doordarshi Z. Office	HP	17	2/6/84	225	42	2,050	33	9	8	21
21		HP	17	7/1/84	5,750	90	170	41	60	13	69
22	Anwarganj, Atherton Text.	TW	87	1/27/83	35	5	570	40	15	12	1,200
23		TW	87	2/6/84	90	60	1,050	41	8	10	3,200
24		TW	87	11/20/84	40	4	700	23	6	19	27,000
25		TW	87	7/7/84	1,690	104	350	23	10	18	27,000
26		TW-2	25	2/6/84	79	55	1,270	37	6	8	200
27	Anwarganj, near Kanpur Chem.	DW	15	6/7/83	305	18	1,120	70	7	9	96
28	Anwarganj, near Kanpur, Hotel	DW	16	2/6/84	490	61	1,740	75	14	11	137
29	Anwarganj, near Mohd. Ka Hatta	DW	13	7/7/84	470	320	975	70	15	14	33
30	Lohia Machines	TW	87	2/6/84	33	9	1,285	37	6	4	200
31	Maharajpur P.S.	DW	13	6/27/84	1,500	86	620	46	19	15	31
32	Anwarganj, near Kanpur Chem.	DW	15	2/6/84	73	18	525	52	9	13	3,100
33	Amrodha	DW	12	7/1/84	215	11	510	26	12	5	49
34	Musanagar, Kanpur	DW	13	7/1/84	340	33	405	24	17	7	35
35	Bura, Kanpur	DW	13	7/1/84	390	110	255	24	12	9	50
36	Pratapganj, Jain Auto.	HP	15	7/7/84	570	170	690	37	7	6	46

HP, handpump; DW, dug well; TW, tubewell; TW-2, Shallow tubewell

Table 3 Groundwater Samples with Anomolous Chromium Content from Other Parts of India

Sample No.	Location	Type	Depth (m)	Sampling Date	Fe	Mn	Sr	Li	Rb	Cr
							(μg L^{-1})			
1	Near Union Carbide, Lucknow	HP	15.5	6/2/82	40	13	265	12	6	22
2	Opposite H.A.L., Lucknow	HP	14.5	2/2/83	260	550	210	23	21	279
3	Hussainganj, Lucknow	HP	14.0	2/2/83	42	17	580	34	24	550
4	Opposite Imambara, Lucknow	DW	10.0	5/18/84	130	12	300	33	115	32
5	Daryayai Masjid, Lucknow	DW	11.3	7/10/84	205	12	210	31	125	57
6	Maharajganj, Barabanki, U.P.	DW	7.9	4/14/84	100	26	215	25	22	43
7	Hathras, Aligarh, U.P.	DW	14.0	4/14/84	84	450	280	28	24	56
8	Hero Cycles, Ludhiana, Punjab	TW	81	4/3/84	92	33	17	37	1	210
9	Pb. Agr. Un. Ludhiana, Punjab	HP	11.3	4/3/84	490	61	1,740	75	14	137
10	Bawa Oil, GT Rd., Ludhiana, Punjab	TW	32	4/3/84	155	19	34	62	4	3,000
11	Orient Packages, Faridabad, Hary.	TW	29	10/1/82	75	95	190	25	43	31,000

HP, handpump; DW, dug well; TW, tubewell.

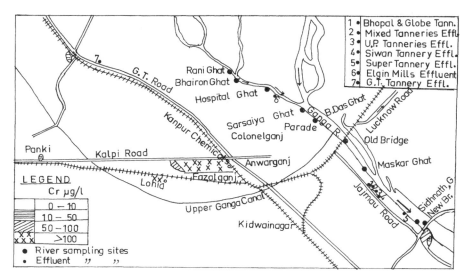

Figure 8. Chromium in groundwater in Kanpur metropolitan city, U.P.

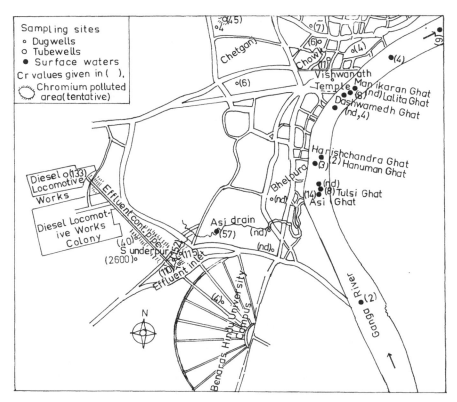

Figure 9. Chromium occurrence in shallow groundwaters in Varanasi. Data for tubewells and surface waters are also given.

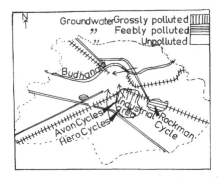

Figure 10. Sketch map of Ludhiana (not to scale).

5.1 Major Anthropogenic Sources of Chromium Contamination

The various anthropogenic sources of chromium may be grouped as follows
(Fig. 11):

Industrial	Agricultural
Mining	Sewage and sludge, manure
Metallurgy	Fertilizer (as impurity)
Chemical	
Dyeing	
Tanning	
Electroplating	
Pigments	
Photography	
Anticorrosive	
Oil well drilling	
Photography, lithography	
Wood preservative	

5.1.1 Ferrochrome Industry

It is estimated that consumption of chromite in India is around 154,315 t (as
compared to chromite production of 344,000 t in 1981). The ferrochrome
industry uses 55% of this chromite, the chemical industry uses 9%, and the rest is
used for other purposes. The important companies manufacturing chromium
compounds are as follows:

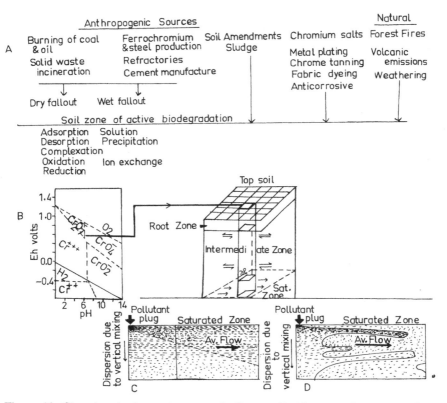

Figure 11. Chromium in the environment. A. Sources. B. Movement in unsaturated zone. C. Movement in uniform granular. D. Movement in heterogeneous materials.

Company	Sodium Dichromate (t)	Basic Chrome Sulfate (t)	Chrome Oxide Green (t)	Chromic Acid (t)
Binny, Madras	1,500	1,200	144	—
Kanpur Chemicals, U.P.	1,800	900	—	216
Sudarshan Textiles and Chemicals (Pune)	1,350	—	—	—
Golden Chemicals, Thane	6,875	6,750	—	—

The chemicals produced are used primarily in the manufacture of pigments, mordants in the textile industry, electroplating, and chromium tanning. The effluents from these industries sometimes have high chromium concentration (Tables 4 and 5).

Table 4 Industrial Waste Effluents Containing High Chromium Concentrations: Kanpur, Varanasi and Bhadohi (U.P.)

Sample No.	Industry Discharging Effluent	Location	Sampling Date	Fe	Mn	Ag	Cu	Zn	Co	Mo	Cd	Sr	Li	Rb	Cs	Cr
									(μg L^{-1})							
1	Siwan Tannery	Kanpur	3/16/83	41	295	13	68	425	nd	30	nd	1,600	25	3,260	68	55,000
2	Siwan Tannery	Kanpur	4/18/83	6,250	645	17	26	525	4	400	0.7	835	28	1,875	63	2,375
3	Siwan Tannery	Kanpur	1/26/84	73	430	13	18	174	83	273	nd	910	23	1,090	1,080	33
4	Super Tannery	Kanpur	1/26/84	80	77	10	29	37	83	185	nd	500	18	141	30	1,525
5	Tannery effluent mixed	Kanpur	3/16/83	340	225	8	31	118	3	10	—	100	16	2,080	31	21,000
6	Tannery effluent mixed	Kanpur	4/18/83	2,000	220	8	114	146	nd	160	3	475	21	131	15	2,140
7	Tannery effluent mixed	Kanpur	1/26/84	97	23	13	14	4	125	265	2	550	22	180	33	108
8	Elgin Mills	Kanpur	1/26/84	920	43	35	78	55	37	115	nd	190	13	59	16	935
9	Standard Tannery	Ianpur	6/28/83	625	95	10	202	136	23	95	nd	510	31	26	13	1,350
10	Super Tannery	Kanpur	6/28/83	660	68	11	25	65	37	125	nd	520	23	63	23	12,500
11	Siwan Tannery	Kanpur	6/20/83	680	94	nd	14	59	10	39	1	300	2	30	9	335
12	G.T. Tannery	Kanpur	6/20/83	2,870	100	nd	51	117	10	9	0.5	102	0.5	4	4	745
13	Diesel Loco. Works	Varanasi	10/26/83	0.6	8	0.5	13	32	10	30	0.5	700	9	80	16	250
14	Diesel Loco. Works	Varanasi	2/20/84	94	38	1	20	29	10	20	nd	240	7	18	6	235
15	Diesel Loco. Works	Varanasi	6/11/84	2,510	165	nd	113	183	nd	nd	1	180	13	10	nd	832
16	Diesel Loco. Works	Varanasi	6/11/84	80	14	nd	2	125	nd	nd	nd	120	8	4	4	1,157
17	Diesel Loco. Works	Varanasi	6/6/84	2,580	215	9	80	105	nd	30	nd	1,200	8	19	86	88,000
18	Bhikam Yarn Dyers	Bhadohi	6/7/84	500	145	5	18	212	nd	nd	nd	840	15	54	8	103
19	Bhadohi Yarn Dyers	Bhadohi	6/7/84	670	106	nd	14	108	nd	20	nd	500	10	34	8	4,500
20	Lal Carpet	Bhadohi	6/10/84	83	13	2	69	83	nd	20	nd	600	10	102	8	475
21	Bharat Dyers	Bhadohi	6/10/84	8	nd	nd	3	nd	nd	30	nd	380	19	15	nd	453
22	Bhadohi Woollens	Bhadohi	6/10/84	3,866	480	4	165	630	6	18	3	390	37	113	8	5,150

nd, not detected.

Table 5 Industrial Waste Effluents from Lucknow (U.P.), Faridabad (Haryana), and Ludhiana (Punjab) Containing High Chromium Concentrations

Sample No.	Industry Discharging Effluent	Location	Sampling Date	Fe	Mn	Ag	Cu	Zn	Mo	Cd	Sr	Li	Rb	Cs	Cr	Pb	Ni
									(μg L^{-1})								
1	Union Carbide	Lucknow	6/21/82	40	43	2	8	5	20	nd	220	14	10	3	261	nd	1,530
2	Union Carbide	Lucknow	2/2/83	1,150	185	5	2,100	4,200	nd	nd	280	9	19	12	1,500	68	1,800
3	Union Carbide	Lucknow	5/17/83	1,220	69	2	276	135	30	nd	200	13	9	2	1,040	68	10,250
4	Union Carbide	Lucknow	2/23/84	23	12	2	51	1,085	15	nd	194	11	10	2	965	nd	1,250
5	Union Carbide	Lucknow	8/7/82	35	115	2	10	35	nd	nd	200	7	8	2	645	10	5,100
6	Union Carbide	Lucknow	10/8/82	250	18	nd	66	526	95	2	46	9	9	nd	3,125	20	3,200
7	Union Carbide	Lucknow	12/13/82	162	34	nd	231	219	10	nd	92	13	7	nd	775	15	10
8	Union Carbide	Lucknow	2/2/84	23	12	2	51	1,085	15	nd	194	11	10	2	965	nd	1,250
9	Gedore Tools Plant	Faridabad	10/1/82	65	80	nd	27	25	15	nd	390	25	39	11	3,850	5	2,300
10	Munneshwari Enterprises	Faridabad	10/1/82	145	13	nd	28	35	35	nd	290	30	142	25	1,130	nd	32
11	Ferro Alloys	Faridabad	10/1/82	255	665	nd	5	38	38	nd	105	67	54	8	22	nd	7,050
12	Ralson Cycles	Luchiana	10/18/85	14,000	530	10	93	3,100	108	nd	15	50	5	9	4,400	6	10,750
13	Hero Cycles	Luchiana	10/18/85	2,900	395	12	95	18,000	120	—	6	38	4	20	15,000	6	11,000
14	Avon Cycles	Luchiana	10/18/85	2,150	120	8	190	8,000	88	0.7	11	36	3	8	1,500	20	12,800
15	Rockman Cycles	Luchiana	10/18/85	37,000	2,700	5	190	700	130	0.3	5	52	3	8	4,500	nd	6,000
16	Ralson Cycles	Luchiana	10/18/85	2,200	2,150	9	100	1,700	10	nd	14	43	6	8	7,570	6	15,000

nd, not detected.

5.1.2 Sewage Sludge

Around 1,700 million t of animal waste is produced in India annually (Handa, 1984), about 20–30% of which is used as manure. No data on the chromium content in these wastes are known. Even the data from other countries show that the chromium concentration varies greatly (Hansen and Chaney, 1984; Page, 1974, Seto and Deanglis, 1978; Freedman and Hutchinson, 1981, Scheffler and Schachtschabel, 1982):

Sweden		England		Germany
Range	Mean	Range	Mean	Max.
(mg/kg)		(mg/kg)		(mg/kg)
20–40,615	872	40–8,800	980	3,000

Michigan		North America	Ontario
Range	Mean	Range	Range
(mg/kg)		(mg/kg)	(mg/kg)
22–30,000	2,031	20–40,000	16–16,000

5.1.3 Phosphate Fertilizers

About 1.5 million t of phosphate fertilizers are used in India annually. Freedman and Hutchinson (1981) recorded 39–92 mgL^{-1} of chromium in phosphatic fertilizers, although no specific data on phosphatic fertilizers used in India are available.

5.1.4 Mining Sources

These sources are likely to be of importance in Orissa and Bihar only.

6.   AREAS WITH HIGH CHROMIUM CONTENT IN GROUNDWATER

6.1   Kanpur

The Kanpur metropolitan town situated in the heart of the Ganga plains, is about 70 km to the south southwest of Lucknow (Fig. 4). It is an important center for chromium tanning of leather. The waste effluents are being released by these industries without any pretreatment. However, the influence of these effluents on groundwater is not known, since these industries are located mostly on Jajmau Road, near the bank of the Ganga river (Fig. 8).

Kanpur Chemicals Ltd. (Fig. 8) is an important company manufacturing chromium salts. The waste effluent from this factory is quite high in chromium. The company did not allow its effluent to be sampled, but one sample collected outside the premises (after having been mixed with other effluents) was found to contain 1,850 μg chromium L^{-1}. In fact, the tubewell water in the Atherton Textile Mill, which is just adjacent to this company's premises, was found to contain as much as 27,000 μg chromium L^{-1}. One dugwell water (Table 2, No. 27), which had only 96 μg chromium L^{-1} in June 1983, had 3,100 μg chromium L^{-1} in February 1984 (Table 2, No. 32), indicating a strong seasonal change in water quality. The dug well is located about 30 m away from the closed drain carrying effluent from Kanpur Chemicals. It is possible that pollution in the Kalpi Road, right up to Lohia Machines (Fig. 8) may be due to the waste effluents from this factory, but further work, including more thorough sampling, is required to substantiate this assumption.

6.2 Sunderpur, Varanasi

The Sunderpur village in Varanasi, bordering the Diesel Locomotive Works, was found to contain groundwater with anomolous chromium concentrations (Table 1, Fig. 9). The highest value recorded was 2,600 μg chromium L^{-1} in one dug well water sample. A study of the chromium content of the waste effluents being discharged by Diesel Locomotive Works revealed that it had varying amounts of chromium, the highest value recorded being 88,000 μg chromium L^{-1}. In fact, even the deeper aquifer zone seems to have become affected, since the tubewell water from the Diesel Locomotive Works premises was found to contain 133 μg chromium L^{-1}.

6.3 Bhadohi, Varanasi District

Bhadohi, in the Varanasi district, is an important center for yarn dyeing as well as for carpet manufacturing. It appears that chromium salts are being used for dyeing purposes, as is evident from the high chromium content in waste effluent samples from industries in this area (Table 4). The highest chromium content observed in shallow groundwater sample was 418 μg chromium L^{-1}.

6.4 Faridabad–Delhi Area

A number of industries in Faridabad are discharging waste effluents with high chromium content (Table 5, Nos. 9 and 10). The occurrence of groundwater with 31,000 μg chromium L^{-1} (Orient Packages, Ltd., Table 3, No. 11) shows that pollution of groundwater is occurring here also.

6.5 Ludhiana–Punjab

There are several manufacturing plants in Ludhiana that discharge their wastes into unlined drains. The analysis of these effluents revealed not only high chromium content but also high cyanide contents and, at times, also nickel

(Handa, 1978a,b; Kakar and Bhatnagar, 1981; Table 5). The soils in Ludhiana are sandy loam in texture and alkaline in reaction. The chromium ions (along with cyanide) have migrated to the saturated zone, thereby polluting the groundwater in this area (Table 3).

6.6 Other Areas

In Lucknow, the waste effluent from Union Carbide was found to contain high chromium content (Table 5), which is affecting the quality of the groundwater in the neighborhood (Table 3, No. 1). Isolated cases with groundwaters containing anomolous values for chromium were detected in other parts of U.P. and Punjab (Table 3).

7. CHROMIUM IN SURFACE WATERS

Förstner and Wittman (1983) have suggested 0.5 μg chromium L^{-1} as the background content of chromium in surface waters. Langård and Norseth (1977) found that stream waters had $1-10$ μg chromium L^{-1}. Over 50% of the stream waters analyzed from different parts of India had less than 2 μg chromium L^{-1} (Fig. 3), the 50% probable frequency value being 5.7 μg chromium L^{-1} (Fig. 3).

7.1 Kerala

Only 13 stream water samples were analyzed from Kerala; the average chromium content was 1.3 μg L^{-1} (SD 3.2). Apparently the heavy rains (ca. 2,500 mm/year) are responsible for this relatively low chromium content.

7.2 The Ganga River

The Ganga River emanates from the Himalayas and after traversing through the plains of U.P., Bihar, and Bengal enters the Bay of Bengal. The analysis of stream waters from Haridwar did not reveal detectable amounts of chromium (Table 6). However, at Kanpur (Fig. 8), 12 of 67 samples analyzed had chromium content exceeding 20 μg L^{-1}, the overall average being 8.4 μg L^{-1} (SD 9.8). Nevertheless the chromium content normally remained below 50 μg L^{-1}. It is obvious that discharge of wastes into the Ganga River at Kanpur is having its effect on the chromium content.

At Varanasi, the average chromium content was found to be 2.2 μg L^{-1} (SD 3.3; Table 6). Dilution seems to be the main factor in such low values of chromium, although adsorption on the bottom sediments may also be exercising some influence.

Table 6 Surface Waters Containing Anomolous Concentrations of Chromium Ions

Sample No.	River/Surface Water	Location	Sampling Date	Fe	Mn (μg L^{-1})	Cr
1	Ganga River, Old Bridge	Kanpur	1/21/84	1,210	135	21
2	Ganga River, Old Bridge	Kanpur	3/21/84	3,750	373	24
3	Ganga River, Old Bridge	Kanpur	6/27/84	—	345	21
4	Ganga River, Sarsaiya Ghat	Kanpur	1/21/84	1,020	109	17
5	Ganga River, Maskar Ghat	Kanpur	1/21/84	1,570	80	41
6	Ganga River, New Bridge	Kanpur	7/7/84	—	385	23
7	Ganga River, New Bridge	Kanpur	9/4/84	—	620	23
8	Ganga River, Dhori Ghat	Kanpur	5/27/84	—	34	21
9	Lower Ganga Canal, Muradnagar	Kanpur	6/28/83	—	52	41
10	Lower Ganga Canal, Muradnagar	Kanpur	6/27/84	—	290	45
11	Assi Drain, Lanka	Varanasi[a]	10/18/81	23,000	1,175	1,300
12	Assi Drain, Lanka	Varanasi	6/2/84	2,490	310	57
13	Gomati River, Hardinge Bridge	Lucknow	5/17/83	770	66	28
14	Gomati River, Gaughat	Lucknow	3/3/84	3,170	95	120
15	Gomati River, Hardinge Bridge	Lucknow	5/18/84	1,700	165	50

[a]Drain entering the Ganga River at Varanasi.

7.3   The Yamuna River

Ajmal et al. (1984) have reported on the chromium content in the Yamuna River water and sediments at various sites:

| | Sites | | | |
Parameter	Delhi	Mathura	Agra	Allahabad
Cr μg L^{-1} (water)	20–37	24–29	3.0–8.8	10–27
Cr mg kg^{-1} (sediments)	98–1,090	190–199	51–59	67–76

In the present studies, the samples were mainly from Delhi, and the average value was found to be 8.7 μg chromium L^{-1} (SD 4.5), the chromium content being much lower than that reported by Ajmal et al. (1984).

7.4   The Gomati River

The Gomati River water at Lucknow was found to contain 2.0 μg chromium L^{-1} (average; SD 6.5). However, at times the chromium content was found to exceed 50 μg L^{-1} (Table 6).

7.5   The Vembanad Lake

The Vembanad Lake in Kerala (Fig. 7) has in its drainage basin laterites, Quilon limestone, Warkallis, and Archaen rocks (charnockites, khondalites, etc.). The lake is under the influence of tides, particularly during the premonsoon dry season, when the water level in the lake is low. The average chromium content in the lake water was found to be 4.5 μg L^{-1} (SD 5.3).

8.   CONCLUSION

The investigations reported here give an idea of the background chromium content in rain, surface water, and groundwater from some parts of India. Areas where groundwater pollution by chromium is quite significant have been delineated. However, much remains to be done, particularly with regard to the movement of polluting plume in the unsaturated and the saturated zones. Furthermore, it is possible that there are other areas in India, particularly in Rajasthan and Tamil Nadu, where chromium pollution may be occurring. The surface waters in general have chromium concentrations less than 50 μg L^{-1}, but regular monitoring is required to see that this limit is not exceeded.

ACKNOWLEDGMENT

I am indebted to Shri B.P.C. Sinha, Chief Hydrogeologist for his permission to publish this paper.

REFERENCES

Ajmal, M., Khan, M.A., and Nomani, A.A. (1984). "Distribution of heavy metals in water and sediments of selected sites of Yamuna River (India)." *Environ. Monit., Assess.* **5**, 205–214.

Bertine, K.K., and Goldberg, E.D. (1971). "Fossil combustion and the major sedimentation cycle." *Science* **173**, 233–235.

Davids, H.W., and Lieber, M. (1951). "Underground water contamination by chromium wastes." *Water Sewage Works* **98**, 525–530.

Förstner, U., and Wittmann, G.T.W. (1983). *Metal Pollution in the Aquatic Environment.* Springer-Verlag, Berlin, 486 pp.

Freedman, B., and Hutchinson, T.C. (1981). "Sources of metal and elemental contamination of terrestrial environments." In N.W. Lepp, Ed., *Effect of Heavy Metal Pollution on Plants.* Vol. 2, Applied Science Publ., London, Vol. 2, pp. 35–94.

Furr, A.K., Parkinson, T.F., Hinrichs, R.A., et al. (1977). "National survey of elements and radio-activity in fly ashes. Absorption of elements by cabbage grown in fly ash soil mixtures." *Environ. Sci. Technol.* **11**, 1194–1201.

Handa, B.K. (1978a). "Occurrence of heavy metals and cyanides in ground water from shallow aquifers in Ludhiana." *IAWPC Tech. Annu.* **5**, 109–115.

Handa, B.K. (1978b). "Trace elements and cyanides as indicators of ground water pollution." *J. Inst. Public Health Eng. (India)* **1978** (1), 30–35.

Handa, B.K. (1983a). "Pollution of natural waters by industrial waste effluents with special reference to pollution in some parts of north and north-west India." *Proc. National Seminar on Assessment, Development and Management of Ground Water Resources.* New Delhi, April 29–30, 1983, pp. 355–365.

Handa, B.K. (1983b). "Occurrence and distribution of trace elements in natural waters of India." *Proc. National Seminar on Assessment, Development and Management of Ground Water Resources.* New Delhi, April 29–30, 1983, pp. 293–304.

Handa, B.K. (1984). "Water quality and water pollution problems on the Indian sub-continent." *Intern. Symp. Hydrochemical Balances of Fresh Water Systems.* Uppsala, Sweden, September 10–14, 1984, pp. 313–322.

Handa, B.K., Kumar, A., Goel, D.K., et al. (1983a). "Pollution of ground waters by chromium in Uttar Pradesh." *IAWPC Tech. Annu.* **10**, 167–176.

Handa, B.K., Kumar, A., Goel, D.K., et al. (1983b). *Hydrochemistry, Water Quality and Water Pollution in Kanpur Metropolis.* Tech. Report No. 7, C.C.L., C.G.W.B., Ministry of Irrigation, 136 pp.

Hansen, L.G., and Chaney, R.L. (1984). "Environmental and food chain effects of the agricultural use of sewage sludges." *Rev. Environ. Toxicol.* **1**, 103–172.

Hem, J.D. (1970). "Study and interpretation of the chemical characteristics of natural water." US Geol. Surv. W.S.P. **1473**, 363 pp.

Horn, M.K., and Adams, J.A.S. (1966). "Computer-derived geochemical balances and element abundances." *Geochim. Cosmochim. Acta* **30**, 279–297.

Kakar, Y.P., and Bhatnagar, N.C. (1981). "Ground water pollution due to industrial effluents in Ludhiana." In W. van Duijenbooden, P. Glasbergen, and H. van Lelyveld, Eds., *Internat. Symp. on Quality of Ground Water.* Noordwigkerhaut, The Netherlands March 23–27, 1981. Elsevier Scientific Publ., Amsterdam, pp. 265–272.

Langård, S., and Norseth, T. (1977). "Chromium." In *Toxicity of Metals.* Permanent Commission & International Assoc. of Occupational Health. Health Effects Research Laboratory, Research Triangle Park, NC, U.S. Dept. of Commerce, NTIS, Vol. 2, pp. 164–187.

Matzat, E., and Shiraki, K. (1978). "Chromium." In K.H. Wedepohl, Ed., *Handbook of Geochemistry.* Springer-Verlag Berlin, Vol. 2.

Natusch, D.F.S., Bauer, C.G., Matusiwicz, H., et al. (1975). "Characterization of trace elements in fly ash." In T.C. Hutchinson, Ed., *Symp. Proc. Intern. Conf. on Heavy Metals in the Environment.* Toronto, Canada, Vol. 2, Part 2, pp. 553–578.

Navaree, J.L., Ronneau, C., and Priest, P. (1980). "Deposition of heavy metals in Belgian agricultural soils." *Water Air Soil Pollut.* **14**, 208–213.

Page, A.L. (1974). *Fate and Effects of Trace Elements in Sewage Sludge when Applied to Agricultural Lands.* EPA 670/2-74-005, Office of Research & Development, U.S. EPA, Cincinnati, OH, 97 pp.

Scheffler, F., and Schachtschabel. (1982). *Lehrbuch der Bodenkunde,* 11th ed. Enke, Stuttgart, 442 pp.

Seto, P., and Deanglis, P. (1978). "Concepts of sludge utilization on agricultural land." In *Sludge Utilization and Disposal Conference Proceedings No. 6.* Environment Canada, Ottawa, pp. 138–155.

Sittig, M. (1980). *Priority Toxic Pollutants.* Noyes Data Corp. Park Ridge, NJ, 368 pp.

Taylor, M.C., Reeder, S.W., and Demayo, A. (1979). "Chromium." In *Guidelines for Surface Water Quality. Inorganic Chemical Substances.* Environment Canada, Vol. 1, 9 pp.

8

CHROMIUM CONTAMINATION OF GROUNDWATER

Lynn M. Calder

Golder Associates
Mississauga, Ontario

1. INTRODUCTION

Groundwater contamination by chromium is a major problem in industrialized areas, particularly those that have large metal-plating industries, such as parts of Japan and the northern United States. Other important industrial sources of chromium are wood treatment and tannery facilities, as well as chromium mining and milling operations. Typically, chromium-containing wastes have been disposed of by discharging them to surface impoundments or lagoons. Leakage from these lagoons into groundwater has been relatively common. Almost all reported incidences of chromium-related groundwater contamination are of industrial origin. Naturally occurring chromium also occurs at potentially harmful concentrations (Robertson, 1975), although such occurrences are thought to be rare.

Chromium has two stable oxidation states in natural environments: hexavalent [Cr(VI)] and trivalent [Cr(III)]. Chromium(VI) is toxic to animals and plants, whereas Cr(III) is considered to be less toxic (Health and Welfare Canada, 1979; see Chapter 19, this volume). Drinking water standards have been set at $0.05 \, mg \, L^{-1}$ ($10^{-6} \, M$) total chromium because of the toxic effects of Cr(VI) and the possibility of oxidation of Cr(III) to toxic Cr(VI) (Health and Welfare Canada, 1979; U.S.EPA, 1976).

Large plumes of chromium-contaminated groundwater in shallow sand and gravel aquifers have been well documented (Deutsch, 1972; French et al., 1985; Perlmutter and Lieber, 1970; Stollenwerk and Grove, 1984; Wiley, 1983). Such plumes have been reported to reach lengths of up to 1,300 m (Perlmutter and Lieber, 1970). Many of the plumes have necessitated the abandonment of local groundwater supplies.

Groundwater contamination by chromium can be extensive in sand and gravel and fractured rock aquifers because groundwater velocities in these materials typically are between about 0.1 and 5 m per day. At the other extreme, groundwater velocities in clayey materials tend to be low, on the order of a few centimeters or less per year, so chromium-contaminated groundwater cannot extend far from the source.

The mobility of chromium in groundwater depends also on its solubility and its tendency to be adsorbed by soil or aquifer materials. These factors, in turn, depend on the groundwater chemistry and the characteristics of soil or aquifer material in contact with the chromium-containing groundwater.

2. SPECIATION OF CHROMIUM

A number of different chromium species exist in groundwater environments. These species differ greatly in their solubility and in their tendency to be adsorbed by soil or aquifer materials. The concentration and mobility of chromium in groundwater are strongly dependent on its speciation characteristics.

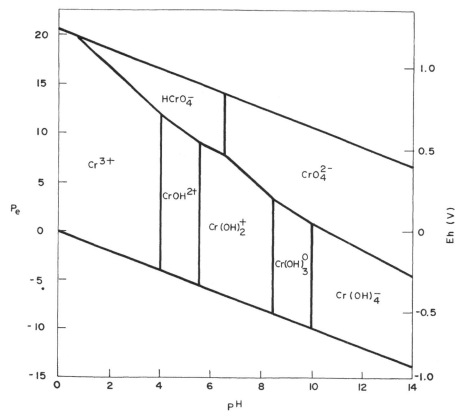

Figure 1. Areas of dominance of dissolved chromium species at equilibrium in the system $Cr + H_2O + O_2$ at 25°C and 1 atm. (Data from Baes and Mesmer, 1977; Hem, 1977)

Chromium speciation in groundwater is affected by pE (redox) and pH conditions (Fig. 1). Chromium(VI) predominates under oxidizing (high redox) conditions, whereas Cr(III) predominates under more reducing (low redox) conditions. Oxidizing conditions generally are found in shallow aquifers within a few meters of the water table, where there is a continual replenishment of oxygen from the atmosphere via the unsaturated zone. Oxygen in groundwater tends to be consumed by hydrochemical and biochemical oxidation reactions, usually involving organic matter. In shallow groundwaters, this oxygen is replaced from the atmosphere. In deeper groundwaters that are isolated from the atmosphere, no replenishment of oxygen occurs, and reducing conditions develop.

Chromium(VI) is extensively hydrolyzed in water. At total chromium concentrations below $500 \, mg \, L^{-1}$ $(0.01 \, M)$, the dominant Cr(VI) species are the oxyanions $HCrO_4^-$ and CrO_4^{2-}. Equilibrium between the two species is dependent on pH:

$$HCrO_4^- \longleftrightarrow CrO_4^{2-} + H^+$$

At low pH, the concentration of H^+ ions is high, so the reaction tends to the left, and $HCrO_4^-$ is the dominant species. At higher pH, where the H^+ concentration is lower, the reaction tends to the right, and CrO_4^{2-} predominates. In natural groundwaters, the pH is typically between 6 and 8, so the CrO_4^{2-} ion is the dominant species (Fig. 1). At Cr(VI) concentrations above 500 mg L^{-1}, the $Cr_2O_7^{2-}$ ion predominates in acidic environments (Baes and Mesmer, 1977). However, Cr(VI) concentrations in polluted groundwaters are generally below 50 mg L^{-1}. Chromium(VI) at concentrations on the order of 10 mg L^{-1} imparts a characteristic yellow color to water (Deutsch, 1972). Chromium(VI) does not commonly form complexes with inorganic or organic ligands (Langård, 1980; National Academy of Sciences, 1974).

The dominant Cr(III) species occurring in groundwater also depend on pH. The governing reactions are:

$$Cr^{3+} + H_2O \longleftrightarrow CrOH^{2+} + H^+$$

$$CrOH^{2+} + H_2O \longleftrightarrow Cr(OH)_2^+ + H^+$$

$$Cr(OH)_2^+ + H_2O \longleftrightarrow Cr(OH)_3^0 + H^+$$

$$Cr(OH)_3^0 + H_2O \longleftrightarrow Cr(OH)_4^- + H^+$$

$Cr(OH)_2^+$ is the dominant species in natural groundwaters with a pH between 6 and 8. $CrOH^{2+}$ and Cr^{3+} predominate in more acidic, waters and $Cr(OH)_3^0$ and $Cr(OH)_4^-$ predominate in more alkaline waters (Fig. 1). Chromium(III) complexes with many organic ligands, as well as fluoride, ammonia, cyanide, thiocyanate, oxalate, and sulfate (Baes and Mesmer, 1977).

3.   SOLUBILITY OF CHROMIUM

There are no significant solubility constraints on the concentrations of Cr(VI) in groundwater. Solubility can, however, significantly limit the concentration of Cr(III) in groundwater at a pH above 4 or 5. The low solubility of the Cr(III) solid phases Cr_2O_3 and $Cr(OH)_3$ (Hem, 1977) is likely the major reason why Cr(III) generally makes up a small percentage of the total chromium concentration in natural or polluted groundwaters and why it rarely occurs at concentrations above the drinking water standard of 0.05 mg L^{-1}. Chromium(III) tends to be essentially immobile in most groundwaters because of its low solubility.

Chromium(III) solution concentrations in equilibrium with Cr_2O_3 are less than 0.005 mg L^{-1} above a pH of 4, whereas Cr(III) solution concentrations in equilibrium with $Cr(OH)_3$ vary from approximately 0.05 to 500 mg L^{-1} between a pH of 5 and 9 (Hem, 1977). Although the theoretical solubility of Cr_2O_3 is much lower than that of $Cr(OH)_3$, Hem (1977) suggests that chromium is more likely to precipitate as $Cr(OH)_3$ in natural aqueous systems where waste has been introduced as $Cr(VI)$ and subsequently been reduced to $Cr(III)$.

When $Cr(VI)$ is transported by groundwater, it may be transformed to and precipitated as $Cr(III)$ if the groundwater enters a low redox zone. Laboratory studies have shown that $Cr(VI)$ can be reduced readily to $Cr(III)$ in the presence of organic matter, especially at low pH (Bartlett and Kimble, 1976b; Bloomfield and Pruden, 1980). Schroeder and Lee (1975) found that $Cr(VI)$ could also be reduced by $Fe(II)$ and dissolved sulfides.

Chromium(III) generally is not transported great distances by groundwater because of its low solubility. However, $Cr(III)$ could be converted to the more soluble $Cr(VI)$ if the redox state at a particular location were to change with time from reducing to oxidizing. Chromium(III) has been found to be oxidized by manganese under natural conditions (Bartlett and James, 1979; Schroeder and Lee, 1975).

Chromium(III) can form highly soluble organic complexes in the laboratory, particularly under acid conditions (Bartlett and Kimble, 1976a; James and Bartlett, 1983a). Thus, if it is in a complexed form, $Cr(III)$ may exist at much higher concentrations in groundwater than if it is uncomplexed. The existence of $Cr(III)$ complexes has not been documented under field conditions.

4. ADSORPTION

4.1 Theory

Adsorption occurs because dissolved ionic species are attracted to mineral surfaces that have a net electrical charge due to imperfections or substitutions in the crystal lattice or chemical dissociation reactions at the particle surface (Freeze and Cherry, 1979). The electrical charge of a mineral surface varies with pH. The surface charge is neutral at the zero point of charge pH (pH_{zpc}). Above the pH_{zpc}, the surface is negatively charged, and cation adsorption occurs. Below the pH_{zpc}, the surface is positively charged, and anion adsorption occurs.

Adsorption of a contaminant by soil or aquifer materials serves to retard the advance of the contaminant front with respect to the groundwater velocity. If the contaminant input is in the form of a slug, adsorption also serves to reduce the peak concentration of the contaminant in the slug. Because adsorption is not irreversible, permanent removal of contaminants from solution does not occur. Instead, adsorbed contaminants will be released back to the solution phase after the center of the slug has passed or after the source of the contamination is removed. This causes solution concentrations to remain above background levels for a longer period of time than if there were no adsorption.

The relative velocity in a porous aquifer of the center of mass of a slug of contamination or the front emanating from a continuous source may be estimated by the following equation:

$$R = \frac{v_{gw}}{v_c} = 1 + \frac{\rho_b}{\theta}\frac{dC}{dS} \qquad (1)$$

where:

R = retardation factor (/)
v_{gw} = velocity of groundwater (L/T)
v_c = velocity of contaminant (L/T)
ρ_b = dry bulk density of aquifer material (M/L^3)
θ = porosity of aquifer material (/)
C = concentration of contaminant adsorbed (M/M)
S = concentration of contaminant in solution (M/L^3)
$\frac{dC}{dS}$ = partitioning ratio (L^3/M)

The partitioning ratio is a theoretical equlibrium ratio between the adsorbed and solution concentrations of a contaminant. This ratio is assumed to be characteristic for a contaminant and a specific adsorbing medium for a specific set of conditions determined by the solution chemistry, usually in terms of the contaminant concentration, pH, pE, and temperature.

The partitioning ratio is determined by means of batch experiments in which known masses of a solid medium (e.g., soil, clay) are equilibrated with a range of known solution concentrations of the contaminant of interest. If the partitioning ratio does not vary with concentration, adsorption is said to be linear, and the partitioning ratio is a constant known as the distribution coefficient or K_d. If the partitioning ratio varies with the concentration of the contaminant, adsorption is described by a nonlinear model.

Equation 1 or its equivalent is commonly incorporated into mathematical models of contaminant transport in groundwater (Grove and Stollenwerk, 1985; Pickens and Lennox, 1976). It is used usually when adsorption is linear or is assumed to be linear and the partitioning ratio is equal to the K_d. A typical aquifer bulk density of 1.9 and a porosity of 0.30 would give a linear retardation factor of:

$$R = 1 + 6.3\,K_d \qquad (2)$$

where the K_d is in mLg^{-1}

K_ds can range from zero if no adsorption occurs to tens of thousands if adsorption is extremely high. If there is no adsorption, the retardation factor is equal to 1, and the contaminant is unretarded. If adsorption is very high, the contaminant is highly retarded and may in some cases be considered immobile for all practical purposes.

For nonlinear adsorption, groundwater transportation equations must be solved iteratively using a concentration-dependent retardation factor because the retardation of the contaminant will vary with time due to changing solution concentrations as the plume or slug of contaminated groundwater passes a particular aquifer segment. This makes comparisons and predictions more difficult than for the linear adsorption model.

Our discussion of adsorption thus far has considered only equilibrium relationships. This is adequate for cases where adsorption reactions are quick relative to groundwater flow rates. In high-velocity groundwater, such as is found in some sand and gravel aquifers, adsorption predicted by equilibrium models may overestimate that found in the field if chemical equilibrium is not instantaneous.

4.2   Chromium Adsorption and Mobility

Table 1 shows examples of Cr(III) and Cr(VI) partitioning ratios obtained directly or calculated from published data. For experiments that were conducted over a wide pH range, adsorption data at two pHs were selected to demonstrate the effect of pH on adsorption and to permit comparison of data among experiments. For experiments that were conducted over a wide range of chromium concentrations, the adsorption data at the minimum and maximum concentrations are included in Table 1. The Cr(III) concentrations used were generally above the 0–50 mg L^{-1} concentration range that might be encountered in contaminated groundwater. The Cr(VI) experiments included some concentrations in this range.

Chromium(III) concentrations in groundwaters are generally low because of the low solubilities of Cr(III) compounds above a pH of 4. However, Cr(III) concentrations in groundwater may be high in very acidic waters with a pH less than 4, which may occur at some tailings impoundments.

Positively charged ions, such as Cr^{3+}, which is the dominant Cr(III) species below a pH of 4, are generally adsorbed by clay minerals, which typically have high cation exchange capacities and low pH_{zpc} in the order of 2–2.5 (Stumm and Morgan, 1981). Cr^{3+} adsorption by soils and clays is generally very high, with adsorption increasing with pH as the clay surfaces become more negatively charged. For example, the batch experiments of Bartlett and Kimble (1976a) using silt loam with distilled water spiked with a Cr(III) concentration of 260 mgL^{-1} gave calculated partitioning ratios of 78 and 290 mLg^{-1} at a pH of 4. The batch experiments of Griffin et al. (1977), using two clays and landfill leachate spiked with Cr(III) concentrations of 92–731 mgL^{-1}, gave even higher partitioning ratios of 215–968 mLg^{-1} at the same pH (Table 1).

These partitioning ratios are very high. Chromium(III) adsorption is nonlinear, linear, so the K_d adsorption model cannot be used for assessing retardation except at the precise concentrations used. If these ratios were equivalent to K_ds they would give retardation factors of 500–6,100 by Equation 2. Such retardation factors indicate that Cr(III) is relatively immobile at a pH of around 4 due to adsorption. Above this pH, it is also relatively immobile because of its low solubility. It is probable that Cr(III) mobility over the entire pH range would be enhanced by the

222

Table 1 Examples of Partitioning Ratios in the Literature

Materials and Methods	pH	Solution Soil Ratio (mLg^{-1})	Initial Cr Concentration (mgL^{-1})	Partitioning Ratio (mLg^{-1})	Reference
Chromium III					
Silt loam and distilled water batch test	2	10	260	0.42	Bartlett and Kimble, 1976a, Figure 2
	4	3.33	260	0.45	
		10	260	78	
		3.33	260	290	
Montmorillonite and landfill leachate batch test	2	500	91	62	Griffin et al., 1977, Figure 6
			731	24	
	4	500	91	968	
			731	215	
Kaolinite and landfill leachate batch test	2	500	27	87	Griffin et al., 1977, Figure 6
			381	76	
	4	500	381	683	
Chromium VI					
Pure Fe(OH)$_3$ and distilled water batch test	7	250	20	886	James and Bartlett, 1983b, Table 1

Description	pH				Reference
38 soils and distilled water batch test	5.4[a]	10	1	5.6	James and Bartlett, 1983b, Table 2
Montmorillonite and landfill leachate batch test	4	16.7	10	59	Griffin et al., 1977, Figure 1
			37	19	
			307	1.8	
	7	16.7	10	12	
			37	4.0	
			307	0.54	
Kaolinite and landfill leachate batch test	4	10.0	4	30	Griffin et al., 1977, Figure 2
			52	2.7	
			191	0.76	
	7	10.0	4	10	
			52	0.83	
			191	0.24	
Alluvial materials and groundwater batch test	6.8	5	0.2	52	Stollenwerk and Grove, 1984, presented data
			100	1.7	
Alluvial materials and groundwater column experiment	6.8		5.0	25[b]	
			50	5[b]	

[a] Average of 38 soils.
[b] Calculated from the retardation factor.

223

formation of complexes due to decreased adsorption or increased solubility of these complexes over the uncomplexed form.

In the anionic form, $HCrO_4^-$ and CrO_4^{2-}, Cr(VI) is attracted to positively charged surfaces, such as iron, manganese, and aluminum oxides and hydroxides. These substances commonly coat aquifer materials and have a high pH_{zpc} of 6.7–8.5 (Stumm and Morgan, 1981). Amorphous iron, with a pH_{zpc} of 8.5, is the adsorbate found at the highest concentrations in most aquifer materials.

Adsorption of Cr(VI) by clays, soils, and natural aquifer materials is low to moderate in pH ranges that would be encountered commonly in groundwaters (Table 1). Adsorption of Cr(VI) characteristically decreases with increasing pH due to the decrease in positive surface charge of the adsorbing medium. Little or no adsorption occurs above a pH of 8.5. The batch experiments of James and Bartlett (1983b) confirm that iron hydroxides are strong adsorbates of Cr(VI). Partitioning ratios in Table 1 vary from 0.24 to 52 mLg^{-1} at a pH of 7 for natural soils, clays, or aquifer materials. These are much lower than those calculated for Cr(III).

Chromium(VI) adsorption has been found to be nonlinear (Griffin et al., 1977; Stollenwerk and Grove, 1984), fitting the Langmuir adsorption model:

$$\frac{C}{S} = \frac{K_l Q}{1 + K_l S} \tag{3}$$

where:

K_l = adsorption equilibrium constant from Langmuir plot (L^3/M)
Q = maximum number of adsorption sites from Langmuir plot (M/M)

If Cr(VI) adsorption were linear, the calculated partitioning ratios would correspond to retardation factors of 2.5–329 by Equation 2. These retardation factors indicate that Cr(VI) mobility at pH 7 could range from high to low. Chromium(VI) mobility would be lower below a pH of 7 and higher above a pH of 7. Above a pH of 8.5, Cr(VI) would be entirely unretarded.

The Cr(III) and Cr(VI) adsorption data described assume the existence of chemical equilibrium. Stollenwerk and Grove (1984) found that equilibrium adsorption of Cr(VI) was achieved within 3 days during batch tests. This may be too slow to allow local equilibrium to occur in fast flowing groundwater.

Laboratory experiments using simulated or natural aquifer materials and groundwaters may be useful for predicting chromium mobility in the field, but extrapolations should be made with caution due to the difficulty of reproducing field conditions in the laboratory and because of the possible existence of nonequilibrium conditions in the field.

5. CASE STUDIES

There are a number of documented cases of extensive chromium contamination of shallow sand and gravel aquifers. The following are accounts of major occurrences of chromium in groundwater, which necessitated abandonment of water supplies in the shallow aquifers where severe chromium contamination was encountered. In general, chromium movement was found to be unretarded or only moderately retarded relative to groundwater velocities. The field data are generally consistent with observed trends in the laboratory.

5.1   Nassau County, New York

The best known and the first major published case study in North America of chromium in water supply wells was in Nassau County, Long Island (Lieber et al., 1964). A number of investigators studied the site between 1942 and 1975, and over 150 wells have been installed to monitor chromium concentrations in groundwater.

The source of chromium was an aircraft plant that used chromium solutions for anodizing and plating metals. Between 1941 and 1949, untreated wastes containing hexavalent chromium at concentrations thought to be in the order of 40 mg L^{-1} (Pinder, 1973) were discharged to a cluster of disposal ponds at the plant.

The site is located on a very permeable sand and gravel aquifer, with groundwater velocities in the order of 0.15–0.5 m per day (Perlmutter and Lieber, 1970). By 1949, the chromium-contaminated groundwater had formed a cigar-shaped plume whose front had travelled 1,200 m from the disposal ponds and whose width was approximately 260 m. The maximum chromium concentration in the groundwater was 40 mgL^{-1}. The chromium appeared to be migrating at approximately the same velocity as the groundwater (Perlmutter and Lieber, 1970).

In 1949, a treatment plant was installed to remove chromium from the wastes disposed in the ponds. This was only partially effective, as evidenced by chromium concentrations of 0.06–35 mg L^{-1} in some samples of treated effluent in 1962 (Perlmutter and Lieber, 1970). Nevertheless, chromium concentrations in the groundwater dropped considerably after installation of the treatment plant.

By 1962, the length of the plume was 1,300 m, and its width was 300 m. The maximum chromium concentration in the groundwater was 10 mg L^{-1}. There appeared to have been much less movement of the plume between 1949 and 1962 than there had been before that time. The groundwater pH was 4.6–6.2, which is the range where Cr(VI) could be significantly adsorbed by the aquifer materials.

Three hypotheses can account for apparent decrease in velocity of the chromium plume after 1949:

1. The lowering of chromium concentrations in the effluent and the groundwater after installation of the treatment plant, which would result in greater adsorption because of the concentration-dependence of chromium adsorption

2. The slow reduction of Cr(VI) to Cr(III), with subsequent precipitation of Cr(III), particularly in deeper groundwaters

3. The slow kinetics of the adsorption process in the relatively fast groundwater regime, such that chemical equilibrium, and therefore maximum adsorption, would not occur until the aquifer had been exposed to chromium-containing groundwater over a long period of time

This case demonstrates the types of uncertainties that complicate predictions of chromium migration in groundwater.

5.2 Telluride, Colorado

Another well-documented occurrence of chromium contamination in groundwater is at the site of a heavy metal mining and milling operation near Telluride, Colorado (Grove, 1985). The source of chromium was the tailings pond, which, since 1977, apparently discharged chromium-containing wastes into the groundwater system. Water in the tailings pond had chromium concentrations of 8.8 mgL^{-1} (Stollenwerk and Grove, 1984). The U.S. Geological Survey initiated a study of the site in October 1978. This study has provided an excellent opportunity for comparison of field observations with laboratory experiments (Stollenwerk and Grove, 1984) and computer simulation (Grove and Stollenwerk, 1985). The shallow aquifer consisted of gravel and sand alluvial material. Calculated groundwater velocities were very high, in the order of 5 m per day.

In 1979, a chromium plume at least 520 m long was observed, with a maximum chromium concentration of 2.7 mg L^{-1} (Grove, 1985). The groundwater pH was approximately 6.8 (Stollenwerk and Grove, 1984). Laboratory and field investigations determined that the chromium was retarded by a factor of 10 relative to the groundwater velocity. Nonlinear, rate-dependent adsorption models were used to simulate the observed plume (Grove, 1985). In this case, the retardation was moderate, but because of the extremely high groundwater velocity, the chromium plume was nevertheless very mobile.

5.3 Michigan

Groundwater contamination by chromium is a problem in the State of Michigan because of its large automotive industry and its associated electroplating industry. Deutsch (1972) discusses several incidents of chromium contamination of groundwater in Michigan during the 1930s to 1950s, many of which were associated with the plating industry.

Wiley (1983) presents a more recent occurrence of groundwater contamination by chromium from a plastic chromeplating facility in southwestern Michigan.

Chromium concentrations of up to 14 mg L^{-1}, most of which was Cr(VI), were encountered in private water supply wells and monitoring wells tapping a shallow permeable glacial outwash aquifer. A contaminant plume front extended approximately 1,000 m from the plating facility. The source of the chromium in the groundwater was not determined. Possible sources were the plant's process water discharge ponds, which were allowed to contain chromium at concentrations up to 0.05 mg L^{-1}, a sludge stockpile adjacent to the disposal ponds, or leaks and spills of process water or waste. A purge well system was designed to reduce chromium concentrations in the shallow aquifer to below drinking water standards.

An incidence of groundwater contamination by chromium from a wood treatment plant was reported in another community in southwest Michigan in 1980 (French et al., 1985). The wood was treated with a 2% solution containing 47.5% chromium trioxide, 34% arsenic pentoxide, and 18.5% copper oxide. Between 1970 and 1979, the lumber was allowed to drip onto the ground and onto an asphalt pad. Excess solution from treatment and runoff from the pad was collected in a sump pit. Effluent from the pit was discharged to the ground adjacent to the treatment building until 1980. An effluent sample from the pit contained 1,600 mg L^{-1} chromium, of which 1,500 mg L^{-1} was Cr(VI). Solid waste, which consisted of sludge, scrap lumber, solution containers, and packing materials also was buried on the plant site.

The site is located on a permeable outwash plain consisting of gravelly sands with up to 17% silt and clay-sized particles. The water table was located at a depth of up to 8 m. The groundwater had a moderately alkaline pH up to 8.4. Groundwater velocities were estimated at 0.15 m per day.

No arsenic or copper plumes were encountered at the site. However a groundwater plume containing total chromum concentrations above 0.05 mgL^{-1} was found in 1980 to extend approximately 600 m from the discharge area. It had a width of approximately 200 m and a thickness of 20 m. Chromium concentrations were highest in the coarser layers. It was not determined whether this was because of preferential movement of chromium-contaminated groundwater in the coarser zones or because chromium adsorption in the finer materials was higher than that in the coarser materials. The highest chromium concentration in groundwater was 6.58 mg L^{-1}. The plant well had a chromium concentration of up to 2.5 mg L^{-1}. The length of the plume was found to be consistent with estimated groundwater velocities, suggesting that chromium was essentially unretarded. It was assumed that the chromium was almost entirely Cr(VI). Chromium mobility in this case was consistent with laboratory evaluations of Cr(VI) mobility in slightly alkaline groundwaters.

Remediation of the site was initiated with the removal of several thousand cubic meters of contaminated soil. The highest chromium concentrations in the soil corresponded to areas of buried waste and to the upper 0.5 m of loamy soil overlying the outwash materials. A purge well–spray irrigation system was established to restore the aquifer to drinking water standards.

6. SUMMARY

There are a number of reported field occurrences of extensive Cr(VI) contamination of shallow sand and gravel aquifers. In the field, Cr(VI) movement has been found to be unretarded or only slightly retarded by adsorption to aquifer materials. This is consistent with laboratory findings.

Chromium(VI) tends to be moderately to highly mobile in most shallow groundwaters due to the lack of solubility constraints and the low to moderate adsorption of Cr(VI) anionic species by iron and other metal oxides and hydroxides in neutral to alkaline waters. Above a pH of 8.5, no Cr(VI) adsorption is expected to occur. Chromium(VI) adsorption generally increases with decreasing pH, so adsorption can be very high in neutral to acidic groundwater. Cromium(VI) adsorption is strongly nonlinear, such that adsorption decreases with increasing Cr(VI) concentration. Adsorption also appears to be rate-dependent, so the kinetics of the adsorption process are very important, particularly in high-velocity groundwater regimes.

Chromium(III) tends to be relatively immobile in most groundwaters because of the precipitation of low-solubility Cr(III) compounds above a pH of 4 and because of the high adsorption of the Cr^{3+} ion by soil clays below a pH of 4. Adsorption of Cr(III) increases with increasing pH. Chromium(III) may be mobile in groundwater if it is in complexed form, although this has not been documented in the field.

ACKNOWLEDGMENTS

I thank J.A. Cherry for his review of the manuscript and his valuable suggestions and encouragement. I also thank my colleagues at the University of Waterloo and at Golder Associates for their comments and assistance and R. Passero and R. Minning for providing useful references.

REFERENCES

Baes, C.F. Jr., and Mesmer, R.E. (1977). *The Hydrolysis of Cations.* John Wiley & Sons, New York, 489 pp.

Bartlett, R., and James, B. (1979). "Behavior of chromium in soils. III. Oxidation." *J. Environ. Qual.* **8**, 31–35.

Bartlett, R.J., and Kimble, J.M. (1976a). "Behavior of chromium in soils. I. Trivalent forms." *J. Environ. Qual.* **5**, 379–383.

Bartlett, R.J., and Kimble, J.M. (1976b). "Behavior of chromium in soils. II. Hexavalent forms." *J. Environ. Qual.* **5**, 383–386.

Bloomfield, C., and Pruden, G. (1980). "The behavior of Cr(VI) in soil under aerobic and anaerobic conditions." *Environ. Pollut. (Series A)* **23**, 103–114.

Deutsch, M. (1972). "Incidents of chromium contamination of groundwaters in Michigan." In W.A. Pettyjohn, Ed., *Water Quality in a Stressed Environment.* Burgess Publishing Co., Minneapolis, pp. 149–159.

Freeze, R.A., and Cherry, J.A. (1979). *Groundwater.* Prentice-Hall, Englewood Cliffs, NJ, 604 pp.

French, W.B., Gallagher, M., Passero, R.N., and Straw, W.T. (1985). "Hydrogeologic investigation for remedial action related to a chromium-arsenic-copper discharge to soil and groundwater." In Glysson, Swan, and Way, Eds., *Innovations in Water and Wastewater Fields.* Ann Arbor Press, Ann Arbor, MI, pp. 209–229.

Griffin, R.A., Au, A.K., and Frost, R.R. (1977). "Effect of pH on adsorption of chromium from landfill-leachate by clay minerals." *J. Environ. Sci. Health* **A12**, 431–449.

Grove, D.B. (1985). "Hexavalent chromium contamination of an alluvial aquifer near Telluride, Colorado." *2nd Annual U.S.G.S. Conference on Hazardous Waste Disposal.* Cape Cod, October, 1985. Extended abstract.

Grove, D.B., and Stollenwerk, K.G. (1985). "Modeling the rate-controlled sorption of hexavalent chromium." *Water Resources Res.* **21**, 1703–1709.

Health and Welfare Canada (1979). *Guidelines for Canadian Drinking Water Quality 1978,* 76 pp.

Hem, J.D. (1977). "Reactions of metal ions at surfaces of hydrous iron oxide." *Geochim. Cosmochim. Acta* **41**, 527–538.

James, B.R., and Bartlett, R.J. (1983a). "Behavior of chromium in soils. V. Fate of organically complexed Cr(III) added to soil." *J. Environ. Qual.* **12**, 169–172.

James, B.R., and Bartlett, R.J. (1983b). "Behavior of chromium in soils. VII. Adsorption and reduction of hexavalent forms." *J. Environ. Qual.* **12**, 177–181.

Langård, S. (1980). "Chromium." In H.A. Waldron, Ed., *Metals in the Environment.* Academic Press, London, pp. 111–132.

Lieber, M., Perlmutter, N.M., and Frauenthal, H.L. (1964). "Cadmium and hexavalent chromium in Nassau County ground water." *J. AWWA* **56**, 739–747.

National Academy of Sciences (NAS). (1974). *Chromium.* National Research Council, Committee on Biologic Effects of Atmospheric Pollutants, Washington, DC, 155 pp.

Perlmutter, N.M., and Lieber, M. (1970). *Dispersal of Plating Wastes and Sewage Contaminants in Ground Water and Surface Water, South Farmingdale Massapequa Area, Nassau County, New York.* U.S. Geological Survey Water Supply Paper 1879-G, 67 pp.

Pickens, J.F., and Lennox, W.C. (1976). "Numerical simulation of waste movement in steady groundwater flow systems." *Water Resources Res.* **12**, 171–180.

Pinder, G.F. (1973). "A Galerkin-finite element simulation of groundwater contamination on Long Island, New York." *Water Resources Res.* **9**, 1657–1669.

Robertson, F.N. (1975). "Hexavalent chromium in the ground water in Paradise Valley, Arizona." *Ground Water* **13**, 516–527.

Schroeder, D.C., and Lee, G.F. (1975). "Potential transformations of chromium in natural waters." *Water Air Soil Pollut.* **4**, 355–365.

Stollenwerk, K.G., and Grove, D.B. (1984). "Adsorption and desorption of hexavalent chromium in an alluvial aquifer near Telluride, Colorado." *J. Environ. Qual.* **14**, 150–155.

Stumm, W., and Morgan, J.J. (1981). *Aquatic Chemistry.* John Wiley & Sons, New York, 780 pp.

US EPA. (1976). "National interim primary drinking water regulations." Washingtn, D.C. Report No. EPA/570/9-76/003, 163 pp.

Wiley, K.G. (1983). *Hydrogeological Investigation Examining the Identification and Cleanup of Chromium Contaminated Groundwater in Richland Township, Kalamazoo County, Michigan.* M.S. Thesis, Wright State University.

9

CHROMIUM REMOVAL FROM INDUSTRIAL WASTEWATERS

Stephen Beszedits

B & L Information Services
Toronto, Ontario, Canada

1. Fate of Chromium in Biologic Treatment Systems
2. Conventional Chemical Precipitation
3. Electrochemical Reduction
4. Sulfide Precipitation
5. The Re-elixirization Process by Nippon Electric Co.
6. Mitsubishi's Electrolytic Ferrite Treatment System
7. Insoluble Starch Xanthate
8. Cellulose Xanthate
9. Complexation with Polygalacturonic Acid
10. Cementation
11. Ion Exchange
12. Reverse Osmosis
13. Electrodialysis
14. Electrolytic Recovery
15. The ChromeNapper System
16. Solvent Extraction
17. Liquid Membranes
18. Evaporation
19. Foam Separation
20. Freeze Separation

Heavy metals is the classification term generally applied to those metals of particular concern in the treatment of industrial wastewaters, namely, copper, silver, zinc, cadmium, mercury, lead, chromium, iron, and nickel (Lanouette, 1977). Other metals that may be considered in this category are tin, arsenic, molybdenum, cobalt, manganese, and aluminum.

Metals can exist in wastewaters in various forms. For example, chromium occurs in aqueous systems in both the trivalent form (Cr^{3+}) and the hexavalent form (Cr^{6+}). Hexavalent chromium present in wastewaters is primarily in the form of chromate (CrO_4^{2-}) and dichromate ($Cr_2O_7^{2-}$).

Because chromium is widely used in a number of industries, chromium-contaminated wastewaters can originate from a multitude of sources (see Chapter 3). In addition to the metal finishing industry, chromium is used in the manufacture of inks, dyes, pigments, glass, ceramics and certain glues. It is also employed in chromium tanning, textile dyeing, and wood preserving. Chromium compounds are used to inhibit corrosion in cooling waters. Consequently, effluents may contain a wide range of concentrations of either hexavalent or trivalent chromium or both. Typical ranges of concentrations of chromium and other heavy metals in various plating effluents are depicted in Tables 1 and 2.

A wide array of technologies are available for the removal of chromium from wastewaters. Some of these are well-established methods that have been in practice for decades; others are more recent innovations. Whereas a number of processes simply remove chromium from wastewaters, emphasis in recent years has been on recovery and use. The following sections describe a host of processes, illustrated with case studies drawn from laboratory experiments and full-scale installations. Additional details on various aspects of metal removal from wastewaters may be found in the articles by Bell (1979), Lanouette (1977), and Stinson (1979) and in the book by Cherry (1982).

In practical applications, the choice of one form of treatment versus another depends on many factors, including the form and concentration of chromium in the wastewater, other constituents present in the effluent stream, extent of removal desired, capital and operating costs associated with a particular treatment technology, environmental regulations governing the discharge of treated wastewater, and the amount of sludges or residues generated and their

Table 1   Composition of Raw Waste Streams from Common Metals Plating

Constituent	Range (mg/L)
Copper	0.032–272.5
Nickel	0.019–2954
Total chromium	0.088–525.9
Hexavalent chromium	0.005–334.5
Zinc	0.112–252
Cadmium	0.007–21.6
Lead	0.663–25.39

Source: U.S. EPA, 1979b.

Table 2   Metal Concentrations in Raw Industrial Waste Streams

Industry	Metal	Concentration Range (mg/L)
Anodizing plants	Total chromium	0.268–79.2
	Cr(VI)	0.005–5.0
Coating plants	Total chromium	0.190–79.2
	Cr(VI)	0.005–5.0
	Iron	0.410–168
Chemical milling and etching	Copper	0.206–272
	Total chromium	0.088–526
	Cr(VI)	0.005–335
	Zinc	0.112–200
	Iron	0.0075–263
Printed board industry	Copper	1.58–536
	Nickel	0.027–8.44
	Lead	0.044–9.70
	Total chromium	0.005–38.5
	Cr(VI)	0.004–3.54

Source: U.S. EPA, 1979b.

disposal costs. Selection of the most appropriate treatment system for any given waste stream, whether consisting of one or more individual metal removal processes, requires accurate determination of the volume and characteristics of the effluent along with bench-scale treatability studies. Advice provided by chemical suppliers and equipment vendors also plays a role in the selection process.

In addition to the techniques described, there are other ways to cope with wastewaters. For example, effluents can often be recycled in a manufacturing operation (Pierce and Thorstensen, 1976) or used by other industries (Golomb,

1974). The growing popularity of waste exchange programs in recent years attests to the fact that one man's waste is often another man's resource (Laughlin, 1980).

1. FATE OF CHROMIUM IN BIOLOGIC TREATMENT SYSTEMS

Because of their effectiveness and favorable capital and operating costs, biologic methods are popular for the treatment of both domestic sewage and a host of industrial wastewaters. Of the various biologic treatment systems, the activated sludge process is widely used. Although the basic function of activated sludge treatment is to remove dissolved and suspended organic matter, it can also tolerate a limited concentration of heavy metals. However, beyond a certain level, heavy metals exert an adverse effect on performance; manifestations of metal toxicity include decreased biochemical oxygen demand (BOD) and chemical oxygen demand (COD), inhibition of nitrification, and deflocculation of sludge (Brown and Lester, 1979; Moore et al., 1961).

Effect of metals on the activated sludge process depends on a multitude of factors. These may be divided into plant-operating parameters, physical or chemical factors, and biologic factors (Brown and Lester, 1979). For example, free ionic metals generally exert a far more deleterious effect than complexed or chelated species. Moreover, acclimated systems are more resistant to metals than are unacclimated ones.

Activated sludge treatment reduces metal concentrations by widely varying and largely unpredictable amounts. Several mechanisms of metal removal in activated sludge have been proposed, including (Brown and Lester, 1979):

1. Physical trapping of precipitated metals in the sludge floc matrix
2. Binding of soluble metal to extracellular polymers
3. Accumulation of soluble metal by the cell
4. Volatilization of metal to the atmosphere

In the case of chromium, the oxidation state has been found to affect removal efficiency. Chromium entering a municipal treatment plant in the hexavalent form may be reduced to the trivalent form during aeration. One source reported typical removal efficiencies for trivalent chromium during activated sludge treatment to be between 70 and 90% and for hexavalent chromium to be about 20% (Brown and Lester, 1979). Hence, it appears that conversion of hexavalent chromium into trivalent chromium is desirable in the activated sludge process.

Since metals cannot be destroyed, when a treatment plant removes metals from the liquid flow, they are concentrated in the much smaller flow of sludge. Consequently, metal concentrations in the sludge can become quite large. The amount of metals detected in digested sludges from 33 municipal treatment works of varying sizes in one survey was (Cohen, 1977):

Metal	Median Value (Dry Sludge Basis) (mg kg^{-1})
Cadmium	31
Chromium	1,100
Copper	1,230
Mercury	6.6
Nickel	410
Lead	830
Zinc	2,780

Heavy metals are known to have an inhibitory effect on the anaerobic digestion of sewage sludge. Reports differ considerably on how much metal may be tolerated. According to the results of one laboratory study, an anaerobic digester was operated satisfactorily with as much as 3.5% chromium in the solids (Moore et al., 1961).

2. CONVENTIONAL CHEMICAL PRECIPITATION

Chemical precipitation is the most common technique for the removal of heavy metals from wastewaters. The chemicals most frequently used for precipitation of metals are lime, caustic, and sodium carbonate. Although most heavy metals are precipitated readily by pH adjustment, hexavalent chromium is highly soluble and does not precipitate out of solution at any pH. Consequently, treatment for chromium usually consists of a two-stage process: first, the reduction of hexavalent chromium to the trivalent form and, second, the precipitation of the trivalent chromium. Hexavalent chromium-bearing streams are segregated and treated separately; the reduced chrome-containing effluent can then be blended with other metal-contaminated streams for further treatment.

Reducing agents most commonly employed are gaseous sulfur dioxide or a solution of sodium bisulfite. Because the reaction proceeds rapidly at a low pH, an acid, for example, sulfuric acid or hydrochloric acid, is added to keep the wastewater pH between 2 and 3. At pH 2, the reduction reaction is almost instantaneous, whereas at pH 3, the reaction takes about 30 minutes Lanouette (1977). The presence of oxidizing agents can prolong treatment time by competing with hexavalent chromium for the reducing agent. Other suitable reducing reagents include sodium sulfite, sodium hydrosulfite, and ferrous sulfate (Kraljik, 1975).

A wide variety of industries have adopted this treatment technology for treating their wastewaters. Based on laboratory studies, Campbell et al. (1977) found sulfur dioxide treatment, using a constant excess SO_2 addition, to be the best practical technology for consistently reducing hexavalent chromium levels to less than 0.5 mg/L in the discharge stream of a multiproducts pigments manufacturing complex.

Treatment of chromium waste arising from plating and metal finishing operations at IBM's Federal Systems Division at Owego, NY, involves reduction of hexavalent chromium to the trivalent state at pH 2–3 with sodium bisulfite, followed by pH adjustment to 8.0 with lime or caustic soda (Forbes, 1979).

A process consisting of hexavalent chromium reduction with ferrous sulfate, neutralization with NaOH, and polyelectrolyte addition to enhance the settling of suspended solids was developed at a major lighting manufacturing facility to treat aluminum anodizing and finishing wastewaters (Olver et al., 1978). According to laboratory and pilot plant results, the treatment system is capable of producing a final effluent containing less than 5.0 mg L^{-1} of aluminum, less than 0.5 mg L^{-1} of copper, and less than 0.5 mg L^{-1} of chromium.

To treat its waterborne wood preservative effluents, a Georgia plant installed a continuous treatment system in which hexavalent chromium is reduced to the trivalent form by the addition of sulfur dioxide, followed by pH adjustment to 8.5 with hydrated lime, yielding settleable precipitates of chromium, copper, fluoride, and arsenic (Teer and Russell, 1972).

In the laboratory, treatability studies conducted by Bishop (1978) on the effluent stream of a New Hampshire tannery containing about 43 mg L^{-1} of trivalent chromium, maximum chromium precipitation occurred at pH 8.5 when sodium hydroxide was used. At this pH, the concentration of chromium in the supernatant was less than 0.01 mg L^{-1}. An Indian publication reported that chromium can be recovered from spent tan liquors by precipitating it with lime at pH 6.6 and then redissolving the precipitated chromium hydroxide in sulfuric acid to give chromium sulfate, which can be recycled for further tanning (Arumugam, 1976).

Precipitation with lime was found to be an effective and economical method for removing heavy metals from the discharge of a medium-sized multiproduct dyestuffs company in Britain (Anderson and Clark, 1978). Typical metal content of the raw wastewater was:

Chromium	18.0 mg L^{-1}
Lead	3.0 mg L^{-1}
Copper	1.5 mg L^{-1}
Zinc	17.0 mg L^{-1}
Iron	19.0 mg L^{-1}
Manganese	26.0 mg L^{-1}

According to the results of pilot plant experiments, the concentration of chromium in the final effluent was consistently less than 1 mg L^{-1}.

3. ELECTROCHEMICAL REDUCTION

Reduction of hexavalent chromium can also be accomplished with electrochemical units. These systems are competitive with treatments that employ chemical-reducing compounds. The electrochemical chromium reduction process uses consumable iron electrodes and an electrical current to generate ferrous ions that react with hexavalent chromium to give trivalent chromium as follows (U.S. EPA, 1979c):

$$3 Fe^{2+} + CrO_4^{2-} + 4 H_2O \rightarrow 3 Fe^{3+} + Cr^{3+} + 8 OH^-$$

The reaction occurs rapidly and requires minimum retention time. Hexavalent chromium in the effluent can be reduced to less than 0.05 mg L^{-1}. If the pH of the wastewater is maintained between 6 and 9, the ferric and trivalent chromium ions will precipitate as hydroxides. The major disadvantage of the electrochemical reduction system is that it results in an increased quantity of sludge; the additional sludge results from the precipitated iron hydroxide.

4. SULFIDE PRECIPITATION

Since most heavy metals form stable sulfides, excellent metal removals can be attained by sulfide precipitation (U.S. EPA, 1980). Not only can sulfide precipitation give lower residual metal concentrations than hydroxide precipitation, but metal sulfides usually settle faster and can be dewatered more readily than hydroxide sludges. Treatment with sulfide is most advantageous when used as a polishing step after conventional hydroxide precipitation or when very high metal removals are required.

Sulfide precipitation processes currently used for wastewater treatment fall into two broad categories: the soluble sulfide process (SSP) and the insoluble sulfide process (ISP). In the SSP, the sulfide is added in the form of a water-soluble reagent, such as sodium sulfide.

Although the SSP can potentially reduce Cr^{6+} and precipitate it as chromium hydroxide in one step, a pilot plant study demonstrated that in actual practice the reduction could be accomplished satisfactorily only if ferrous ions or some other suitable secondary metals were present (U.S. EPA, 1980). In these reactions, the ferrous ion acts primarily as a catalyst for chromium reduction.

In the proprietary ISP process called Sulfex, developed by the Permutit Co., a freshly prepared sulfide slurry (made by reacting $FeSO_4$ and NaHS) serves as the source of the sulfide ions (Feigenbaum, 1978; U.S. EPA, 1980). It operates on the principle that the resultant FeS will dissociate into ferrous ions and sulfide ions to the degree predicted by its solubility product. As sulfide ions are consumed, additional FeS will dissociate to maintain the equilibrium concentration of ions. Sulfide and ferrous ions reduce hexavalent chromium to the trivalent state, thereby eliminating the need to segregate and treat the chromium wastes separately. Under alkaline conditions, the chromium will then precipitate as chromium hydroxide.

5. THE RE-ELIXIRIZATION PROCESS BY NIPPON ELECTRIC CO.

Developed by Japan's Nippon Electric Co., the so-called re-elixirization process is a ferrite coprecipitation technique (Iammartino, 1975b; Okuda et al., 1975). In this treatment method, the wastewater is first mixed with a divalent ferrous salt, usually ferrous sulfate. Approximately 2 mol or more of salt are required per mol of metal. Once the divalent iron ion coexists with the nonferrous metal ion in aqueous solution, an equivalent amount of alkali is added to neutralize the waste stream, yielding a dark green hydroxide mixture. When the hydroxide is oxidized with air under specific conditions, redissolution and complex formation take place, eventually forming a black spinel ferrite.

According to Nippon Electric Co., the re-elixirization process offers several advantages over conventional alkali precipitation for heavy metals removal. For example, whereas in conventional processes, the most suitable conditions for precipitation vary, depending on the type of heavy metal involved, in the re-elixirization process, all the heavy metals, including hexavalent chromium, can be treated simultaneously once the most appropriate condition for precipitating ferrite has been established. The hexavalent chromium is removed by reduction to the trivalent form followed by precipitation. Unlike simple metal hydroxides, ferrite precipitates do not redissolve. Because of their large particle size and ferromagnetic properties, the precipitates can be easily separated and recovered from solution either by conventional filtration or magnetically.

Not only synthetic samples but actual industrial discharges have been treated successfully by this process. Treatment of wastewater from Nippon's own Central Research Laboratory yielded the following results (Okuda et al., 1975):

Metal	Concentration, mg L^{-1}	
	Influent	Effluent
Mercury	7.4	0.001
Cadmium	240	0.008
Copper	10	0.01
Zinc	18	0.016
Chromium	10	0.01
Nickel	1,000	0.2
Manganese	12	0.007
Iron	600	0.06
Bismuth	240	0.1
Lead	475	0.01

6. MITSUBISHI'S ELECTROLYTIC FERRITE TREATMENT SYSTEM

The electrolytic ferrite formation treatment system devised by another Japanese firm, Mitsubishi Petrochemical Co., is quite similar to Nippon Electric's re-elixirization technique (Anon., 1978a). However, in the Mitsubishi process, iron dissolved in the waste stream by electrolysis of iron anodes supplies the hydroxide that coagulates the pollutants. After adding an alkali for pH control and a flocculant, the waste stream is fed to a thickener from which the sludge is withdrawn to a proprietary oxidation vessel. Aeration completes the formation of the crystalline ferrite precipitate, which is separated from the liquid phase by magnetic or convection techniques.

With this technique, by supplying iron ions to chromium-containing waste-waters in excess of the stoichiometric requirement for the reduction of Cr^{6+}, hexavalent chromium can be reduced in the whole pH range without controlling pH during reduction, and the resultant trivalent chromium can be removed completely as a spinel-type chromite, that is, chromium ferrite. In order to reduce Cr^{6+} to the trivalent form and completely incorporate it into the ferrite sludge, the atomic ratio of iron to chromium must be greater than 5 (Nojiri et al., 1980).

The process is capable of very high chromium removals. For example, the level of hexavalent chromium in electroplating wastewater was reduced from 200 mg L^{-1} to less than 0.05 mg L^{-1} at one site (Anon., 1978a).

7. INSOLUBLE STARCH XANTHATE

Insoluble starch xanthate (ISX) is made from commercial crosslinked starch by reacting it with sodium hydroxide and carbon disulfide. To give the product stability and to improve the sludge settling rate, magnesium sulfate is added also. ISX works like an ion exchanger, removing the heavy metals from the wastewater and replacing them with sodium and magnesium. Average capacity is 1.1–1.5 mEq of metal ion per gram of ISX (Anon., 1978b). The production and use of ISX are illustrated schematically in Figure 1. The most cost-effective use of ISX is realized when it is applied as a polishing step after conventional hydroxide precipitation.

ISX is most commonly used by adding it to the wastewater as a slurry for continuous flow operations or in the solid form for batch treatments. It should be added to the effluent at pH $\geq$ 3. Then the pH should be allowed to rise above 7 for optimum metal removal (Wing, 1978). Stoichiometric quantities of ISX at pH above 7 will in most cases reduce heavy metal concentrations below discharge limits. In some cases, less than stoichiometric amounts of ISX can give excellent results. Heavy metal removal is virtually instantaneous. However, longer contact times are not detrimental and in most cases increase removal. The resultant ISX–metal sludge settles quickly and dewaters to 30–90% solids after filtration or

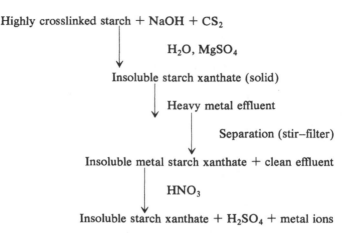

Highly crosslinked starch + NaOH + CS$_2$

H$_2$O, MgSO$_4$

Insoluble starch xanthate (solid)

Heavy metal effluent

Separation (stir–filter)

Insoluble metal starch xanthate + clean effluent

HNO$_3$

Insoluble starch xanthate + H$_2$SO$_4$ + metal ions

Figure 1. Preparation and use of insoluble starch xanthine. (From Wing, 1978)

centrifugation. However, due to the viscous nature of ISX, the sludge is somewhat difficult to handle and dewater.

Treatment with ISX is very effective, not only for the removal of common heavy metal ions but also for the removal of hexavalent chromium and some complexed metals. When ISX is used for the removal of hexavalent chromium, the pH is first lowered to 3 and then raised to 8 (Wing, 1978). In the acidic environment, ISX reduces Cr^{6+} to the trivalent form, and in the alkaline solution the chromium is removed as Cr^{3+}–starch xanthate or chromic hydroxide.

Treatment with ISX generates a considerable amount of sludge. Although metals can be reclaimed from ISX sludge by treatment with 4N nitric acid or by incineration, which yields metal oxides, recovery has been found to be economically viable only for gold (Anon., 1978b). The usual method of disposal of the ISX–metal sludge is landfilling.

When 1,000 ml solutions containing the individual metals at the concentrations shown in Table 3 were treated with the indicated amounts of ISX, the residual concentrations depicted in Table 3 were obtained. In these experiments, the solutions were stirred for 50–60 minutes at a final pH of 8.9. Treatment of a 1,000 ml solution containing a mixture of heavy metals at the concentrations indicated in Table 4 with 0.32 g of ISX to a final pH of 8.9 yielded the residual metal levels shown.

8.   CELLULOSE XANTHATE

Treatment with ISX added to wastewaters as a solid or slurry suffers from several shortcomings, including:

Table 3 Removal of Heavy Metal Cations from Water with Insoluble Starch Xanthate (ISX)

Metal	Initial Metal Concentration $(mg\ L^{-1})$	ISX (g)	Residual Metal Concentration $(mg\ L^{-1})$
Ag^+	53.94	0.32	0.016
Au^{3+}	30.00	0.50	0.010
Cd^{2+}	56.20	0.64	0.012
Co^{2+}	29.48	0.64	0.090
Cr^{3+}	26.00	0.64	0.024
Cu^{2+}	31.77	0.32	0.008
Fe^{2+}	27.92	0.32	0.015
Hg^{2+}	100.00	0.64	0.001
Mn^+	27.47	0.64	0.015
Ni^{2+}	29.35	0.64	0.160
Pb^{2+}	103.60	0.64	0.035
Zn^{2+}	32.69	0.32	0.294

Source: Anon., 1976.

Table 4 Removal of Metals from Dilute Solution with Insoluble Starch Xanthate (ISX)

Metal	Initial Concentration $(mg\ L^{-1})$	Residual Concentration $(mg\ L^{-1})$
Cd^{2+}	5.62	0.001
Co^{2+}	2.95	0.010
Cr^{3+}	2.60	0.026
Cu^{2+}	3.18	0.005
Fe^{2+}	2.79	0.001
Hg^{2+}	10.00	0.0007
Mn^{2+}	2.75	0.010
Ni^{2+}	2.93	0.050
Pb^{2+}	10.36	0.031
Zn^{2+}	3.27	0.007

Source: Anon., 1976.

1. Slow sedimentation of suspended solids
2. High turbidity and cloudy appearance of the supernatant liquid
3. Formation of deposit on the filter medium when the resultant sludge is filtered
4. Difficulties in handling the sticky and gummy filter cake

Researchers at General Motors Corporation developed a treatment process using cellulose xanthate, since the fibrous nature of cellulose xanthate makes dewatering and handling of the spent cellulose metal sludge much easier (Hanway et al., 1979).

Batchwise treatment of four industrial wastewater samples containing lead, nickel, copper, chromium, and iron with cellulose xanthate in one study indicated that the pH needed to achieve minimum levels of residual metals was in the range 6.5–9.5. These experiments did not provide a conclusive basis for establishing optimum dosage. Indeed, in subsequent continuous pilot scale runs, the amount of cellulose xanthate required for effective metal removal was significantly less (by a factor of 5 to 10) than the dosage employed in the bench-scale experiments.

Subjecting the cellulose xanthate metal precipitate to leaching tests with water indicated that only minimal leaching would occur; the metal ion concentrations detected in the filtrates were:

Lead	$0.05–0.12$ mg L^{-1}
Copper	$0.02–0.46$ mg L^{-1}
Nickel	$0.05–2.40$ mg L^{-1}
Iron	$0.03–0.06$ mg L^{-1}
Chromium	$0.05–4.5$ mg L^{-1}

According to the investigators, these results imply that no contamination of groundwater supplies would occur from the leaching of spent cellulose xanthate placed in landfills, provided proper disposal procedures are adhered to.

9. COMPLEXATION WITH POLYGALACTURONIC ACID

A number of naturally occurring polyelectrolytes readily complex with heavy metals. Of these substances, polygalacturonic acid (PGA), a derivative of pectins, in particular has received much attention. This natural polyacid forms sparingly soluble complexes with metals present in the form of cations, for example, Cr^{3+} (Jellinek and Sangal, 1972). Metals present in anionic form, for example, hexavalent chromium, can also be precipitated with PGA by using it in conjunction with a polybase, such as polyethylene immine.

10. CEMENTATION

Cementation, that is, the displacement of a metal from solution by a metal higher in the electromotive series, offers an attractive possibility for treating any waste stream containing reducible metallic ions. The fundamental principles of cementation have been known from time immemorial. Iron cementation in launders, for example, is the oldest and most common technique for winning copper from its ore.

Strictly speaking, the term *cementation* is restricted to cases where the displaced metal is precipitated from solution. However, the term may be extended to cover the reduction of any ionic species by contact with a metal of higher oxidation potential whether the reduced species is precipitated or not, for example, the reduction of ferric iron to the ferrous form by contact with metallic iron or zinc or the reduction of hexavalent chromium to the trivalent state by contact with iron (Case, 1974).

In practice, a considerable spread in the electromotive force between metals is necessary to ensure adequate cementation capability. Because of its low cost and ready availability, scrap iron is the sacrificial metal used most often.

Cementation is especially suited for small wastewater flows. It is not very practical for large flows because of the long contact times required. The sacrificial metal can be packed into columns or into rotating reactors, for example, rotating or tumbling barrels.

Some common examples of cementation in wastewater treatment include the precipitation of silver from photoprocessing discharges, the precipitation of copper from printed etching solutions, and the reduction of Cr^{6+} in chromium plating, iriditing, and chromate-inhibited cooling water discharges (Case, 1974). In order to cope with the hexavalent chromium and other heavy metals in its wastewater stream, a Connecticut manufacturer of consumer and automotive products installed a treatment system featuring cementation some years ago, primarily on its merits as a continuous hexavalent chromium-reduction process (Jester and Taylor, 1973). In the presence of sufficient clean scrap steel, the system is capable of rapidly converting chromium to the trivalent form at pH less than 2.5 and can reclaim a considerable amount of copper.

11. ION EXCHANGE

Ion exchange is a stoichiometric and reversible chemical reaction wherein an ion from solution is exchanged for a similarly charged ion attached to an immobile solid particle (U.S. EPA, 1981). Although there are numerous inorganic materials possessing ion exchange capability, the synthetic organic resins are the predominant type used today because their characteristics can be tailored to specific applications. Each resin has a distinct number of mobile ion sites that set the maximum quantity of exchange per unit of resin.

Cation exchange resins are those that have positively charged mobile ion exchange, whereas anion exchangers are those whose exchangeable ions are negatively charged. By using a cation exchange resin and an anion exchange resin in combination, virtually all the mineral contaminants can be removed from a solution. Resins can be broadly classified as strong or weak cation exchangers or strong or weak anion exchangers. Chelating resins behave similarly to weak acid cation resins but exhibit a high degree of selectivity for metal cations over sodium, calcium, or magnesium ions.

Ion exchange processing can be accomplished by either a batch method or a column method. Most industrial applications of ion exchange use fixed bed column systems. Another system that deserves more than a passing mention is the reciprocating flow ion exchanger—a proprietary unit especially developed for purifying the bleed stream of a large volume solution, such as overflow from a plating rinse tank. It operates on the principle that, for the short period of time the unit goes off stream for regeneration, the buildup of contaminants in the rinse system is negligible. These units are more attractive than fixed bed systems for plating chemical recovery because the columns use smaller resin volumes. Consequently, capital costs and space requirement are usually lower.

After the feed solution is processed to the extent that the resin becomes exhausted and cannot accomplish any further ion exchange, the resin must be regenerated. If the spent regenerant solution cannot be used, it must be processed for recycling or recovery of the metals, or it must be treated or disposed of in an environmentally acceptable manner.

There are a vast number of case studies cited in the literature on the use of ion exchange technology for the removal of chromium from wastewaters. One study demonstrated that strongly basic ion exchange resins could remove chromate from the effluents generated by a zinc yellow pigment manufacturing plant (Robinson et al., 1974). Furthermore, the recovered chromate solution could be recycled into product manufacture without sacrificing product quality.

Ellis and Kunin (1977) reported that Amberlite IRA-94 weak basic anion exchange resin can effectively and economically remove hexavalent chromium from cooling tower blowdowns. During regeneration, the chromate can be recovered in a form suitable for direct recycling into the cooling tower. By using this resin in conjunction with the weakly acidic cation exchange resin Amberlite DP-1, such other metals as trivalent chromium and zinc also present in the blowdown can also be removed.

In the ion exchange treatment system installed some years ago on the chromeplating line of an Illinois firm, the wastewater first passes through a column containing Amberlite IR-120 cation exchange resin to remove iron, nickel, trivalent chromium, and various other cations and then through a second column packed with Amberlite IRA-402 anion exchange resin, which removes hexavalent chromium, fluoride, and other anions (Anon., 1973b).

The development and application of a continuous ion exchange system for recovering chromic acid from spent chromeplating baths was the subject of an article by Raman and Karlson (1977). In this system, treatment and resin regeneration occur simultaneously in separate sections of the unit. Constituting the heart of the system is an endless fluid-transfusible belt containing the ion exchange resin in narrow transverse pockets.

12. REVERSE OSMOSIS

Reverse osmosis (RO) is a pressure-driven membrane process in which a feed stream under pressure is separated into a purified permeate stream and a concentrate stream by selective permeation of water through a semipermeable membrane (Applegate, 1984; Gooding, 1985). To obtain reasonable flow rates through the membranes, RO systems are operated at $21-105$ kg cm^{-2}.

The earliest practical RO membrane was cellulose acetate. Asymmetric cellulose acetate is still widely used in RO applications; also available are asymmetric polyamide and thin film composites of polyamide and several other polymers. Differing considerably from conventional polymeric barriers are dynamic membranes. These are formed by passing a dilute solution of membrane-forming material over a clean porous substrate at high pressure. Although several membrane materials are possible, most of the work has involved composites of polyacrylic acid and hydrous zirconium oxide deposited on porous stainless steel or ceramic tubes. There are three types of RO configurations used in commercial applications: tubular, spiral wound, and hollow fiber.

Since RO is not a thermal process and performs a separation without a phase change, energy requirements are low, and it can be used to concentrate temperature-sensitive materials without loss of quality. Another advantage is that no sludge is produced. Reverse osmosis systems are compact, and space requirements are modest. Capital costs are relatively low, and the process is simple to operate and can be fully automated, requiring little operator attendance. Moreover, because RO plants are modular in design, maintenance can be performed without shutting down the entire plant. The modular design also makes expansion an easy option.

Reverse osmosis enjoys widespread popularity in the treatment of numerous diverse wastewaters. For example, in the metal finishing industry, it is used for recovery of plating chemicals for rinsewater as well as purification of mixed wastewater to allow its reuse (Crampton, 1982; Stinson, 1979). In plating chemical recovery applications, RO units separate the valuable metal salts from rinse solutions, yielding a concentrated metal solution, which can be recycled to the plating bath, and water of sufficient purity for use in rinsing. Reverse osmosis has also been successfully demonstrated for the removal of chromium, lead, iron, nickel, copper, and zinc from vehicle washrack wastewater (Chian, 1976), the removal of chromium and other metals from acid mine drainage (Wilmoth et al., 1978), and the removal of chromium, iron, zinc, and various other constituents from cooling tower blowdown (Chian and Fang, 1976).

13. ELECTRODIALYSIS

Electrodialysis (ED) is a membrane process that can be used for the separation, removal, or concentration of ionized species in water solutions (Applegate, 1984; Schoeman, 1985). These operations are accomplished by the selective transport of ions through ion exchange membranes under the influence of an electrical potential across the membrane. Ion exchange membranes, permeable to either anions or cations but not both, are thin sheets of ion exchange material normally reinforced by forming on a synthetic fabric backing. Cation selective membranes are usually prepared from crosslinked polystyrene that has been sulfonated to produce sulfonate groups attached to the polymer. Anion membranes can be crosslinked polystyrene containing quaternary ammonium groups. Aliphatic anion membranes are also widely used because of their lower electrical resistance. In the usual configuration employed for ED, hundreds of alternating anion selective and cation selective membranes are arrayed in parallel between two electrodes to form an ED multicell or stack. The passage of a direct current through the ED stack causes the anions and cations in the raw process solution fed to the diluting cells (the feed) to move in the direction of the anode and cathode, respectively. Because of the alternating membrane arrangement, they leave the diluting cells and accumulate in the concentrating cells. Water exiting from the ion-depleted cells is the desalted product, and the stream leaving the ion-concentrated cells is the brine.

In recent years, conventional ED has been modified by the incorporation of polarity reversal, giving the so-called electrodialysis reversal (EDR) process (Applegate, 1984). The purpose of the EDR process is to remove deposits from the membrane. When the cell function is reversed, scale in the concentrating cells tends to redissolve. With the exception of a few key features, the EDR process is nearly the same as the ED process.

Eisenmann (1979) carried out laboratory ED experiments with a five cell-pair stack on simulated chromic acid rinse solutions prepared by dissolving chromium trioxide in tapwater. Electrodialysis of these solutions over a range of current densities yielded the following results:

Run	Current Density $(ma\ cm^{-2})$	Feed Concentration $(g\ L^{-1}\ CrO_3)$	Product Concentration $(g\ L^{-1}\ CrO_3)$
1	10	0.37	61
2	12	0.32	106
3	14	1.24	143
4	16	0.98	167
5	18	0.70	174

According to the data, chromic acid can be concentrated from dilute aqueous solutions at least to about 70% of the strength of many chromeplate solutions and may be useful for direct return to the plate tanks. During the course of these experiments, the membranes were in contact with varying concentrations of chromic acid over a period of 2–3 months without obvious deterioration or loss of selectivity.

Subsequent to the laboratory studies, a 50 cell-pair demonstration unit was installed on an actual chromeplating line of a Connecticut metal finisher. Despite some operational problems, the results were encouraging, indicating that the use of ED for chromic acid recovery is promising.

14. ELECTROLYTIC RECOVERY

Solutions containing Cr^{6+} and sulfuric acid are used in brass finishing, printed circuit board etching, preparation of plastic for plating, anodizing, and various other surface treatments. As the solutions are used, Cr^{6+} is reduced to Cr^{3+}, the dissolved solids content increases, and the acid concentration decreases. For example, a spent printed circuit etchant would contain considerable amounts of Cu^{2+}, Cr^{3+}, and Cr^{6+}, whereas a spent brass etchant would contain zinc in addition to these constituents. When the etchant no longer performs properly despite replenishment with sodium dichromate, the entire tank is dumped.

In-plant recycling of such chromium-bearing process solutions not only reduces the volume of effluents requiring treatment and disposal but also conserves chromium, thus lowering process costs by reducing chromium purchases. Several electrolysis methods have been proposed and developed for recycling. One of these methods, developed by the U.S. Bureau of Mines, is an oxidation technique involving an electrolytic diaphragm cell that oxidizes trivalent chromium back to the hexavalent state and at the same time removes copper or other metal contaminants that gradually accumulate in etching solutions (Basta, 1983; George et al., 1981; Soboroff et al., 1978).

The process uses a diaphragm cell fitted with a Nafion 427 cation-selective membrane. When the spent solution is placed in the anode chamber, most of the Cr^{3+} produced during the etching operations is oxidized to Cr^{6+}, thus allowing the recycle of the regenerated solution back to the etching tank. Impurity metals dissolved during the etching operations, for example, Cu^{2+}, Zn^{2+}, Cr^{3+}, are transferred to the catholyte. Copper plates out on the cathode, where it can be easily recovered as a powder containing 85–90% copper, whereas Zn^{2+} and Cr^{3+} are precipitated from the catholyte by the addition of a base.

15. THE CHROMENAPPER SYSTEM

Developed by Innova, Inc., the ChromeNapper system is a novel electrolytic method designed to reduce the cost of chromium recovery (Militello, 1981). An electrolytic ion transfer membrane, a proprietary substance that requires no implanting of ion exchange resin as in electrodialysis membranes, constitutes the key feature of the process. Furthermore, instead of using thin membranes separating three compartments as in the conventional electrodialysis cell, this system uses a single, thick ion-permeable membrane that separates two compartments. The membrane surrounds an inner compartment, and platinum-plated titanium anodes are inserted through the top of the module that contains the recovered chromic acid–sulfuric acid anolyte. The outside of the membrane is wrapped in a stainless steel mesh cathode. Rinsewater is the catholyte solution. Ion transfer and concentration of the chromic acid are accomplished by applying a direct current between the anodes and the mesh cathodes on the outside of each cell. Chromic acid concentrates in the anode compartment of the cell, whereas treated dilute rinsewater is returned to the rinse tanks.

Several articles appearing in trade publications have described the implementation of this treatment technique in a number of plating shops (Anon., 1980; Anon., 1981).

At one of these sites where the performance of the ChromeNapper was monitored for a 5-day period by the U.S. EPA, the ChromeNapper was dedicated to the chromium line in a closed loop mode for the purpose of recycle reuse (Militello, 1981). The objective of the system was to maintain a relatively constant chromium concentration in final rinse, with no discharge of rinsewater to the waste. Hence, as the chromium concentration increased from dragout, the ChromeNapper removed the excess.

Sampling of the influent and effluent streams showed the following average values:

	Influent	Effluent
Total chrome	11.5 mg L^{-1}	9.9 mg L^{-1}
Hexavalent chrome	6.2 mg L^{-1}	5.4 mg L^{-1}
Nickel	3.1 mg L^{-1}	2.9 mg L^{-1}
pH	7.8	8.0

Influent to the recovery system was from the final rinse tank. Effluent from the recovery unit was returned to the final rinse. The flow rate through the ChromeNapper system was chosen to be such a value that the chromium concentration into and out of the unit was held nearly constant. As shown in the table, this goal was accomplished.

16. SOLVENT EXTRACTION

Liquid–liquid extraction (also frequently referred to as solvent extraction) of metals from solutions on a large scale has experienced a phenomenal growth in recent years due to the introduction of selective complexing agents. In addition to hydrometallurgical applications, solvent extraction has gained widespread usage for waste reprocessing and effluent treatment.

Solvent extraction involves an organic and an aqueous phase. The aqueous solution containing the metal or metals of interest is mixed intimately with the appropriate organic solvent, and the metal passes into the organic phase. In order to recover the extracted metal, the organic solvent is contacted with an aqueous solution whose composition is such that the metal is stripped from the organic phase and is re-extracted into the stripping solution. The concentration of the metal in the strip liquor may be increased, often 10 to 100 times over that of the original feed solution. Once the metal of interest has been removed, the organic solvent is recycled either directly or after a fraction of it has been treated to remove impurities. There are several ways to cope with the metal-laden strip solution. However, the best approach is one that regenerates the stripping solution and recovers the metal directly, electrolytically, or by direct reduction, or as an insoluble salt.

The applicable solvent is often made up of an extractant dissolved in a diluent, since many extractants are very viscous materials in the pure state (Bailes et al., 1976). In some cases, a third component, a modifier, may be added to the organic phase to prevent salt precipitation or third-phase formation.

There are many different extractants available. The accepted classification of these into groups is based on reaction type and includes inert extractants, solvating extractants, acidic extractants, chelating extractants and ionic extractants (Bailes et al., 1976). For example, tributyl phosphate is a versatile solvating extractant, whereas di-2-ethylhexyl phosphoric acid (D2EHPA) is a common acidic extractant. LIX 64N is a well-known chelating agent, and Alamine 336, a tertiary amine, is an ionic extractant.

In addition to lowering the viscosity of the organic phase, the main function of the diluent, or carrier solvent, is to facilitate contact between the organic and aqueous phases. Commercial diluents are cuts from certain processes adopted during the production of chemicals from petroleum feedstocks and, as such, are complex mixtures. Because of its low cost, kerosene is a widely used diluent.

Of the various compounds suitable for use as modifiers, long-chain alcohols are frequently employed in solvent extraction.

The extractant evaluated for the removal of Cr^{6+}, Cr^{3+}, Co^{2+}, and Ni^{2+} from aqueous solutions in the studies of Reddy and Sayi (1977) was di-n-pentyl sulfoxide (DPSO). Although nickel could not be extracted to any appreciable extent, appropriate conditions could be developed for the removal of the other three species with this reagent.

McDonald and Bajwa (1977) found a 25% Alamine 336–xylene solution to be the most effective solvent for the recovery of chromium, cadmium, and zinc from metal finishing wastewaters taken from the Dixie Metal Finishing Plant of Houston, Texas. The three metals could be extracted selectively or simultaneously. Stripping agents evaluated included sulfuric acid, EDTA, and sodium hydroxide. When 4 M NaOH was used for stripping, more than 99.5% recovery of the three metals was attained, and the regenerated solvent could be recycled and reused without any loss of extraction efficiency.

17. LIQUID MEMBRANES

A liquid membrane is a thin liquid film that selectively permits the passage of a specific constituent from a mixture. Unlike solid membranes, however, liquid membranes separate by chemistry rather than size, and thus in many ways liquid membrane technology is similar to solvent extraction.

There are basically two types of liquid membranes: supported and unsupported. With supported membranes, the liquid is impregnated into the pores of a solid membrane. Unsupported membranes, also known as emulsion membranes or liquid surfactant membranes, are usually in the form of double emulsion drops. Both types of liquid membranes operate on the same principles. A thorough discussion on the fundamental aspects of these membranes may be found in the publications by Parkinson (1983) and Stroeve and Varanasi (1982).

Since liquid membrane technology is a fairly recent development, a number of problems remain to be solved. A major issue with the use of supported membranes is the long-term stability of the membranes, whereas the efficient breakup of microspheres for product recovery is one of the difficulties encountered frequently with emulsion membranes.

In the laboratory and pilot plant experiments conducted by Kitigawa et al. (1974) with liquid surfactant membranes, the concentrations of Cr^{6+}, Cu^{2+}, Hg^{2+}, and Cd^{2+} in synthetic solutions were rapidly reduced from several hundred milligrams per liter to less than $1\ mg\ L^{-1}$ in batch and continuous flow operations. The ion carrier added to the oil phase for chromium and mercury extraction was Alamine 336 (a C_8–C_{10} tertiary amine), LIX 64N (a mixture of two oxime compounds) for the separation of copper, and Aliquat 336 (methyl trioctyl ammonium chloride) for the removal of cadmium.

In the process developed by Bend Research, Inc., proprietary microporous hollow fibers hold an oil by capillary action (Anon., 1983). The oil contains a tertiary amine that serves as a complexing agent for metal ions. The process has been successfully tested for recovering chromium from electroplating rinsewaters, reducing chromium concentration from $300\ mg\ L^{-1}$ to less than $10\ mg\ L^{-1}$, and yielding a concentrated stream containing 5 wt% chromium.

18. EVAPORATION

In the electroplating industry, evaporators are used chiefly to concentrate and recover valuable plating chemicals. Recovery is accomplished by boiling off sufficient water from the collected rinse stream to allow the concentrate to be returned to the plating bath. Many of the evaporators in use also permit the recovery of the condensed steam for recycle as rinsewater. Evaporation has been demonstrated successfully on all types of plating effluents, and consequently there are numerous units in operation at North American plating facilities.

Four types of evaporators are used throughout the electroplating industry (U.S. EPA, 1979a):

1. Rising film evaporators
2. Flash evaporators using waste heat
3. Submerged tube evaporators
4. Atmospheric evaporators

Site-specific conditions and the mode of operation determine construction materials and influence the selection of one system over another.

Ideally, evaporation yields a concentrate and a deionized solution. However, carryover can result in metal concentrations as high as 10 mg L^{-1} in the condensate, although the usual level is less than 3 mg L^{-1}, pure enough for most final rinses. The buildup of impurities, which can cause serious quality problems, can be prevented by the incorporation of treatment techniques, such as ion exchange, filtration, precipitation, and activated carbon adsorption. For example, most of the plating establishments using evaporators to recover Cr^{6+} use cation exchangers to control the buildup of trivalent chromium.

Commercial evaporators are available from a host of vendors. Developed by Corning Glass Works, the climbing film evaporator (CFE) is one of the more recent recovery techniques for use in the plating industries (Bhatia and Jump, 1977). It has a high efficiency for removing plating chemicals, and the rinsewater purified by distillation can be recycled to the rinse tanks. Low pressure steam provides the energy by which rinsewater vaporizes and drives the climbing film of concentrate into the vapor–liquid separator where the heavier metallic materials settle out while the water vapor flows through a filter into a condenser. When the recovered solution reaches the desired concentration, it is drained into a storage vessel and retained for reuse.

Results of a 6-month study of chromeplating operations centered around a CFE unit at Advanced Plating Co., Cleveland, Ohio, showed that the system could be accommodated with little impact on existing operations and that the recovered chromic acid could be recycled back into the bath without any adverse effects on product quality (Elicker and Lacy, 1978).

19. FOAM SEPARATION

Foam separation processes rely on the adsorption of a surface-active agent (ionic or nonionic) at the gas–aqueous solution interfaces of generated bubbles. In some cases, a surface-active agent already present in the waste can be used as a collector, whereas in other cases, a surface-active agent must be added to the waste, with the required charge of the long-chain surfactant ion determined by the nature and charge of the species to be separated. The surfactant is concentrated in a foam or froth formed on top of the bulk solution. Separation is accomplished by physically separating the foam and the bulk solution phases in batch operations and by taking off steady-state foam (overflow) and underflow streams in continuous operations.

Foam separations may be subdivided into several groups of processes, depending on the nature of the interaction between the surface-active agent and the contaminant of interest. These include foam fractionation, ion flotation, adsorbing colloid flotation, and microflotation (Grieves, 1970, 1975).

McIntyre et al. (1981, 1983) employed adsorbing colloid foam flotation to remove copper and zinc, mixtures of copper and zinc, and chromium from solutions. The test samples were prepared by dissolving $CuSO_4 \cdot 5H_2O$, $ZnSO_4 \cdot 7H_2O$, or $Cr(NO_3)_3 \cdot 9H_2O$, alone or in combinations in tapwater. Fe^{3+} and Al^{3+} were used as coagulant adsorbents. Sodium lauryl sulfate (NLS) served as the surfactant throughout the experiment, and its recovery from the foamate was also investigated.

Excellent results were achieved in a reasonably wide pH range working at $100-114$ mg L^{-1} Fe^{3+} and $75-100$ mg L^{-1} Al^{3+} and adjusting the NLS dosage as a function of the total floc concentration in the system when solutions containing 20 mg L^{-1} each of copper, zinc, and chromium were treated. For example, when a sample was treated with 150 mg L^{-1} Fe^{3+}, 100 mg L^{-1} Al^{3+}, and 85 mg L^{-1} NLS at pH 7.3, residual copper, zinc, and chromium concentrations were 0.06, 0.40, and 0.12 mg L^{-1}, respectively.

Treatment of the foamates with NaOH or Na_2CO_3 solution gave a readily settleable sludge and a liquid supernatant containing virtually all the NLS initially present in the foamate. The surfactant recovered from the foamate could be recycled, thereby realizing considerable savings in surfactant consumption.

To evaluate the effectiveness of a pilot plant-scale dissolved air ion flotation unit for the removal and concentration of hexavalent chromium, Grieves et al. (1968) fed a solution of water, potassium dichromate, and ethyl hexadecyl dimethyl ammonium (EHDA) bromide to the treatment apparatus. After treatment, chromium was concentrated into a liquid volume less than 1% of the volume of the original waste stream.

The surfactant was liberated by slightly depressing the pH, followed by reduction of the hexavalent chromium with sodium bisulfite. Once freed, the surfactant could be extracted readily by organic solvents and then recovered by vacuum distillation for reuse.

20. FREEZE SEPARATION

All freeze separation processes are based on difference in component concentration between solid and liquid phases that are in equilibrium (Heist, 1979). Although the largest uses of freeze crystallization are to fractionate solutions in specialized applications and to desalinate brackish water and seawaters, it has also received considerable attention for wastewater treatment. For example, it can be employed for the regeneration of spent sulfuric acid pickle liquor, and it has been found to be a feasible technique for recovering metals from spent plating and etching baths and rinsewaters.

Advantages cited for freeze separation in the treatment of metal finishing effluents include (Crumpler, 1977):

1. Energy costs are lower than for evaporative recovery.
2. Heat-sensitive plating solutions are not damaged.

3. Volume of water used for rinsing is reduced, since ice is recovered.
4. The plating chemicals can be recovered and recycled.
5. There is no or only a minimum amount of sludge to contend with.
6. Pretreatment of the waste stream is unnecessary.
7. Due to low operating temperatures, corrosion problems are minimized and hence inexpensive materials can be used for piping, pumps, valves, and so on.

One of the more recent innovations in this type of treatment is the so-called Crystalex process (Campbell and Emmerman, 1972; Iammartino, 1975a). Developed by Avco's Systems Division, it has been successfully tested on a 10,000 L per day electroplating waste, cooling tower blowdowns, and various other effluents. Treatment of electroplating rinsewater by a pilot plant unit yielded the following results (Campbell and Emmerman, 1972):

Constituent	Feed Stream (mg L^{-1})	Product Water (mg L^{-1})
Nickel	105	0.44
Cadmium	105	0.40
Chromium	110	0.225
Zinc	100	0.34
NaCl	30,000	120

21. ADSORPTION BY ACTIVATED CARBON

Activated carbon adsorption, in both the granular and powdered form, is one of the most popular physical–chemical treatment techniques for the removal of dissolved organics from wastewaters. Regeneration of the spent carbon can be accomplished by several methods; of these thermal regeneration is the most common. About 5–10% of the carbon is lost during thermal regeneration. Despite the rather steep rise in the price of virgin carbon during the past few years, activated carbon adsorption has experienced a rapid growth in the purification of municipal and industrial effluents. Due to increasingly stringent environmental regulations, activated carbon will likely remain the favorite treatment choice for the removal of toxic and hazardous organics in the years ahead. Since activated carbon also possesses an affinity for heavy metals, considerable attention has been focused on the use of carbon for the removal of hexavalent chromium, complexed cyanides, and metals present in various other forms from wastewaters.

The mechanism of removal of hexavalent and trivalent chromium from synthetic solutions and electroplating effluents has been extensively studied by a number of researchers. According to some investigators, the removal of Cr^{6+} occurs through several steps of interfacial reactions (Huang and Bowers, 1979):

1. The direct adsorption of Cr^{6+} onto the carbon surface
2. The reduction of Cr^{6+} species to Cr^{3+} by carbon on the surface
3. The adsorption of the Cr^{3+} species produced, which occurs to a much lesser extent than the adsorption of the Cr^{6+} species

Huang and Bowers (1979) found that reduction and adsorption occurred simultaneously when the kinetics of Cr^{6+} removal onto Filtrasorb 400 activated carbon were examined in batch experiments. Maximum Cr^{6+} adsorptive capacity occurred at pH 2.5 and decreased rapidly between 2.5 and 7.1, largely due to the decreasing electrostatic attraction between the positively charged carbon surface and the anionic hexavalent chromium in solution. At pH < 2.5, Cr^{6+} adsorptive capacity decreased due to the rapid reduction of hexavalent chromium species and the subsequent dominance of the cationic trivalent chromium species at low pH.

Since the practical application of activated carbon adsorption for the treatment of wastewaters depends not only on the extent of adsorption, but also on the feasibility of regenerating the spent carbon, a substantial amount of work has been devoted to identifying viable methods for regenerating spent carbon used in the treatment of hexavalent chromium-containing effluents. Wu et al. (1976) and Huang and Bowers (1979) evaluated several regeneration methods:

1. Thermal regeneration ($550°C$ in air or $950°C$ in CO_2)
2. Caustic regeneration with NaOH
3. Combined caustic–thermal regeneration (caustic and $950°C$ in CO_2)
4. Acid regeneration with HCl

The best results were obtained with the combined caustic–thermal process; more than 97% of the original removal efficiency was maintained through three regeneration cycles. Thermal regeneration at $550°C$ resulted in only partial recovery of the adsorptive capacity, whereas about 60% of the chromium adsorbed was desorbed with simple caustic regeneration. Only Cr^{3+} was stripped from the carbon surface with HCl.

22. POWDERED ACTIVATED CARBON–ACTIVATED SLUDGE

Adding powdered activated carbon (PAC) to the activated sludge process in the treatment of municipal and industrial wastewaters can impart a number of benefits, including (Adams, 1974):

1. More uniform operation and effluent quality
2. Improved BOD, COD, and TOC removals
3. Better removal of phosphorus and nitrogen

4. Less tendency for foaming in the aeration tank due to the adsorption of detergents
5. Adsorption of refractory organics and protection of the biologic system from such toxics as heavy metals
6. Enhanced color removal in the case of highly colored effluents
7. Decreased effluent solids and thicker sludges
8. Greater treatment flexibility and increased effective plant capacity at little or no added capital investment.

Carbon added to the activated sludge aids by direct adsorption of pollutant and by providing a more favorable environment for the microorganisms to propagate (Sublette et al., 1982). It may be introduced to the aeration basin in dry or slurry form at any convenient point in the plant, and it may be added continuously or batchwise. Optimum carbon dosage is best established by laboratory studies or field trials. The sludge–carbon mixture is removed in the secondary clarifier, and most of the settled sludge is recycled to the aeration tank. Like spent granular activated carbon, powdered carbon can be regenerated thermally. In recent years, wet air oxidation has become a well-established regeneration technique for PAC used in conjunction with the activated sludge process.

A wide variety of wastewaters have been treated successfully by this method, and there are numerous full-scale treatment plants using this technology. One of the in-house studies initiated at the U.S. Industrial Environmental Research Laboratory in Cincinnati, Ohio, was designed to evaluate the treatability of a dyes and pigments processing wastewater using an activated sludge process both with and without PAC (Shaul et al., 1983).

Raw wastewater was collected from a municipal sewer downstream from the discharge of the plant. The raw wastewater could be characterized as a highly variable, highly colored effluent containing significant amounts of organics and metals. Concentration of metals varied substantially during the collection of the samples. Total chromium ranged from a minimum of 0.58 mg L^{-1} to a maximum of 13.9 mg L^{-1}, total lead from 1.5 to 13.6 mg L^{-1}, total copper from 0.23 to 2.4 mg L^{-1}, and total zinc from 0.27 to 12.2 mg L^{-1}.

Influent feed rate to the continuous flow activated sludge pilot plants was maintained at about 173 L day^{-1}, resulting in a 1.2-day hydraulic detention time in the aeration basin (based on influent rate only). The activated sludge return rate from the secondary clarifier to the aeration basin was also about 173 L day^{-1}. To obtain the desired solids retention times (SRT), the rate of wasting of the mixed liquor was varied. Temperature was maintained at 22–$27°C$ and pH at 7.0–8.5. The dissolved oxygen concentration was kept at or above 2 mg L^{-1}. Powdered activated carbon dosages ranged from 850 to $14,000$ mg L^{-1}.

BOD removal was very good in all systems whether PAC was present or not. However, the units containing PAC attained higher percentages of removals of COD, TOC, and color than the units operated without PAC.

Removal trends of all metals were similar. Primary clarification reduced total lead and chromium concentrations by 50–75%. Aeration and secondary clarification further reduced total lead and chromium concentrations by an additional 50–80%. Generally speaking, the secondary effluent concentrations for total lead and chromium of the PAC–activated sludge plants was similar to or slightly lower than those from the units where PAC was absent.

23. MISCELLANEOUS ADSORBENTS

Besides activated carbon, a myriad of other substances, both natural and synthetic, possess a marked affinity for chromium and other heavy metals. Because of their low cost and ready availability, much interest has been focused especially on the use of various industrial solid wastes, agricultural byproducts and discards, and similar products as inexpensive sorbents. Peat, zeolites, clay, and soil have also been tested with varying degrees of success for removing metals from wastewaters.

Adsorption of metals by peat in the laboratory experiments of Moo et al. (1976) was found to be optimum at pH 5. When peat was added to synthetic metal solutions, 5 g of peat adsorbed 95% of the lead, 95% of the copper, 70% of the tin, and 45% of the hexavalent chromium, respectively, from 100 ml samples containing 100 mg L^{-1} metal.

Contacting actual textile dye house wastewaters with peat reduced the concentrations of such heavy metals as cadmium, copper, iron, lead, nickel, zinc, mercury, and hexavalent and trivalent chromium by about 99% at one site (Leslie, 1974).

According to an Australian article, an efficient ion exchange medium can be obtained by treating peat with a solution of calcium hydroxide (Cullen and Siviour, 1982). With calcium loadings of 12% on the peat, such metals as copper, nickel, chromium, cadmium, lead, zinc, mercury, iron, and silver were readily extracted from 1,000 mg L^{-1} solutions.

Fly ash samples obtained from three Texas power plants fuelled with lignite coal were used to remove heavy metals, including chromium, from 100 mg L^{-1} solutions in a recent study conducted at the University of Texas (Netzer and Donahue, 1983). Although metal removals of about 99% were obtained under most conditions, large amounts of fly ash were required.

Better than 99.5% removals were attained for most of the metals when aqueous solutions of cadmium, chromium, cobalt, copper, iron, lead, mercury, silver, nickel, and zinc were treated by a combination of lime and shredded automobile tires in the studies of Netzer et al. (1974).

Australian researchers have prepared a crosslinked, insoluble matrix from casein possessing an excellent affinity for heavy metals (Anon., 1979). The granular material functions like an ion exchange resin—adsorbing cations at pH 4 and anions at pH below 3—but is considerably less costly than synthetic ion exchange resins and is also biodegradable. Mining, tanning, and electroplating wastewaters are some of the effluents successfully treated with this product.

Adsorption onto silica gel removed 50–98% of the metal cations from solution of cadmium, chromium, copper, lead, manganese, mercury, nickel, and zinc in a study conducted by Rigo (1974) at the Canada Centre for Inland Waters.

The capability of soil to retain metals from solution is markedly influenced by its clay content. One publication, for example, reported that the affinity of three soils tested for the removal of copper, chromium, and zinc ions was directly proportional to the clay mineral content of the soils (Wentink and Etzel, 1972).

24. WATER HYACINTHS

Water hyacinths are ubiquitious nuisance plants throughout many tropical and subtropical parts of the world. However, water hyacinths can be used for pollution control because, with their extensive root system, they can rapidly and effectively extract a wide variety of pollutants, including heavy metals, from surrounding wastewaters (Anon., 1975).

Most of the work involving the use of water hyacinths for wastewater treatment has been conducted at the National Aeronautics and Space Administration's (NASA) National Space Technology Laboratories (NSTL) at Bay St. Louis, Mississippi. To provide a relatively inexpensive and effective means of treating a mixture of chemical and photographic wastes containing a multitude of organics, silver, cadmium, and chromium, a water hyacinth filtration system was installed at NSTL some years ago (NASA, undated). The specially designed lagoon containing the hyacinths was constructed in a zigzag shape to maximize the lagoon's length within a relatively small area. The lagoon, receiving approximately 95,000 L of wastewater per day while providing a minimum detention time of 20 days, was stocked with sufficient hyacinths to cover 20% of its surface area. The water hyacinths maintained effluent pH between 6.8 and 7.8, and the dissolved oxygen level generally remained above 5 mg L^{-1}. Reductions in dissolved solids and COD levels ranged from 29 to 75% and 83 to 92%, respectively, and BOD removals attained were approximately 98%. Over a 6-week period, the hyacinths accumulated heavy metals to concentrations several hundred times the initial levels.

Using water hyacinths for the removal of metals from industrial and municipal wastewaters has been investigated by other. Dinges (1978), for example, reported that nickel, mercury, chromium, lead, iron, zinc, and copper were readily assimilated by water hyacinths when these plants were used to upgrade stabilization pond effluents in Texas.

25. SLUDGE HANDLING AND ULTIMATE DISPOSAL

Because in many treatment methods, metals removed from wastewater end up in sludges, sludge handling and disposal constitute important segments of the overall treatment procedure. A comprehensive discussion of sludge dewatering technology may be found in an EPA publication (U.S. EPA, 1982), as well as in articles by Okey et al. (1979) and Lanouette (1977). Sanitary landfilling, chemical landfilling, landfarming, ocean disposal, deep well injection, and placement into abandoned mines are some of the leading ultimate disposal methods. Choice of one versus the others is governed by many factors, including the metal content of the sludge, disposal costs, and, most important, prevailing environmental regulations. Of the various disposal methods, chemical landfilling is regarded as the most appropriate for sludges containing high amounts of metals, that is, metal finishing wastewater treatment sludges.

Although metals can be recovered from sludges by a variety of techniques, recovery is usually practiced only when the metal is of high value or when one species is present. Separation is neither simple nor economical when there are several metals in the sludge. Techniques and case studies describing the reclamation of metal values from municipal and industrial wastewater treatment sludges have been discussed in numerous publications (Browning et al., 1979; Mueller, 1976; Mehta, 1981; Kalinske, 1981; Jenkins et al., 1981). According to a Czechoslovakian patent, chromium can be separated and recovered from mixed sludge by making moist sludge alkaline with lime and then treating the suspension with ozone until trivalent chromium is oxidized to the hexavalent form (Drokos and Bahensky, 1969). The soluble $CaCrO_4$ is then filtered out and separated from the residue.

Metal-containing sludges can be used in the manufacture of various products. For example, at an Italian factory, sludges arising from the treatment of metal finishing wastewaters are dewatered, dried, and then used for making ceramics, roof tiles, and other building products (Anon., 1974). According to a Polish article, sludges containing hydroxides of nickel, chromium, zinc, cadmium, copper, and aluminum have been found suitable as additives to low-grade cement mortars and concrete (Tuznik and Kieszkowski, 1972).

There are basically two designs for a secure chemical landfill (U.S. EPA, 1982). The first takes advantage of natural geologic barriers created by impervious clays. The second adds a flexible elastomer liner as further protection against leaching of pollutants into the groundwater. In both cases, disposal involves direct burial of wastes in cells designed to avoid contamination of the surrounding environment.

The wastes to be buried are classified and segregated, and their positions within a burial cell are noted. Sludges are solidified with various reagents before burial. Only compatible wastes are placed in a given disposal cell. When a cell is full, a compacted clay cover is placed over the top to prevent precipitation from filtering into the cell, thereby minimizing the formation of leachate.

A piping system for leachate collection is buried in a permeable bottom layer at the center of each cell. All leachate is recovered and is periodically pumped out of the cell through a standpipe connected to the piping system. The recovered leachate is solidified, then buried in the landfill. A monitoring well system is placed outside the landfill cells for early detection of any leachate that may leak out of the area. A properly operated secure chemical landfill does not usually experience leachate in its monitoring wells.

The term *solidification* collectively defines disposal technologies that fixate or encapsulate waste in a solid matrix end product (Pojasek, 1979). Fixation processes chemically and physically bind hazardous wastes with a solidification agent to form an inert material, whereas encapsulation methods physically surround the waste within a given agent or material (Serper, 1981). Encapsulation can be subdivided into two categories: microencapsulation and macroencapsulation. Microencapsulation attempts to seal each waste within a matrix, whereas macroencapsulation is the process by which the matrix containing the waste is encapsulated. Both fixation and encapsulation techniques reduce waste permeability and yield an endproduct with significant shear strength. Leading additives used in fixation processes are portland cement, fly ash, and lime (Pojasek, 1979; Serper, 1981). Polyurethane, polyethylene, asphalt, and concrete are some of the materials used for encapsulation (Fields and Lindsey, 1975).

There are many solidification processes to choose from. Most of the available techniques are fairly recent developments. The effectiveness of solidification methods is usually assessed by contacting the endproduct with a leachate solution for a predetermined time and then analyzing the metals present in the extractant. The amount of metals leached from solidified products of proven fixation–encapsulation processes is usually negligible.

The Sealosafe process, developed by Crossford Pollution Services Ltd. Great Britain, converts sludges into an environmentally stable, rocklike solid (Helsing, 1980; Schoefield, 1978). After pretreatment to ensure ease of handling, the wastes are contacted with special reactants. Ionic pollutants react with the process chemicals to form strong chemical bonds while waste constituents present in insoluble form are dispersed and trapped. The resultant slurry is transported to the designated landfill site where it forms into a hard, inert mass in about 3 days. The compressive strength of the solidified product is less than that of concrete and mortar but is comparable to industrial grouts. Mine tailings and drosses, electroplating residues, organic sludges, metal finishing chromium sludges, tin production wastes, and latex and paint sludges are some of the wastes successfully solidified by the Sealosafe process.

The process of Chemfix Inc. uses a two-part inorganic system that reacts with all polyvalent metal ions and various other waste constituents to yield a chemically and mechanically inert soillike product (Anon., 1973a; Conner, 1974; Gray, 1977). The complex chemical reactions basically evolve from the reactions between soluble silicates and silicate setting agents that interact to yield a pseudomineral solid matrix. Pollutants are retained by physical entrapment as well as by physical and chemical bonds. Brine sludges, steel plant sludges, and chemical manufacturing residues are some of the myriad of wastes disposed of by this technique.

Developed by the Belgian firm, Cemstobel SA, the Soliroc process treats waste sludges with a proprietary dry ingredient that, when added to the waste, forms a monomer (Rousseaux and Craig, 1981). To obtain the monomer from the dry ingredient, acid must first be added to the waste if it is not already present. The resultant mixture subsequently turns into an insoluble bulk mass when lime or some other base is added. To harden the resultant mass, cement is also usually added. The final product is a rigid and highly impervious solid that can be used for landfilling or as a construction material. Organic as well as inorganic wastes are readily handled by this process. The reagent employed does not merely encapsulate metallic pollutants but actually reacts with them, thus retaining them very strongly.

Dewatered and dried wastes are cemented into aggregates of 250–500 kg with polybutadiene and then encapsulated with a high density, flexible polyethylene jacket in the disposal process developed by the TRW Systems Group, Inc., under the sponsorship of the U.S. EPA (Lubowitz et al., 1977). The resultant endproduct can be stored indefinitely or used in landfill operations. Many wastes, for example, electroplating sludges, chlorine brine sludge, pigment manufacturing sludges, and nickel–cadmium production sludge, have been treated by this process. When leaching tests were performed on typical encapsulated samples with a variety of solutions, the amount of heavy metals extracted was negligible even when the samples were contacted with citric acid and hydrochloric acid, two powerful solvents for heavy metals.

IU Conversion Systems' Poz-O-Tec process involves the treatment of sludges with fly ash and one or more additive(s) (Mullen et al., 1978; Samanta, 1977). Pozzolanic reactions yield structurally stable and environmentally acceptable cementitious material suitable for landfilling. The solidified waste may also be used as a construction material for land base, berms, pond liners, and dikes. For example, Poz-O-Tec material was used for the preparation of a 120-acre roadbed surface at the Dulles International Airport in Virginia some years ago.

REFERENCES

Adams, A.D. (1974). "Powdered carbon: Is it really that good?" *Water Wastes Eng.* **11**(3), B8–B11.

Anderson, D., and Clark, R. (1978). "Development of a pretreatment process for toxic metals removal." Effluent and Water Treatment Convention, Birmingham, Great Britain.

Anon. (1973a). "Chemfix treatment system for liquid and sludge effluents." *Materials Reclamation Weekly* **122**(20), 26.

Anon. (1973b). "Water treatment saves 45,000 gallons daily." *Plant Operation Management* **93**(3), 43.

Anon. (1974). "Waste water treatment at Zanussi." *Electroplating Metal Finishing* **27**(11), 35, 37.

Anon. (1975). "Floating aquatic plants remove chemicals from polluted waters." *Water and Pollut. Control* **113**(6), 23, 25.

Anon. (1976). "Insoluble starch xanthate (ISX)—Preparation and use in heavy metal recovery." *CA–NRRC–41 (Rev. 2)*. Northern Research Center, U.S. Dept. of Agriculture, Peoria, IL.

Anon. (1978a). "Chementator." *Chem. Eng.* **85**(3), 47–48.

Anon. (1978b). "Heavy metal removal? Try starch xanthate." *Prod. Finishing* **31**(9), 72–74.

Anon. (1979). "Chementator." *Chem. Eng.* **86**(11), 83–84.

Anon. (1980). "Bumper plater recovers chromic acid with new process." *Plating Surface Finishing* **67**(4), 34–35, 37.

Anon. (1981). "Ion transfer recovers chrome." *Ind. Finishing* **57**(3), 34–36.

Anon. (1983). "Chementator." *Chem. Eng.* **90**(8), 9–10.

Applegate, L.E. (1984). "Membrane separation processes." *Chem. Eng.* **91**(12), 64–89.

Arumugam, V. (1976). "Recovery of chromium from spent chrome tan liquor by chemical precipitation." *Ind. J. Environ. Health* **18**(1), 47–57.

Bailes, P.J., Hanson, C., and Hughes, M.A. (1976). "Liquid–liquid extraction: Metals." *Chem. Eng.* **83**(18), 86–94.

Basta, N. (1983). "Total metals recycle is metal finishers' goal." *Chem. Eng.* **90**(16), 16–17, 19.

Bell, J.P. (1979). "How to remove metals from plating rinse water." *Prod. Finishing* **43**(11), 40–47.

Bhatia, S., and Jump, R. (1977). "Metal recovery makes good sense!" *Environ. Sci. Technol.* **11**(8), 752–755.

Bishop, P.L. (1978). "Physicochemical treatment of wastes from a secondary tanner." In *Proc. 33rd Ind. Waste Conf., Purdue Univ.,* pp. 64–72.

Brown, M.J., and Lester, J.N. (1979). "Metal removal in activated sludge: The role of bacterial extracellular polymers." *Water Res.* **13**(9), 817–837.

Browning, M.E., Kraljic, J., and Santini, G.S. (1979). "Metal finishing sludge disposal: Economic, legislative and technical considerations for 1979." In *Proc. 2nd Conf. on Advanced Pollut. Control for the Metal Finishing Industry.* EPA 600/8-79-014, Cincinnati, OH, pp. 26–31.

Campbell, R.J., and Emmerman, D.K. (1972). "Freezing and recycling of plating rinsewater." *Ind. Water Eng.* **9**(4), 38–39.

Campbell, H.J. Jr., Scrivner, N.C., Batzar, K., and White, R.F. (1977). "Evaluation of chromium removal from a highly variable wastewater stream." In *Proc. 32nd Ind. Waste Conf., Purdue Univ.,* pp. 102–115.

Case, O.P. (1974). *Metallic Recovery from Waste Waters Utilizing Cementation.* EPA-670/2-74-008, U.S. Govt. Printing Office, Washington, DC.

Cherry, K.F. (1982). *Plating Waste Treatment.* Ann Arbor Science Publishers, Inc., Ann Arbor, MI.

Chian, E.S.K. (1976). "Renovation of vehicle washrack wastewater for reuse." *Water–1975, AIChE Symp. Ser.* **71**(151), 87–92.

Chian, E.S.K., and Fang, H.H.P. (1976). "RO treatment of power plant cooling tower blowdown for reuse." *Water–1975, AIChE Symp. Ser.* **71**(151), 82–86.

Cohen, J.M. (1977). "Trace metal removal by wastewater treatment." *Technol. Trans.* January, U.S. EPA, Cincinnati, OH.

Conner, J.R. (1974). "Ultimate disposal of liquid wastes by chemical fixation." In *Proc. 29th Ind. Waste Conf., Purdue Univ.,* pp. 906–922.

Crampton, P. (1982). "Reverse osmosis in the metal finishing industry." *Metal Finishing* **80**(3), 21–27.

Crumpler, E.P. Jr. (1977). *Management of Metal-Finishing Sludge.* PB-263 946, Natl. Tech. Info. Serv., Springfield, VA.

Cullen, G.V., and Siviour, N.G. (1982). "Removing metals from waste solutions with low rank coals and related materials." *Water Res.* **16**(8), 1357–1366.

Dinges, R. (1978). "Upgrading stabilization pond effluent by water hyacinth culture." *J. Water Pollut. Control Fed.* **50**(5), 833–845.

Drokos, F., and Bahensky, V. (1969). "Recovery of chromium compounds from waste sludge after disposal of spent electroplating baths." Czech. Patent 130, 794 (CA 72:45599).

Eisenmann, J.L. (1979). "Membrane processes for metal recovery from electroplating rinse water." In *Proc. 2nd Conf. on Advanced Pollut. Control for the Metal Finishing Industry*. EPA-600/8-79-014, Cincinnati, OH, pp. 99–105.

Elicker, L.N., and Lacy, R.W. (1978). *Evaporative Recovery of Chromium Plating Rinse Waters*. EPA-600/2-78/127, Cincinnati, OH.

Ellis, M.J., and Kunin, R. (1977). "New technology for the recovery of chromates from cooling tower blowdown." In *Proc. 37th Int. Water Conf.*, Pittsburgh, PA, pp. 41–49.

Feigenbaum, H.N. (1978). "Process for removal of heavy metals from textile waste streams." *Am. Dyestuff Rep.* **67**(3), 43–44, 46.

Fields, T. Jr., and Lindsey, A.W. (1975). *Landfill Disposal of Hazardous Wastes: A Review of Literature and Known Approaches*. SW-165, U.S. EPA, Cincinnati, OH.

Forbes, J.M. Jr. (1979). "IBM Owego give metal finishing wastes total treatment." *Pollut. Eng.* **11**(3), 46–49.

George, L.C., Soboroff, D.M., and Cochran, A.A. (1981). "Regeneration of waste chromic acid etching solutions in an industrial scale research unit." In *Proc. 3rd. Conf. on Advanced Pollut. Control. for the Metal Finishing Industry*. EPA-600/2-81-028, Cincinnati, OH, pp. 33–36.

Golomb, A. (1974). "Application of reverse osmosis to electroplating waste treatment. 4. Potential reutilization of chromium plating wastes by other industries." *Plating* **61**(10), 931–934.

Gooding, C.H. (1985). "Reverse osmosis and ultrafiltration solve separation problems." *Chem. Eng.* **92**(1), 56–62.

Gray, R. (1977). "Toxic waste disposal." *Water Pollut. Control* **76**(1), 30–38.

Grieves, R.B. (1970). "Foam separations for industrial wastes: Process selection." *J. Water Pollut. Control Fed.* **42**(8), Part 2, R336–R344.

Grieves, R.B. (1975). "Foam separations: A review." *Chem. Eng. J.* **9**(1), 93–106.

Grieves, R.B., Ettelt, G.A., Schrodt, J.T., and Bhattacharyya, D. (1968). "Dissolved-air ion flotation of industrial wastes hexavalent chromium." In *Proc. 23rd Ind. Waste Conf. Purdue Univ.* pp. 154–164.

Hanway, J.E. Jr., Mumford, R.G., and Mishra, P.N. (1979). "Treatment of industrial effluents for heavy metals removal using the cellulose xanthate process." *Water–1978, AIChE Symp. Ser.* **75**(190), 306–314.

Heist, J.A. (1979). "Freeze crystallization." *Chem. Eng.* **86**(10), 72–82.

Helsing, L.D. (1980). "Methods and systems currently available for the safe management of chemical wastes." *Chem. Times and Trends* **3**(3), 16–23.

Huang, C.P., and Bowers, A.R. (1979). "The development of an activated carbon process for the treatment of chromium VI-containing plating wastewater." In *2nd Conf. on Advanced Pollut. Control for the Metal Finishing Ind.* EPA-600/8-79-014, Cincinnati, OH, pp. 114–122.

Iammartino, N.R. (1975a). "Freeze crystalization: New water-processing tool." *Chem. Eng.* **82**(13), 92–93.

Iammartino, N.R. (1975b). "Mercury cleanup routes—II." *Chem. Eng.* **82**(3), 36–37.

Jellinek, H.H.G., and Sangal, S.P. (1972). "Complexation of metal ions with natural polyelectrolytes (removal and recovery of metal ions from polluted waters)." *Water Res.* **6**(3), 305–314.

Jenkins, R.L., Scheybeler, B.J., Smith, M.L., et al. (1981). "Metals removal and recovery from municipal sludge." *J. Water Pollut. Control Fed.* **53**(1), 25–32.

Jester, T.L., and Taylor, T.H. (1973). "Industrial waste treatment at Scovill Manufacturing Co., Waterbury, Connecticut." In *Proc. 28th Ind. Waste Conf., Purdue Univ.*, pp. 129–137.

Kalinske, A.A. (1981). "Extracting heavy metals and toxic organics from sludge." *Water Eng. Mgt.* R140–R151.

Kitigawa, T., Nishikawa, Y., Frankenfeld, J.W., and Li, N.N. (1977). "Wastewater treatment by liquid membrane process." *Environ. Sci. Technol.* **11**(6), 602–605.

Kraljik, J. (1975). "Practical guide to the treatment of chromium waste liquors." *Metal Finishing* **73**(10), 49–55.

Lanouette, K.H. (1977). "Heavy metals removal." *Chem. Eng.* **84**(22), 73–80.

Laughlin, R.G.W. (1980). "Canadian waste exchange program: Successes and failures." Presented at National Conf. on Hazardous and Toxic Waste Management, New Jersey Institute of Technology.

Leslie, M.E. (1974). "Peat: A new medium for treating dye house effluent." *Am. Dyestuff Rep.* **63**(8), 15–16, 18.

Lubowitz, H.R., Derham, R.L., Ryan, L.E., and Zakrzewski, G.A. (1977). "Development of a polymeric cementing and encapsulating process for managing hazardous wastes." EPA-600/2-77-045, Cincinnati, OH.

McDonald, C.W., and Bajwa, R.S. (1977). "Removal of toxic metal ions from metal finishing wastewater by solvent extraction." *Sep. Sci.* **12**(4), 435–445.

McIntyre, G., Rodriguez, J.J., Thackston, E.L., and Wilson, D.J. (1981). "Inexpensive heavy metal removal by foam flotation." In *Proc. 36th Ind. Waste Conf. Purdue Univ.*, pp. 564–572.

McIntyre, G., Rodriguez, J.J., Thackston, E.L. and Wilson, D.J. (1983). "Inexpensive heavy metal removal by foam flotation." *J. Water Pollut. Control Fed.* **55**(9), 1144–1149.

Mehta, A. (1981). "Routes to metals recovery from metal finishing sludges." In *Proc. 3rd. Conf. on Advanced Pollut. Control for the Metal Finishing Industry.* EPA-600/2-81-028, Cincinnati, OH, pp. 76–79.

Militello, P. (1981). "Assessment of emerging technologies for metal finishing pollution control: Three case studies." EPA-600/S2-81-153, Cincinnati, OH.

Moo, C.T., Chung, F.Z., and Lung, T.N. (1976). "Treatment of metal-bearing solutions with peat. A preliminary investigation." *K'ung Yeh Chi Shu* **14**(5), 166–172 (CA 86:60118e).

Moore, W.A., McDermott, G.N., Post, M.A., et al. (1961). "Effects of chromium on the activated sludge process." *J. Water Pollut. Control Fed.* **33**(1), 54–72.

Mueller, W. (1976). "Possibilities of recycling electroplating sludges." *Galvanotechnik* **67**(5), 381–383.

Mullen, H., Ruggiano, L., and Taub, S.I. (1978). "Converting scrubber sludge and flyash into landfill material." *Pollut Eng.* **10**(5), 71–74.

National Aeronautics and Space Administration (undated). *Compiled Data on the Vascular Aquatic Program: 1975–1977. National Space Technology Laboratories.* NSTL Station, MS.

Netzer, A., and Donahue, M.R. (1983). *Heavy Metal Removal from Wastewater by Lignite Flyash.* University of Texas at Dallas, Richardson, TX.

Netzer, A., Wilkinson, P., and Beszedits, S. (1974). "Removal of trace metals from wastewater by treatment with lime and discarded automotive tires." *Water Res.* **8**(10), 813–817.

Nojiri, N., Tanaka, N., Sato, K., and Sakai, Y., (1980). "Electrolytic ferrite formation system for heavy metal removal." *J. Water Pollut. Control Fed.* **52**(7), 1898–1906.

Okey, R.W., Digregorio, D., and Kominek, E.G. (1979). "Waste sludge treatment in the CPI." *Chem. Eng.* **86**(3), 86–100.

Okuda, T., Sugano, I., and Tsuji, T. (1975). "Removal of heavy metals from wastewater by ferrite co-precipitation." *Filtr. Sep.* **12**(5), 472, 475–476, 478.

Olver, J.W., Kreye, W.C., Michelsen, D.L., and Sutton, H.C. (1978). "Treatment and disposal of anodizing wastewater with options for water reuse." *In Proc. 6th Annual Water and Wastewater Equipment Manufacturers Assoc. Ind. Pollut. Conf., McLean, VA*, pp. 441–456.

Parkinson, G. (1983). "Liquid membranes: Are they ready?" *Chem. Eng.* **90**(17), 22–23, 25, 27.

Pierce, R., and Thorstensen, T.C. (1976). "The recycling of chrome tanning liquors." *J. Am. Leather Chem. Assoc.* **71**(4), 161–166.

Pojasek, R.B. (1979). "Solid-waste disposal." *Chem. Eng.* **86**(17), 141–145.

Raman, R., and Karlson, E.L. (1977). "Reclamation of chromic acid using continuous ion exchange." *Plating Surface Finishing* **64**(8), 40–42.

Reddy, A.S., and Sayi, Y.S. (1977). "Solvent extraction separation of chromium VI, iron III, cobalt II and nickel II with di-n-pentyl sulfoxide." *Sep. Sci.* **12**(6), 645–648.

Ricci, L.J. (1975). "Heavy metals recovery promises to pare water cleanup bills." *Chem. Eng.* **82**(27), 29–31.

Rigo, L.C. (1974). *Removal of Heavy Metals from Aqueous Solution by Adsorption onto Silica Gel.* Unpublished report, Canada Centre for Inland Waters, Burlington, Ontario.

Robinson, D.J., Weisberg, H.E., Chase, G.I., et al. (1974). *An Ion Exchange Process for Recovery of Chromate from Pigment Manufacturing.*" EPA-670/2-74-044, Cincinnati, OH.

Rousseaux, J.M., and Craig, A.B. Jr. (1981). "Stabilization of heavy metal wastes by the Soliroc process." *Proc. 3rd Conf. on Advanced Pollut. Control for the Metal Finishing Industry.* EPA-600/2-81-028, Cincinnati, OH.

Samanta, S.C. (1977). "Physical and chemical characteristics of stabilized SO_2 scrubber sludges." In *Sixth Environ. Eng. and Science Conf., University of Louisville.*

Schoeman, J.J. (1985). "The status of electrodialysis technology for brackish and industrial water treatment." *Water SA* **11**(2), 79–86.

Schofield, J.T. (1978). "Getting rid of toxic waste." *Industrial World,* March.

Serper, A. (1981). "Consider alternatives to landfilling for disposing of hazardous wastes." *Solid Wastes Mgt.* **24**(2), 62, 64, 66, 163–164.

Shaul, G.M., Barnett, M.W., Neiheisel, T.W., and Dostal, K.A. (1983). "Activated sludge with powdered activated carbon treatment of a dyes and pigments processing wastewater." In *Proc. 38th Ind. Waste Conf., Purdue Univ.,* pp. 659–671.

Soboroff, D.M., Troyer, J.D., and Cochran, A.A. (1978). "A one-step method for recycling waste chromic acid–sulfuric acid etching solutions." In *Proc. 33rd Ind. Waste Conf., Purdue Univ.,* pp. 758–763.

Stinson, M.K. (1979). "Emerging technologies for treatment of electroplating wastewaters." *Water–1978, AIChE Symp. Ser.* **75**(190), 270–284.

Stroeve, P., and Varanasi, P.P. (1982). "Transport processes in liquid membranes: Double emulsion separation system." *Sep. Purif. Methods* **11**(1), 29–69.

Sublette, K.L., Snider, E.H., and Sylvester, N.D. (1982). "A review on the mechanism of powdered activated carbon enhancement of activated sludge treatment." *Water Res.* **16**(7), 1075–1082.

Teer, E.H., and Russell, L.V. (1972). "Heavy metal removal from wood preserving wastewater." In *Proc. 27th Ind. Waste Conf., Purdue Univ.,* pp. 281–286.

Tuznik, F., and Kieszkowski, M. (1972). "Preliminary studies on complete neutralization and utilization of sludge from plating effluent and treatment processes. *Electroplating Metal Finishing* **25**(7), 10–11, 13–17.

U.S. EPA (1979a). *Control Technology for the Metal Finishing Industry—Evaporators.* EPA-625/8-79-002, Cincinnati, OH.

U.S. EPA (1979b). *Development Document for Existing Source Pretreatment Standards for the Electroplating Point Source Category.* EPA-440/1-79/003, Washington, DC.

U.S. EPA (1979c). *Environmental Pollution Control Alternatives: Economics of Wastewater Treatment Alternatives for the Electroplating Industry.* EPA-625/5-79-016, Cincinnati, OH.

U.S. EPA (1980). *Control and Treatment Technology for the Metal Finishing Industry: Sulfide Precipitation.* EPA-625/8-80-003, Cincinnati, OH.

U.S. EPA (1981). *Control and Treatment Technology for the Metal Finishing Industry: Ion Exchange.* EPA-625/8-81-007, Cincinnati, OH.

U.S. EPA (1982). *Sludge Handling, Dewatering, and Disposal Alternatives for the Metal Finishing Industry.* EPA-625/5-82-018, Cincinnati, OH.

Wentink, G.R., and Etzel, J.E. (1972). "Removal of metal ions by soil." *J. Water Pollut. Control Fed.* **44**(8), 1561–1574.

Wilmoth, R.C., Baugh, T.L., and Decker, D.W. (1978). "Removal of selected trace elements from acid mine drainage using existing technology." In *Proc. 33rd Ind. Waste Conf., Purdue Univ.*, pp. 886–894.

Wing, R.E. (1978). "Process for heavy metal removal from plating wastewaters." In *Proc. 1st Annual EPA3AES Conf. on Advanced Pollut. Control for the Metal Finishing Industry, Lake Buena Vista, FL.*

Wu, M.H., Hsu, D.Y., and Huang, C.P. (1976). "The regeneration of activated carbon for chromium removal." In *Proc. 31st Ind. Waste Conf., Purdue Univ.*, pp. 409–419.

10

MOBILITY AND BIOAVAILABILITY OF CHROMIUM IN SOILS

R.J. Bartlett and B.R. James*

The University of Vermont
Burlington, Vermont

*B.R. James is at The University of Maryland, College Park, MD.

1. INTRODUCTION

Chromium ranks as the tenth metal in the earth's crust, ahead of nickel, copper, lead, and zinc in abundance and economically higher still. It appears to find its place in the world mainly where man puts it—into steel, into leather, or on top of metal to protect or decorate it, and into the environment as a pollutant. In sludge, fly ash, slag, industrial waste, and even in its natural forms, it emerges as a contaminant.

Contaminating Cr(III) substitutes for small amounts of octohedral aluminum in clay minerals, and in chromite chromium ore, it borrows the familiar magnetite structure by substituting for two atoms of Fe(III), forming $FeCr_2O_4$. The green of the emerald and the red of the ruby are examples of primeval pollution, with an occasional Cr(III) substituting for an aluminum. Thus, chromium plays a key role in the well-known ruby laser, which can emit intense bursts of monochromatic in-phase radiation (Cotton and Wilkinson, 1980). Small amounts of chromium are essential in the human diet, yet the appearance of chromium in our food seems to be another example of random contamination.

There is a dearth of literature dealing with mobility and bioavailability of polluting chromium in soils. The first and most basic reason for the dearth is the kinetic inertness of Cr(III) and its complexes. Most forms of Cr(III) present in or added to soils are low in solubility and reactivity and are oxidized to Cr(VI) only when environmental circumstances fit together in a rather narrow and delicately balanced optimum.

The second reason is the high instability of Cr(VI) in soils. Ready leaching, plant uptake, and reduction make Cr(VI) difficult to find or to follow, if found, in field soils. There is good reason for not writing about Cr(VI) if it is not there.

The third reason is the propensity of soil chemists to treat samples of soils as reagents by drying, pulverizing, sieving, and storing them before studying them. This practice has retarded the study of chromium redox behavior in the laboratory because, unlike most moist field samples, dry soils if long-stored (designated for emphasis "lab dirt") will not oxidize Cr(III) (Bartlett and James, 1979).

The preceding reasons lead to the logical assumption that all chromium added to soils soon becomes permanently entrenched as Cr(III), explaining why analysis of Cr(VI) in soil and water often has not been included in the objectives of major studies of chromium waste disposal. The last reason, then, is that not looking for Cr(VI) has stacked the odds against finding this mobile and toxic species.

2. SPECIATION AND MOBILITY OF CHROMIUM (III)

2.1 The Aluminum Analog

Wherever Cr(III) is found in nature, it will be in close association with Al(III) and Fe(III), mainly because of the overwhelming prevalence of the latter two elements in the earth's crust. The atomic radii of chromium, aluminum, and iron (64, 67, and 54 pm, respectively) and, therefore, the ionic potentials of the trivalent ions are close enough together that substitution of one for another occurs in many mineral and organic structures. In spite of the greater ionic potential of Al(III), Cr(III) and Fe(III) are stronger Lewis acids because of their need to share electrons in order to acquire the noble gas (krypton) configuration, whereas aluminum has only to lose three electrons to acquire the stable neon core. Chromium(III) forms a multitude of complexes, both organic and inorganic, in a variety of colors. However, it is slow in forming them compared with aluminum. The result is stronger bonding, and hence chromium has replaced aluminum for tanning leather. In spite of similarities, we know of course that Al(III) and Cr(III) are fundamentally very different. Chromium(III) has three unpaired electrons, and in giving up these electrons, Cr(III) becomes Cr(VI). Aluminum(III), having none to give, does not oxidize.

Both aluminum and Cr(III) form the hexaqua ion $[M(H_2O)_6]^{3+}$, and both form hexacoordinate complexes with oxygen and soluble polynucleate complexes by olation, that is, bridging by –OH groups. Both Al(III) and Cr(III) become anions above pH 7, and both form strong organic complexes and complexes with fluorine and phosphate (Stunzi and Marty, 1983; Mertz, 1969; Bartlett and Kimble, 1976a). A principal characteristic of Cr(III) complexes is their relative kinetic inertness (Cotton and Wilkinson, 1980; Baes and Mesmer, 1976). Hamm (1958) used polarography to determine rates of formation of acetate, glycolate, lactate, phthalate, citrate, and tartrate complexes and concluded that essentially the only slow step in complex formation was dissociation of water from the hydration shell.

The slowness of complex formation makes it possible to use catechol violet to sensitively measure Cr(III) in the presence of iron and aluminum and to use rates of reaction with catechol violet to categorize soluble Cr(III) species according to kinetic behavior (Bartlett, unpublished data). The colored aluminum complex forms at about 20 times the rate of the chromium complex. Detailed procedures have not been worked out for chromium.

Bartlett and Kimble (1976a) found that behavior of Cr(III) added at 10 mmol kg^{-1} to a spodic horizon, a Spodosol Ap horizon, and a clay Alfisol A horizon was strikingly similar in quality to that of aluminum already present in the soils. The effects of increasing pH levels adjusted with $CaCO_3$ and of added phosphorus produced parallel results between Cr(III) and aluminum, as characterized by extraction with pH 4.8 NH_4OAc, 1 M HCl, 1 M NaF, and 0.12 M $Na_4P_2O_7$. Added chromium appeared to become part of the aluminum pool in these soils.

Only HCl and $Na_4P_2O_7$ extracted significant portions of the Cr(III) added. The NH_4OAc and NaF removed very small fractions of the amounts present. These fractions could be considered intensity measurements. The intensities of Cr(III) in the two Spodosols, as characterized by the NH_4OAc and NaF extracts, were decreased as the pHs were increased by additions of $CaCO_3$. The effect of pH in the clay was similar with NH_4OAc but not NaF. Fluoride, which coordinates with Cr(III) to form a complex similar to that of aluminum and Fe(III), extracted very little Cr(III) from the clay, regardless of pH.

Decreases in aluminum extracted by NH_4OAc were associated with $CaCO_3$ and phosphorus treatments in the same way as those of Cr(III). Amounts of Cr(III) extracted by $Na_4P_2O_7$, as they varied with $CaCO_3$ and phosphorus treatments, generally were parallel to those extracted by the weaker NH_4OAc and NaF extractants, especially in the Spodosols. It seemed that the milder extractants were removing reactive quantities of Cr(III) in equilibrium with the much greater quantity represented by the pyrophosphate extraction.

Probably all of the Cr(III) removed by pyrophosphate in this study was held as part of the organic fraction of the soil. Aleksandrova (1960) found that $Na_4P_2O_7$ extracted calcium, aluminum, and iron humates completely without extracting aluminum or iron from the parent material or from nonsilicated forms of sesquioxides. The effects of $CaCO_3$ and phosphorus in lowering the chromium extractable with the pyrophosphate were much greater in all soils where the lime or phosphorus or both were equilibrated with the soil before the addition of $CrCl_3$ than where the chromium was added and equilibrated first. For example, 65% of the chromium added was removed by pyrophosphate from the Adams B2ir without lime or phosphorus, 36% was removed where chromium was applied before lime and phosphorus, and only 6% where chromium was added after equilibration with lime and phosphorus. Probably treatment with lime and/or phosphorus prevented the formation of organic complexes by immediately tying up the Cr(III) as inorganic hydroxides or phosphates. Where chromium was added first, stable chromium–organic complexes may have been formed that resisted precipitation by hydroxyl or phosphate groups.

Similar reasoning would lead to the conclusion that the NH_4OAc and the NaF extractions also represent organic–Cr(III) fractions. Hydrochloric acid, on the other hand, appeared to extract the inorganic forms of Cr(III). In response to added $CaCO_3$ and phosphorus, the amounts extracted by HCl were inversely related to amounts extracted by pyrophosphate. Apparently in soils receiving high levels of lime and/or phosphorus, a significant proportion of the chromium forms inorganic hydroxides or phosphates, rather than organic complexes, and these inorganic compounds are soluble in HCl. Thus, behavior of organic and inorganic fractions of Cr(III) with different extractants in soils was quite similar to that expected or found for aluminum in response to varying amounts and sequences of applied lime and phosphorus. Past research dealing with aluminum should be helpful in predicting solubility and mobility of Cr(III) in waste added to soils.

2.2 Mobility of Natural Forms

Naturally occurring Cr(III) in Spodosols appears to be mobilized by the podsolization process similarly to aluminum. In a study of background levels of heavy metals in Vermont soils, the chromium extractable by 1 M HCl appeared to have accumulated in spodic horizons (Bartlett, 1982). Levels of chromium extracted were significantly greater in 46 Typic Haplorthod Bhs or Bs horizons than in the Ap or C horizons of the same soils (3.7 ± 0.46 versus 2.38 ± 0.36 versus 2.09 ± 0.24 mg kg^{-1}), respectively. The E horizons contained only 0.4 ± 0.23 mg kg^{-1}. Although levels of extractable aluminum were orders of magnitude higher, the aluminum relationships among horizons were similar. It would be interesting to learn whether Cr(III) is taken up by acid forest soil vegetation and recycled back to the soil in the manner of aluminum.

Even though most of the 492 horizon samples examined were from agricultural soils, all but 1 of the 18 horizons with detectable amounts of Cr(III) extractable by pH 4.8 NH_4OAc were acid forest soil horizons. Of these, 12 were spodic horizons, either Bhs or Bs.

Lindau and Hossner (1982) found in salt marsh sediments in Texas that approximately 87% of the total chromium was fixed within the crystalline lattice of clay and silicate minerals, and only 7% could be accounted for as associated with organic matter plus sulfides.

2.3   Addition of Organic Complexes to Soil

James and Bartlett (1983a) demonstrated that presence of citric acid, diethylene-triaminepentaacetate acid (DTPA), fulvic acids, and a water-soluble extract of an air-dried soil all formed soluble Cr(III) complexes. These complexes remained soluble as the pH was increased above 5.5, at which pH all of the uncomplexed Cr(III) was precipitated. After adding soil to these solutions until the suspension pH reached 6.7, half or more than half of the chromium complexed in these four ligand systems remained in solution. In another study, 24-hour equilibration of soil suspensions with alfalfa meal allowed about 20% of added Cr(III) to remain soluble (Bartlett and Kimble, 1976a).

Chaberek and Martell (1959) presented diagrams of probable chelated forms of Cr(III) with hydroxy carboxylic acids in which the cation is bonded to an –OH group and to two –COOH groups rather than to three carboxyl groups. This means that in hydroxy acids, such as citric, chelated forms of Cr(III) could be anionic in soils at pHs greater than the pKas of uncomplexed carboxyl groups. Above pH 4.7, the pKa for the second carboxyl group of citric acid, Cr(III) could be chelated as $Cr(OH)^{2+}$ or as an olated polymer carrying a 2+ charge (Rengasamy and Oades, 1978).

During the first week of equilibration, solubilities of chromium added as either citrate or as chloride to Hapludalf samples at pHs of 5.1, 6.0, and 6.7 decreased markedly, but during the next 7 weeks, soluble chromium in the citrate treatments decreased slowly compared with the chromium in treatments receiving inorganic chromium (James and Bartlett, 1983a). After 1 year of moist incubation, 750 μmol kg^{-1} still remained water soluble in the citrate-treated samples. The very slow decomposition of the citrate complexed by Cr(III) suggests that the Cr(III) bound to it made citrate less available to microbes as an energy source.

2.4   Chromium/Aluminum Mobility

To continue the analogy with aluminum, it seems safe to assume the ionic trivalent chromium in solution or on acid soil exchange sites (below pH 5) will be the most available form of inorganic Cr(III), chemically and biologically, even though exchangeable chromium would be held tightly compared with other cations. However, the most mobile Cr(III), by diffusion or mass flow, will be soluble organic (and some fluoride) complexes and dispersable colloidal sized inorganic hydroxy polymers, silicates, carbonates, and crystalline $FeCr_2O_4$ and clay minerals.

Dispersability will depend on wetting and drying, freezing and thawing, ionic strength of soil solution, and biologically mediated surface interactions of cations, anions, H_2O, and organic substances. Perhaps the biogeochemical models being developed under the sponsorship of acid rain dollars for predicting aluminum mobility can be applied to predict mobility of Cr(III).

3. OXIDATION OF CHROMIUM(III)

3.1 Occurrence

Thermodynamically, Cr(VI) should be the stable form of chromium in soils, more stable than nitrate, if we are considering equilibrium with atmospheric oxygen. If equilibrium with organic carbon is governing, then, of course, Cr(III) will be the stable form. Early work in detecting Cr(III) oxidation in soils was convincingly negative. Bartlett and Kimble (1976a) reported no Cr(VI) detectable in any soil extract, regardless of level of Cr(III) added, original pH, $CaCO_3$ or phosphorus treatments, or aeration, moisture, time, and temperature conditions imposed.

Subsequently, however, Bartlett and James (1979) discovered that rapid oxidation of a portion of Cr(III) salts or hydroxides added to almost any soil with a pH above 5 took place readily, provided the soil sample was fresh and moist and directly from the field. They showed that oxidized manganese, present in most fresh moist field soil samples, served as the electron link between the added Cr(III) and oxygen of the atmosphere. The amount of Cr(III) oxidized to Cr(VI) was proportional to the manganese reduced (and exchangeable) and also to the amount of manganese reducible by hydroquinone before adding Cr(III). These findings were verified by Amacher and Baker (1982).

3.2 Laboratory Dirt

Soil oxidation of chromium was not demonstrated earlier because dried, stored, "lab dirt" samples had been studied instead of fresh, moist, field soil samples. Both processes of drying field-moist soils and rewetting them later have drastic effects on their redox behavior (Bartlett and James, 1980; Bartlett, 1981). Drying alters surface chemical characteristics so that the behavior of the dried sample will change erratically with time during dry storage and while it returns to its metastable moist state when water is added back.

Drying in the laboratory or natural drying in the field topples the fragile portion of the poise in the metastable soil system. Changes occur instantly. Reverse changes after rewetting are slow, requiring biologic activity. The most obvious effects of drying are greater solubility and reducing ability of soil organic matter. Surface acidity increases, and manganese is reduced, becoming exchangeable and soluble. Both lower pH and higher solubility of organic matter probably result from the increased polarity of surface-oriented H_2O as drying proceeds. Manganese oxide reduction on drying is associated with partial oxidation of the disrupted colloidal organic matrix (Hammes and Berger, 1960). The net effect is a soil with greater potential for oxidation. That is, the soil pe is lowered by drying (Bartlett, 1981).

Flooding soils dried for storage or dried naturally in the field results in rapid reduction of any manganese oxides remaining and intensifies development of anaerobic conditions. This effect is related to reduced organic matter made more available by soil drying. Flooding a dried soil greatly lowered the net chromium oxidation by soil manganese oxides the first day, and it eliminated it altogether within 3 days. However, in the same soil stored in the field-moist state, chromium oxidizing capacity was not significantly affected by flooding in open cups for 10 days (Bartlett, 1981).

3.3 Kinetics and Mechanisms

Amacher and Baker studied the kinetics of chromium oxidation by Hagerstown silt loam (Typic Hapludalf) in dilute suspension comparable to a sediment-laden stream contaminated with chromium and also by suspensions of a delta-MnO_2 preparation. The amount of soil used was such that its reducible manganese content was just enough to oxidize all of the added chromium, but because of adsorption of chromium and lack of accessibility, amounts of each that reacted were less than the maximum. At $2°C$, the oxidation of chromium followed zero-order kinetics, that is, the rate did not depend on concentrations of the reactants, which is not unusual for reactions involving surfaces. At low temperature, the oxidation reaction was sufficiently slowed so that the manganese oxide surface was "saturated" with chromium undergoing oxidation, and higher concentrations of Cr(III) could not be expected to speed up the reaction. At higher temperatures, an initial rapid increase in Cr(VI) was quickly followed by a fall-off in reaction rate. Reduction at higher temperatures of some of the Cr(VI) formed could explain the fall-off. Another explanation offered by the authors was that not all of the manganese oxide was available for oxidizing chromium and the reaction slowed as the active part became reduced by chromium.

A similar phenomenon was observed by Amacher and Baker (1982) in measuring the chromium oxidizing abilities of incubated whole-soil samples. At the lowest level of added Cr(III), virtually all of the chromium in immiscibly displaced soil solution was Cr(VI), but as increasing levels of Cr(III) were added to the soil, the fraction of soil solution Cr(VI) decreased. Even though there was more than enough manganese oxide in the soil to oxidize all of the soil solution chromium, it did not, indicating either that a major portion of the manganese oxide was inaccessible to the chromium or that reduction of Cr(VI) by the soil organic matter also was occurring.

When these researchers studied the effects of temperature on the kinetics of chromium oxidation in suspensions of prepared hydrated amorphous MnO_2, they found the general shapes of the kinetic curves to be the same as those in soil suspensions, with an initial rapid oxidation of chromium followed by a rapid fall-off in rate. Even though in large excess, the MnO_2 did not oxidize all of the Cr(III), indicating that the reactive part of the MnO_2 surface had been depleted.

Reduction of Cr(VI) could not be an alternative explanation because of the absence of an electron donor in the synthetic oxide system.

Based on the observation that Mn(II) in solution did not increase as the manganese was reduced, they concluded that the Mn(II) formed during chromium oxidation was retained on the manganese oxide surfaces and thus blocked access to the Mn(IV) below it. At all times during the reaction, all of the added chromium was found to be in solution. This meant that oxidation of Cr(III) must have occurred almost instantly without prior adsorption onto the manganese oxide surface. Lack of adsorption of Cr(III) by MnO_2 is in marked contrast to the high affinity of MnO_2 for cobalt, nickel, copper, lead, cadmium, zinc, and especially Mn(II) (Jenne, 1968; Loganathan et al., 1977; Posselt et al., 1968; Amacher and Baker, 1982; and Ross and Bartlett, unpublished data). The levels of "adsorption" of 1 mol of Mn(II) per mol of MnO_2 reported by Morgan and Stumm (1964) certainly suggest a chemical reaction at the MnO_2 surface rather than simple ion exchange.

Ross and Bartlett (1981) showed that the increase in chromium oxidation six days after addition of Mn(II) to 14 widely differing soils was well correlated with the original oxidation test, suggesting a relationship between Mn(II) oxidation and preexisting oxide surfaces. Rates of oxidation of added manganese, as evidenced by disappearance of pH 4.8 NH_4OAc extractable manganese, were first-order reactions. The relationship ($r = -.974$) between extractable manganese and chromium oxidized for all points in a rate/temperature trial showed that chromium was oxidized more rapidly as extractable manganese decreased. This finding indicated that oxidation of chromium was blocked by added Mn(II) and that the blocking effect decreased as the adsorbed Mn(II) became oxidized. Lowered extractable Mn(II) indicated that more reducible manganese oxide was available for accepting electrons from Cr(III) and that more oxide surface was free of Mn(II) and therefore accessible to Cr(III).

Adsorption of Mn(II) and its oxidation appeared to be related to the experimental demonstration by Loganathan et al. (1977) of the formation of a positively charged MnO_2 surface above pH 5–6 following adsorption of cobalt and zinc. Ross and Bartlett (1981) suggested that Mn(II) adsorption might also lower negative charge on a manganese oxide surface and that concentration of –OH ions near plus charges on the surface would increase the surface pH and promote the autooxidation of Mn(II) by atmospheric oxygen. By contributing to positive charges, adsorption of Mn(II) would prevent close approach of Cr(III) and thus temporarily "poison" the surface for oxidation of chromium. As the Mn(II) became autooxidized, surface charge again would become more negative and protons would replace hydroxyl ions.

3.4 Oxidation of Organic Chromium

Low molecular weight organic acids such as citrate and gallic acids increase solubility and mobility of Cr(III) and thereby facilitate its oxidation. Organic acids may be present in organic waste materials added to soils, may be synthesized during their decomposition or that of other organic residues in soils, or may form from plant root exudates (Hale et al., 1978). According to a laboratory study of James and Bartlett(1983b), initial oxidation rates of different species of $Cr(III)$ added to an Aquic Udorthent, and the maximum $Cr(VI)$ levels observed, decreased in the order: freshly precipitated $Cr(OH)_3$ > chromium-citrate > aged $Cr(OH)_3$ in citrate > aged $Cr(OH)_3$. The oxidation of $Cr(III)$ in tannery sewage sludge and in tannery effluent added to Typic Eutrochrept samples in a continuously aerated slurry was studied as a function of time and loading rate. Amounts of $Cr(VI)$ increased with increasing additions of $Cr(III)$, added as high-chromium sewage sludge or tannery effluent.

3.5   Potential Chromium(VI) Formation in Field Soils

Results of research on the biohazards associated with land disposal of chromium wastes have been summarized by Chaney et al. (1981). Land application of refuse, compost, sewage sludge, tannery sludge, cooling tower blowdown, and sewage effluent has yielded negative results in showing chromium buildup in soil below the zone of tillage, and it has not been shown to increase chromium in plants beyond the first crop year (Chaney et al., 1981; Cunningham et al., 1975; Dowdy and Ham, 1977; Grove and Ellis, 1980; Mortvedt and Giordano, 1975; Wickliff et al., 1982).

Lack of demonstrated change in total chromium in subsoil after application of $Cr(III)$-containing waste (Chaney et al., 1981; Cunningham et al., 1975; Dowdy and Ham, 1977; Grove and Ellis, 1980; Mortvedt and Giordano, 1975) does not prove that chromium oxidation did not occur. Chromium(VI) was not analyzed in these studies. Behavior of $Cr(III)$ in an organic waste is somewhat analogous to the organic nitrogen in such a material and soluble and mobile $Cr(VI)$ behaves similarly to nitrate, which is released slowly by mineralization and does not accumulate in humid region subsoils. An ion may pass through, and its level still will be zero in a particular subsoil. Neither $Cr(VI)$ nor nitrate is readily reduced in subsoils in the absence of available organic energy. However, unlike nitrate, $Cr(VI)$ may become slowly reduced by mineral $Fe(II)$ in subsoils, and $Cr(VI)$ is much more strongly adsorbed than nitrate by many subsoils. Enriched groundwater levels of $Cr(VI)$ above water quality standards have been identified in some studies (e.g., Deutsch, 1972; Dutt and McCreary, 1970).

Under optimum conditions for oxidation, chromium in sludge could release low levels of $Cr(VI)$ over a period of months or years. Possibly, the concentration of $Cr(VI)$ released could be at an environmentally acceptable level. Just as slow-release organic nitrogen fertilizer efficiently provides nitrate for plant use, slow-release sludge could produce polluting chromate at a level too low for groundwater pollution but optimum for food plants supplying human and animal nutrition needs.

There is evidence that $Cr(III)$ complexed by high molecular weight organic ligands is not readily dissociated and oxidized. Amacher and Baker (1982) reported that most of the sludge-borne chromium was still in the top 20 cm of a

Hagerstown soil profile four years after the last sludge addition. They did not find any Cr(VI) in KH_2PO_4 extracts of this soil or of soil incubated in the laboratory with chromium-containing leather dust. However, continued moist incubation of some of these same samples in the Vermont laboratory for an additional year produced positive Cr(VI) tests in phosphate extracts in two of the leather dust soil samples (Bartlett and Amacher, unpublished data).

3.6 Oxidation of Old Soil Chromium

The test for potential oxidizable soil chromium (Section 3.7.2) was devised to measure by equilibration citrate extractable and oxidizable "old" soil Cr(III). Chromium previously added to soils as sludge, worn-out moccasins, or inorganic native chromium in a high chromium soil, as well as in fresh tannery sludge or lagoon waste, was demonstrated to become partially oxidized to Cr(VI) under the conditions of the test (Table 1, Bartlett, 1985). Potential oxidizable chromium was measured in soil samples from 30 research plots in eight states that had received chromium-containing sludge applications. (Samples were donated by M.C. Amacher, R.H. Dowdy, P.M. Giordano, T.D. Hinsley, L.W. Jacobs, D.R. Keeney, C. Lue-Ling, A.L. Page, and V.V. Volk.) When these values were compared with the total soil chromium extractable by the hypochlorite method (Section 3.7.3), the following relationship was obtained:

$$\text{Cr oxidized} = 0.02 \times \text{NaOCl-Cr} + 1.7 \tag{1}$$

$$r^2 = 0.86$$

Table 1 Chromium Oxidized by the Potential Oxidizable Soil Cr Test

	Cr(VI), mg kg^{-1}	
	By NaOCl	Oxidization Test
Tannery sludge 0.1 cmol kg^{-1} (Eutrochrept, VT)	72	1.4
Tannery sludge 1. cmol kg^{-1} (Eutrochrept, VT)	556	18.2
Tannery sludge 10 cmol kg^{-1} (Eutrochrept, VT)	5000	76.8
Eutrochrept, VT, 1% moccasin	240	6.4
Eutrochrept, VT, 5% moccasin	1280	32.8
Hapludalf, PA, 2% leather dust	600	4.6
Hapludalf, PA, 5% leather dust	1230	26.8
Cr(OH)$_3$, aged 1 year, pH 7, 1 cmol kg^{-1}	520	4.0
Cr(OH)$_3$, aged 1 year, pH 7, 2.5 cmol kg^{-1}	1300	9.6
Chrome soil, PA, 0–5 cm	80	1.6
Chrome soil, PA, 5.30 cm	72	1.5
Chrome soil, PA, 30–60 cm	60	1.2

Although the overall regression was indicative of a significant correlation between quantity of chromium in the soils and its oxidizability, there were notable individual soils that strayed from the regression—soils with either especially high or especially low fractions of the chromium in them becoming oxidized. Oxidized chromium values based on soil weight ranged from 0 to 77 mg kg^{-1} with a mean of 11.0 mg kg^{-1}; NaOCl extracted values were from 20 to 5000 mg kg^{-1}, and the mean was 505 mg kg^{-1}.

In another study, a 1% chromium tannery sludge incubated with Typic Eutrochrept A horizon samples (2 g kg^{-1}) maintained at field capacity moisture inside polyethylene bags produced Cr(VI) levels of about 1 mg kg^{-1} after three weeks and 2 to 3 mg kg^{-1} after four months (Bartlett, 1985). Thirteen of these sludge- and leather-treated samples were triple bagged in thin polyethylene to prevent moisture loss and stored for four more years (Bartlett, 1986a). At the end of four years, the samples were extracted with 10 mM KH$_2$PO$_4$·K$_2$HPO$_4$ and were found to contain Cr(VI) in amounts varying from 0 to 41 μmol L^{-1} of soil.

In 1 m^2 field plots, applications of 1000 mg kg^{-1} chromium as tannery sludge produced 0.4 mg kg^{-1} Cr(VI) after one month. This Cr(VI) was gone one week later, either lost by "dechromification," that is, reduction, or by leaching.

Formation of manganese oxides, optimal for oxidation of chromium, appears to be favored by a somewhat restricted oxygen supply along with moderate levels of organic reducing substances. Ten times the concentration of Cr(VI) (50 versus 0.5 mg L^{-1}) was produced in a soil suspension incubated nine days with MnSO$_4$ and Cr(OH)$_3$ in a stoppered flask as compared with the same volume of suspension swirled in an open one. Vigorous aeration by shaking a high organic matter soil with Cr(OH)$_3$ for 10 days resulted in formation of high levels of Cr(VI) but none at all when a low organic matter soil was treated in the same way. However, with no shaking of the suspension during the 10-day period, there were high levels of Cr(VI) associated with the low organic matter soil, but no Cr(VI) formed in the high organic matter soil suspension. When the two soils were incubated in polyethylene bags at field capacity moisture, only trace amounts of Cr(VI) were formed from the Cr(OH)$_3$ in each after 10 days.

3.7   Procedures

The mechanistic bases for these methods are discussed in Section 6.1.5.

3.7.1   Determination of Soil Chromium Oxidized by Hypochlorite

This test, modified from the method of Amacher and Baker (1982), measures most of the organic chromium in soils, especially that chromium added as sludge or leather waste.

1. Add 40 mL of undiluted laundry bleach solution (5.25% or about 0.7 M NaOCl) adjusted to pH 9.5 to 1.0 g dry chromium-containing soil in a 100 ml glass test tube. Mix thoroughly, using a vortex mixer.

2. Place tube in a boiling water bath for 20 minutes (hood), mix again (caution: Wear goggles), and obtain a clear extract by centrifuging a portion of the suspension.

3. Dilute as necessary and determine chromium in the extract by atomic absorption, using an air-acetylene flame and standards containing the same concentrations of NaOCl as the unknowns.

3.7.2 Standard Chromium Oxidizing Test

This text evaluates the potential of a soil for oxidizing available chromium.

Shake 2.5 g of soil (dry weight basis) 15 minutes with 25 mL $10^{-3} M$ CrCl$_3$, add 0.25 mL 1 M KH$_2$PO$_4 \cdot$ K$_2$HPO$_4$, shake 30 seconds more, and filter or centrifuge. Determine Cr(VI) by adding 1 mL azide reagent to 8 mL of extract, mix, let stand 40 minutes, and compare color with that in standards at 540 nm. (Prepare azide reagent by adding 120 mL of 85% H$_3$PO$_4$, diluted with 280 mL of distilled water, to 0.4 g of s-diphenylcarbazide dissolved in 100 mL 95% ethanol.)

Bloomfield and Pruden (1980) concluded that Cr(VI) was reduced during the s-diphenylcarbazide determination, which, if correct, would mean that this test underestimates the amount of chromium oxidation by a soil. However, Bloomfield and Pruden erred in equating the reduction of Cr(VI) in extracts they treated with H$_3$PO$_4$ alone with potential reduction under the conditions of the foregoing s-diphenylcarbazide method in which both H$_3$PO$_4$ and azide indicator are mixed with the aliquot at exactly the same time. In studying the kinetics of the complexation of Cr(VI) by s-diphenylcarbazide, Amacher (personal communication) showed that the complex was formed extremely rapidly compared with the rate of reduction and that the complex color was not at all affected by levels of organic matter that did not precipitate. We found that the main problem with high organic matter is caused by turbidity when humic substances are flocculated by acidification. This turbidity can be easily removed by filtration (0.2 μm) before reading colors.

Yamazaki's (1980) procedure of preconcentration of Cr(VI) with BaSO$_4$ greatly increases the sensitivity of the diphenylcarbazide method for low concentrations in water, although the standard method is sensitive enough for most soil extracts.

3.7.3 Test for Potential Oxidizable Chromium

A *Chemical Estimation of "Available" Soil Cr* is made by intermittently shaking 1 g of soil with 50 mL of 10 mM K$_2$H-citrate [or (NH$_4$)H-oxalate] for 18 hours and determining by atomic absorption spectrophotometry the chromium in the supernatant obtained by filtration or centrifugation.

The *Equilibration by Shaking Test* measures oxidizability of soil chromium in a well-aerated suspension in the presence of freshly oxidized manganese oxides and a soluble chelating agent.

1. Weigh 5 g (dry weight basis) of moist or dry chromium-containing soil to be tested into a 125 or 250 ml conical flask. Add 5 cm^3 (a packed and leveled teaspoon measure) of moist garden soil that has a pH of 6.2 or greater. The soil must be sampled fresh from the garden and maintained in the moist state before use. The sample should be from the A horizon, but material from the surface that has been dry at any time during a two-week period before sampling should be avoided.

2. Add 0.2 mmol $MnSO_4$, 0.5 mmol K_2H-citrate (made from KOH and citric acid), and enough water to bring the total solution volume to 50 ml.

3. Stopper the flask loosely with a foam plug and place on a rotating shaker providing vigorous horizontal swirling motion.

4. After three weeks of shaking, filter or centrifuge some of the suspension and determine Cr(VI) in the clear filtrate by the s-diphenylcarbazide method (Section 3.7.2).

The *Test of Time* is simply a long-term incubation of Cr(III) containing soil at field capacity moisture and at a fairly constant temperature (15 to 20°C) (Bartlett, 1986a). Half-L soil samples are a convenient size. Thoroughly mix each sample after adding distilled water or evaporating some of the water in it to ensure the soil is near its field capacity moisture, estimated by the proper "feeling of friability," indicating that the large soil pores are air-filled and the small ones water-filled. Water should not move out of the soil by capillarity when the soil is sprinkled onto a piece of filter paper. Triple bag each sample separately using 4-L thin polyethylene bags (about 2.6 mg cm^{-2}). Thin polyethylene is permeable to oxygen, carbon dioxide, and water vapor. Add 1–2 mL of water occasionally to the space between the outer two bags to prevent loss of soil moisture. If the Cr(III) in the soil is low in amount, is largely inorganic, or if several months have elapsed since the addition of an organic form, spike the soil with an organic energy source, namely, 5 mmol of citric acid thoroughly mixed with a one-half L sample.

Probably one year is a long enough incubation period for determining the likelihood of chromium oxidation. Checking for Cr(VI) once a month (after gentle mixing) for the first year may show Cr(VI) formation and then disappearance early in the incubation. At testing time, shake 5 cm^3 (one teaspoon packed and leveled) with 25 mL of 10 mM $KH_2PO_4 \cdot K_2HPO_4$ for one minute, centrifuge or filter, and determine Cr(VI) in the solution (Section 3.7.2).

4.   BEHAVIOR OF HEXAVALENT CHROMIUM

4.1   Fates

Chromium(VI) added to or formed in soils may be leached, reduced, adsorbed, precipitated, or taken up by a living organism. Occasional reports of Cr(VI) in groundwater is circumstantial evidence that leaching occurs, but the leaching process has not been documented. Overall fates of chromium are reviewed by Cary (1982).

Griffin et al. (1977) studied the removal of Cr(VI) in landfill leachate by clay minerals and found that low pH favored Cr(VI) removal. They assumed, however, that adsorption of Cr(VI) was the only reaction taking place. In contrast, Grove and Ellis (1980) suggested that reduction alone accounted for the disappearance of soluble Cr(VI) added to three soils, but they did not measure Cr(VI) in various extracting solutions used to characterize soil chromium. Both reduction and adsorption occur simultaneously in many soils. Even though it is often impossible to affix the cause for Cr(VI) disappearance as being one or the other, an understanding of the separate mechanisms of reduction and adsorption should help in evaluating the importance of each in a given soil environment.

4.2   Reduction

4.2.1   Occurrence and Mechanisms

Presence of soil organic matter brought about spontaneous reduction of Cr(VI), without regard to pH, in soils incubated with Cr(VI) for four weeks at field capacity moisture but dried to lab dirt before extraction and analysis (Bartlett and Kimble, 1976b). Presumably Bloomfield and Pruden (1980) also used dried soil in observing losses of Cr(VI) inversely proportional to soil pH and attributed to reduction and adsorption.

Bartlett and Kimble (1976b) found no reduction of Cr(VI) in acidified organic-free Hapludult B horizon samples. Only 2% of the Cr(VI) was lost in 48 hours by reduction in samples of the same soil containing added cow manure when the pH was above 6; but lowering the pH to 3 with HCl resulted in rapid reduction of all of the Cr(VI). Bloomfield and Pruden (1980) observed a similar effect of pH on loss of Cr(VI) with water-soluble organic matter.

Stollenwerk and Grove (1985b) concluded that water samples collected for analysis of Cr(VI) should not be acidified because of possible reduction of Cr(VI) by reducing impurities, even in filtered samples. Use of the extraction procedure for the EP toxicity test described in the Federal Register(45[98], May 19, 1980) practically eliminates the possibility of finding Cr(VI) even if it is present. The procedure calls for acidifying solid material with 0.5 N acetic acid and maintaining it at 20–40°C for a 24-hour period. Even if the soil is devoid of organics, the acetic acid itself can reduce Cr(VI) under these conditions.

The following equation was developed by Bartlett and Kimble (1976b) to explain the pH changes measured during reduction of Cr(VI):

$$HOHC = R = CHOH + K_2Cr_2O_7 + 6HCl = \tag{2}$$
$$KOOC\text{-}R\text{-}COOK + 2CrCl_3 + 5H_2O$$

When hydrochloric acid was added at three times the molar quantity of Cr(VI), the pH after reduction was found to exactly equal the pH of the system receiving Cr(III) directly. Thus, the equation developed appeared to be a good approximation of the reaction taking place. But more chromium was extracted by hydrochloric acid, $Na_4P_2O_7$, and NH_4OAc where chromium was applied in the reduced form than where it became reduced from the Cr(VI) form. The tighter binding as a result of the reduction may mean that the Cr(III) was held by the oxidized half of the redox couple, an organic coordination compound.

Related to this finding, the reduction of soluble Cr(VI) added to unlimed (pH 5.3) and limed (pH 6.5) suspensions of an Ultic Hapludalf resulted in the formation of soluble organic complexes of Cr(III) if citric acid was added with the Cr(VI) (James and Bartlett, 1983b). During a nine-day incubation period, more reduction of Cr(VI) occurred in unlimed than in limed samples, but higher levels of soluble Cr(III)-chelate were measured in the limed treatments. Hexavalent chromium removed from solution by reduction and then adsorption was increased by adding citric acid with unlimed or limed samples.

Compared with oxidation kinetics, Amacher and Baker (1982) found the kinetics of Cr(VI) reduction to be straightforward and simple and reduction to be a much slower process than oxidation. Reduction of 96.2 μmol L^{-1} of Cr(VI) added to Hagerstown loam (Typic Hapludalf) was a first-order reaction at 26°C at pH 5.0, with a rate constant of 0.012 day^{-1} and a half-life of 55 days. Reduction by a soil fulvic acid was slow. It was increased somewhat when the temperature was raised and was dramatically faster when the pH was lowered from 5.5 to 1.0.

Light can bring about photooxidation of soil organic matter accompanied by reduction of Cr(VI) (Bartlett, unpublished data). A 1 mg L^{-1} Cr(VI) solution was added to a sample of pH 7 phosphate-extracted soil organic matter solution, and the mixture in a cuvet was placed in the lamp beam of a Perkin-Elmer 403 atomic absorption spectrophotometer. Lamps and wavelength settings were varied. Visible light ranging from 423 to 671 nm appeared to be the most effective in increasing the amount of Cr(VI) reduced in the light compared with that in the dark during the same 15-minute time period.

4.2.2 Predicting Reduction

The following tests are suggested for predicting the Cr(VI) reducing capacities of soils and the fates of Cr(VI) added to them.

Total Cr(VI) reducing capacity. This is most easily estimated by the classic Walkley-Black (1934) soil organic matter determination in which carbon oxidizable by $K_2Cr_2O_7$ is measured by titrating with $Fe(NH_4)_2(SO_4)_2$ the Cr(VI) not reduced by a soil sample in suspension with concentrated H_2SO_4.

Available reducing capacity. Shake 2.5 cm^3 of moist soil 18 hours with 25 mL of 0.1 to 10mM chromium as $K_2Cr_2O_7$ in 10 mM H_3PO_4, filter or centrifuge, and determine Cr(VI) not reduced in the extract by the s-diphenylcarbazide method (Section 3.7.1). Try 0.1 mM chromium first. If all of the Cr(VI) is lost, repeat with increased concentration until the Cr(VI) remaining is measurable but below 0.1 mM.

Reducing intensity. The procedure is the same as that used in the available capacity determination except that 10 mM KH_2PO_4 should be used in the matrix solution in place of H_3PO_4.

4.3 Adsorption and Precipitation

4.3.1 Effects of pH

Soil pH will determine both the speciation of Cr(VI) and the charge characteristic of the colloidal surface with which it reacts. Above pH 6.4, $HCrO_4^-$ dissociates to $CrO_4^=$ as the dominant form of Cr(VI) in dilute aqueous systems (Deltombe et al., 1966). Divalent $CrO_4^=$ may be adsorbed in similar fashion to $SO_4^=$ and $HPO_4^=$, forming binuclear bridged complexes on goethite, Fe $\cdot$ OCr(00)O $\cdot$ Fe, or on aluminum oxides and other soil colloids with positively charged surfaces. Chromate also may be adsorbed by ligand exchange as $MoO_4^=$ is on Fe(III) and Al(III) oxides (Russell, 1973; Parfitt, 1978). Depending on the colloid, $HCrO_4^-$ may be similar to $H_2PO_4^-$ and be tightly held in soils, or it may be similar to HCO_3^-, Cl^-, and NO_3^- and remain soluble.

Soil pH also affects the quantities of positive and negative charges on soil colloids, especially on organic matter and on Fe(III), Al(III), and Mn(III, IV) oxides (McKenzie, 1977; Parfitt, 1978). Binding of Cr(VI) species in soils will depend on soil mineralogy and on the relation of soil pH to the pH of zero point of charge of the colloids involved.

4.3.2 Exchangeable Chromium(VI) Versus Immobilized Chromium

Study of adsorption versus precipitation of phosphate by soil aluminum species exposes a broad gray area in which the two processes overlap, and their separation for practical purpose is impossible. So it is with chromate.

The solubilities of both Cr(VI) and orthophosphate in the presence of excess aluminum change similarly with pH (Bartlett and Kimble, 1976b). They showed that orthophosphate prevented adsorption of chromate, presumably by competition for the same sites, and that adsorption and desorption behavior of chromate was analogous to that of phosphate in Spodosol horizons as pH was increased with $CaCO_3$. Consistent with this, KH_2PO_4 was the best extracting agent for Cr(VI). Amacher and Baker (1982) found that 10 mM KH_2PO_4 at a 3:1 solution:soil ratio was optimum for extraction of chromate. James and Bartlett (1983c) used 10 mM $(KH)_{1.5}PO_4$ buffered at pH 7.2 to minimize reduction. They considered the Cr(VI) removed *exchangeable*.

The Cr(VI) not extracted by the 10 mM phosphate buffer was referred to as nonexchangeable. In retrospect, a better term would be *immobilized chromium*. After Cr(VI) is added to a soil, a portion of it cannot be exchanged, either because it is bonded tightly to a colloidal surface (adsorbed or precipitated), because it has been reduced to Cr(III) and then adsorbed or precipitated or, more likely, because it is complexed by the organic functional groups that were implicated in its reduction. The term *exchangeable* is applicable to any soil from which Cr(VI) can be extracted by pH 7.2 10 mM phosphate, whereas *immobilized* chromium is a fraction of the Cr(VI) added to a particular soil.

4.3.3   Removal by Hydroxides

The fractions of bound Cr(VI) can be more effectively separated in oxide materials not containing organic matter than in whole soils. About three fourths of the Cr(VI) removed from solutions by Fe(OH)$_3$ was exchangeable by phosphate (James and Bartlett, 1983c). An almost exactly equal quantity was reduced to Cr(III) and complexed by Tamm's acid oxalate at pH 3.3. This similarity in quantities of chromium removed by the two solutions suggests that the phosphate-exchangeable Cr(VI) was adsorbed on the amorphous fraction of the Fe(OH)$_3$ removed by oxalate (Schwertmann and Fischer, 1973).

Sodium hydroxide extracted all the Cr(VI) from the oxide, but because it would disperse and dissolve huge quantities of organic matter, NaOH is not satisfactory for removing strongly bound Cr(VI) from whole soils.

Hexavalent chromium also was shaken with freshly prepared Cr(OH)$_3$ and with Cr(OH)$_3$ aged for 15 days. Both NaOH and phosphate extracted a greater percentage of Cr(VI) from aged than from fresh Cr(OH)$_3$, indicating that precipitation predominated in the fresh material and adsorption in the aged hydroxide. The Cr(VI) incorporated into fresh Cr(OH)$_3$ probably was part of the amorphous or crystalline structure, for example, as Cr(OH)$_2$HCrO$_4$, and was not easily removed by other anions. This type of precipitate may have been similar to the coprecipitate of Al(OH)$_3$ and Cr(VI) noted by Bartlett and Kimble (1976b) or to certain aluminum-phosphates in soils with pHs above 5.5 (Lindsay, 1979).

4.3.4   Removal by Whole Soils

An average of 36% of added Cr(VI) was removed from solution by 38 soil horizons (19 A and 19 B horizons from the U.S. Northeast) having a mean pH of 5.4, whereas 13% was removed by limed samples with a mean pH of 7.0 (James and Bartlett, 1983c). Liming the soils decreased both the exchangeable Cr(VI) and the immobilized chromium, with each form accounting for about 50% of the total lost. Liming decreased exchangeable Cr(VI) the most in soils having high levels of iron and aluminum sesquioxides and kaolinite. This effect probably resulted from a decrease in positive charge on soil colloids as pH increased.

In the A horizons that contained organic matter, the Cr(VI) removed by soils and not exchangeable probably was reduced to Cr(III) and complexed by humic acids or precipitated as Cr(OH)$_3$. Some reduction of Cr(VI) by Fe^{2+} may have occurred in B horizons. Presence of Fe^{2+} was shown by extraction with 1 M hydrochloric acid. However, precipitation with aluminum, Cr(III), or barium seemed more likely here. The solubility product of BaCrO$_4$ (K$_{SP}$ 3 $\times$ 10^{-10}) is low enough that Ba^{2+} might control the solubility of Cr(VI) in certain soils (Trotman-Dickenson, 1973). Ammonium acetate (pH 4.8, 1.25 N) and 1 M hydrochloric acid extracted Ba^{2+} from almost all the 38 soils, the highest value being 350 ppm from Vergennes B horizon (Glossaquic Hapludalf). Therefore, Ba^{2+} precipitation can be an important mechanism for removal of chromium not exchangeable by the phosphate buffer.

Sulfate and phosphate added with Cr(VI) to two A and two B horizon soils decreased Cr(VI) removal by the soils, with phosphate having a greater effect than sulfate. Although liming decreased Cr(VI) removal by the soils, even less Cr(VI) was held in exchangeable or immobilized forms in limed treatments containing the competing anions.

Stollenwerk and Grove (1985a) explained adsorption of Cr(VI) anions by a Colorado alluvium as being driven by positive charges on coatings of iron oxide, which had a pH_{zpc} of 8.3. This nonspecifically adsorbed chromium was readily desorbed by chromium-free water. They hypothesized that the more slowly released chromium was specifically adsorbed by complexing directly with iron oxide and hydroxide coatings and that chromium not released was either incorporated as Cr(VI) into the structure of the oxides or was reduced to Cr(III) and precipitated.

Aoki and Munemori (1982) showed that adsorption of chromate by iron oxides was accompanied by release of $-OH$ and was increased in the presence of cadmium, lead, copper, and zinc divalent cations. Sulfate and SCN^- competed with chromate for the sites. In a field experiment, Jaiswal and Misra (1984) showed that increases in available iron, up to 750%, correlated with Cr(VI) addition, with unrecoverable Cr(VI), soluble Cr(VI), pH decrease, decrease in available phosphorus, and decrease in available manganese.

4.3.5 Reduction of Adsorbed Chromium(VI)

Reducibility of adsorbed Cr(VI) was studied in six soil samples chosen because of their high levels of amorphous, organically complexed aluminum and iron and because of reputed strong phosphorus fixation by soils with this mineralogy (James and Bartlett, 1983c). Samples included A and B horizons of a Typic Fragiochrept and a Typic Haplorthod, a Hydric Dystrandept Ap, and a Typic Haplorthod Bs horizon. These samples were limed to near-neutral pH and equilibrated overnight with $K_2Cr_2O_7$ in 10 mM $NaNO_3$ and then with 200 μM gallic acid, except for controls. In the B horizons and the Andept Ap, about half or more than half, and in the other As, 15 to 18%, of the adsorbed and exchangeable Cr(VI) was still exchangeable by 10 mM phosphate after the gallic acid treatment, indicating that the Cr(VI) was protected from reduction by adsorption. Such protection was not observed with limed A and B horizon samples of a Typic Paleudult and a Typic Hapludult, even though these soils removed more Cr(VI) than the amorphous aluminum-dominated materials. Apparently, these more highly weathered iron oxide-rich colloids did not bind the Cr(VI) in such a way as to prevent its reduction.

5. BIOAVAILABILITY OF CHROMIUM

5.1 Uptake and Translocation of Chromium by Higher Plants

The toxicity of Cr(VI) to living cells and the essentiality of Cr(III) in human nutrition are opposing concerns that find common ground in the soil chemistry of this heavy metal. The subject of chromium bioavailability to plants has been reviewed by Cary (1982) and by the National Research Council of the NAS (1974). As a soluble anion, Cr(VI) readily penetrates cell membranes and is toxic as an oxidizing agent; hence the drinking water standard maximum of 50 μgL^{-1} chromium (U.S. EPA, 1976). In contrast, Cr(III) is soluble at biological pHs only when organically complexed in low molecular weight organic complexes and, therefore, soil forms probably do not penetrate membranes. An understanding of the dynamic balance governing chromium redox is essential for explaining how plant roots and soil microorganisms respond to absolute concentrations and relative levels of Cr(VI) and Cr(III) in their environments.

Plants growing in soils amended with Cr(III) as soluble salts, Cr(OH)$_3$, tannery wastes, or sewage sludge often do not contain higher concentrations of chromium in the above-ground portion than do plants in control treatments (Bolton, 1975; Cunningham et al., 1975; Cary et al., 1977a,b; Kelling et al., 1977; Hinsley, 1976; Dowdy and Ham, 1977; Lahouti and Peterson, 1979; Silva and Beghi, 1979; Sykes and Earl, 1981; Cary et al., 1983; Sheppard et al., 1984; Stomberg et al., 1984). Soil Cr(VI) was not determined in these studies.

In a pot study, mustard (*Brassica rapa* L.), barley (*Hordeum vulgare* L.), and alfalfa (*Medicago sativa* L.) were severely stunted, misshapen, and chlorotic after growing two weeks in fresh field-moist Eldridge loamy fine sand (Aquic Udorthent) treated with 30 mmol kg^{-1} CrCl$_3$ (Bartlett and James, 1979). In contrast were plants grown in identical samples of the same soil, except that the CrCl$_3$ was added after the samples had been air-dried for one week. These plants, in the rewet soil, showed some injury, though considerably less than those in the continuously moist soil, and the dry weights were at least three times as great. The differences were attributed to Cr(VI) injury. Forty hours after CrCl$_3$ was added, the mean value of exchangeable Cr(VI) was 260 μmol kg^{-1} in the continuously moist soil versus 20 μmol kg^{-1} in the rewet lab dirt.

The same soil was treated with increasing levels of CrCl$_3$ and planted with Swiss chard (*Beta vulgaris* L.). Because of plant damage and death caused by Cr(VI) at the higher levels of Cr(III), plants in the first harvest were not analyzed. Results of the second crop, planted five weeks after addition of Cr(III) and harvested eight weeks after that, are shown in Table 2 (Bartlett, unpublished data). Manganese taken up by the plants, presumably made available by reduction as Cr(III) was oxidized, was directly proportional to the Cr(III) added to the soil ($r^2 = 0.99$). Chromium in the plant tops was related to the manganese in the tops ($r^2 = 0.83$) and to the Cr(VI) in the soil one week after planting ($r^2 = 0.68$).

James and Bartlett (1984b) found that levels of chromium in *Phaseolus*

Table 2   Chromium and Manganese in Tops of Swiss Chard Plants Grown in an Aquic Udorthent Receiving Increasing Amounts of CrCl₃

CrCl$_3$ Added to Soil	Cr(VI) in Soil at Six Weeks	Chromium in Plant Tops	Manganese in Plant Tops
mmol kg^{-1}		*μmol kg^{-1}*	*mmol kg^{-1}*
0	0	8	1.0
.5	0	101	2.0
1.5	2.5	186	2.5
5.0	4.8	250	3.2
10.0	9.0	303	5.9
20.0	38.5	416	10.6

vulgaris L. roots after 46 days of growth were highly correlated ($r^2 = 0.99$) with soluble Cr(VI) concentrations measured in unplanted field-moist soil amended with the same levels of Cr(III) as Cr(OH)$_3$ or tannery wastes. Despite this linear relationship between soil Cr(VI) and root chromium, levels of chromium in bean tops were only increased above those of the control in the highest (10 mmol Cr(OH)$_3$ kg^{-1} soil) chromium treatment. This difference between root and shoot levels of chromium suggests that oxidation of Cr(III) enabled anionic Cr(VI) formed in the soil to move to the roots and be absorbed by them. It appears that most of the Cr(VI) was reduced again to Cr(III) and retained by the roots in a tightly bound or insoluble form or in a soluble organic complex that was not translocated to the tops.

Shewry and Peterson (1974) found little translocation of chromium in barley seedlings grown in nutrient solutions. Pickrell and Ellis (1980) used ^{51}Cr(III) to show that 4% of chromium placed on the midrib of soybean leaves was translocated to other parts of the leaf. Parr and Taylor (1980) applied soluble Cr(VI) (9 mg L^{-1}) to soil and found that chromium concentration in soybean seedling roots was 100 times that in leaves. Foliarly applied Cr(VI) was not translocated from the leaves. Leaves of bush beans grown by Barcelo et al. (1985) in nutrient solution contained 12 mg kg^{-1} compared with 510 mg kg^{-1} in the roots.

James and Bartlett (1984b) observed that addition of 2 mmol citric acid kg^{-1} with the 10 mmol Cr(OH)$_3$ kg^{-1} increased shoot levels of chromium significantly ($p < .05$) from 25 to 42 mg chromium kg^{-1} dry plant tissue, while citrate did not increase chromium content of shoots in control treatments without chromium added, or in any treatments with high chromium sewage sludge or tannery wastes added. It was concluded that the increased chromium uptake with citrate was a Cr(VI) effect that resulted from enhanced solubility of Cr(OH)$_3$ by citrate in the soil, which in turn led to increased oxidation of Cr(III) by soil manganese oxides, as it did in the unplanted soils. Wallace et al. (1976) found that ethylenediaminetetraacetic acid added to soil increased the toxicity to bush beans of chromium added as Cr(III) but very little chromium was translocated to the leaves. There was no test for Cr(VI).

Citrate was found to increase both soluble Cr(III) and Cr(VI) in water extracts of the Cr(OH)$_3$ treatments of unplanted, the bulk portion of the planted soil, and rhizosphere portion of the planted soil (James and Bartlett, 1984b). We also observed that soluble Cr(VI) decreased in the order: unplanted soil > bulk of the planted soil > rhizosphere of the planted soil with or without citrate added. In contrast, soluble Cr(III) increased in the order: unplanted soil < bulk of the planted soil < rhizosphere of the planted soil in the treatments in which citrate was added.

Organic acids exuded by roots might be expected to have a similar effect on rhizosphere soil as observed with citrate in the unplanted soil, but no such effects could be found in comparing the rhizosphere soil with the bulk planted soil. However, when the total planted soil was compared with the unplanted soil, the former was shown to have higher soluble Cr(III) than the latter, which is the same effect observed with citrate compared with no citrate in the unplanted soil. Roots appeared to have a more reducing effect on Cr(VI) than citrate. Citrate increased both soluble Cr(III) and Cr(VI); roots increased soluble Cr(III) apparently at the expense of Cr(VI).

If Cr(III) were to be translocated to the leaves of plants, it is not unreasonable to think that it might be oxidized to Cr(VI) in the powerful oxidative environment within the chloroplasts (if it could get in there) where water is oxidized to O^{2-}. Skeffington et al. (1976) showed that 0.5% of the Cr(III) mixed with ground fresh barley roots was oxidized to Cr(VI). Likely electron acceptors here would be superoxide or the hydroxyl free radical or perhaps manganese, which trades electrons with oxygen and its free radicals, sometimes becoming a free radical itself (Bartlett, 1986b). Theoretically, N_2O, produced by microbial reduction of nitrate, could serve as an electron acceptor in the oxidation of Cr(III) as it becomes reduced to N_2.

It seems likely that plant growth responses, occasionally observed with Cr additions to soils or nutrient solutions (National Resource Council, 1974; Mertz, 1969; Pratt, 1966) could have been caused by indirect effects of Cr(III), added or formed by reduction, in increasing availability of organically bound trace elements. Adding Cr^{3+} or Al^{3+} to nutrient solutions has been shown to increase availability of iron from chelated forms (Bartlett, unpublished data). Another possibility is improved manganese availability caused by Cr(III) reduction of manganese oxides. This effect was demonstrated with barley grown in a limed soil sample of an acid highly leached manganese-depleted Vermont Udifluvent.

5.2   Forms of Chromium in Plants

The bioavailability of chromium to animals eating the plant tissue may be related to the form of chromium in the plant (National Research Council, 1974). Recent research has shown that the chromium complex in plants is different from the active glucose tolerance factor in humans. Starich and Blincoe (1982) isolated a chromium complex from alfalfa (*Medicago sativa* L.) that was distinct from that in brewer's yeast. The valence or mode of incorporation of the chromium into the plant did not affect the form of the complex, and it was extremely stable once formed. In contrast to the cationic GTF, the alfalfa complex of chromium was anionic and contained sugars and primary amines instead of amino acids. The plant complex had a molecular weight of < 2000 daltons, while the GTF molecular weight is < 400 daltons. The plant chromium complex also was found in crested wheatgrass (*Agropyron cristatum* L.), beans (*Phaseolus* sp.), and wheat (*Triticum* sp.). Further investigations of the transformations and structure of the plant complex of chromium and the GTF may elucidate mechanisms governing translocation of chromium in plants and perhaps forms of Cr(III) in soil that may be absorbed by roots or formed in them and translocated to shoots.

5.3   Effects of Chromium on Microbial Systems

Availability and toxicity relationships in soil microbial systems appear to be at least superficially similar to those in higher plants.

Ross et al. (1981) found that 100 mg Cr(III) kg^{-1} soil or 10 mg Cr(VI) kg^{-1} decreased respiration in soils. In axenic broth cultures, gram-negative bacteria were more sensitive to 10 to 12 mg Cr(VI) L^{-1} than were gram-positive organisms. In a general survey of the effects of heavy metals on nitrification in soils, Liang and Tabatabai (1978) found that 5.0 mmol $CrCl_3$ kg^{-1} soil inhibited this process during a 10-day period. They did not treat soils with Cr(VI) or measure Cr(VI) that may have formed from the Cr(III). Zibilski and Wagner (1982) enriched sewage sludge with Cr(III) before adding it to soil at pH 6 and observed some inhibitory effects on bacterial populations when 556 mg kg^{-1} chromium were added to soil. Again, since Cr(VI) was not measured, the form of chromium causing the inhibition is not known.

Babich et al. (1982a) found that mycelial growth rates of a number of fungi were inhibited more by Cr(VI) than by Cr(III), although they reported that autoclaving of media could reduce Cr(VI) in it to Cr(III) and thereby lower the toxicity of the metal to microorganisms (Babich et al., 1982b).

James and Bartlett (1984a) observed the effects of different concentrations of Cr(III) and Cr(VI) on nitrification in soil suspensions amended with soluble salts and with chromium-rich tannery wastes and also compared the effects of Cr(VI) on NH_4^+ and NO_2^- oxidation in solution culture. Results indicated that both NH_4^+ and NO_2^- oxidizing bacteria were at first partially inhibited by Cr(VI) in culture solution at pH 7.2–7.6, but then both groups of microbes adapted to the extent that all of the added NH_4^+ was oxidized to NO_3^- after 35 days, even though Cr(VI) was still present. The autotrophs appeared to have lower sensitivity to Cr(VI) than the heterotrophs studied by Ross et al. (1981), who observed that respiration not only was lowered in soils by Cr(VI) but also that the effects persisted after Cr(VI) was no longer detected after reduction to Cr(III). Perhaps low sensitivity to Cr(VI) is related to the abilities of *Nitrosomonas* and

Nitrobacter species to produce oxyanions (NO_2^- and NO_3^-) in their normal respiratory and metabolic functions. In contrast, heterotrophs that reduce NO_3^- may be sensitive to the easily reducible Cr(VI) oxyanions. Therefore, the balance between autotrophic and heterotrophic microbial activity might be shifted in favor of autotrophs in soils containing Cr(VI).

In the chromium and rhizosphere study previously discussed, James and Bartlett (1984b) noted differences in root nodulation among treatments of bean plants. The root systems growing in the control, 1.0 mmol Cr(OH)$_3$, and sewage sludge treatments had 10 to 100 spherical, pink nodules in each pot, indicating production of leghemaglobin and fixation of N_2 (Alexander, 1977). In contrast, there were fewer than 10 nodules per root system in the 10 mmol Cr(OH)$_3$ treatments, and many of the nodules appeared as elliptical, swollen tissue rather than as distinct spheres. Nodules on the roots grown in soil treated with tannery effluent at both levels also resembled ellipses, were white, and numbered fewer than 10 per root system. These roots had fewer branches of fine roots than did the roots of the other treatments. Although the roots grown in tannery effluent-treated soil were poorly nodulated, there were no significant differences in Kjeldahl nitrogen content of shoots or roots in the various treatments. These effects on nodulation may have been caused by nitrogen in the wastes or by effects of chromium on *Rhizobium*, as shown for other bacteria (Ross et al., 1981).

Effects of chromium on microbial cells appear to be similar to effects of aluminum toxicity on root cells growing in acid soils, effects that have been attributed to malfunctioning proteins and DNA synthesis (Petrilli and deFlora, 1977). In other research with chromium and *Salmonella typhimurium*, Petrilli and deFlora (1978) found that mutagenicity only occurred if KMnO$_4$ was added to culture plates with Cr(III) salts, and oxidation of Cr(III) to Cr(VI) was observed. The mutagenic effects of chromium were mitigated by addition of ascorbic acid to the growth medium, which undoubtedly reduced Cr(VI) formed by KMnO$_4$.

Ushikuko et al. (1985) found that Cr(VI) at 0.1–1.0 mg L^{-1} decreased activities of urease, phosphatase, aryl sulfatase, and B-glucosidase 20 to 100% in soils, but inhibitions were less marked than those for cadmium or mercury. In contrast, Chang and Broadbent (1982) found that chromium inhibited microbial nitrogen transformations in soil more than did cadmium, copper, zinc, manganese, or lead. They added up to 400 mg Cr(III) kg^{-1} as CrCl$_3$ to a neutral Yolo silt loam soil (Typic Xerothent). Although they recovered only 10 to 13% of added chromium from the soil by a series of extractants, a small percentage of added Cr(III) could have been oxidized to Cr(VI) and have been responsible for the toxicity to the microbes.

The soil reactions and intracellular transformations responsible for bioavailability and toxicity of chromium appear similar for higher plants and microorganisms, except for the lower sensitivity to Cr(VI) shown by nitrifying autotrophic bacteria. Microbial population genetics in relation to forms and transformations of chromium and other redox metals in soils and water is an unexplored area for research.

6. CHROMIUM REDOX MECHANISMS SUMMARIZED

6.1 Oxidation

6.1.1 Interacting Factors

In this section, we attempt to synthesize many of the facts discussed earlier that deal with mechanisms of chromium transformations and to fit chromium redox into the overall redox framework of soils and water. Here and there we chinked a few of the fissures in the framework with suppositions and conjectures, but drew the line at the oxymoronic educated surmise.

The barriers to Cr(III) oxidation in soils and water are kinetic barriers. The oxidative behavior of Cr(III) is the resultant interaction among the following factors, each of which has been demonstrated to contribute to the rates and quantities of Cr(VI) formation:

1. Solubility, speciation, pH, presence of binding ligands, and all other factors that determine the activity of Cr(III) and its mobility by diffusion and mass flow.
2. Availability of receptive surfaces of recently formed manganese oxides, relatively free of organic substances and adsorbed cations, especially Mn(II).
3. Geometry between reduced organic species such that activities of easily oxidizable substances are lower in the region of the manganese oxide surfaces than the activities of reactive oxidized manganese.

Solubility and speciation of Cr(III) in solutions and dilute colloidal suspensions could be determined operationally by measuring rates of oxidation of chromium by a standard MnO_2. The hexaqua ion $[Cr(H_2O)_6]^{3+}$ will oxidize faster than exchangeable Cr^{3+} adsorbed on colloidal surfaces or than olated polymers such as $[Cr_6(OH)_2]^{6+}$. Organically chelated Cr(III) will oxidize more slowly in a suspension than chromium freshly precipitated by hydroxyls, phosphates, or silicates but faster than precipitated Cr(III) that has been aged or chromium in well-ordered minerals.

6.1.2 Effects of pH

According to Equation 3, Cr^{3+} releases a proton during oxidation, suggesting that low pH would inhibit oxidation. On the other hand, $Cr(OH)_3$ should be more easily oxidized at low pH since two protons are consumed during its oxidation (Equation 4).

$$Cr^{3+}+1.5\ MnO_2+H_2O=HCrO_4^-+1.5\ Mn^{2+}+H^+\ [Log\ K=-1.0] \qquad (3)$$

$$Cr(OH)_3+1.5\ MnO_2+2H^+ = HCrO_4^-+1.5\ Mn^{2+}+2H_2O\ [Log\ K=+8.9] \qquad (4)$$

At intermediate pHs, the degree of hydroxylation of olated chromium should determine effects of pH on oxidation. Bartlett and James (1979) demonstrated this, showing that as the pH of a 1 μM Cr(III) solution was lowered, amount of oxidation by a dilute soil suspension ranged from 20% at pH 7.5 to 100% at pH 3.2.

In some systems, effects of pH on charge characteristics and surface behavior of manganese oxides can mask effects of pH on chromium speciation and solubility. Amacher and Baker (1982) observed no differences in rate of oxidation of Cr(III) by a MnO_2 preparation at pH 5.5 compared with 7.5, some slowing at pH 3, and considerable slowing at pH 1. They concluded that the pH effects resulted from lowering of negative surface charge at pH 3 and a reversal of charge to positive at pH 1.

In addition to pH dependent charges characteristic of oxide minerals, it seems that manganese oxide surfaces might also develop so-called permanent negative charges as substitution of Mn(II) and Mn(III) for Mn(IV) occurs during their oxidative formation. Because of the rapid redox transformations and specific adsorption continually taking place on manganese oxide surfaces, these "permanent" charges will be quite temporary.

Exchangeable Cr^{3+} will not be found in soils with pHs greater than 4.5 to 5.

6.1.3 Oxidation of Inorganic Chromium(III)

The manganese oxide surfaces must be clean for optimum oxidation of Cr(III) by manganese oxides, that is, relatively free of specifically adsorbed Mn(II) and other heavy metals. Reduction and reoxidation are required to rid the surface of the foreign metals; oxidation alone takes care of Mn(II), the number one contaminant. Chromium(III) appears not to be adsorbed (Amacher and Baker, 1982). As a willing chromium gets close to a receptive surface, it is quickly oxidized to the anionic form and then repelled by the like negative charges.

However, unless the Cr(III) is in a mobile low molecular weight organic complex, there is a major kinetic problem with this mechanism in a soil, namely getting that chromium close to the manganese oxide surface. It is easy to see how it could happen in a reciprocating test tube on a shaker, in a leaching column, or even in a plastic bag intimately mixed and kneaded. We know that it does happen. But in an unmixed field soil, contacts will be less frequent and more random. The manganese surfaces are by no means continuous and do not blanket the soil particles in the manner of iron oxide in an Ultisol (Bartlett, 1986). Furthermore, as soil water films become discontinuous at field capacity moisture and below, movement of ions from one surface to another is restricted, even for a highly diffusable species, which inorganic Cr(III) is not. Thus, oxidation of inorganic chromium in field soils will be slow and, given opportunities for its reduction, accumulated Cr(VI) from inorganic sources may rarely be measurable.

6.1.4 Manganese Oxide Surfaces and Organic Chromium(III)

Oxidation of Cr(III) held by low molecular weight organic acids that can move by diffusion or mass flow appears to be more kinetically feasible in unmixed samples or field soils than oxidation of inorganic Cr(III). Still, the necessary and complex side reactions required will tend to make oxidation even by this mechanism very slow unless conditions are at a precisely defined optimum. Conjecture concerning the mechanism must take into account the difficulty of moving a negatively charged ligand close to a negatively charged manganese oxide surface.

Carboxyl groups or phosphates (particularly pyrophosphates) form strong complexes with Mn(III) and will cause reverse dismutation to take place according to the following equation:

$$Mn^{2+} + MnO_2 + 4H^+ = 2Mn^{3+} + 2H_2O \qquad Log\ K = -7.4 \qquad (5)$$

Although uphill thermodynamically, the reverse dismutation is favored by acidity and is driven energetically by the attraction of the ligands for Mn(III). While bound by organic functional groups, the Mn(III) will not oxidize chromium. Instead, it will oxidize the organic carbon that is holding it and become reduced to Mn(II). As organic acids become oxidized, the pH rises and the redox reaction slows down. Once the pH of Mn(III)-citrate reaches about 7.7, the dark orange Mn(III) solution can be shown to remain stable in a test tube for months. With Mn(III)-oxalate, on the other hand, the redox reaction continues at a slow rate, regardless of pH increase, until all of the oxalate is oxidized or else all of the Mn(III) is reduced to Mn(II) (Bartlett, 1986a).

Surplus Mn(III)-organics create an unfavorable environment for oxidation of chromium (Bartlett, 1986a). Even though an excess of MnO_2 is there, their presence either prevents formation of Cr(VI) or reduces it shortly after it forms. Manganese(III)-citrate or oxalate will prevent chromium oxidation or cause reduction at much lower concentrations than are required of metal-free citrate or oxalate for similar effects. Perhaps, by lowering the negative charge on the organic, the Mn(III) may allow closer approach to the negatively charged MnO_2, lowering the pe in the vicinity of the surface. Thus, a soil containing high levels of oxidized manganese is an effective Cr(VI) reducing soil, while at the same time being a good Cr(III) oxidizing soil by the standard oxidizing test (Section 3.7.1) (Bartlett, 1986a).

The Mn(II) formed by reduction has a double effect in inhibiting chromium oxidation. It has a direct effect in blocking adsorption and access of Cr(III) to the manganese oxides surface as discussed in Section 3.3. It also helps drive by mass action the reverse dismutation reaction (Equation 5) causing an increased rate of Mn(III)-organic complex formation which tends to hold Cr(III) in the reduced state. In spite of this evidence that chromium reduction is favored by the overall process of manganese reverse dismutation, it also appears to be true that Mn(III) formed by reverse dismutation is the number one electron parking place in the oxidation of organically held Cr(III) to Cr(VI). If the Cr(III) is occupying the site ordinarily reserved for Mn(III) in the dismutation reaction, and if there are not many other well-suited organic sites for the Mn(III) ions, the Cr(III) appears to be displaced by the manganese and at the same time oxidized to the Cr(VI) anion. This is to say, for Cr(VI) to form, the organic sites must be in low supply relative to the potential Mn(III), and Cr(III) must occupy a significant percentage of these sites (Bartlett, 1986a).

An excess of 20 mM Cr(III)-citrate or Cr(III)-oxalate, adjusted to pH 4.8, was allowed to react with a small amount of synthetic amorphous MnO_2 until all of the MnO_2 had been reduced (Bartlett, 1986a). Determination of the Cr(VI) found in the solutions indicated that about three fourths of the MnO_2 was required to be reduced to Mn(II) to account for the Cr(VI) produced. Only one fourth of the MnO_2 was used in oxidizing organic carbon. That is, a significant portion of the Cr(III) bound by the organic acids was oxidized preferentially by the MnO_2 as opposed to the organic carrier. Thus, chromium can be thought of as temporarily protecting the organic matter from oxidation.

We need to explain why a somewhat restricted O_2 environment sometimes increases chromium oxidation compared with a highly aerated soil system (Section 3.6). A reasonable hypothesis is that oxygen free radicals are created by partial reduction of O_2 in a slightly reducing environment and that these free radicals provide the mechanism for the oxidation of Mn(II) or Mn(III) and thereby the freshening of the MnO_2 surfaces. An additional effect, which might be expected if the pe is lowered gently, is the reduction of Mn(III) associated with organic compounds near the oxide surfaces. The freed organics will be less inhibiting of Cr oxidation and could even facilitate it by forming complexes with Cr(III) (Bartlett, 1986a). Undoubtedly, the phenomenon is related in some way to the establishment of a more favorable redox balance among electron donors, acceptors, and carriers (Bartlett, 1986b). The participatory roles of soil microorganisms in all of the transformations we have discussed should be considered.

6.1.5   Models for Chromium(III) Oxidizability in Natural Soils

Chromium oxidation by a natural soil depends upon equiponderance among the chemical availabilities of Cr(III), manganese oxides, and reducing organic substances. The first model also is a standard test for evaluating the net chromium oxidizing power of a soil as fixed by the natural or uncontrolled balance between manganese oxides and reducing organics when Cr(III) is supplied at a normalized level of availability. Three other models are representations of the limited availabilities of Cr(III) for oxidation as determined in soils where the balance between manganese oxides and reducing organic substances is controlled by different mechanistic approaches.

Standard Chromium Oxidizing Test. This test measures the potential of a soil for oxidizing available Cr(III). Ample opportunity for maximum contact between available Cr(III) and manganese surfaces is provided by the 15-minute shaking. The test evaluates the freshly formed chemically receptive manganese oxide sites, which are responsible for oxidation, as they are balanced against available reducing organics, which reduce some of the Cr(VI) formed during the test. By measuring net oxidation, the test evaluates the extent that oxidizing behavior exceeds reducing behavior under standardized conditions of Cr(III) availability.

Bartlett (1981) showed that severity of sample drying was reflected in decreases in chromium oxidation test values. Drying at 40°C for 100 hours was more severe in terms of reducing manganese oxides, but 105°C was more severe in releasing easily oxidized soluble organic compounds. By removing the soluble organics, leaching dried samples with 10 mM KH_2PO_4 before the test restored much of the net chromium oxidizing ability lost by drying.

The Equilibration by Shaking Test. This test measures potential oxidizable Cr(III) under controlled soil redox conditions (Section 3.7.2). The model is one of equilibration by shaking to control the oxidation versus the reduction capacity of the soil. The goal of the test is to predict the oxidizability of Cr(III) having unknown availability. The system is continuously agitated and aerated for equilibration. The ingredients, to be swirled together as a slurry for three weeks, consist of a resistant chromium source for testing, a "healthy" medium-manganese soil, $MnSO_4$, and pH 5 citrate. The citrate functions to solubilize, extract, and mobilize immobile or bound chromium and also to provide an electron source for balancing redox.

Formed by reverse dismutation (Equation 5), Mn(III) catalyzes the decomposition of excess citrate, and, at first, any Cr(VI) formed is reduced by the Mn(III)-organic complex. Oxidation of the excess organic is catalyzed by the Mn(III) and the Cr(VI) until it is low enough to coexist with Cr(VI). Freed of surplus citrate, the Cr(III) is available to be oxidized by Mn(III) competing with it for citrate or by MnO_2 at the oxide surface. While this redox poise balance is being achieved, newly formed and surface adsorbed Mn(II) is reoxidized to MnO_2 by O_2 and the surfaces are becoming continually freshened.

The Redox Interface Model. A pond and paddy model was proposed to represent a mechanism for oxidizing quantities of soluble organic ligands along with Cr(III) bound to them in the manganese oxide layer that tends to form at the interface between anerobic soil and water with dissolved O_2 above it (Bartlett, 1986a). The geometry of a redox interface provides for tight separation of the manganese oxide electron acceptors from the huge pool of organic electron donors so that the latter are met by manganese oxides in very small portions as they gradually diffuse across the interface. Because the capacity of the system to reduce greatly exceeds the capacity to oxidize, mixing across an interface would eliminate the oxidized phase and any possibility for oxidation of chromium.

Work with lab lagoons (tubs containing soil and tannery sludge) and with test tube paddies, containing soil and sludge saturated with dilute oxalate and a layer of synthetic MnO_2 on top, led to the conclusion that, while the possibility of Cr(VI) formation at a redox interface is a valid hypothesis, a pond loaded with tannery sludge would require many years, perhaps centuries, before a suitable redox balance was reached for it to begin pumping great quantities of Cr(VI) into the drinking water (Bartlett, 1986a). The obstacle appears to be the amount of organic matter preserved in the flooded situation. The oxidizing manganese interface did form, but it was accompanied by an overwhelming abundance of apparent Mn(III) complexes that inhibited chromium oxidation or reduced the occasional Cr(VI) that we could sometimes detect slipping through. Nitrate formed at rice paddy interfaces is quickly reduced as it diffuses back through the interface (Reddy et al., 1980).

The Test of Time, or the Waiting with Patience Model. Time tends to cure the redox imbalance between manganese oxides and available easily oxidizable organic matter, and hence the time test is another approach to evaluating soil Cr(III) oxidizability while the manganese and organic redox parameters are controlled. Patience is the watchword if you want to find out whether a given chromium sludge is likely to release Cr(VI) in a natural soil. The essence of the time test is *waiting*, to see if indeed it does. Justification for the test is the haphazardness of field sampling for predicting the possibility that a particular soil having received a loading of Cr(III) will form Cr(VI). Because of probable hour-to-hour redox changes in the field brought about by moisture, aeration, temperature changes, leaching, and root and microbial metabolic activities, sampling would have to be almost daily for a period of many years.

What happens in a soil kept aerated at field capacity moisture while we are waiting? Each soil manganese oxide surface forms an interface or interfaces with the soil near it. The fact that the soil is not continuously churned allows the chromium, manganese, and organic phases to get together only a little bit at a time with small contacts by slow diffusion or minor physical disturbances. Incompatible species and phases persist side by side that could not exist if they touched.

Meanwhile, the manganese oxide surfaces are slowly but continually changing. Through slow cycle after slow cycle of reverse dismutation, reduction, and reoxidation, the oxide surfaces play an active role in the cleanup of the extraneous organic matter and in their own fouling and refreshening. In the absence of new additions of superfluous organic matter, the manganese surfaces are deprived of their dynamism, and they gradually become encrusted and occluded by Mn(II) and old organic refuse [possibly stable Mn(III) compounds]. Even a soil sample kept in the refrigerator, moist and well aerated, will little by little lose its chromium oxidizing capacity. Although some reduction of the MnO_2 has occurred, it can be shown by an "electron demand" titration with NH_2OH (Bartlett, 1986b) that most of the manganese is still oxidized. Drying, warming, or freezing and thawing will bring about considerably greater deterioration of the oxide surfaces than time alone (Bartlett, 1981; 1986a). Still, following any of these treatments (except for long storage of dried soil), the chromium oxidizing ability can be at least partially restored by simply leaching the soil with water or 10 mM pH 7 phosphate solution to remove some of the soluble or dispersed colloidal organic substances interfering with oxidation of chromium by manganese oxide surfaces or causing reduction after oxidation.

An interesting restorative phenomenon was noted in a pH 3.4 Oe horizon of a forested Typic Haplorthod in northeastern Vermont (Bartlett, 1986a). After leaching by pH 4 melting snow, the horizon material regained an ability to oxidize chromium that had been totally absent the preceding summer and fall (Bartlett, 1986a). It was hypothesized that as the snow melted, organic matter, dispersed by freezing, was leached downward. Removal of interfering organic substances allowed the highly organic soil material to oxidize Cr(III) added at only one tenth the concentration used in the standard oxidation test. In most soils, reducing organics prevent oxidation of $CrCl_3$ added at 0.1 mM.

The main reason for waiting with patience is that the results are not predicted well by chemical extractions. At the end of the four-year room temperature incubation (Section 3.6), there was only a rough correlation ($r^2 = 0.44$) between Cr(III) extracted by K_2H-citrate and Cr(VI) in the incubated soils (Bartlett, 1986a).

Dilution is another way to lower the effectiveness of soluble Cr(VI) reducing organics. As increasing volumes of water were added to a soil suspension equilibrated on a reciprocating shaker after treatment with Cr(III), the concentration of Cr(VI) in solution was decreased with each increment of water. However, the total amount of Cr(VI) formed increased as solution:soil ratio was increased until, at a 2000:1 ratio, 100% of the chromium was in the oxidized form (Bartlett, unpublished data).

Finding natural levels of Cr(VI) in soils may be a matter of sampling at the right moment in history. On the day it arrived from California, we extracted a sample of Maxwell clay (a Typic Pelloxerert formed from an ultramafic parent material known to contain natural chromium sampled by John Kimble) and discovered that phosphate removed about 1 mg kg^{-1} of Cr(VI). Two weeks later, another extraction showed the Cr(VI) had disappeared. It was never found again, even though the soil was stored in its original well-aerated moist state (Bartlett, unpublished data).

6.2 Reduction

Reduction of Cr(VI) added to or formed in soil or water hinges on whether Cr(VI) is in a more favorable or less favorable position than the usual oxygen, nitrogen, or manganese electron sinks to accept electrons from a reactive organic (or iron or sulfur) electron source. A pe and pH balance tilted toward reduction of Cr(VI) means that activities of the usual electron acceptors will be low and those of the available electron donors will be high. In addition, reduction of Cr(VI) is contingent upon the availability of the Cr(VI), not only chemical availability but also availability from the geometric perspective implicating the Cr(VI) and the reductant.

Adding Cr(VI) to a soil sample kept moist and aerated until the organic matter has reached a steady state of slow oxidation (Bartlett, 1986b) will result in extremely slow reduction of the Cr(VI), even if the soil is quite acid. Reducing organic substances such as might be exuded by growing plant roots in the field lower soil-pe and stimulate reduction. Natural drying or freezing of the soil surface in the field releases soluble reducing compounds that leaching waters carry downward where they can furnish electrons for reducing otherwise stable Cr(VI). Interestingly, soils that are good chromium oxidizers are also good reducers. Addition of organic residues to soils containing high levels of oxidized manganese will result in formation of unstable Mn(III) organic complexes that not only temporarily prevent Cr(III) oxidation but also catalyze the reduction of Cr(VI).

Adsorption or precipitation of Cr(VI) can bind it strongly enough to regulate its rate of reduction and also prevent its reduction altogether (Section 4.3.5). As a general rule, any environmental factor affecting adsorption/desorption of phosphate in soils can be considered to affect the availability of Cr(VI). Thus, increasing soil pH by liming, addition of phosphate fertilizers, and supplying organic residues that decompose to form chelating ligands that lower the plus charges on aluminum and iron mineral surfaces all will increase availability of soil phosphate to plant roots, and also can be expected to increase the availability of Cr(VI) to be reduced, or its availability to plants as well.

Because it is a proton consuming process, Cr(VI) reduction will tend to slow down as pH is increased, but this tendency will be counteracted by the increase in solubilities of both Cr(VI) and the reducing organic matter as pH rises. Conversely, in acid soils, above pH 3, Cr(VI) may be partially protected against reduction by subdued solubility of organic substances along with its own enhanced adsorption.

Adsorption of Cr(VI) by the reducing surface itself may occur on organic aluminum or iron sites. Of course, even without adsorption, proximity of the reductant and the Cr(VI) is required for the redox reaction. Chromate may escape reduction by remaining in solution until diffusion or mass flow eventually brings it into contact with a reactive reducing surface.

After reduction of Cr(VI) has begun, loss of Cr(VI) from solution can become accelerated by adsorption onto plus charged surfaces of $Cr(OH)_n$ newly formed by reduction. Precipitated Cr(III) and adsorbed Cr(VI) will cover up active sites on organic matter surfaces inhibiting its reactivity and preventing microbial activity that would tend to metabolize carbon and facilitate reduction. Thus, the reduction process is self limiting and conservative of the reactants, and the self-preserving stable state of soil nonequilibrium perseveres.

Both oxidation of Cr(III) and reduction of Cr(VI) are thermodynamically favorable in soils and, because of the nonequilibrium between O_2 and organic matter, both take place at the same time in the same soils. Because oxidation so strikingly increases the mobility of chromium in soils and because oxidation is kinetically fast compared with reduction, reduction of Cr(VI) can be expected to occur at an indefinitely later time and perhaps at some point far from its place of formation. Therein lies the possibility of chromium pollution of ground or surface water.

6.3 The Chromium Cycle in Soils and Water

Like any circle, this cycle has no beginning and no ending (Fig. 1). A logical starting place might be Cr(VI), the thermodynamically stable chromium form in a system in equilibrium with the air. But we know that in real soils, Cr(VI) is the least stable form, and this tells us something about soils. The $HCrO_4^-$ anion can

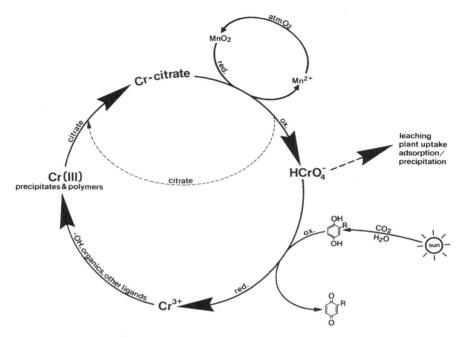

Figure 1. The chromium cycle in soils and water.

temporarily depart from the cycle by being adsorbed or precipitated, leached into streams or groundwater, or taken up by living organisms. Eventually the sidetracked chromium should find its way back to the circle, probably in a reduced form.

The next step is "dechromification", or Cr(VI) reduction by carbon reduced by the sun's energy through photosynthesis. An intermediate species, such as Fe^{2+} or S^{2-}, reduced by carbon, can serve as the direct electron donor. Analogous to denitrification, the most important part of the nitrogen cycle because it preserves O_2, "dechromification" is the vital segment of the chromium cycle. Without it, all atmospheric O_2 theoretically could end up as chromate, even if each chromium in the earth's crust only took on one more oxygen. However, all life probably would be poisoned by Cr(VI) long before all of the crust's chromite could meet with manganese. Indeed, oxidation processes are a threat to life on earth.

Following reduction of Cr(VI), Cr(III) is bound by a variety of ligands that render it insoluble, immobile, and unreactive, and chromium reciprocates by stabilizing the ligand matrices, for example, leather, or soil humus. Mobile ligands, such as citrate, extract Cr(III) and deliver it to manganese oxide surfaces where both the organic and the chromium become oxidized. Sometimes the ligand is recycled because the Mn(III) formed by reverse dismutation picks up its electrons from the Cr(III) in "preference" to those from the organic ligand, and only the chromium is oxidized. When organic ligands are excessive in relation to small amounts of Cr(III) held, the surplus organic sites tend to induce reverse dismutation of MnO_2 by binding the Mn(III). The Mn(III)-organic may prevent formation of Cr(VI) or reduce it as fast as it forms, short-circuiting the cycle or speeding up the cycling gyrations.

It seems reasonable to conclude that the chromium cycle operates in all soils that contain chromium in the biological zone. In other words, Cr(VI) is present in most soils—some Cr(VI), occasionally.

ACKNOWLEDGMENT

The core of research discussed in this chapter was supported by the Vermont Agricultural Experiment Station, Burlington. We are grateful for the immeasurable contributions of Martha Bartlett to this work.

REFERENCES

Aleksandrova, L.N. (1960). "The use of sodium pyrophosphate for separating free humic substances and their organo-mineral compounds from soil." *Sov. Soil Sci.* (English transl.) 2, 190–197.

Alexander, M. (1977). *Introduction to Soil Microbiology*. John Wiley & Sons, New York, p. 326.

Amacher, M.C., and Baker, D.E. (1982). *Redox Reactions Involving Chromium, Plutonium, and Manganese in Soils*. DOE/DP/04515-1. Institute for Research on Land & Water Resources, Pennsylvania State University and U.S. Department of Energy, Las Vegas, Nevada, 166 pp.

Aoki, T., and Munemori, M. (1982). "Recovery of chromium(VI) from wastewaters with iron(III) hydroxide. I. Adsorption mechanism of chromium(VI) on iron(III) hydroxide." *Water Res.* 16, 793–797.

Babich, H., Schiffenbauer, M., and Stotzky, G.S. (1982a). "Comparative toxicity of trivalent and hexavalent chromium to fungi." *Bull. Environ. Contam. Toxicol.*, 28, 452–459.

Babich, H., Schiffenbauer, M., and Stotzky, G.S. (1982b). "Effect of sterilizaiton method on Cr^{3+} and Cr^{6+} to fungi." *Microbios Lett.* 20, 55–64.

Baes, C.F., Jr., and Mesmer, R.E. (1976). *The Hydrolysis of Cations*. Wiley-Interscience, New York, pp. 211–220.

Barcelo, J., Poschenreider, C., and Gunse, B. (1985). "Effect of chromium(VI) on mineral element composition of bush beans." *J. Plant Nutr.* 8, 211–218.

Bartlett, R.J. (1981). "Oxidation-reduction status of aerobic soils." In R.H. Dowdy, Ed., *Chemistry in the Soil Environment*. ASA Special Publ. No. 40, Am. Soc. of Agron., Madison, Wisconsin, Chap. 4.

Bartlett, R.J. (1982). "Background levels of metals in Vermont soils." *Vt. Agr. Exp. Sta. RR 29*, Burlington, Vermont.

Bartlett, R.J. (1985). "Criteria for land spreading of the sludges in the Northeast: Chromium." In *Criteria and Recommendations for Land Application of Sludges in the Northeast*. NE Regional Publication, Bulletin 851, The Pennsylvania State University, pp. 49–52.

Bartlett, R.J. (1986a). "Chromium oxidation in soils and water: Measurements and mechanisms." In D.M. Serrone, Ed., *Proceedings of the Chromium Symposium: Update, 1986*. Industrial Health Foundation, Pittsburgh, pp. 310–330.

Bartlett, R.J. (1986b). "Soil redox behavior." In D.L. Sparks, Ed., *Soil Physical Chemistry*. CRC Press, Inc., Boca Raton, FL, pp. 179–207.

Bartlett, R.J., and James, B.R. (1979). "Behavior of chromium in soils: III. Oxidation." *J. Environ. Qual.* **8**, 31–35.

Bartlett, R.J., and James, B.R. (1980). "Studying air-dried, stored soil samples—Some pitfalls." *Soil Sci. Soc. Am. J.* **44**, 721–724.

Bartlett, R.J., and Kimble, J.M. (1976a). "Behavior of chromium in soils: I. Trivalent forms." *J. Environ. Qual.* **5**, 379–383.

Bartlett, R.J., and Kimble, J.M. (1976b). "Behavior of chromium in soils: II. Hexavalent forms." *J. Environ. Qual.* **5**, 383–386.

Bloomfield, C., and Pruden, G. (1980). "The behavior of Cr(VI) in soil under aerobic and anaerobic conditions." *Environ. Pollution Series A* **23**, 103–114.

Bolton, J. (1975). "Liming effects on the toxicity to perennial ryegrass for a sewage sludge contaminated with zinc, nickel, copper and chromium." *Environ. Pollution* **9**, 295–304.

Cary, E.E. (1982). "Chromium in air, soil, and natural waters." In S. Langård, Ed., *Biological and Environmental Aspects of Chromium*. Elsevier Biomedical Press, Amsterdam, pp. 49–64.

Cary, E.E., Allaway, W.H., and Olson, O.E. (1977a). "Control of chromium in food plants. I. Adsorption and translation of chromium by plants." *J. Agr. Food Chem.* **25**, 300–304.

Cary, E.E., Allaway, W.H., and Olson, O.E. (1977b). "Control of chromium in food plants. II. Chemistry of chromium in soils and its availability to plants." *J. Agr. Food Chem.* **25**, 305–309.

Cary, E.E., Gilbert, M., Bache, C.A., et al. (1983). "Elemental composition of potted vegetables and millet grown on hard coal bottom ash-amended soil." *Bull. Environ. Contam. Toxicol.* **31**, 418–423.

Chaberek, S., and Martell, A.E. (1959). *Organic Sequestering Agents*. John Wiley & Sons, New York, p. 363.

Chaney, R.L., Hornick, S.B., and Sikora, L.J. (1981). "Review and preliminary studies of industrial land treatment practices." *Proc. Seventh Annual USEPA Research Symposium on Land Disposal*, Philadelphia, PA, EPA-600/9-81-002b, pp. 200–212.

Chang, F.-H., and Broadbent, F.E. (1982). "Influence of trace metals on some soil nitrogen transformations." *J. Environ. Qual.* **11**, 1–4.

Cotton, F.A., and Wilkinson, G. (1980). *Advanced Inorganic Chemistry*. John Wiley & Sons, New York, pp. 719–749.

Cunningham, J.D., Keeney, D.R., and Ryan, J.A. (1975). "Yield and metal composition of corn and rye grown on sewage sludge-amended soil." *J. Environ. Qual.* **4**, 448–454.

Deltombe, E., deZoubov, N., and Pourbaix, M. (1966). "Chromium." In M. Pourbaix, Ed., *Atlas of Electrochemical Equilibria in Aqueous Solutions*. (J.A. Franklin, Eng. transl.). Pergamon Press, Oxford, pp. 256–271.

Deutsch, M. (1972). "Incidents of chromium contamination of ground waters in Michigan." In W.A. Pettyjohn, Ed., *Water Quality in a Stressed Environment*. Burgess Publ. Co., Minneapolis, MN, pp. 149–159.

Dowdy, R.H., and Ham, G.E. (1977). "Soybean growth and elemental content as influenced by soil amendments of sewage sludge and heavy metals: Seedling studies." *Agron. J.* **69**, 300–303.

Dutt, G.R., and McCreary, T. (1970). "The quality of Arizona's domestic, agricultural, and industrial waters." *Arizona Agr. Exp. Sta. Rep. No. 256* pp. 1–83.

Griffin, R.A., Au, A.K., and Frost, R.R. (1977). "Effect of pH on adsorption of chromium from landfill leachate by clay minerals." *J. Environ. Sci. Health* **A12**, 431–449.

Grove, J.H., and Ellis, B.G. (1980). "Extractable chromium as related to pH and applied chromium." *Soil Sci. Soc. Am. J.* **44**, 238–242.

Hale, M.G., Moore, L.D., and Griffin, G.J. (1978). "Root exudates and exudation." In

Y.R. Dommergues and S.V. Krupa, Eds., *Interactions Between Non-Pathogenic Soil Micro-organisms and Plants*. Elsevier, Amsterdam, pp. 163–203.

Hamm. R.E. (1958). "Complex ions of chromium. VIII. Mechanism of reaction of organic acid anions with chromium(III)." *J. Am. Chem. Soc.* **80**, 4469–4471.

Hammes, J.K., and Berger, K.C. (1960). "Chemical extraction and crop removal of Mn from air-dried and moist soils." *Soil Sci. Soc. Am. Proc.* **24**, 361–364.

Hinesly, T.D. (1976). "Data. Univ. Illinois." In Council for Agricultural Science and Technology, *Application of Sewage Sludge to Cropland*. USEPA, Washington, DC, p. 47.

Jaiswal, P.C., and Misra, S.G. (1984). "Available Fe in soil as influenced by chromium(VI) application." *J. Plant Nutr.* **7**, 541–546.

James, B.R., and Bartlett, R.J. (1983a). "Behavior of chromium in soils. V. Fate of organically complexed Cr(III) added to soil." *J. Environ. Qual.* **12**, 169–172.

James, B.R., and Bartlett, R.J. (1983b). "Behavior of chromium in soils. VI. Interactions between oxidation-reduction and organic complexation." *J. Environ. Qual.* **12**, 173–176.

James, B.R., and Bartlett, R.J. (1983c). "Behavior of chromium in soils. VII. Adsorption and reduction of hexavalent forms." *J. Environ. Qual.* **12**, 177–181.

James, B.R., and Bartlett, R.J. (1984a). "Nitrification in soil suspensions treated with chromium (III, VI) salts or tannery wastes." *Soil Biol. Biochem.* **16**, 293–295.

James, B.R., and Bartlett, R.J. (1984b). "Plant-soil interactions of chromium." *J. Environ. Qual.* **13**, 67–70.

Jenne, E.A. (1968). "Controls on Mn, Fe, Co, Ni, Cu, and Zn concentrations in soils and water: The significant role of hydrous Mn and Fe oxides." In R.F. Gould, Ed., *Trace Inorganics in Waters*. *Advan. Chem. Series* pp. 337–389.

Kelling, K.A., Keeney, D.R., Walsh, L.M., and Ryan, J.A. (1977). "A field study of the agricultural use of sewage sludge. III. Effect on uptake and extractability of sludge-borne metals." *J. Environ. Qual.* **6**, 352–360.

Lahauti, M., and Peterson, P.J. (1979). "Chromium accumulation and distribution in crop plants." *J. Sci. Food Agr.* **30**, 136–142.

Liang, C.N., and Tabatabai, M.A. (1978). "Effects of trace elements on nitrification in soils." *J. Environ. Qual.* **7**, 291–293.

Lindau, C.W., and Hossner, L.R. (1982). "Sediment fractionation of Cu, Ni, Zn, Cr, Mn, and Fe in one experimental and three natural marshes." *J. Environ. Qual.* **11**, 540–545.

Lindsay, W.L. (1979). *Chemical Equilibria in Soils*. John Wiley & Sons, New York, pp. 169.

Loganathan, P., Burau, R.G., and Fuerstenau, D.W. (1977). "Influence of pH on the sorption of Co^{2+} and Ca^{2+} by a hydrous manganese axide." *Soil Sci. Soc. Am. J.* **41**, 57–62.

McKenzie, R.M. (1977). "Manganese oxides and hydroxides." In J.B. Dixon and S.B. Weed, Eds., *Minerals in Soil Environments*. Soil Sci. Soc. Am., Madison, WI, pp. 181–193.

Mertz, W. (1969). "Chromium occurrence and function in biological systems." *Physiol. Rev.* **49**, 163–239.

Morgan, J.J., and Stumm, W. (1964). "Colloid-chemical properties of Mn dioxide." *J. Colloid Sci.* **19**, 347–359.

Mortvedt, J.J., and Giordano, P.M. (1975). "Response of corn to zinc and chromium in municipal wastes applied to soil." *J. Environ. Qual.* **4**, 170–174.

National Research Council, Committee on Biological Effects of Atmospheric Pollutants. (1974). *Chromium*. National Acad. Sci., Washington, DC, 155 pp.

Parfitt, R.L. (1978). "Anion adsorption by soils and soil materials." *Adv. Agron.* **20**, 1–50.

Parr, P.D., and Taylor, F.G., Jr. (1980). "Incorporation of chromium in vegetation through root uptake and foliar absorption pathways." *Environ. Exp. Bot.* **20**, 157–160.

Petrilli, F.L., and deFlora, S. (1977). "Toxicity and mutagenicity of hexavalent chromium on *Salmonella typhimurium.*" *Appl. Environ. Microbiol.* **33**, 805–809.

Petrilli, F.L., and deFlora, S. (1978). "Oxidation of inactive trivalent chromium to the mutagenic hexavalent form." *Mutation Res.* **58**, 167–174.

Pickrell, D.J., and Ellis, B.G. (1980). "Absorption and translocation of chromium through the surface of a soybean leaf." *Agron. J.* **72**, 854–855.

Posselt, H.S., Anderson, F.J., and Weber, W.J., Jr. (1968). "Cation sorption on colloidal hydrous manganese dioxide." *Environ. Sci. Tech.* **2**, 1087–1093.

Pratt, P.F. (1966). "Chromium." In H.D. Chapman, Ed., *Diagnostic Criteria for Plant and Soils.* Univ. of California, Riverside, Div. of Agr. Sci., pp. 136–141.

Reddy, K.R., Patrick, W.H., Jr., and Phillips, R.E. (1980). "Evaluation of selection processes controlling nitrogen loss in a flooded soil." *Soil Sci. Soc. Am. J.* **44**, 1241–1246.

Rengasamy, P., and Oades, J.M. (1978). "Interaction of monomeric and polymeric species of metal ions with clay surfaces. III. Aluminum(III) and chromium(III)." *Aust. J. Soil Res.* **16**, 53–66.

Ross, D.S., and Bartlett, R.J. (1981). "Evidence for nonmicrobial oxidation of manganese in soil." *Soil Sci.* **132**, 153–160.

Ross, D.S., Sjögren, R.E., and Bartlett, R.J. (1981). "Behavior of chromium in soils. IV. Toxicity to microorganisms." *J. Environ. Qual.* **10**, 145–148.

Russell, E.W. (1973). *Soil Conditions and Plant Growth.* Longmans, London, p. 657.

Schroeder, D.C., and Lee, G.F. (1975). "Potential transformations of chromium in natural waters." *Water Air Soil Pollution* **4**, 355–365.

Schwertmann, U., and Fischer, W.R. (1973). "Natural 'amorphous' ferric hydroxide." *Geoderma* **10**, 237–247.

Sheppard, M.I., Sheppard, S.C., and Thibault, D.H. (1984). "Uptake by plants and migration of uranium and chromium in field lysimeters." *J. Environ. Qual.* **13**, 357–361.

Shewry, P.R., and Peterson, P.J. (1974). "The uptake and transport of chromium by barley seedlings (*Hordeum vulgare* L.)." *J. Exp. Bot.* **25**, 785–797.

Silva, S., and Beghi, B. (1979). "The use of chromium-containing organic manures in rice fields." *Riso.* **28**, 105–113; *Abstr. Soils Fert.* **43**, 4100.

Skeffington, R.A., Shewry, P.R., and Peterson, P.J. (1976). "Chromium uptake and transport in barley seedlings (*Hordeum vulgare* L.)." *Planta* **132**, 209–214.

Starich, G.H., and Blincoe, C. (1982). "Properties of a chromium complex from higher plants." *J. Agr. Food Chem.* **30**, 458–462.

Stollenwerk, K.G., and Grove, D.B. (1985a). "Adsorption and desorption of hexavalent chromium in an alluvial aquifer near Telluride, Colorado." *J. Environ. Qual.* **14**, 150–155.

Stollenwerk, K.G., and Grove, D.B. (1985b). "Reduction of hexavalent chromium in water samples acidified for preservation." *J. Environ. Qual.* **14**, 396–399.

Stomberg, A.L., Hemphill, D.D., Jr., and Volk, V.V. (1984). "Yield and elemental concentration of sweet corn grown on tannery waste-amended soil." *J. Environ. Qual.* **13**, 162–165.

Stunzi, H., and Marty, W. (1983). "Early stages of the hydrolysis of chromium(III) in aqueous solution. I. Characterization of a tetrameric species." *Inorganic Chem.* **22**, 2145–2150.

Sykes, R.L., and Earl, N.J. (1981). "The effect of soil-chromium(III) on the growth and chromium absorption of various plants." *J. Am. Leather Chem. Assoc.* **76**, 102–125.

Trotman-Dickenson, A.F., Ed. (1973). *Comprehensive Inorganic Chemistry.* Pergamon Press, Oxford, Vol. 3, pp. 664–700.

U.S. Environmental Protection Agency. (1976). "Chromium." In *Quality Criteria for Water.* Washington, DC, pp. 37–41.

U.S. Environmental Protection Agency. (1980). "Hazardous waste management system: Identification and listing of hazardous wastes." *Fed. Reg.* **45**(212), 72029–72039.

Ushikuko, A., Oyama, G., Ishimani, K., and D'Istri, F.M. (1985). "Effects of heavy metal concentration on microbial activity in paddy soil." *Nohaku Shuho (Tokyo Nogyo Daigaku)* **30**, 44–59; *Chem. Abstr.* (1986). **104**, 50712.

Walkley, A., and Black, I.A. (1934). "An examination of the Degtjareff method for determining soil organic matter and a proposed modification of the chromic acid titration method." *Soil Sci.* **37**, 29–38.

Wallace, A., Soufi, S.M., Cha, J.W., and Romney, E.M. (1976). "Some effects of chromium toxicity on bush bean plants grown in soil." *Plant Soil* **44**, 471–473.

Wickliff, C., Volk, V.V., Tingley, D.T., et al. (1982). "Reactions of chrome tannery sludge with organic and mineral soils." *Water Air Soil Pollution* **17**, 61–74.

Yamazaki, H. (1980). "Preconcentration and spectrophotometric determination of Cr(VI) in natural waters by coprecipitation with barium sulfate." *Anal. Chim. Acta* **113**, 131–137.

Zibilski, L.M., and Wagner, G.H. (1982). "Bacterial growth and fungal genera distribution in soil amended with sewage sludge containing cadmium, chromium, and copper." *Soil Sci.* **134**, 364–370.

11

CHROMIUM TOXICITY TO ALGAE AND BACTERIA

P.T.S. Wong

Fisheries and Oceans
Greak Lakes Laboratory for Fisheries and Aquatic Sciences
Canada Centre for Inland Waters
Burlington, Ontario, Canada

J.T. Trevors

Department of Environmental Biology
Ontario Agricultural College
University of Guelph
Guelph, Ontario, Canada

1. INTRODUCTION

Chromium (atomic weight 51.996; atomic number 24) can exist in oxidation states from -2 to $+6$, with 0, $+3$, and $+6$ being the most common oxidation states (Merck Index, 1976). Its reported abundance in the earth's crust varies from 100 to 300 mg/kg. Chromium is primarily used in the manufacture of chrome-steel or stainless steel (chrome-nickel-steel alloys) for increasing the resistance and durability of metals, or for chromeplating other metals (Chapter 3). Chromium is also used in relatively large amounts in the pulp and paper, automobile, dye, leather, and tanning industries (NRC, 1976). The average chromium levels in waters of the Great Lakes are from 1 to 2 μg/L (Weiler and Chawla, 1969). However, the levels were reported as high as 500 μg/L in certain streams and rivers in Canada polluted by domestic and industrial sources (NRC, 1976).

Chromium can exist in a variety of oxidation states. However, only Cr(III) and Cr(VI) are important in natural water systems. Most of the Cr(III) present in water is adsorbed on particulate matter whereas Cr(VI) remains in solution. The Cr(VI), a powerful oxidizing agent, readily oxidizes organic material and is reduced to Cr(III). Chromium(VI) has long been recognized as a toxic substance whereas Cr(III) is considered to be relatively innocuous and even essential to human health in minute quantities (NAS, 1980).

Since large amounts of chromium are released into the environment through industrial processes (Shepherd and Jones, 1971), it is important to evaluate its impact on aquatic organisms. A substantial body of literature has evolved dealing with the toxicity of chromium in fish and invertebrates (Towill et al., 1978; API, 1982). Limited information is available on the effects of chromium in algae and bacteria, organisms central to nutrient and mineral cycling in the biosphere.

The objectives of this review are to examine recent available information on chromium toxicity in algae and bacteria with discussion on various abiotic and biotic factors that may influence the toxicity. In addition, plasmid-encoded resistance to chromium in bacteria is dicussed.

2. CHROMIUM TOXICITY TO ALGAE

Information on chromium in algae is rather sparse. The few papers published before 1978 have been adequately reviewed (NRC, 1976; Towill et al., 1978; API, 1982). This chapter focuses mainly on information available since that time.

No studies have conclusively demonstrated that chromium is an essential element in algae. On the other hand, chromium is shown to be quite toxic to algae (Table 1). Of the commonly occurring forms of chromium [(III) and (VI)], the Cr(VI) form seems to be the most toxic (Towill et al., 1978). Hence, almost all the studies on the effects of chromium on algae have been conducted with Cr(VI). The range of Cr(VI) concentrations necessary to cause effects varies from 20

Table 1   Some Recent Examples of Cr(VI) Toxicity to Algae

Algal Species	Toxic Concentrations (μg/L)	Effects	Reference
Chlorella sp.	10,000	Growth inhibition	Petria, 1978
	100	Photosynthesis inhibition	
Chlorella pyrenoidosa	20	Growth reduced	Schroll, 1978
Ulothrix fimbriata	150	Growth inhibition	Bharti et al., 1979
Cladophora glomerata	250	Growth inhibition	
Stigeoclonium tenure	250	Growth inhibition	
Scenedesmus sp.	1000	Growth inhibition	Filip et al., 1979
Chlorella sp.	1000	Growth inhibition	
Microsistis sp.	1000	Growth inhibition	
Lemna paucicostata	500	Inhibited frond multiplication	Nasu and Kugimoto, 1981
Thalassiosira pseudonana	<20	Growth inhibition with 0.03 ppt salinity	Frey et al., 1983
Skeletonema costatum	980	Growth inhibition at 32.5 ppt salinity	
Selenastrum capricornutum	600	Growth inhibition	Michnowicz and Weaks, 1984
Natural phytoplankton	98	Growth inhibition at 0.03 ppt salinity	Riedel, 1984
	>980	Growth inhibition at 32.5 ppt salinity	

μg/L to 10,000 μg/L. Several factors have contributed to this variability, many of which are related to experimental design. For example, a number of different algal species have been examined under different exposure conditions and a variety of parameters have been measured, for example, growth rate, photosynthesis, morphology, and so on. Very few, if any, studies have followed the experimental protocol of the standard algal toxicity test (APHA, 1980) with regard to recommended test organisms, culture techniques, temperature, illumination, and parameter to measure. Consequently, it is difficult to compare the results from different investigators.

Most of the studies on the toxicity of chromium in algae have concerned the effects on growth (Table 1). Algal growth appears to vary considerably in its sensitivity to chromium. Growth of *Ulothrix fimbriata* was inhibited at 150 μgCr(VI)/L, while *Cladophora glomerata* and *Stigeoclonium tenure* were at 250 μg/L (Bharti et al., 1979). Patrick (1978) studied the effect of Cr(VI) on the species composition of freshwater algal communities. Addition of 0.4 mg Cr(VI)/L caused a shift in community dominance from diatoms of high prey value to green and blue-green algae of low prey value. Filip et al. (1979) exposed the indigenous algae from wastewater lagoons to Cr(VI) (1 to 40 mg/L). Only one alga (*Oscillatoria* sp.) grew while other algal species (*Scenedesmus*, *Chlorella*, *Microsistis* sp.) were virtually absent in the chromium-amended media. Hollibaugh et al., (1980) examined the toxicity of 10 metal ions, including chromium, on species composition of natural marine assemblages of phytoplankton. Shifts in the species composition with *Skeletonema costatum* as the dominant diatom were observed when the natural assemblages were exposed to toxic levels of metals. In the estuarine environment, the addition of 98 μg Cr(VI)/L greatly reduced the growth of *Surirella ovata*, *Cyclotella* sp., and *Detonula confervacea* (Frey et al., 1983). The mechanism for the differences in algal sensitivity to chromium is not known. However, in other metals, the differences in sensitivity are apparently related to the size of the algal species (Munawar and Munawar, 1982), rate of uptake of the metal (Conway and Williams, 1979), sites of metal-binding (Hart, 1975), or genetic determinant (Singh et al., 1978).

Besides growth, other parameters such as photosynthesis, morphology, and enzyme activities have been used for chromium toxicity studies. Petria (1978) reported that chromium was more inhibitory to photosynthesis than growth in a *Chlorella* species. Addition of 50–100 μg chromium/L to the medium greatly reduced the photosynthesis while 10 mg/L was required to inhibit the growth. Concentrations of chromium less than 50 μg/L stimulated these processes. In contrast, Schroll (1978) observed that the growth of *Chlorella pyrenoidosa* was more sensitive to chromium than the photosynthesis. Thomas et al., (1980) studied the effects of 11 metal ions, including chromium, on the morphology of several marine phytoplankton. The main effects of metals on phytoplankton were granular, yellow cytoplasm; a disruption of chloroplast integrity and dispersion; and more delicate spines extruding from the marginal processes. This change in cell size or morphology has been observed in a variety of algae exposed to metals (Rosko and Rachlin, 1977; Silverberg et al., 1977) and has been suggested as the uncoupling of cell growth and cell division by the metals (Davies, 1976).

Very few studies on the effects of environmental factors on chromium toxicity in algae are published. The salinity, sulfur, pH, and organic ligands have been identified as affecting the chromium toxicity. The limited number of studies investigating the interaction of salinity and chromium toxicity have found that chromium toxicity in estuarine algae is inversely related to salinity.

Frey et al., (1983) found that the inhibitory concentration of chromium to natural phytoplankton populations varied from 98μg/L in water of 0.03 ppt salinity to greater than 980 μg/L in seawater of 32.5 ppt salinity. For pure culture of *Thalassiosira pseudonana*, addition of 98 μg/L of chromium completely inhibited the growth of the culture at 0.03 ppt salinity and slightly reduced the growth in water of 2.1 ppt salinity. Similarly, Riedel (1984) reported that the sensitivity of the growth rate of *T. pseudonana* to chromium was directly related to the salinity in the media. Organisms other than algae have also been shown to decrease their sensitivity to chromium in response to increased salinity (Fales, 1978).

The mechanism of the detoxification of chromium by increasing salinity was examined by Riedel (1984). The inhibition by chromium was found to be a function of the ratio of chromate to sulfate. Inhibition occurred when this ratio exceeded about 500:1. Since sulfate is a major constituent of seawater, the concentration of sulfate would be varied as a function of salinity. The mechanism for chromium toxicity would involve a competition with sulfate for the sulfate uptake transport system. Other studies have also shown that chromium inhibits sulfate uptake by algae and higher plants (Jeanjean and Broda, 1977; Smith, 1976).

Michnowicz and Weaks (1984) exposed *Selenastrum capricornutum* to sublethal concentrations of arsenic, copper, chromium, nickel, and zinc over a broad range of pH levels. At pH 4, growth was depressed for all metal ions tested. The adjustment of pH to higher levels (6, 8, 10, or 12) resulted in increased growth. Toxicity was least at the optimum pH range (8–10) for growth of the alga. In contrast, chromium was found to be more toxic at high pH than at low pH to the growth and multiplication of duckweed, *Lemna paucicostata* (Nasu and Kugimoto, 1981). More studies should be carried out to relate the pH values and chromium effects before generalization can be made.

While most ionic forms of metals studied to date are cationic (e.g., copper, cadmium, lead, mercury), Cr(VI) in water is anionic (e.g., CrO_4^{-2} and $HCrO_4^{-1}$). Thus, unlike the cationic metal ions, Cr(VI) will not be regulated and detoxified by organic ligands and standard culturing chelators such as ethylenediaminetetraacetic acid (Frey et al., 1983; Riedel, 1984).

Other factors, such as temperature, water hardness, and light, that may influence the toxicity of chromium on algae have not been reported in the literature since 1978.

3.   CHROMIUM TOXICITY TO BACTERIA

Chromium has been shown to exert a toxic effect on a wide variety of bacterial strains and bacterial processes (Table 2). Chromium toxicity has been investigated in soil by Drucker et al. (1979). Chromium(III) added to soil at concentrations of 10 and 100 mg/kg reduced the number of aerobic and anaerobic bacteria, whereas fungal numbers were decreased by only the higher 100 mg/L Cr(III) treatment.

James and Bartlett (1984) reported on the effect of chromium salts and tannery wastes (containing chromium) on nitrification in soil. Solutions of CrK $(SO_4)_2$ or $K_2Cr_2O_7$ were applied to soil at three levels (520, 5200, or 52,000 μg chromium/kg soil). Tannery effluents were also added at the same chromium

Table 2   Examples of the Toxic Effects of Chromium on Bacteria and Bacterial Processes

Effect	Reference
Plasmid-encoded chromium resistance	Bopp et al., 1983 Summers and Jacoby, 1978
Pseudomonas ambigua G-1 resistant to 4000 mg/L Cr(VI)	Horitsu et al., 1983
10 and 100 mg/L Cr(III) inhibits aerobic and anaerobic bacteria in soil	Drucker et al., 1979
Tannery effluents rich in chromium as well as 5200 μg/kg CrK(SO$_4$)$_2$ and K$_2$Cr$_2$O$_7$ do not inhibit soil nitrification	James and Bartlett, 1984
Chromium(VI) at concentrations between 10–12 mg/L inhibited bacterial growth	Ross et al., 1981
Gram-negative bacteria more sensitive to chromium than gram-positive bacteria	Ross et al., 1981
Chromium at 556 μg/g inhibited bacterial growth, whereas 1 μg/g chromium caused initial stimulation	Zibilske and Wagner, 1982
Electroplating waste rich in chromium was toxic to saprophytic and nitrifying bacteria	Ajmal et al., 1984
Aerobic heterotrophic bacteria from chromium-polluted marine sediment more capable of tolerating Cr(III) than bacteria from previously nonpolluted sediment	Aislabie and Loutit, 1984
Chromium(VI) (as CrO$_3$) but not as CrCl$_2$ inhibited rumen microbiota	Forsberg, 1978
K$_2$CrO$_4$ mutagenic to *E. coli*	Venitt and Levy, 1974
K$_2$CrO$_4$ and K$_2$Cr$_2$O$_7$ mutagenic to *B. subtilis*	Nishioka, 1975
Na$_2$Cr$_2$O$_7$, CrO$_3$ and K$_2$CrO$_4$ mutagenic to *Salmonella typhimurium*	Petrilli and DeFlora, 1977

concentrations. Nitrite (NO$_2^-$) and nitrate (NO$_3^-$) were determined using colorimetric assays (James and Bartlett, 1984). Nitrate formation during the initial 23 days of incubation was reduced (compared with control soils) in soils challenged with 52,000 μg chromium/kg. After 42 days, NO$_3^-$ levels were about equivalent in all treatments. The 520 and 5200 μg chromium/kg and all the tannery waste effluents tested did not significantly reduce nitrification.

Chromium toxicity to soil bacterial isolates has been evaluated by measuring cell turbidity in broth cultures containing Cr(VI) or Cr(III) (Ross et al., 1981). Chromium(VI) at concentrations from 10 to 12 mg/L was observed to inhibit most bacterial isolates growing in a soil-extract medium or a semisynthetic medium. In addition, gram-negative bacteria were more sensitive to chromium than gram-positives (Ross et al., 1981). The authors suggested that because of the toxicity to such low levels of Cr(VI), soil microbial processes such as nitrification may be temporarily or permanently affected. The high degree of toxicity exerted by such relatively low levels of this metal also illustrates the need for caution in handling and disposal of chromium-containing wastes.

In another study by Zibilske and Wagner (1982), a 1 $\mu g/g$ chromium treatment caused initial stimulation in soil bacterial numbers. However, at a high chromium concentration (556 $\mu g/g$), an inhibition of bacterial growth was observed. Measurement of microbial biomass using the adenosine triphosphate (ATP) method produced trends similar to those observed in the enumeration studies; an initial stimulation of ATP levels, followed by lower ATP levels (when compared with a nonpolluted control soil) after two weeks.

A mixed continuous culture (chemostat) grown at a low dilution rate was used by Lester et al. (1979) to assess the toxic effect of cadmium, chromium, copper, and lead at concentrations of 50 mg/L. The most toxic metal ion was copper, followed by cadmium, lead, and chromium. Other studies on chromium toxicity in soil have focused on investigating the adsorption of chromate and Cr(VI) on a clay-sand mixture, and the acute toxicity of electroplating wastes on microorganisms (Ajmal et al., 1984). The researchers found the electroplating wastes were rich in sulfate, chloride, silica, nitrogen, phosphate, chromate, and Cr(VI). The studies indicated that the electroplating waste was toxic to both saprophytic and nitrifying bacteria. Clay proved to be a suitable absorbent for chromate and Cr(VI) because of its high cation exchange capacity (CEC) and strong binding capability (Ajmal et al., 1984).

Only a few reports exist on chromium toxicity to heterotrophic marine sediment bacteria (Aislabie and Loutit, 1984) and organisms responsible for anaerobic digestion processes to treat chromium-polluted tannery waste (Maeda et al., 1984). A survey of aerobic heterotrophic bacteria found in marine sediments at Sawyers Bay (New Zealand) was conducted to assess the impact of effluent high in chromium. The bacterial population present in the polluted sediment was more capable of tolerating Cr(III) at levels less than 10 mg/L. In addition, the types and frequency of aerobic heterotrophs differed between the control and polluted sediment. For example, gram-positive bacteria comprised from 32 to 92% of the population in the polluted sediment. In addition, if 5 mg/L Cr(III) was added to the agar medium, a larger number of bacteria was isolated from the sediment than if no chromium was present in the agar (Aislabie and Loutit, 1984). This finding did not occur with heterotrophic bacteria isolated from the nonpolluted control sediment.

Chromium(VI) (as CrO_3) but not Cr(II) (as $CrCl_2$) is known to inhibit the rate of fermentation of rumen microbiota (Forsberg, 1978). Chromium in specific forms has also been reported to exert a mutagenic effect on *Escherichia coli, Bacillus subtilis,* and *Salmonella typhimurium.* Chromium(VI) as K_2CrO_4, but not Cr(III) [(as $CrK(SO_4)_2$] was found to be mutagenic to *E. coli* (Venitt and Levy, 1974). Nishioka (1975) reported that K_2CrO_4 and $K_2Cr_2O_7$, but not $CrCl_3$, was mutagenic to *B. subtilus.* Mutagenicity bioassays with *S. typhimurium* revealed that Cr(VI) (as $Na_2Cr_2O_7$, CrO_3, and K_2CrO_4) but not Cr(III) [(as $CrK(SO_4)_2$] or Cr(II) (as $CrCl_2$) was mutagenic to this organism (Petrilli and DeFlora, 1977).

Very few investigations have been conducted on chromium resistance in bacteria and yeasts (Trevors et al., 1985; Luli et al., 1983; Chapter 14). The paucity of information clearly shows that additional research is required. To date, two reports have been published that prove that chromium resistance can be plasmid-encoded in pseudomonads. Summers and Jacoby (1978) studied gram-negative bacteria (with known resistances to ions of arsenic, mercury, silver, and tellurium) to determine their tolerances to previously untested metals. No new resistances were observed in enterobacterial plasmids, but some plasmids in *Pseudomonas aeruginosa* belonging to the Inc P-2 group did specify resistance to both chromium and boron. In addition, these plasmids also conferred resistance to metaborate, but not perborate. The plasmid-encoded resistance mechanisms allowed the organisms to tolerate a two- to four-fold increase in borate concentrations, and a two- to five-fold increase in metaborate levels.

Another pseudomonad, identified as *Pseudomonas fluorescens* LB300, has also been found to be chromate-resistant. This organism was isolated from chromium-polluted sediment in the upper Hudson River, United States (Bopp et al., 1983). The organism harbored a plasmid (designated pLHB1) that conferred chromate resistance, as evidenced by the simultaneous loss of resistance when the plasmid was cured using mitomycin C, or deleted by spontaneous segregation. The pLHB1 plasmid was transferrable to *E. coli* by conjugation, where it still conferred chromate resistance (resistance being defined as the capability to tolerate 40- to 200-fold higher levels of chromate than pseudomonads isolated from other environments, or other bacterial strains taken from the same chromium contaminated sediment).

Other research on chromium resistance and sensitivity in bacteria has been conducted by Horitsu et al., (1983). These researchers isolated a strain of *Pseudomonas ambigua* (G-1), resistant to 4000 mg/L Cr(VI), 250 mg/L Cd(II), and 200 mg/L Cu(II). Mutagenesis with N-methyl-N-nitro-N-nitroso-quanidine (MNNG) at 37°C for two hours was used to obtain Cr(VI)-sensitive mutants from the *P. ambigua* parent strain. Most of the sensitive mutants were unable to grow on nutrient agar amended with any of the three previously used metal ions [Cr(VI), Cd(II), Cu(II)]. Two sensitive mutants, S-1 and S-3, could only grow on nutrient agar amended with 200 mg/L and 50 mg/L Cu(II), respectively. Since S-1 was a tolerant to Cu(II) as the parental G-1 isolate, it may be highly probable that the mechanism for Cr(VI) resistance is completely different from that for Cu(II) (Horitsu et al., 1983).

Cr(VI) uptake experiments using both a sensitive and resistant strain showed that the Cr(VI) sensitive mutant took up more Cr(VI) into the cells than the parent G-1 strain (Horitsu et al., 1983). The authors suggested that the membrane barrier for Cr(VI) permeation may be nonfunctioning or damaged in the sensitive mutant. In addition, there was a significant difference in the fatty acid composition between the parent G-1 and the sensitive S-1 strain. In the parent G-1 strain, the major lipid component was a saturated C16 acid. However, in the sensitive mutant, a monounsaturated C18 acid was predominant. Cell envelope differences were observed using transmission electron microscopy. The parent G-1 strain was surrounded by a thick envelope that was absent in the sensitive S-1 mutant.

4. SUMMARY

Compared with some of the other metals (such as lead, mercury, tin), relatively little information is available on chromium toxicity to algae and bacteria. The observations summarized in Tables 1 and 2 clearly show that additional research is needed on the chromium toxicity in the physiology and genetics of these organisms. In addition, it is not known how different forms of chromium influence the microbial ecology of sediments and soils over extended periods of time, and in the chromium-polluted environments whether the microorganisms have the capability to recover from such pollution. Other areas requiring further research include the effects of environmental factors such as pH, temperature, salinity, and hardness on chromium toxicity.

REFERENCES

Aislabie, J., and Loutit, M.W. (1984). "The effect of effluent high in chromium on marine sediment aerobic heterotrophic bacteria." *Marine Environ. Res.* 13, 69–79.

Ajmal, M., Nomani, A.A., and Ahmad, A. (1984). "Acute toxicity of chrome electroplating wastes to microorganisms: Adsorption of chromate and chromium(VI) on a mixture of clay and sand." *Water Air Soil Pollution* 23, 119–127.

American Petroleum Institute (API). (1982). *The Sources, Chemistry, Fate and Effects of Chromium in Aquatic Environment.* American Petroleum Institute, Washington, DC, 179 pp.

American Public Health Association (APHA), American Water Works Association and Water Pollution Control Federation. (1980). "Bioassay methods for aquatic organisms." In *Standard Method for the Examination of Water and Wastewater*, 15th Ed., APHA, Washington, DC.

Bharti, A., Saxena, R.P., and Pandey, G.N. (1979). "Physiological imbalances due to hexavalent chromium in fresh water algae." *Indian J. Environ. Health* 21, 234–243.

Bopp, L.H., Chakrabarty, A.M., and Ehrlich, H.L. (1983). "Chromate resistance plasmid in *Pseudomonas fluorescens*." *J. Bacteriol.* 155, 1105–1109.

Conway, H.L., and Williams, S.C. (1979). "Sorption of cadmium and its effect on growth and the utilization of inorganic carbon and phosphorus of two freshwater diatoms." *J. Fish. Res. Board Can,* 36, 579–586.

Davies, A.G. (1976). "An assessment of the basis of mercury tolerance in *Dunaliedla tertiolecta*." *J. Mar. Biol. Ass. UK* 56, 39–57.

Drucker, H., Garland, T.R., and Wildung, R.E. (1979). "Metabolic response of microbiota to chromium and other metals." In N. Kharasch, Ed., *Trace Metals in Health and Disease*. Raven Press, New York.

Fales, R.R. (1978). "The influence of temperature and salinity on the toxicity of hexavalent chromium to the grass shrimp, *Palaemonetes pugis* (Holthuis)." *Bull. Environ. Contam. Toxicol.* 20, 447–450.

Filip, D.S., Peters, T., Adams, V.D., and Middlebrooks, E.J. (1979). "Residual heavy metal removal by an algae—Intermittent and filtration system." *Water Res.* 13, 305–313.

Forsberg, C.W. (1978). "Effects of heavy metals and other trace elements on the fermentative activity of the rumen microflora and growth of functionally important rumen bacteria." *Can. J. Microbiol.* 24, 298–306.

Frey, B.E., Riedel, G.F., Bass, A.E., and Small, L.F. (1983). "Sensitivity of estuarine phytoplankton to hexavalent chromium." *Estuarine Costal Shelf Sci.* 17, 181–187.

Hart, B.A. (1975). "Bioconcentration and toxicity of cadmium in *Chlorella pyrenoidosa*." In *The Effect of Cadmium on Freshwater Phytoplankton*. Office of Water Research and Technology, Washington, DC, PB 257-547, pp. 1–31.

Hollibaugh, J.T., Seibert, D.L.R., and Thomas, W.H. (1980). "A comparison of the acute toxicities of ten heavy metals to phytoplankton from Saanich Inlet, B.C., Canada." *Estuarine Costal Marine Sci.* **10**, 93–105.

Horitsu, H., Futo, S., Ozawa, K., and Kawai, K. (1983). "Comparison of characteristics of hexavalent chromium-tolerant bacterium, *Pseudomonas ambigua* G-1, and its hexavalent chromium-sensitive mutant." *Agric. Biol. Chem.* **47**, 2907–2908.

James, B.R., and Bartlett, R.J. (1984). "Nitrification in soil suspensions treated with chromium(III, VI) salts or tannery wastes." *Soil Biol. Biochem.* **16**, 293–295.

Jeanjean, R., and Broda, E. (1977). "Dependence of sulfate uptake by *Anacystis nidulans* on energy, on osmotic shock and on sulfate starvation." *Arch. Microbiol.* **114**, 19–24.

Lester, J.N., Perry, R., and Dadd, A.H. (1979). "The influence of heavy metals on a mixed bacterial population of sewage origin in the chemostat." *Water Res.* **13**, 1055–1063.

Luli, G.W., Talnagi, J.W., Strohl, W.R., and Pfister, R.M. (1983). "Hexavalent chromium-resistant bacteria isolated from river sediments." *Appl. Environ. Microbiol.* **46**, 846–854.

Maeda, Y., Shoji, Y., Yoneda, A., and Azumi, T. (1984). "Preliminary studies on treatment of chromium tannery waste sludge by anaerobic digestion." *J. Ferment. Technol.* **62**, 421–427.

Merck Index. (1976). Merck and Co., Inc., Rahway, NJ.

Michnowicz, C.J., and Weaks, T.E. (1984). "Effects of pH on toxicity of As, Cr, Cu, Ni and Zn to *Selenastrum capricornutum* Printz." *Hydrobiologia* **118**, 299–305.

Munawar, M., and Munawar, I.F. (1982). "Phycological studies in Lakes Ontario, Erie, Huron and Superior." *Can. J. Botany* **60**, 1837–1858.

Nasu, Y., and Kugimoto, M. (1981). "*Lemna* (duckweed) as an indicator of water pollution. I. The sensitivity of *Lemna paucicostata* to heavy metals." *Arch. Environ. Contam. Toxicol.* **10**, 159–169.

National Academy of Sciences (NAS). (1980). *Drinking Water and Health*. Safe Drinking Water Committee, National Academy Press, Washington, DC, Vol. 3, pp. 364–369.

National Research Council (NRC). (1976). *Effects of Chromium in the Canadian Environment*. Associate Committee on Scientific Criteria for Environmental Quality. NRCC No. 15017, Ottawa, 168 pp.

Nishioka, H. (1975). "Mutagenic activities of metal compounds in bacteria." *Mutation Res.* **31**, 185–189.

Nriagu, J.O. (1980). *Cadmium in the Environment*. John Wiley & Sons, New York, 682 pp.

Patrick, R. (1978). "Effects of trace metals in the aquatic ecosystem." *Am. Sci.* **66**, 185–191.

Petria, V. (1978). "Effect of chromium salts from water sediments on physiological processes in the alga *Chlorella vulgaris*." *Rev. Roum. Biol. Ser. Biol. Veg.* **23**, 55–57.

Petrilli, F.L., and DeFlora, S. (1977). "Toxicity and mutagenicity of hexavalent chromium on *Salmonella typhimurium*." *Appl. Environ. Microbiol.* **33**, 805–809.

Riedel, G.F. (1984). "Influence of salinity and sulfate on the toxicity of chromium(VI) to the estuarine diatom *Thalassiosira pseudonana*." *J. Phycol.* **20**, 496–500.

Rosko, J.J., and Rachlin, J.W. (1977). "The effect of cadmium, copper, mercury, zinc, and lead on cell division, growth, and chlorophyll—A content of the chlorophyte *Chlorella vulgaris*." *Bull. Torrey Bot. Club* **104**, 226–233.

Ross, D.S., Sjögren, R.E., and Bartlett, R.J. (1981). "Behavior of chromium in soils. IV. Toxicity to microorganisms." *J. Environ. Qual.* **10**, 145–148.

Schroll, H. (1978). "Determination of the absorption of Cr^{+6} and Cr^{+3} in an algal culture of *Chlorella pyrenoidosa* using ^{51}Cr." *Bull. Environ. Contam. Toxicol.* **20**, 721–724.

Shepherd, C.M., and Jones, R.L. (1971). *Hexavalent Chromium—Toxicological Effects and Means for Removal from Aqueous Solution*. U.S. Clearing house Fed. Sci. Tech. Info. AD71-7348.

Silverberg, B.A., Wong, P.T.S., and Chau, Y.K. (1977). "Effect of tetramethyllead on freshwater green algae." *Arch. Environ. Toxicol. Contam.* 5, 305–313.

Singh, H.N., Vaishampayan, A., and Singh, R.K. (1978) "Evidence for the involvement of a genetic determinant controlling functional specificity of Group VI B elements in the metabolism of nitrogen and nitrate in the blue-green alga *Nostoc muscorum*." *Biochem. Biophys. Res. Commun.* 81, 67–74.

Smith, I.K. (1976). "Characterization of sulfate transport in cultured tobacco cells." *Plant Physiol.* 58, 358–362.

Summers, A.O., and Jacoby, G.A. (1978). "Plasmid-determined resistance to boron and chromium compounds in *Pseudomonas aeruginosa*." *J. Bacteriol.* 13, 637–640.

Thomas, W.H., Hollibaugh, J.T., and Seibert, D.L.R. (1980). "Effects of heavy metals on the morphology of some marine phytoplankton." *Phycologia* 19, 202–209.

Towill, L.E., Shriner, C.R., Drury, J.S., et al. (1978). *Reviews of the Environmental Effects of Pollutants. III Chromium*. EPA-600/1-78-023. United States Environmental Protection Agency, Cincinnati, OH.

Trevors, J.T., Oddie, K.M., and Belliveau, B.H. (1985). "Metal resistance in bacteria." *FEMS Microbiol. Rev.* 32, 39–54.

Venitt, S., and Levy, L.S. (1974). "Mutagenicity of chromates in bacteria and its relevance to chromate carcinogenesis." *Nature (Lond)* 250, 493–495.

Weiler, R.R., and Chawla, V.K. (1969). "Dissolved mineral quality of Great Lakes waters." In *Proc. 12th Conference Great Lakes Research*, pp. 801–808.

Zibilske, L.M., and Wagner, G.H. (1982). "Bacterial growth and fungal genera distribution in soil amended with sewage sludge containing cadmium, chromium and copper." *Soil Sci.* 134, 364–370.

12

THE SIGNIFICANCE OF THE INTERACTIONS OF CHROMIUM AND BACTERIA IN AQUATIC HABITATS

M. Loutit, P. Bremer, and J. Aislabie

Microbiology Department
University of Otago
Dunedin, New Zealand

1. INTRODUCTION

Reports that Cr(III) may affect the metabolism and the young of certain aquatic organisms (Calabrese et al., 1973; Muramoto, 1981; Tacon and Beveridge, 1982) emphasize the need to determine the effects of discharging chromium into the environment. Even when low concentrations of Cr(III) are present in effluents (Muir, 1983; Rossin et al., 1983), the effects on fish and their food may have economic repercussions for man. It seems desirable therefore to learn more of the fate, effects, and partitioning of Cr(III) and Cr(VI) in the aquatic environment. In particular, the interactions of bacteria and chromium and their significance are discussed.

2. SOURCES OF CHROMIUM IN AQUATIC HABITATS

2.1 Natural Sources

Chromium is found in the earth's surface and may enter streams, rivers, lakes, and the sea through run-off after rain, land subsidence, wind-blown dust, and volcanic debris following eruptions. Natural sources once accounted for all of the chromium entering the aquatic environment, but the rapidly increasing world population with its associated urbanization and industrialization means that in some areas anthropogenic sources now contribute more chromium to aquatic habitats than natural sources (Leland et al., 1978; Cary, 1982; Duedall et al., 1983; Buat-Menard, 1984; Forstner, 1984; Moore and Ramamoorthy, 1984) (see Chapter 5).

2.2 Anthropogenic Sources

Chromium is one of the most widely used metals in industry (NAS, 1974; Stern, 1982; Kimbell and Panulas, 1984; Moore and Ramamoorthy, 1984) (see Chapters 3 and 4) and is discharged in liquid, solid, and gaseous wastes from industries, either directly or indirectly, into fresh water and marine habitats (Eggett and Thorpe, 1978; Leland et al., 1978; Cary, 1982; Duedall et al., 1983; Buat-Menard, 1984; Forstner, 1984; Moore and Ramamoorthy, 1984). Although input of Cr seems to be declining in some areas (Forstner, 1984), the estimated input from sewage, tanneries, and chrome and steel works (Duedall et al., 1983; Moore and Ramamoorthy, 1984) is still vast.

3. FATE OF CHROMIUM

The chromium in most effluents is not diluted and dispersed upon entering the water, but rather a proportion is deposited into the sediments, the amount varying with the form of the chromium, the presence of other substances in the effluents, and the conditions of the receiving waters (Curl et al., 1965; Elderfield, 1970; Capuzzo and Anderson, 1973; Schroeder and Lee, 1975; Cranston and Murray, 1978; Pankow et al., 1977; Jan and Young, 1978; Leland et al., 1978; Zhou et al., 1979; Pfeiffer et al., 1980; Forstner and Wittman, 1981; Katz and Kaplan, 1981; Smillie et al., 1981; Duedall et al., 1983; Buat-Menard, 1984; Moore and Ramamoorthy, 1984). In some instances little chromium is deposited, particularly if no organic or particulate matter is present. For example, when soluble Cr^{51} in effluent from nuclear reactors was discharged into the Columbia River, and eventually into the sea, the soluble Cr^{51} remained mostly as Cr(VI) and was carried many kilometers from the discharge point (Osterberg et al., 1965; Cutshall et al., 1966). In the same studies, the only indication that chromium might have been deposited in sediment was given by Curl et al. (1965) who stated that any Cr(VI) reduced to Cr(III) would have been adsorbed rapidly to particulate matter and as such would have been subject to deposition.

At the other extreme is the fate of chromium in sewage discharged off the Southern Californian coast. The chromium was found to be in the Cr(III) form and a considerable proportion was deposited in the sediment within a relatively short distance from the outlet (Mearns and Young, 1977; Jan and Young, 1978; Katz and Kaplan, 1981; Duedall et al., 1983). Chromium in the water column away from the discharge point was mainly associated with particulate matter being carried down-current in the effluent plume (Jan and Young, 1978). Rapid deposition of chromium into marine sediments was also noted in a New Zealand study by Smillie et al. (1981), who found that Cr(III), the dominant form in tannery effluent, was largely deposited within 100 m of the discharge point (3000 μg chromium g^{-1} dry weight sediment at the outfall decreasing to 100 μg chromium g^{-1} dry weight, 100 m out). Similar results have been reported for chromium in tannery effluent discharged into freshwaters and estuaries (Capuzzo and Anderson, 1973; Mayer and Fink, 1980).

Gradations between these extremes of partitioning of chromium in the overlying water and the sediments have been demonstrated by Mayer and Schick (1981) in laboratory experiments and by Pfeiffer et al. (1980) in an extended study of the Iraja River.

No complete balance sheets for chromium in effluents entering aquatic habitats have been published. Of the studies on the fate of chromium entering such habitats, those on sewage effluent discharged off the Southern Californian coast are among the most detailed so far (Mearns and Young, 1977; Jan and Young, 1978; Hershelman et al., 1981). An interesting point to emerge from this work was that even an effluent such as sewage having a low chromium concentration (average 0.03 μg chromium ml^{-1}) can, because of the large volume discharged, contribute 600 to 700 metric tons of chromium per year to a relatively localized area (Mearns and Young, 1977; Jan and Young, 1978; Duedall et al., 1983). A considerable proportion of the chromium ends up in the sediment. The result is sediments in an area of several square kilometers with a chromium concentration of up to 1000 μg chromium g^{-1} dry weight.

Where various industries such as tanneries contribute high concentrations of chromium but smaller volumes of effluent, the result is also a sediment with a high chromium concentration which is distributed over a small area (Johnson et al., 1981; Smillie et al., 1981). It is also of interest that metals can be added to aquatic sediments through particulate matter in gaseous emissions from smelters (Neverauskas, 1983) and the burning of fossil fuels (Buat-Menard, 1984).

Clearly, a consequence of urbanization and industrialization is that a number of point sources of effluents contribute metals including chromium to rivers, estuaries, and off-shore areas (Duedall et al., 1983).

4. CHROMIUM AND BIOTA

4.1 Biotic Accumulation in Water Column

The idea that chromium is ingested with food was proposed by Elwood et al. (1980) and is supported by Moore and Ramamoorthy (1984) who state "it is possible that food is probably a more significant source of chromium than water." For filter feeders, chromium could be taken up with either nonbiological or biological particulates during feeding (Karbe et al., 1977). That chromium binds to particulates was indicated as far back as 1965 by Curl et al. Some of these particulates, however, could be small biota. Any soluble chromium either in or adsorbed to the surface of these small organisms could then be passed to the higher animals (Kostic and Draskovic, 1982). There is, however, surprisingly little information on chromium concentration in microbiota (Fukai and Broquet, 1965; Mearns and Young, 1977; Schroll, 1978). Most studies have concentrated on the toxicity of chromium to such organisms (Hervey, 1949; Wium-Andersen, 1974; Subba Rao, 1981; Fasulo et al., 1982; 1983; Kostic and Draskovic, 1982; Frey et al., 1983). Further, these investigations have not considered the possibility that bacteria on particles could concentrate soluble chromium from the water (Paerl, 1977; Leland et al., 1978).

While this suggestion has not been addressed directly, there is evidence that bacteria on surfaces in aquatic habitats do concentrate soluble chromium. Epiphytic bacteria on algae (Lee et al., 1975) and plants (Patrick and Loutit, 1977) have been shown to concentrate a variety of soluble heavy metals, including chromium. Grazing of these epiphytic bacteria (Lee et al., 1975) provides a mechanism by which soluble chromium enters a foodchain.

Periphytic bacteria growing on the gills and carapaces of crabs (*Helice crassa*) have been shown to concentrate chromium (Johnson et al., 1981). This could result in entry of chromium to the crab and to consumers of the crabs. It is of interest that Riva et al. (1981) reported that soluble chromium was adsorbed to the gills of crabs and had toxic effects on the animals, but they did not suggest involvement of periphytic bacteria in the process. Further, animals that feed directly on algae and plants may take in varying amounts of chromium. The chromium content of macrophytes has been shown to vary with species and location (Fukai and Broquet, 1965; Saenko et al., 1976). Animals feeding on such macrophytes would therefore ingest different amounts of chromium in different locations. The work of Capone et al., (1983) supports the contention that animals take in chromium with food; they showed that animals that consumed primary producers in a polluted lagoon had elevated chromium concentrations.

While soluble and particulate chromium in the water column are clearly of interest and have stimulated considerable research, certain aspects of the topic have been neglected. For example, few studies have attempted to assess the importance to the biota of chromium bound to dissolved organic molecules. Nriagu and Coker (1980) demonstrated that chromium attaches to fulvic and humic acids and Nakayama et al. (1981, 1981a,b,c) also reported binding of chromium to organic molecules. Whether these compounds affect the biota has not been established. Most attention has focused on the effects of inorganic substances on plant and animals, but in certain habitats the chromium bound to organic molecules may be of considerable importance (Nakayama et al., 1981; 1981a,b,c).

4.2 Sediment Chromium

The effects on the biota of soluble and suspended particulate chromium have been extensively studied. However, the higher concentrations of chromium in the sediments compared with the overlying waters, both in fresh and marine habitats (Capuzzo and Anderson, 1973; Phillips, 1977; Forstner, 1977; Mearns and Young, 1977; Leland et al., 1978; Bryan, 1979; Mayer and Fink, 1980; Hershelman et al., 1981; Smillie et al., 1981; Young et al., 1981; Duedall et al., 1983; Buat-Menard, 1984; Forstner, 1984; Moore and Ramamoorthy, 1984), and their effect on the biota have been largely ignored (Bryan, 1979). Perhaps the lack of interest in sediment chromium and its interactions with the biota has been due to evidence that suggests that such chromium is biologically inactive. For example, Loring (1979), in a study of the estuary and the open Gulf of St. Lawrence, found that at least 76 to 98% (depending on the place from which the samples were taken) of the chromium in the sediments was not directly available to the biota.

Even in highly polluted sediments such as those receiving tannery discharges, the chromium concentration in the overlying water (OLW) (Smillie et al., 1981) and the interstitial waters (ISW) is low (<20 ng ml^{-1}) (Pillidge, 1985), suggesting that little of the chromium is soluble and hence bioavailable. Yet animals in and around polluted sediments often have elevated chromium concentrations, the amount concentrated depending on the species of animal, its size, age, and the conditions prevailing in the sediment (Patrick and Loutit, 1976, Mearns and Young, 1977; Phillips, 1977; Leland et al., 1978; Hershelman et al., 1981, Johnson et al., 1981; Young et al., 1981; Kostic and Draskovic, 1982; Moore and Ramamoorthy, 1984).

Our interest in sediments and metals began with freshwater studies and led us to postulate that bacteria in sediments are a means by which chromium and other metals enter foodchains (Loutit et al., 1973; Patrick and Loutit, 1976).

5.   SEDIMENT BACTERIA AND THE PASSAGE OF METALS THROUGH FOODCHAINS

5.1   Sediment Bacteria as Food for Higher Trophic Levels

A variety of aquatic animals consume bacteria to wholly or partly satisfy their nutritional needs (Zobell and Feltham, 1938; Fenchel and Jorgensen, 1977; Rieper, 1978; Hobbie and Lee, 1980; Phillips, 1984), but sediment bacteria are mentioned in only a few studies (Ustach, 1982). Whether bacteria are the sole source of nutrients for the organisms that consume them is not the issue here. The question is can bacteria be the vehicle by which chromium and perhaps other metals are transferred to higher trophic levels? Some evidence suggests bacteria do mediate this transfer (Loutit et al., 1973; Lee et al., 1975, Patrick and Loutit, 1976, 1977; Johnson et al., 1981; Bremer, 1984; Bremer and Loutit, 1986a).

5.2   Sediment Bacteria and Metals

If bacteria are involved in the transfer of metals to foodchains, their abundance in sediments compared with the water column may be significant. While it is recognized that the higher organic matter content of sediments supports this abundant bacterial population (Zobell, 1946; Walker and Colwell, 1975), there is a lack of information about many of the sediment bacteria and their functions. To some extent this deficiency is due to their pleomorphism and the absence of suitable tests for confirming their specific activities (Austin et al., 1977; Aislabie and Loutit, 1984). In particular, there is a paucity of information on the interaction of sediment bacteria. Most studies have involved use of pure cultures and have focused on requirements for metals (Diekart et al., 1981) and their toxic effects (Babich and Stotzky, 1977, 1983; Petrilli and DeFlora, 1977; Gadd and Griffiths, 1978; Seyfried, 1980; Wong et al., 1982), although some attention has been paid to the ability of bacteria to transform metals (Jernelov and Martin, 1975; Austin et al., 1977; Summers and Silver, 1978; Timoney et al., 1978). The interactions with bacteria of the metals mercury, lead, arsenic, zinc, copper, and cadmium have received the most attention. However, in spite of its wide use in industry and the elevated concentrations reported in sediments, chromium has been largely neglected.

5.3   Bacteria in Sediments High in Chromium

The observation that sediment bacteria could act as a vehicle for the entry of chromium to a freshwater foodchain (Loutit et al., 1973; Patrick and Loutit, 1976) prompted us to investigate this interaction in an area where tannery effluent had been discharged for over 100 years. Studies so far have concentrated on aerobic heterotrophic bacteria (Aislabie, 1984; Aislabie and Loutit, 1984) and facultative anaerobes and sulphate- reducing bacteria (Pillidge, 1985). The numbers and distribution of the bacteria in the sediment have been found to reflect oxygen tension, organic matter content, Eh, and pH at different depths in the sediment (Pillidge, 1985) and the aerobic heterotrophic population has been shown to exhibit a seasonal fluctuation (Aislabie, 1984; Aislabie and Loutit, 1984). From these studies several points have emerged. Although the tannery ceased to discharge chromium in effluent in 1982, the concentration in the sediment in 1985 was still about 2000 μg chromium g^{-1} dry weight compared with up to 3000 μg chromium g^{-1} in 1981 (Johnson et al., 1981; Smillie et al., 1981). Analysis of the sediment indicates that bacterial numbers are highest in the organic fraction (Pillidge, 1985) and most of the chromium is associated with organic matter (Pillidge, 1985). Capuzzo and Anderson (1973), Monaco (1977), and Mayer and Fink (1980) also found that chromium in sediment was associated with the organic fraction and iron oxide. Mogensen and Jorgensen (1978) suggested that chromium is deposited as Cr(OH)$_3$ into sediments and this also appears to apply to our study area, a suggestion previously made by Smillie et al. (1981). It is possible that because chromium is now removed from the tannery effluent before discharge, although discharge of organic matter has continued, there has been a change in the partitioning of chromium in the sediment in the intervening period. Whether such a change does take place requires experimental confirmation.

Laboratory experiments have confirmed an in vivo observation of an association between facultative anaerobic sediment bacteria and the release of soluble chromium (Pillidge, 1985). In spite of the involvement of sediment bacteria in chromium solubilization, the observed chromium concentration in the ISW and the OLW in vivo was never greater than 20 ng ml^{-1}. The low chromium concentration in the ISW and the OLW, together with the fact that isolates from the polluted sediment were more mucoid than those from a control site (Aislabie, 1984; Aislabie and Loutit, 1986), led us to investigate whether extracellular bacterial polymer could concentrate chromium in a manner similar to that reported for molybdenum (Tan and Loutit, 1976; 1977). We postulated that the polymers around the bacteria in sediment could bind the chromium and prevent it appearing in the ISW. Binding of chromium in this way could offer an explanation for the tolerance of bacteria to chromium in polluted sediments and might facilitate the entry of chromium into foodchains.

5.4   Concentration of Chromium by Sediment Bacteria

In laboratory experiments, sediment bacteria have been shown to concentrate 80% of their total Cr(III) outside the cell. Much of it is located in the polysaccharide fraction of the extracellular polymer material (Aislabie, 1984; Aislabie and Loutit, 1986). Binding of metals to extracellular bacterial polysaccharides has been reported previously by Corpe (1975) and Tan and Loutit (1976). Further, Tan and Loutit (1977) reported that molybdenum was bound to the uronic acids in the polysaccharide. Chromium may also bind to these uronic acids, because an analysis of the extracellular polysaccharides of certain sediment isolates have been shown to have an uronic acid content of 24 to 30% (Bremer, 1984; Bremer and Loutit, 1986a).

The accumulation of chromium by bacteria is known to be affected by the strain of the organism, chromium speciation, chromium concentration in the medium, and the incubation temperature (Aislabie, 1984; Aislabie and Loutit, 1985). The observation that chromium stimulates polysaccharide production is of particular interest (Aislabie, 1984; Aislabie and Loutit, 1984) because of the contention of Costerton et al. (1981) that polysaccharide production by bacteria is necessary for survival in nature. Further, Geesey (1982) has stated that polysaccharide conditions the cell environment, a view supported by Bitton and Freihofer (1978), who showed that polysaccharide reduced the toxicity of metals to bacteria. Aislabie (1984) and Aislabie and Loutit (1986) also proposed that production of extracellular polysaccharide by sediment bacteria was a mechanism of tolerance to metals.

5.5   Tolerance of Sediment Bacteria to Chromium

Some authors use the term *resistance* when discussing whether microorganisms are capable of growing in the presence of heavy metals, but we prefer the term *tolerance* unless the mechanism for tolerance has been proven to be due to the presence of a plasmid or to be chromosomally mediated. Resistance to Cr(VI) has been reported in *Streptococcus lactis* (Efstathiou and McKay, 1977) and *Pseudomonas* strains (Summers and Jacoby, 1978; Bopp et al., 1983). Resistance to Cr(III) is discussed in Chapters 11 and 14.

Bacteria are known to carry out various chemical transformations of heavy metals which, while they may be incidental to metabolism, may offer the organisms a mechanism for tolerating metals (Silver, 1984). Little work, however, has appeared on chromium. Smillie et al. (1981) reported that bacterially produced H_2S in a marine sediment was capable of reducing Cr(VI) to Cr(III), thus rendering it less soluble. Such a transformation of chromium could be interpreted as a mechanism by which sulphate-reducing bacteria tolerate Cr(VI) which is reported to be toxic to their enzyme systems (Postgate, 1979).

In attempting to study tolerance of sediment bacteria to Cr(VI) and Cr(III), we have encountered a number of difficulties. Ramamoorthy and Kushner (1975) and Tilton and Rosenberg (1978) have reported that metals bind to organic components in media, thus making it difficult to assess the concentration of added metal available to bacterial cells. We have attempted to find a synthetic liquid medium in which to grow sediment bacteria. However, many isolates would not grow in liquid media and when the Pope and Skermans Mineral Salts Medium (Skerman, 1967) was found to be appropriate, addition of chromium caused precipitation of various components (Aislabie, 1984). Solid media therefore had to be used to study tolerance. Adjustment of the pH to correct for acidity created by the addition of Cr(III) and Cr(VI) salts again put in doubt the amount of chromium that was available. Nevertheless, some information has been obtained on the degree of tolerance of sediment bacteria to chromium (Aislabie, 1984; Bremer, 1984; Aislabie and Loutit, 1984; Bremer and Loutit, 1986a).

The production of extracellular polysaccharide appears to offer a mechanism by which sediment bacteria keep Cr(III) from binding to the cell surface. The importance of this protective mechanism for some bacteria is illustrated by the work of Wong et al. (1982) who showed that Cr^{3+} competed with Fe^{2+} in binding to *Thiobacillus ferrooxidans*. The binding of Cr(III) to extracellular polysaccharide produced by sediment bacteria has other possible consequences. As indicated earlier, such binding could contribute to the low concentration of chromium detected in ISW. Perhaps chromium, which does form sulphides in sediments, may be mobilized by biological or chemical means and then bind to polysaccharide. As solubilization of the chromium by bacteria takes place at a microsite, the binding of chromium to polysaccharide must occur immediately after solubilization (Pillidge, 1985).

5.6 Extracellular Polysaccharide as a Vehicle for Entry of Chromium to Foodchains

The addition of certain effluents to bodies of water, particularly those containing organic matter, causes an increase in the activity of heterotrophic bacteria (Goulder et al., 1980). Heavy metals in such effluents have been reported to have detrimental effects on sediment bacteria (Goulder et al., 1980, Nugent et al., 1980). Other workers, however, have found that many bacteria do occur in metal-rich sediments (Loutit et al., 1973; Mills and Colwell, 1977; Wong et al., 1982; Pillidge, 1985). These bacteria are available for ingestion by the sediment-dwelling biota (Fenchel and Jorgensen, 1977). To establish whether chromium enters animals bound to extracellular polysaccharide ingested with bacteria, it is necessary to distinguish ingested chromium from that which is bound to the outside of the animal living in a chromium-rich sediment. Bremer (1984) and Bremer and Loutit, (1986a), using a technique similar to that described by Fleming and Richards (1981) for zinc, were able to show that 61% of the total chromium in polychaetes had been ingested. This result differed from that of Fleming and Richards (1982) who found that 86 to 90% of zinc was bound to the outside of tubicifids, while Ireland and Richards (1981) showed that 68% of cadmium was bound to the outer layers of earthworms. Further, Bremer (Bremer, 1984; Bremer and Loutit, 1986a) was able to show that the chromium was ingested attached to polysaccharide.

The difference in the amount bound to the outside of an animal, compared with that ingested, may have considerable importance when considering the subsequent effects of the chromium on the animals. Rice and Windom (1982) have implicated bacteria in the transfer of metals from detritus to seawater by "production of extracellular materials which may form metal complexes" and by "uptake and binding of metals within living microbial biomass." They also implied that metals may be transferred to animals feeding on metal-rich detritus (Rice et al., 1981; Windom et al., 1982), but they did not include chromium in their study nor did they suggest that ingestion of metals bound to bacterial extracellular polysaccharide in detritus was a mechanism of entry of metals to a foodchain.

It has been our experience that molybdenum and chromium bind in greater quantities and more strongly to bacterial polysaccharide than other metals (Loutit, unpublished results). This observation may be significant in explaining whether chromium is ingested or adsorbed to the outside of sediment-dwelling biota.

The binding of chromium to bacterial exopolysaccharide produces some interesting changes in the polysaccharide (Bremer, 1984; Bremer and Loutit, 1986b), particularly in those which render it less degradable (Bremer, 1984; Bremer and Loutit, 1986b). If binding of chromium to polysaccharide results in an increased persistence in the sediment, the chance of ingestion of the polysaccharide with its bound chromium increases.

Elevated chromium concentrations have been found in polychaetes and *Amphibola* kept in the laboratory on sediment collected from near a tannery discharge compared to animals kept on a similar sediment from a nonpolluted site (Aislabie, 1984; Bremer, 1984). Further, polychaetes and *Amphibola* fed sediment bacteria with chromium bound to their extracellular polysaccharides had elevated chromium levels relative to animals fed bacteria without chromium (Bremer, 1984; Bremer and Loutit, 1986a). Moore and Ramamoorthy (1984) have quoted results indicating that chromium is not bioaccumulated, a result that is not in question. Our evidence, however, suggests that bacterial polysaccharides are a vehicle for the entry of chromium to foodchains from sediments where the chromium is apparently bound so that concentration in the ISW remains low. Higher trophic levels feeding on biota that have ingested bacteria with high chromium concentrations may then take up the chromium as shown for studies on a freshwater system (Patrick and Loutit, 1976).

6. CONSEQUENCES OF ENTRY OF CHROMIUM TO FOODCHAINS

Certain animals, depending on the species, age, and condition, can eliminate Cr(III) readily from their bodies. Chromium does not accumulate in particular tissues (Stary et al., 1982; Zaroogian and Johnson, 1983; Moore and Ramamoorthy, 1984; van Weerelt et al., 1984). However, few sediment dwellers or bottom feeders have been studied (Oshida and Wright, 1976; Oshida et al., 1981). Where they have, chromium has been added to the test vessel in soluble form (Oshida et al., 1981). Whether ingested chromium enters specific tissues and has an effect on metabolism, reproduction, or the larval or embryonic stages or has genetic effects has yet to be established.

7. INTERCONVERSION OF CHROMIUM(III) AND CHROMIUM(VI)

If Cr(III) enters the tissues or remains bound to surfaces of an animal, or is accumulated in epiphytic or periphytic bacteria, we need to know if its oxidation to Cr(VI) is possible. Reports indicate that interconversion of Cr(III) and Cr(VI) can occur in aquatic habitats (Schroeder and Lee, 1975; Nakayama et al., 1981b) (see Chapters 6, 8, 10). Most attention has been paid to the reduction of toxic Cr(VI) to Cr(III), which can be achieved, for example, by production of H_2S by bacteria in sediments (Smillie et al., 1981) and in the gut of animals (Edwards and Ewing, 1972).

The conversion of Cr(III) to Cr(VI) in animal tissues has been suggested by Petrilli and DeFlora (1978), but they could not find evidence to support this hypothesis. Bacteria producing hydrogen peroxide (Gottschalk, 1981) could be responsible for converting Cr(III) to Cr(VI) at microsites in sediments, or oxidation could occur in bacterial colonies on the gills of animals (Johnson et al., 1981). Subsequent entry to the animal may have toxic consequences (Riva et al., 1981). Chromium(VI) produced at such microsites may interfere with enzymatic and biochemical processes in fish (Sastry and Sunita, 1983).

8. CONCLUSIONS

Chromium is present at different concentrations in a variety of effluents which are disposed of into aquatic habitats (Duedall et al., 1983; Buat-Menard, 1984; Moore and Ramamoorthy, 1984). Depending on the composition of these discharges and that of the receiving waters, the proportion of chromium remaining in solution as suspended particulate matter or deposited into sediments varies (Forstner and Wittman, 1981). The effects of soluble and particulate chromium on biota have been extensively studied (Moore and Ramamoorthy, 1984). Most studies have involved assessing the effects (usually the LD_{50}) of inorganic chromium salts on a particular species in a culture vessel. While such an approach has been useful, more information is required on the effects on the biota of chromium bound to organic moieties, on the amount of added chromium that is biologically available, on whether the presence of other substances is able to modify the effects of chromium (Ajmal et al., 1984) or act synergistically with it (Moraitou-Apostolopoulou and Verriopoulos, 1982), and if sublethal concentrations of chromium affect the metabolism or reproduction of the organism. Further, the possibility of conversion of Cr(III) to the more toxic Cr(VI) on the surface of biota requires investigation.

What has been ignored in assessing bioavailability of chromium is the interaction of microorganisms and chromium in sediments. Sediment bacteria can solubilize chromium (Pillidge, 1985), concentrate it in extracellular polysaccharides (Aislabie, 1985; Aislabie and Loutit, 1986), and facilitate its entry to foodchains (Bremer, 1984; Bremer and Loutit, 1986a). Because of their ability to concentrate metals, fungi may play a similar role (Gadd and Griffiths, 1978). It is our opinion that bacteria act in microsites in the sediment, and only under certain conditions is chromium released in sufficient quantities to be detectable in overlying water (Pillidge, 1985).

The role of bacteria–chromium interactions in the water column has also been poorly studied. Periphytic bacteria on algae (Lee et al., 1975), plants (Patrick and Loutit, 1977), and animals (Johnson et al., 1981) have been shown to concentrate chromium. Whether the bacterial population is grazed or the whole organism is ingested, the chromium enters the animal, resulting in elevated chromium levels (Aislabie, 1984; Bremer, 1984; Bremer and Loutit, 1986a). The effect of this ingested chromium has yet to be established.

ACKNOWLEDGMENT

We are grateful to Dr. Gillian Lewis for her willingness to read this manuscript.

REFERENCES

Aislabie, J. (1984). *Aerobic Heterotrophic Bacteria in a Marine Sediment Polluted with Chromium*. PhD. Thesis. University of Otago Library.

Aislabie, J., and Loutit, M.W. (1984). "The effect of effluent high in chromium on marine sediment aerobic heterotrophic bacteria." *Mar. Environ. Res.* **13**, 69–79.

Aislabie, J., and Loutit, M.W. (1986). "Accumulation of Cr(III) by bacteria isolated from polluted sediment." *Mar. Environ. Res.* **20**, 221–232.

Ajmal, M., Nomani, A.A., and Ahmad, A. (1984). "Acute toxicity of chrome electroplating wastes to microorganisms; adsorption of chromate and chromium(VI) on a mixture of clay and sand." *Water, Air, Soil Pollut.* **23**, 119–127.

Austin, B., Allen, D.A., Mills, A.L., and Colwell, R.R. (1977). "Numerical taxonomy of heavy metal tolerant bacteria isolated from an estuary." *Can. J. Microbiol.* **23**, 1433–1447.

Babich, H., and Stotzky, G. (1977). "Sensitivity of various bacteria, including actinomycetes, and fungi to cadmium and the influence of pH on sensitivity." *Appl. Environ. Microbiol.* **33**, 681–695.

Babich, H., and Stotzky, G. (1983). "Further studies on environmental factors that modify the toxicity of nickel to microbes." *Regu. Toxicol. Pharmacol.* **3**, 82–99.

Bitton, G., and Freihofer, V. (1978). "Influence of extracellular polysaccharides on the toxicity of copper and cadmium towards *Klebsiella aerogenes. Microb. Ecol.* **4**, 119–125.

Bopp, L.H. Chakrabarty, A.M., and Ehrlich, H.L. (1983). "Chromate resistance plasmid in *Pseudomonas fluorescens." J. Bacteriol.* **155**, 1105–1109.

Bremer, P.J. (1984). *The Role of Polysaccharide Produced by Bacteria in the Accumulation of Chromium (Cr) by Marine Bacteria.* MSc. thesis. University of Otago Library.

Bremer, P.J., and Loutit, M.W. (1986a). "Bacterial polysaccharide as a vehicle for the entry of Cr(III) to a food chain." *Mar. Environ. Res.* **20**, 235–248.

Bremer, P.J., and Loutit, M.W. (1986b). "The effect of Cr(III) on the form and degradability of a polysaccharide produced by a bacterium isolated from a marine sediment." *Mar. Environ. Res.* **20**, 249–260.

Bryan, G.W. (1979). "Bioaccumulation of marine pollutants." *Philos. Trans. R. Soc. London B.* **286**, 483–505.

Buat-Menard, P.E. (1984). "Fluxes of metals through the atmosphere and ocean." In J.O. Nriagu, Ed., *Changing Metal Cycles and Human Health.*, Dahlem Konferenzen, Berlin, Springer Verlag, New York, pp. 43–69.

Calabrese, A., Collier, R.S., Nelson, D.A., and MacInnes, J.R. (1973). "The toxicity of heavy metals to embryos of the American oyster *Crassostrea virginica." Mar. Biol.* **18**, 162–166.

Capone, W., Mascia, C., Porcu, M., Tagliasacchi Masala, M.L. (1983). "Uptake of lead and chromium by primary producers and consumers in a polluted lagoon." *Mar. Pollut. Bull.* **14**, 97–102.

Capuzzo, J.McD., and Anderson, F.E. (1973). "The use of modern chromium accumulations to determine estuarine sedimentation rates." *Mar. Geol.* **14**, 225–235.

Cary, E.E. (1982). "Chromium in air, soil and natural waters." In S. Langård, Ed., *Biological and Environmental Aspects of Chromium.* Elsevier Biomedical Press, New York, Chapter 3.

Corpe, W.A. (1975). "Metal binding properties of surface materials from marine bacteria." *Dev. Ind. Microbiol.* **16**, 249–255.

Costerton, J.W., Irvin, R.T., and Cheng, K.J. (1981). "The bacterial glycocalyx in nature and disease." *Annu. Rev. Microbiol.* **35**, 299–324.

Cranston, R.E., and Murray, J.W. (1978). "The determination of chromium species in natural waters." *Anal. Chim. Acta.,* **99**, 275–282.

Curl, H., Cutshall, N., and Osterberg, C. (1965). "Uptake of Chromium(III) by particles in sea-water." *Nature.* **205**, 275–276.

Cutshall, N., Johnson, V., and Osterberg, C. (1966). "Chromium- 51 in seawater: chemistry." *Science* **152**, 202–203.

Diekart, G., Konheiser, U., Piechulla, K., and Thaven, R.K. (1981). "Nickel requirements and Factor F 430 content of methanogenic bacteria." *J. Bacteriol.* **148**, 459–464.

Duedall, I.W., Ketchum, B.H., Park, P.K., and Kester, D.R. (1983). "Global inputs, characteristics and fates of ocean dumped industrial and sewage wastes: An overview." In I.W. Duedall, B.H.

Ketchum, K.P. Park, and D.R. Kester, Eds., *Wastes in the Ocean 1. Industrial and Sewage Wastes in the Ocean.*, Wiley International Science Series. John Wiley, New York, pp. 3–45.

Edwards, P.R., and Ewing, W.H. (1972). *Identification of Enterobacteriaceae.* 3rd. Ed. Burgess Publishing Company, Minneapolis.

Efstathiou, J.D., and McKay, L.L. (1977). "Inorganic salts resistance associated with a lactose-fermenting plasmid in *Streptococcus lactis.*" *J. Bacteriol.* **130**, 257–265.

Eggett, J.M., and Thorpe, T.M. (1978). "Mobilization of chromium from fly ash particulates by aqueous systems modeling natural waters." *J. Environ. Sci. Health Part A* **13** (4), 295– 313.

Elderfield, H. (1970). "Chromium speciation in sea water." *Earth Planet. Sci. Lett.* **9**, 10–16.

Elwood, J.W., Beauchamp, J.J., and Allen, C.P. (1980). "Chromium levels in fish from a lake chronically contaminated with chromates from cooling towers." *Int. J. Environ. Stud.* **14**, 289–298.

Fasulo, M.P., Bassi, M., and Donini, A. (1982). "Cytotoxic effects of hexavalent chromium in *Euglena gracilis.* I. First observations." *Protoplasma* **110**, 39–47.

Fasulo, M.P., Bassi, M., and Donini, A. (1983). "Cytotoxic effects of hexavalent chromium in *Euglena gracilis.* II. Physiological and ultrastructural studies." *Protoplasma* **114**, 35–43.

Fenchel, T.M., and Jorgensen, B.B. (1977). "Detritus foodchains in aquatic ecosystems: the role of bacteria." *Adv. Microb. Ecol.* **1**, 1–58.

Fleming, T.P., and Richards, K.S. (1981). "A technique to quantify surface adsorption of heavy metals by soft-bodied invertebrates." *Comp. Biochem. Physiol.* **69**, 391–394.

Fleming, T.P., and Richards, K.S. (1982). "Uptake and surface adsorption of zinc by the freshwater tubificid oligochaete *Tubifex tubifex.*" *Comp. Biochem. Physiol.* **71**, 69–75.

Forstner, U. (1977). "Metal concentrations in freshwater sediments—natural background and cultural effects." In H.L. Golterman, Ed., *Interactions Between Sediments and Fresh Water.* Proceedings of the International Symposium held at Amsterdam, the Netherlands, September 1976. Dr. W. Junk B.V. Publishers, The Hague, pp. 94–103.

Forstner, U. (1984). "Metal pollution of terrestrial waters." In J.O. Nriagu, Ed., *Changing Metal Cycles and Human Health.* Dahlem Konferenzen, Berlin, Springer-Verlag, New York, pp. 71–94.

Forstner, U., and Wittman, G.T.W. (1981). *Metal Pollution in the Aquatic Environment.* 2nd Ed. Springer Verlag, New York.

Frey, B.E., Riedel, G.F., Bass, A.E., and Small, L.F. (1983). "Sensitivity of estuarine phytoplankton to hexavalent chromium." *Estuarine Coastal Shelf Sci.* **17**, 181–187.

Fukai, R., and Broquet, D. (1965). "Distribution of chromium in marine organisms." *Bull. Insitut. Oceanogr. Monaco.* **65** (1336), 3–19.

Gadd, G.M., and Griffiths, A.J. (1978). "Microorganisms and heavy metals toxicity." *Microb. Ecol.* **4**, 303–317.

Geesey, G.G. (1982). "Microbial exopolymers; ecological and economic considerations." *ASM News* **48**, 9–14.

Gottschalk, G. (1981). "The anaerobic way of life of prokaryotes." In M.P. Starr, H. Stolp, H.G. Truper, A. Balows, and H.G. Schlegel, Eds., *The Prokaryotes.*, II. 3rd Ed. Springer-Verlag, Berlin, pp. 1415–1424.

Goulder, R., Blanchard, A.S., Sanderson, P.L., and Wright, B. (1980). "Relationship between heterotrophic bacteria and pollution in an industrialized estuary." *Water Res.* **14**, 591–601.

Hershelman, G.P., Schafer, H.A., Jan, T.K., and Young, D.R. (1981). "Metals in marine sediments near a large California municipal outfall." *Mar. Pollut. Bull.* **12**, 131–134.

Hervey, R.J. (1949). "Effect of chromium on the growth of unicellular Chlorophyceae and diatoms." *Bot. Gaz.* **3** (1), 1–11.

Hobbie, J.E., and Lee, C. (1980). "Microbial production of extracellular material: Importance in benthic ecology." In K.R. Tenore, and B.C. Coull, Eds., *Marine Benthic Dynamics.*" The Belle W. Baruch Library in Marine Science, University of South Carolina Press, pp. 341–346.

Ireland, M.P., and Richards, K.S. (1981). "Metal content, after exposure to cadmium, of two species of earthworms of known differing calcium metabolic activity." *Environ. Pollut. Series A* **26**, 69–78.

Jan, T.K., and Young, D.R. (1978). "Chromium speciation in municipal wastewater and seawater." *J. Water Pollut. Control Fed.* **50**, 2327–2336.

Jernelov, A., and Martin, A.C. (1975). "Ecological implications of metal metabolism by micro-organisms." *Annu. Rev. Microbiol.* **29**, 61–75.

Johnson, I., Flower, N., and Loutit, M.W. (1981). "Contribution of periphytic bacteria to the concentration of chromium in the crab *Helice crassa.*" *Microb. Ecol.* **7**, 245–252.

Karbe, L., Schnier, C.H., and Siewers, H.O. (1977). "Trace elements in mussels (*Mytilus edulis*) from coastal areas of the North Sea and the Baltic. Multielement analysis using instrumental neutron activation analysis." *J. Radioanalyt. Chem.* **37**, 927–943.

Katz, A., and Kaplan, I.R. (1981). "Heavy metals behaviour in coastal sediments of Southern California: A critical review and synthesis." *Mar. Chem.* **10**, 261–299.

Kimbell, C.L., and Panulas, J. (1984). "Minerals in the world economy." In *Minerals Yearbook 1982, III.* Area Reports International Bureau of Mines, U.S. Department of the Interior, pp. 1–35.

Kostic, K., and Draskovic, R.J. (1982). "Studies of iron, cobalt and chromium distribution in some continental aquatic ecosystems and biological materials." *J. Radioanal. Chem.* **69**, 417–426.

Lee, Y., Patrick, F.M., and Loutit, M.W. (1975). "Concentration of metals by a marine bacterium *Leucothrix* and a periwinkle *Melarapha.*" *Proc. Univ. Otago Med. Sch.* **53** (1), 17–18.

Leland, H.V., Luoma, S.N., Elder, J.F., and Wilkes, D.J. (1978). "Heavy metals and related trace elements." *J. Water Pollut. Control Fed.* **June**, 1469–1514.

Loring, D.H. (1979). "Geochemistry of cobalt, nickel, chromium and vanadium in the sediments of the estuary and open Gulf of St. Lawrence." *Can. J. Earth Sci.* **16**, 1196–1209.

Loutit, M.W., Patrick, F.M., and Malthus, R.S. (1973). "The role of metal concentrating bacteria in a food chain in a river receiving effluent." *Proc. Univ. Otago Med. Sch.* **51**, 37–38.

Mayer, L.M., and Fink, L.K. (1980). "Granulometric dependence of chromium accumulation in estuarine sediments in Maine." *Estuarine Coastal Mar. Sci.* **11**, 491–503.

Mayer, L.M., and Schick, L.L. (1981). "Removal of hexavalent chromium from estuarine waters by model substrates and natural sediments." *Environ. Sci. Technol.* **15**, 1482–1484.

Mearns, A.J., and Young, D.R. (1977). "Chromium in the Southern California marine environment." In C.S. Giam, Ed., *Pollution Effects on Marine Organisms.* Lexington Books, Lexington, Chapter 8, pp. 125–142.

Mills, A.L., and Colwell, R.R. (1977). "Microbiological effects of metal ions in Chesapeake Bay water and sediments." *Bull. Environ. Contam. Toxicol.* **18** (1), 99–103.

Mogensen, B.B., and Jorgensen, S.E. (1978). "Modelling the distribution of chromium in a Danish Firth." In S.E. Jorgensen, Ed., *State of the Art in Ecological Modelling.* Proceedings of a Conference, Copenhagen, pp. 367–374.

Monaco, A. (1977). "Geochimie des milieux d' estuaire: Comparaison entre les suspensions fluviatiles et les depots prodeltaiques de L' Aude (Languedoc)." *Chem. Geol.* **20**, 45–55.

Moore, J.W., and Ramamoorthy, S. (1984). *Heavy Metals in Natural Waters Applied Monitoring and Impact Assessment.* Springer Verlag, New York, Chapter 4, pp. 58–76.

Moraitou-Apostolopoulou, M., and Verriopoulos, G. (1982). "Individual and combined toxicity of three metals, Cu, Cd and Cr for the marine copepod *Tisbe holothuriae.*" *Hydrobioogia.* **87**, 83–87.

Muir, W.C. (1983). "History of ocean disposal in the Mid-Atlantic Bight." In I.W. Duedall, B.H.

Ketchum, P.K. Park, and D.R. Kester, Eds., *Waste in the Ocean 1, Industrial and Sewage Wastes in the Ocean.*" Wiley International Series, John Wiley, New York, 273–291.

Muramoto, S. (1981). "Influence of complexans (NTA, EDTA) on the toxicity of trivalent chromium (chromium chloride, sulphate) at levels lethal to fish." *J. Environ. Sci. Health A* 16 (6), 605–610.

Nakayama, E., Tokoro, H., Kuwamoto, T., and Fujinaga, T. (1981). "Dissolved state of chromium in seawater." *Nature* 290, 768–770.

Nakayama, E., Kuwamoto, T., Tsurubo, S., Tokoro, H. and Fujinaga, T. (1981a). "Chemical speciation of chromium in seawater: Part 1 Effect of naturally occurring organic materials on the complex formation of chromium(III)." *Anal. Chim. Acta.,* 30, 289–294.

Nakayama, E., Kuwamoto, T., Tsurubo, S. and Fujinaga, T. (1981b). "Chemical speciation of chromium in seawater: Part 2 Effects of manganese oxide and reducible organic materials on the redox processes of chromium." *Anal. Chim. Acta,* 130, 401–404.

Nakayama, E., Kuwamoto, T., Tokoro, H. and Fujinaga, T. (1981c). "Chemical speciation of Cr in seawater: Part 3 the determination of chromium species." *Anal. Chim. Acta,* 131, 247–254.

NAS (1974), *Medical and Biologic Effects of Environmental Pollutants, Chromium.* Committee on the Biologic Effects of Atmospheric Pollutants, Medical Sciences National Research Council, Washington, 155 pp.

Neverauskas, V.P. (1983). "The effect of heavy metals on marine life around Port Pirie." *Aust. Fisheries.,* December, 6–9.

Nugent, C.E., Atchison, G.J., Nelson, D.W., and McIntosh, A.W. (1980). "The effects of heavy metals on microbial biomass in sediments of Palestine lake." *Hydrobiologia* 70, 69–73.

Oshida, P.S., Word, L.S., and Mearns, A.J. (1981). "Effects of hexavalent and trivalent chromium on the reproduction of *Neanthes arenaceodentata* (Polychaeta)." *Mar. Environ. Res.* 5, 41–49.

Oshida, P.S., and Wright, J.L. (1976). "Acute responses of marine invertebrates to chromium." In *Annual Report Southern California Coastal Waters Research Project for 1976.* El Segundo, California, pp. 155–159.

Osterberg, C., Cutshall, N., and Cronin, J. (1965). "Chromium-51 as a radioactive tracer of Columbia River water at sea." *Science* 150, 1585–1587.

Paerl, H.W. (1977). "Bacterial sediment formation in lakes: Trophic implications. In H.L. Golterman, Ed., *Interactions Between Sediments and Freshwaters.*" Proceedings of an International Symposium held at Amsterdam, the Netherlands, September 1976. Dr. W. Junk, B.V. Publishers The Hague, pp. 40–47.

Pankow, J.F., Leta, D.P., et al. (1977). "Analysis for chromium traces in the aquatic ecosystem. II. A study of Cr(III) and Cr(VI) in the Susquehanna River basin of New York and Pennsylvania." *Sci. Total Environ.* 7, 17–26.

Patrick, F.M., and Loutit, M.W. (1976). "Passage of metals in effluents, through bacteria to higher organisms." *Water Res.* 10, 333–335.

Patrick, F.M., and Loutit, M.W. (1977). "The uptake of heavy metals by epiphytic bacteria on *Alisma-plantago-aquatica.*" *Water Res.* 11, 699–703.

Petrilli, F.L., and DeFlora, S. (1977). "Toxicity and mutagenicity of hexavalent chromium on *Salmonella typhimurium.*" *Appl. Environ. Microbiol.* 33, 805–809.

Petrilli, F.L., and DeFlora, S. (1978). "Oxidation of inactive trivalent chromium to the mutagenic hexavalent form." *Mutat. Res.* 58, 167–173.

Pfeiffer, W.C., Fiszman, M., and Carbonell, N. (1980). "Fate of chromium in a tributary of the Iraja River, Rio de Janeiro." *Environ. Pollut. (Series B)* 1, 117–126.

Phillips, D.J.H. (1977). "The use of biological indicator organisms to monitor trace metal pollution in marine and estuarine environments—a review." *Environ. Pollut.* 13, 281–317.

Phillips, N.W. (1984). "Role of different microbes and substrates as potential suppliers of specific, essential nutrients to marine detritivores." *Bull. Mar. Sci.* 35 (3), 283–298.

Pillidge, C.J. (1985). *Bacterial Mobilisation of Chromium(III) in a Polluted Marine Sediment*. PhD. thesis. University of Otago Library.

Postgate, J.R. (1979). *The Sulphate Reducing Bacteria*., Cambridge University Press, Cambridge.

Ramamoorthy, S., and Kushner, D.J. (1975). "Binding of mercuric and other heavy metal ions by microbial growth media." *Microb. Ecol.* 2, 162–176.

Rice, D.L., Tenore, K.R., and Windom, H.L. (1981). "The effect of detritus ration on metal transfer to deposit feeding benthos." *Mar. Ecol. Prog. Ser.* 5, 135–140.

Rice, D.L., and Windom, H.L. (1982). "Trace metal transfer associated with the decomposition of detritus derived from estuarine macrophytes." *Bot. Mar.* 25, 213–223.

Rieper, M. (1978). "Bacteria as food for marine harpacticoid copepods." *Mar. Biol.,* 45, 337–345.

Riva, M.C., Flos, R., Crespi, M. and Balasch, J. (1981). "Lethal potassium dichromate and whitening (Blankophor) exposure of goldfish (*Carassius auratus*): Chromium levels in gills." *Comp. Biochem. Physiol.,* 68, 161–165.

Rossin, A.C., Sterritt, R.M. and Lester, J.N. (1983). "The influence of flow conditions on the removal of heavy metals in the primary sedimentation process." *Water, Air Soil Pollut.,* 19, 105–121.

Saenko, G.N., Koryakova, M.D., Makienko, V.F. and Dobrosmyslova, I.G. (1976). "Concentration of polyvalent metals by seaweeds in Vostok Bay, Sea of Japan." *Mar. Biol.,* 34, 169–176.

Sastry, K.V. and Sunita, Km. (1983). "Enzymological and biochemical changes produced by chronic chromium exposure in a teleost fish, *Channa punctatus.*" *Toxicol. Lett.,* 16, 9–15.

Schroeder, D.C., and Lee, G.F. (1975). "Potential transformations of chromium in natural waters." *Water, Air Soil Pollut.* 4, 355–365.

Schroll, H. (1978). "Determination of the adsorption of Cr^{+6} and Cr^{+3} in an algal culture of *Chlorella pyrenoidosa* using ^{51}Cr." *Bull. Environ. Contam. Toxicol.* 20, 721–724.

Seyfried, P.L. (1980). "Heavy metal resistance in aquatic bacteria in lakes." In J.G. Eaton, P.R. Parrish, and A.C. Hendricks, Eds., *Aquatic Toxicology*, ASTM STP 707, pp. 224–232.

Silver, S. (1984). "Bacterial transformations of and resistance to heavy metals." In J.O. Nriagu, Ed., *Metal Cycles and Human Health*. Dahlem Konferenzen, Berlin, Springer Verlag, New York, pp. 199–223.

Skerman, V.B.D. (1967). *A Guide to the Identification of the Genera of Bacteria*. 2nd Ed. Williams and Wilkins, Baltimore.

Smillie, R.H., Hunter, K., and Loutit, M.W. (1981). "Reduction of Chromium(VI) by bacterially produced hydrogen sulphide in a marine environment." *Water Res.* 15, 1351–1354.

Stary, J., Kratzer, K., Prasilova, J., and Vrbska, T. (1982). "The cumulation of chromium and arsenic species in fish (*Poecilia reticulata*)." *Intern. J. Environ. Anal. Chem.* 12, 253–257.

Stern, R.M. (1982). "Chromium compounds: Production and occupational exposure." In S. Langård, Ed., *Biological and Environmental Aspects of Chromium*, Elsevier Biomedical Press, New York, Chapter 2.

Subba Rao, D.V. (1981). "Growth response of marine phytoplankters to selected concentrations of trace metals." *Bot. Mar.* 24, 369–379.

Summers, A.O., and Jacoby, G.A. (1978). "Plasmid-determined resistance to boron and chromium compounds in *Pseudomonas aeruginosa*." *Antimicrob. Agents Chemother.* 13, 637–640.

Summers, A.O., and Silver, S. (1978). "Microbial transformations of metals." *Annu. Rev. Microbiol.* 32, 637–672.

Tacon, A.G.J., and Beveridge, M.M. (1982). "Effects of dietary trivalent chromium on rainbow trout." *Nutr. Rep. Int.* 25, 49–56.

Tan, E.L., and Loutit, M.W. (1976). "Concentration of molybdenum by extracellular material produced by rhizosphere bacteria." *Soil Biol. Biochem.* 8, 461–464.

Tan, E.L., and Loutit, M.W. (1977). "Effect of extracellular polysaccharide of rhizosphere bacteria on the concentration of molybdenum in plants." *Soil Biol. Biochem.* 9, 411–415.

Tilton, R.C., and Rosenberg, B. (1978). "Reversal of the silver inhibition of microorganisms by agar." *Appl. Environ. Microbiol.* **35**, 1116–1120.

Timoney, J.F., Port, J., Giles, J., and Spanier, J. (1978). "Heavy metal and antibiotic resistance in the bacterial flora of sediments of New York Bight." *Appl. Environ. Microbiol.* **36**, 465–472.

Ustach, J.F. (1982). "Algae, bacteria and detritus as food for the harpacticoid copepod, *Heteropsyllus pseudonunni. J. Exp. Mar. Biol. Ecol.* **64**, 203–214.

van Weerelt, M., Pfeiffer, W.C., and Fiszman, M. (1984). Uptake and release of ^{51}Cr(VI) and ^{51}Cr(III) by barnacles (*Balanus* sp.)." *Mar. Environ. Res.* **11**, 201–211.

Walker, J.D., and Colwell, R.R. (1975). "Factors affecting enumeration and isolation of Actinomycetes from Chesapeake Bay and southeastern Atlantic ocean sediments." *Mar. Biol.* **30**, 193–201.

Windom, H.L., Tenore, K.T., and Rice, D.L. (1982). "Metal accumulation by the polychaete *Capitella capitata*: Influence of metal content and nutritional quality of detritus." *Can. J. Fish. Aquatic. Sci.* **39**, 191–196.

Wium-Andersen, S. (1974). "The effect of chromium on the photosynthesis and growth of diatoms and green algae." *Physiol. Plant.* **32**, 308–310.

Wong, C., Silver, M., and Kushner, D.J. (1982). "Effects of chromium and manganese on *Thiobacillus ferrooxidans.*" *Can. J. Microbiol.* **28**, 536–544.

Young, D.R., Moore, M.D., Jan, T.K., and Eganhouse, R.P. (1981). "Metals in seafood organisms near a large California municipal outfall." *Mar. Pollut. Bull.* **12**, 134–138.

Zaroogian, G.E., and Johnson, M. (1983). "Chromium uptake and loss in the bivalves *Crassostrea virginica* and *Mytilus edulis.*" *Mar-Ecol. Prog. Ser.* **12**, 167–173.

Zhou, J., Chein, W., Liu, M., et al. (1979). Paper delivered at the 17th General Assembly of the International Union of Geodesy and Geophysics in Canberra, December 1979.

Zobell, C.E., and Feltham, C.B. (1938). "Bacteria as food for certain marine invertebrates." *J. Mar. Res.* **1**, 312–327.

Zobell, C.E. (1946). *Marine Microbiology, A Monograph on Hydrobacteriology.* Chronica Botanica Press, Waltham, MA. p. 240.

13

CHROMIUM TOXICITY: EFFECTS ON MICROORGANISMS WITH SPECIAL REFERENCE TO THE SOIL MATRIX

R. N. Coleman

Microbiology Group
Alberta Environmental Centre
Vegreville, Alberta, Canada

1. INTRODUCTION

Chromium was first isolated in 1798 (Jaworski, 1976) (See Chapter 1). Since then, the metal and its salts have been ever increasingly used in a wide range of industrial and manufacturing applications. In relative abundance, chromium ranks 21st of all elements in the earth's crust (Taylor et al., 1979). It commonly exists in three valence states in a wide variety of compounds (Sittig, 1985). However, in environmental terms, only (III) and (VI) valence states of chromium are of major significance. Normally, Cr(III) exists as the cation or as an oxyanion, whereas Cr(VI) exists primarily in two oxyanionic forms. The toxic nature of chromium has the potential to affect many environments and concern exists that chromium salts may be released to the environment. Chromium may be released either in treated wastewater discharged to surface waters or in produced sludges applied to land as soil amending agents (MacNaughton, 1977; Rundle et al., 1984). These areas of concern are mainly due to the fact that Cr(VI) is a strong oxidant (Taylor et al., 1979), a mutagen (Petrilli and DeFlora, 1977; Tso and Fung, 1981), and a carcinogen (Norseth, 1981).

Chromium toxicity effects may be lethal to many different trophic levels and documented cases of human poisoning with chromium have been reported in Japan (Forstner and Wittman, 1981). In this instance, building blocks were manufactured from Cr(VI)-containing wastes that had been stored in heaps and were used in fortifying reclaimed lands, recreational and school grounds. Contact with this material was reported to be responsible for 30 deaths and over 200 incurable conditions. In addition, Tokyo drinking water, which was obtained from groundwater beneath the chromium waste heaps, contained 2,000 times the allowable chromium concentration limit.

Although chromium may be toxic under certain conditions it has significant biological importance as an essential element for potentiation of the insulin effect in animals (Forbes and Erdman, 1983). It is also important in yeast metabolism (Anderson et al., 1977).

Very little is known regarding the toxic effects of chromium on microbial cells, especially in the soil system. This chapter deals with certain influences of chromium on microbial life, especially in the soil matrix, and will present information on the toxic effects of chromium in soil and interactions of microbes, chromium, and soil.

2. ASPECTS OF CHROMIUM IN CERTAIN ENVIRONMENTS

Chromium found in the environment is usually derived from natural sources and anthropogenic origins such as industrial and manufacturing activities.

2.1 Distribution

2.1.1 Soil

The total chromium content of soil usually ranges from 100–300 mg/kg, but may vary from as low as traces to as high as 4 g/kg. The observed soil level is usually related to the level found in the underlying rocks and parent material forming the soil (Aubert and Pinta, 1977). The quantity of chromium "available" to plants and other soil flora is usually low. For example, using weak reagents (2.5% acetic acid to simulate "availability") only 0.01–0.4% of the total chromium was extracted from a brown podzolic soil in Scotland. As well, little chromium (0.1–1.0% of the total chromium) was extracted from other soils using 1M NH_4CH_3COO (Aubert and Pinta, 1977). However, there is little data to allow for valid conclusions to be made as to the "availability" of chromium in most other soils. Therefore, although the total soil chromium content may be known by chemical methods and its distribution determined, the availability could vary from soil to soil. This variability makes it difficult to understand and quantitate chromium mobility and transfer through the foodchain.

2.1.2 Sediments

Although sediments are not considered soils, they are a solid matrix and show similarity to soils in terms of chromium interaction. Sediments become enriched with many metals by continual deposition of suspended solids from the overlying water (Forstner and Wittman, 1981). Sources of chromium found in sediment usually result from industrial activity and wastes from these industries and are treated via conventional domestic sewage treatment systems or are discharged directly to receiving waters (Smillie et al., 1981). Continual discharge tends to accumulate chromium in sediments in the aquatic environment. The chromium concentration in sediment varies greatly and can range from 10 mg/g in sediment at a tannery waste outfall to 438 mg/g in sediment taken beneath receiving waters of an electroplating plant (Nugent et al., 1980), whereas, background sediment levels of natural waters are typically 48 μg/g and 37 μg/g, as detected in the sediments of Lakes Superior and Huron, respectively (Kemp et al., 1978).

Chromium speciation is variable and depends on physico-chemical conditions of the water column. Chromium(VI) has been found to be the predominant species in water off the California coast, whereas in municipal wastewater 97 to 99% of the dissolved chromium ion was found to be Cr(III), mostly associated with the particulate fraction (Jan and Young, 1978). Chromium(VI) in overlying water undergoes reduction in the sediment or at its interface with the water. Microbially produced hydrogen sulphide from sediments can reduce Cr(VI) to Cr(III) in a seawater/sediment system (Smillie et al., 1981). After reduction, Cr(III) may easily be sorbed by organic material (Pfeiffer et al., 1980). Some of these observed activities also occur in soils and are further discussed in Section 3.5.

2.2 Sources

The major source of chromium is chromite ore ($FeO \cdot Cr_2O_3$) found in recoverable quantities in Rhodesia (Zimbabwe), U.S.S.R., Republic of South Africa, New Caledonia, and the Philipines; in 1971, estimates for world chromite ore production were 6.3 million tons (Fishbein, 1981) (see Chapter 2). The ore is used in production of stainless steel, refractory material, chromium metal, plating, alloy manufacture, chromates, and dichromates (used in dye manufacture, leather tanning, wool preservation, anticorrosion agents, etc.) (Aislabie and Loutit, 1980; Fishbein, 1980; 1981). Wastes from these industrial operations are typically disposed of by way of domestic sewage treatment systems or directly to receiving waters with little treatment; in many cases the chromium levels discharged may be significantly high. During the tanning process, leather shavings are produced; although they are potential fertilizer material, the product may contain in excess of 1% chromium which exhibits toxicity in the soil system (Thorstensen and Shah, 1979).

Point source chromium contamination problems are evident and much of the metal ends up in sediments (especially by direct discharge of industrial waste waters to receiving waters) and land (the increasing use of sewage sludges and tannery wastes as soil-amending agents). Another significant source is domestic sewage discharge that may contain effluents from plating and other industries where dilution is often a key operating factor in allowing the discharge of sewage effluent. However, chromium concentrates differentially on (in) the sediment even though the incoming chromium concentration is low (Gray and Clarke, 1984).

3. CHROMIUM AND MICROBIAL SYSTEMS

Many microbial cells, when in contact with chromium-containing compounds, exhibit negative responses that may be concentration-dependent with the notable exception of certain yeasts (Anderson, et al., 1977). Negative responses from microbial cells are usually considered to be a toxic effect and toxic manifestations may take several forms, from changes in morphology, to reactions with cell components, to reproducible physiological reactions such as changes in growth rate. Accumulation of chromium might also result in the cell sequestering the metal in such a way that its effects on the cell are minimized.

3.1 Effects on Morphology, Cells, and Cellular Components

Morphological changes have been observed in cells in contact with chromium-containing salts. For example, Bondarenko and Ctarodoobova (1981) observed three classes of responses in certain microorganisms exposed to various concentrations of $K_2Cr_2O_7$. In the case of *Staphylococcus aureus, S. epidermidis, Bacillus cereus*, and *B. subtilus* a few colonies were formed with "degenerate" cells and the colonies were usually small in the presence of 1×10^{-2} M $K_2Cr_2O_7$. Gram negative cells such as *Shigella sonnei, S. flexneri, Salmonella typhosa, Proteus mirablis*, and *Escherichia coli* exposed to 3.16×10^{-2} M $K_2Cr_2O_7$ produced the same morphological effects as observed in the gram positive microorganisms. In all the bacteria tested, cell division slowed as the $K_2Cr_2O_7$ concentration increased and at 1×10^{-1} M division ceased in the gram negative isolates tested (see previously). Although no visible morphological effects were reported in *S. aureus, S. epidermidis, B. cereus* and *B. subtilus*, all other isolates tested exhibited cell elongation, cell enlargement, filamentous forms, and spherical forms, as well as normal appearing cells. In addition, changes in motility, pigment, and spore generation were also observed in isolates that exhibited cellular morphological changes (Bondarenko and Ctarodoobova, 1981). Chromium(III) did not produce these changes (Bondarenko and Ctarodoobova, 1981). In contrast to the foregoing, other workers have shown that Cr(VI) complexes produce filamentous forms in *E. coli* (NCIB 10097) (Theodotou et al., 1976).

Chromium(VI) depressed the growth rate of some bacterial cells. As an example, the mean generation time (MGT) of an *Arthrobacter* sp. increased from 2.1 to 10.2 hours over a Cr(VI) concentration range of 0–200 mg/L [expressed as Cr(VI) ion concentration using $K_2Cr_2O_7$], whereas, in an *Agrobacterium* sp., the MGT increased from 1.0 to 3.0 hours over a concentration range of 0–100 mg/L. Above these concentrations growth was inhibited in both bacteria (Coleman and Paran, 1983). When the same *Arthrobacter* sp. was incubated with 2000 mg/L of Cr(VI) and examined using transmission electronmicroscopy, few morphological differences were seen in either whole or sectioned cells compared with unexposed control cells. However, a significant chromium signal was emitted from the surface of the chromium exposed cells when excited with x-rays (energy dispersive x-ray microanalysis) (Bhatnagar and Coleman, 1984).

Chromium effects on cellular components have been described and many of these significantly affect the cells activity. Chromous nitrate inhibited prodigiosin production in *Serratia marcescens* at a concentration of 0.37 mM (Furman et al., 1984). Although the valence of chromium in this salt is II, it is possible that Cr(II) when in liquid medium could oxidize to Cr(III) although no analytical evidence was presented. Chromous ion (as $CrCl_2$) also appears to react with cell components where it is has been shown to reduce *Pseudomonas aeruginosa* cytochrome c_{555} oxidase in vitro (Barber et al., 1977). In 1977, Romanenko and Koren'kov reported a bacterium, tentatively named (by the authors) *Pseudomonas dechromaticans*, as being able to utilize chromates and dichromates as electron acceptors when grown anaerobically. Subsequent Cr(III) hydroxides were reported to be produced and were implicated in binding numerous bacteria into cell clumps, whereas, in the absence of chromates, the cells were observed as single rods. Chromate reduction has also been shown to occur in *P. aeruginosa*; chromate resistance in this microorganism was shown to be plasmid-associated and transmissable to a restricted host range (Bopp, 1980; Bopp et al., 1983).

Yeast cells may also be affected on exposure to chromium and although chromium initiated larger cell formation in *Saccharomyces cerevisiae*, growth was reduced at the same time (Loveless et al., 1954). The fungus *Fusarium oxysporum* f. sp. *lycopersici* tolerated levels of chromium in the 0.8 mM range (as dichromate) whereas *Cunninghamella echinulata* was found to tolerate 0.2 mM chromium (Naguib et al., 1984). Dichromate was reported to be less toxic than chromate which, in turn, was less toxic than chromic ion. At low concentration (0.005 and 0.01 mM), however, chromate and chromic ion stimulated the growth of the *Fusarium oxysporum* f.sp. *lycopersici*.

3.2 Toxicity Detection

Chromium(VI) is reported to be the most toxic of all commonly occurring valence states of chromium (Williams and Silver, 1984), and its toxicity is expressed either as genotoxic effects or the more general metabolic toxic effects.

Genotoxic effects are mainly produced by Cr(VI) in microbial cells and result in frameshift mutations and base pair substitutions (Petrilli and DeFlora, 1977) as well as a more general genetic effect of unbalanced nucleotide pools (DeFlora et al., 1984). The toxic outcome of Cr(VI) is lethal DNA damage and mutagenic damage as well (see Chapter 16). Chromium(III), on the other hand, has been shown to produce DNA-DNA crosslinks, thus decreasing the fidelity of DNA replication (DeFlora et al., 1984). To be effective as a mutagen chromium must cross the cell membrane; it has been shown that Cr(VI) will diffuse across the cell membrane, whereas Cr(III) will do so only under extreme conditions (long incubations and high concentrations).

Metabolic toxic effects in microbial cells produced by chromium are often exhibited by changes in electron transport and similar manifestations. Studies using the Microtox test (Beckman Instruments, Inc., Carlsbad, CA) have shown that Cr(VI) is approximately twice as toxic as Cr(III) in the *Photobacterium phosphoreum* based test at the optimum pH of 6.5 (Qureshi et al., 1984). Chromium(VI) and (III) levels of 26.9 and 15.8 mg/L, respectively, produced the EC50 effect (that concentration lowering the *P. phosphoreum* light output by 50%). However, a different bacterial system (*Spirillum volutans* test) showed the opposite effect occurred where Cr(III) [CrK(SO$_4$)$_2$] was six times as toxic as Cr(VI) (K$_2$Cr$_2$O$_7$). The pH in the latter was measured as 6.0 in the Cr(VI) salt and 3.3 in the Cr(III) salt (Goatcher et al., 1984). Perhaps if the pH had been raised in the latter, the toxic effect might have been reduced.

Very little has been reported on the assessment of toxicity associated with solid matrices. Although not directly related to chromium toxicity and disposal, methods have recently been reported for the assessment of toxicity associated with solid material before disposal by landfill (Matthews et al., 1985). The protocol involved water extraction of the solid material followed by toxicity assessment of the water extract using the Microtox test. These procedures may be adaptable to the assessment of chromium toxicity in soils.

3.3 Chromium Accumulation

Both chromium(III) and (VI) have clearly been shown to be accumulated by microbial cells. Brewers yeast (*Saccharomyces carlsbergensis*), probably the best known example, forms organic complexes with Cr(III). In this system, Cr(III) appears to passively diffuse into the cell; however, the chromium concentration within the cell was always lower than that of the supporting medium (Kumpulainen and Koivistoinen, 1978). Marques et al. (1982) have shown that an S128 strain of *P. aeruginosa* was able to grow in the presence of 1600 μg/L K_2CrO_4 (428 mg/L as chromium(VI)) and presumptive evidence of chromium accumulation by cells was reported using transmission electron microscopy (electron dense deposits as compared to a control).

Coleman and Paran (1983) reported that a species of *Arthrobacter* when incubated (at 30°C) with 400 mg/L of Cr(VI) (using $K_2Cr_2O_7$) accumulated chromium up to 0.4% of its weight (dry weight). In the same report, an *Agrobacterium* sp. accumulated chromium to 0.16% of its dry weight in the presence of either 50 or 100 mg/L of Cr(VI). The *Arthrobacter* sp. could tolerate high levels of Cr(VI) (400 mg/L), whereas the *Agrobacterium* could tolerate up to 100 mg/L. From the appearance of the uptake curves and the response of uptake to various levels of Cr(VI), it was speculated that perhaps two different uptake mechanisms were in effect: an equilibrium mechanism for the *Arthrobacter* sp. and a biologically mediated mechanism for the *Agrobacterium* sp. Chromium has been specifically visualized on the latter *Arthrobacter* sp. where x-ray-excited chromium emissions were detected from the cell surface but not from any intracellular site examined (Bhatnagar and Coleman, 1984).

In natural systems, chromium can accumulate in the microbial populations and often those populations are associated with higher life forms. Bacteria periphytic to the crab, *Helice crassa*, have been shown to concentrate chromium and were implicated in contributing to the passage of chromium to the foodchain, since these crabs are eaten by fish (Johnson et al., 1981). During low tide this crab burrows into sand and sediments and is presumably exposed to chromium. These researchers showed that if the periphytic bacteria were removed from the carapace of the crab, the chromium content of that area of the carapace was likewise reduced. Sewage fungus or heterotrophic slimes present on the sediment in rivers are reported to concentrate chromium. For example, Gray and Clarke (1984) indicated that these slimes concentrate chromium to 2.9×10^4 times as much chromium compared with the concentration in the overlying water. The actual chromium concentration was reported to be up to 97 μg/g dry weight (Table 1). While observing the effects of copper and mercury on the uptake of other elements, Guthrie et al. (1977) stated that bacterial cells appeared to concentrate chromium (up to 2.6 times) from their suspending medium (water from ash settling ponds). In Table 1, results of chromium concentration by bacterial cells from various sources are listed. In order to make comparisons between data in Table 1, the first three entries are shown both by wet packed cell volume (as published) and by estimated dry weight. The reported values range from a "small quantity" to 3.6 μg/mg or almost 0.4% of the dry cell weight. Other values appear higher (e.g., 260 μg/mg) but were calculated by this author and must be considered approximations at best.

Table 1   Chromium Concentrations Found in Certain Microbial Cells and Systems

Cell Chromium Concentration	Calculated[a] Cell Chromium Concentration	Chromium[b] Concentration of Medium	System	Reference
0.1 mg/L[c]	40.0 μg/mg	0.4 mg/L	Mixed culture	Guthrie et al., 1977
0.4 mg/L[c]	160.0 μg/mg	0.2 mg/L	Mixed culture	Guthrie et al., 1977
0.65 mg/L[c]	260.0 μg/mg	0.25 mg/L	Mixed culture	Guthrie et al., 1977
Small quantity	—	1104.7 mg/L	*P. aeruginosa*	Marques et al., 1982
3.6 μg/mg[d]	—	400 mg/L	*Arthrobacter* sp.	Coleman and Paran, 1983
1.6 μg/mg[d]	—	50 mg/L	*Agrobacterium* sp.	Coleman and Paran, 1983
97 μg/mg[d]	—	NR[e]	Sewage fungus	Gray and Clarke, 1984

[a]A calculation was made by this author to convert the reported chromium concentration per wet packed cell volume to per dry weight based on the assumption that bacterial cells contain at least 75% water and the specific gravity of cells is approximately 1.00. [b]Concentration as Cr(VI). [c]Reported value per wet volume of packed cells (14,000 $\times$ G; 10 minutes). [d]Reported value per dry weight of cells. [e]Not reported.

3.4 Chromium Tolerance and Resistance

Tolerance to chromium may be considered as the ability of a microbial cell to undergo binary fission and grow in the presence of chromium. Since chromium is toxic, its effects on a cell will ultimately modify growth to the point where growth ceases. Many bacteria are sensitive to low chromium levels [5 mg/L as Cr(VI)]; however, others may be quite resistant to a wide range of chromium concentration. Data has been published on the levels of Cr(VI) which bacteria can tolerate and some of them are compiled in Table 2. Certain of these data suggest that bacterial cells tolerate as much as 5.3 g/L [Cr(VI)], whereas other resistance levels fall in the range of 100–450 mg/L [Cr(VI)]. In the case of the *Arthrobacter* sp., resistance may be due to the gram positive nature of the cell (Coleman and Paran, 1983). The mean inhibitory concentration (MIC) of 127 mg/L reported for *E. coli* will be lower in terms of Cr(VI); if the assumption is made that CrO_3 is hydrolysed to chromic acid, then less than half of the MIC value of CrO_3 will be Cr(VI) by weight, or approximaely 60 mg/L.

In Table 3, a listing of some bacterial isolates and their unpublished chromium

Table 2   A Compilation of Published Levels of Bacterial Chromium VI Tolerance

Bacterial Species	Level of Chromium Salt Added	Calculated Level of Chromium VI Added	Reference
Pseudomonas K21	20,000 mg/L[a]	5,356 mg/L	Shimada, 1979
P. fluorescens	>1,500 mg/L[a]	>401.7 mg/L	Bopp, 1980
P. aeruginosa	>1,600 mg/L[a]	>428.2 mg/L	Marques et al., 1982
Arthrobacter sp.	1271.2 mg/L[b]	450 mg/L	Coleman and Paran, 1983
Agrobacterium sp.	282.5 mg/L[b]	100 mg/L	Coleman and Paran, 1983
E. coli	127 mg/L[c]	66 mg/L	Thompson and Watling, 1984

[a] As K_2CrO_4; [b] as $K_2Cr_2O_7$; [c] as CrO_3.

sensitivity levels in liquid and solid media are compared. The isolates were made in our laboratory from lysimeters that were batch fed and continuously perfused with nutrient broth and increasing levels of chromium [up to 500 mg/L of Cr(VI)]. Two lysimeters were used, one held at a temperature of 50°C and the other held at 22°C. Bacteria were isolated (on plate count agar) from the effluent, identified using the in-house computer-assisted scheme (Coleman, 1980), and tested for Cr(VI) sensitivity in liquid [nutrient broth, 1% yeast extract, and Cr(VI)] and solid media (for methods, see Coleman and Paran, 1983). The results showed bacterial resistance on solid media to chromium levels of up to 300 mg/L, whereas generally higher resistance levels were observed in liquid media. Of these resistant colonies (resistant to 50 mg/L or higher), only 28% were gram negative and they tolerated intermediate (approximately 50–100 mg/L) Cr(VI) levels; the remaining were gram positive and tolerated significantly higher levels of Cr(VI). Those bacteria, considered thermophilic (able to grow at 50°C), of the genus *Bacillus* were able to tolerate higher chromium levels in solid media than the other mesophilic *Bacillus*. This is an initial observation but may be significant in chromium removal from waste waters since the temperature of waste waters can be higher than ambient water temperatures.

Genetic factors play a role in resistance to Cr(VI) (see Chapter 14). A plasmid has been isolated that confers Cr(VI) resistance to *P. fluorescens* (Bopp et al., 1983). Resistance to chromium and antibiotics has also been shown to occur and the possibility of selection of antibiotic/metal resistant strains in highly polluted areas concerned certain researchers (Simon-Pujol et al., 1978), whereas others indicated little difference in antibiotic resistance and chromium-sensitive or resistant strains (Luli et al., 1983).

Table 3   Chromium IV Tolerance of Lysimeter Bacterial Isolates

| Genus and Species | Selection[a] Conditions | Chromium VI[b] Tolerance (mg/L) on Growth Medium | |
		Solid	Liquid
Bacillus circulans	T	300	ND[c]
B. cereus	M	100	150
B. polymyxa	M	150	300
B. circulans	T	150	ND
B. megaterium	M	100	200
B. lichenformis	T	250	200
B. circulans	T	250	ND
B. lichenformis	T	150	ND
B. lichenformis	T	200	ND
B. fastidiosis	M	100	100
B. coagulans	T	200	ND
B. stearothermophilus	T	150	ND
B. brevis	M	100	150
Arthrobacter globiformis	M	200	200
A. globiformis	M	200	350
A. globiformis	M	200	450
A. globiformis	M	150	ND
Alcaligenes paradoxus	M	100	150
A. denitrificans	M	100	150
Enterobacter sakazaki	M	50	250
Pseudomonas stutzeri	M	100	ND
P. maltophila	M	100	150
P. mendocina	M	50	250
P. maltophila	M	200	ND
P. pseudoflava	M	100	ND

[a]Selection temperature in Cr(VI) perfused lysimeters, M=mesophilic, 22°C and T=thermophilic, 50°C; [b]visible growth (colony formation or turbidity within 14 days); [c]ND=not determined.

3.5   Interactions of Chromium with Microorganisms in Soil

Chromium in soil may have toxic and inhibitory effects on microflora and these effects may vary according to the valence state of the metal. Certain valence states of chromium will be considered briefly. Since addition of Cr(VI) to soils may result in its reduction to Cr(III) (Bartlett and Kimble, 1976b) followed by formation of complexes with organic material, toxic effects vary with chromium speciation even though the total chromium content remains unchanged (Cary, 1982). Conversely, the oxidation of Cr(III) or Cr(VI) could theoretically occur in well aerated soils; however, Bartlett and Kimble (1976a) could not detect it in such soils. Thermodynamically, Cr(VI) may exist in soils for extended periods of time, especially in low organic soils (Cary, 1982). These observations appear contradictory, but, in fact, may be a measure of the complexity of the soil ecosystem, which in turn decreases the simplicity of explanation of the toxic effects of chromium in soil (see Chapter 10).

Varying degrees of chromium sensitivity have been observed in soil bacteria and soil bacterial activities. The effect on soil amended with sewage sludge containing 556 μg total chromium per g of soil was to depress the bacterial counts and this was paralleled with a decrease in ATP levels. At 1 μg total chromium per g of soil chromium, the counts at first increased and then returned to pretreatment levels (Zibilske and Wagner, 1982). Others have observed the effect of Cr(VI) on bacteria grown in soil extract medium compared with bacteria grown in soil and noted less toxic effects in the soil grown cells (Ross et al., 1981). Observed tolerance patterns of these soil bacteria fell into two categories; in general, the gram negative bacteria were less tolerant than the gram positive cells. Some of the gram positive cells (*Corynebacterium* sp.) grew almost as vigorously at 11 mg/L of Cr(VI) as they did at 0 mg/L, whereas some gram negative cells could not grow at 1 mg/L.

Microbial carbon metabolism has been reported to be significantly affected by the presence of Cr(VI). In glucose-amended soils treated with 100 mg/kg Cr(VI), the molar ratio of CO_2 to soil glucose was 2.4, whereas at 0, 1, and 10 mg/kg the ratio was 4.5 (Drucker et al., 1979). These authors indicated that as the Cr(VI) concentration increased, metabolic carbon routing moved from oxidative to fermentative patterns. They also reported a significant depression of CO_2 (derived from starch) in the presence of 100 mg/kg of Cr(VI). Lesser effects, however, were observed at 1 and 10 mg/kg (Drucker et al., 1979).

Dehydrogenase activity in soil and alfalfa-amended soil was shown to be affected by Cr(III) additions. Activity was totally inhibited in soil containing 1000 mg/kg Cr(III), whereas in the alfalfa-amended soil, activity was not entirely inhibited at 5000 mg/kg (3% of control) (Rogers and Li, 1985). Soil nitrification was inhibited in the presence of soluble Cr(VI). However, the NH_4^+ and NO_2^- oxidizing bacteria adapted to the presence of Cr(VI) (at 100 uM $K_2Cr_2O_{10}$) and following a short period of inhibition, all added soluble nitrogen was oxidized (James and Bartlett, 1984).

Although soil exerts an attenuating effect on the toxicity of chromium to the microbial flora (Beck, 1981), little is known of the fate and the effect of the metal ions and its salts when applied to soils. Some researchers speculate that because of the mobility of the zincate anion in soil and its similarity to the chromate anion, the chromate anion should be highly mobile (Giordano and Mortvedt, 1976). However, unpublished results from our laboratory indicate that when $K_2Cr_2O_7$ is applied to a 1 m^2 plot of soil (solodized Solonetzic soil; organic material, 9.5–9.8%; sand, 37–40%; silt, 34–39%; clay, 24–26%) to give a Cr(VI) concentration of 100 mg/L (assuming a penetration depth of 30 cm), little penetration occurred. Four months after the application less than 33% of the added chromium penetrated past a depth of 5 cm. Concentration of Cr(VI) in the surface 5 cm varied up to 1878 mg/kg. When the effects of added chromium were assessed, initial results of heterotrophic bacterial counts indicated that soil attenuated chromium's toxic effect.

Total heterotrophic counts in the soil from the spill site did not significantly differ (p= 0.05, t test) from those of immediately adjacent control sites. Bacterial counts [heterotrophic count on PCA containing 100 mg/L of Cr(VI)] increased slightly after the spill but then returned to prespill levels and control site levels. A differential count of the pigmented colonies associated with the total heterotrophic count increased in numbers in response to chromium. Previously, in our laboratory, it was observed that the pigmented count increased in response to Cr(VI) in soil lysimeters.

Soil fungi have been reported to be 10 to 50 times as resistant as soil bacteria to the effects of soil Cr(VI) (Beck, 1981). However, little data could be located by this author with respect to the soil fungal population interactions with chromium, whereas in vitro data indicates that Cr(VI) depresses sporulation and mycelial growth as well (Babich et al., 1982).

Clearly, soil attenuates the toxic effect of Cr(VI) or sequesters by reacting with Cr(VI). This may, in some cases, be due to reduction of Cr(III).

4. SUMMARY

Chromium pervades all environments but does so from both geological and anthropogenic sources. Chromium also occupies a peculiar position in that it may be toxic at low levels but at the same time be required for a few physiological functions such as insulin potentiation. By contrast, many other metals, such as zinc, are required for physiological function and are toxic only at very high levels.

The distribution of chromium is very wide and is found in most soils in concentrations ranging up to 4 g/kg; however, concentrations are usually much lower, but even in high chromium soils much of the chromium will be unavailable to soil microflora. Chromium is toxic especially in its Cr(VI) valence state and toxicity is exhibited in altered genetic material and in altered metabolic and physiological reactions (such as growth rates, normal light output for luminescent bacteria, and certain morphological effects). Certain bacterial cells can accumulate chromium to about 0.4% (on a dry weight basis) and accumulation may be a biologically mediated process. Chromium accumulation has been shown to occur in the natural environment, in bacteria periphytic to a crab's carapace and in sewage fungus. Tolerance to chromium has been demonstrated and is partly due to the presence of genetic material expressing the production of chromate reductase and partly due to the gram positive character of certain cells. Fungi can tolerate 10 to 20 times the chromium concentration compared to bacteria.

Chromium effects on soil are not well understood but it is known that the toxic response is attenuated, and that soil bacterial counts change very little in the presence of high chromium concentrations. Chromium has been shown to affect the metabolism of glucose and flow of carbon in soil bacteria; as well, certain soil enzymes (dehydrogenase and nitrogenase) are depressed by the presence of chromium.

Much of the anthropogenically derived chromium could be controlled if industrial effluents high in chromium were treated before they were discharged to receiving waters or domestic sewage treatment facilities. A microbiological system for chromium removal may allow treatment in an energy efficient manner with a minimum of technological development. Research into areas such as microbiological chromium removal and the assessment of the toxicity of contaminated solid material should be actively encouraged.

ACKNOWLEDGMENTS

The author wishes to thank Drs. J.S. Davies, A. Khan, A.A. Qureshi, and S. Ramamoorthy for helpful and critical evaluation of this manuscript, T. Kazmierczak for translation of the Russian, V. Bohaychuk, I. Gaudet, and J.H. Paran for technical assistance, the AEC Chemistry Wing Secretarial personnel for manuscript production, and the AEC Library personnel for literature search and document acquisition.

REFERENCES

Aislabie, J., and Loutit, M.W. (1984). "The effect of effluent high in chromium on marine sediment aerobic heterotrophic bacteria." *Mar. Environ. Res.* **13**, 69–79.

Anderson, R.A., Polansky, M.M., Brantner, J.H., and Roginski, E.E. (1977). "Chemical and biological properties of biologically active chromium." In *International Symposium on Trace Element Metabolism in Man and Animals.* Freising, pp. 269–271.

Aubert, H., and Pinta, M. (1977). *Trace Elements in Soils.* Elsevier, Amsterdam, Oxford, New York, pp. 13–17.

Babich, H., Schiffenbauer, M., and Stotsky, G. (1982). "Comparative toxicity of trivalent and hexavalent chromium to fungi." *Bull. Environ. Contam. Toxicol.* **28**, 452–459.

Barber, D., Parr, S.R., and Greenwood, C. (1977). "The reduction of *Pseudomonas* cytochrome c_{551} oxidase by chromous ions." *Biochem. J.* **163**, 629–632.

Bartlett, R.J., and Kimble, J.M. (1976a). "Behavior of chromium in soils I. Trivalent forms." *J. Environ. Qual.* **5**, 379–383.

Bartlett, R.J., and Kimble, J.M. (1976b). "Behavior of chromium in soils II. Hexavelent forms." *J. Environ. Qual.* **5**, 383–386.

Beck, T. (1981). "Untersuchungen uber die toxische wirkung der in siedlung-sabfallen haufigen schwermetalle auf die bodenmikroflora." *Z. pflanzenernaehr bodenk* **144**, 613–627.

Bhatnagar, R., and Coleman, R.N. (1984). "Hexavalent chromium uptake by an *Arthrobacter* species: Electron Microscope studies." *Proceedings of the 42nd Annual Meeting of the Electron Microscope Society*, San Francisco Press Inc., San Francisco, pp. 680–681.

Bondarenko, B.M., and Ctarodoobova, A.T. (1981). "Morphological and cultural changes in bacteria under the effect of chromium salts." *Zhurnal Microbiologii Epidermiolgii Immunobiology* **4**, 99–100.

Bopp, L.H. (1980). *Chromate resistance and chromate reduction in bacteria.* Ph.D. thesis, Rensselaer Polytechnic Institute, New York. p. 165.

Bopp, L.H., Chakrabarty, A.M., and Ehrlich, H.L. (1983). "Chromate resistance in *Pseudomonas fluorescens.*" *J. Bacteriol.* **155**, 1105–1109.

Cary, E.E. (1982). "Chromium in air, soil and natural waters." In *Biological and Environmental Aspects of Chromium.* Elsevier Biomedical Press, Amsterdam, pp. 49–64.

Coleman, R.N., Ed. (1980). Microbiological identification service computer program. Alberta Environmental Centre, Vegreville, Alberta, Canada. 2 data files (298 records); 1 program file; 1 object program (Fortran).

Coleman, R.N., and Paran, J.H. (1983). "Accumulation of hexavalent chromium by selected bacteria." *Environmental Technol.* Lett **4**, 149–156.

DeFlora, S., Bianchi, V., and Levis, A.G. (1984). "Distinctive mechanisms for interaction of hexavalent and trivalent chromium with DNA?" *Toxicol. Environ. Chem.* **8**, 287–294.

Drucker, H., Garland, T.R., and Wildung, R.E. (1979). Metabolic response of microbiota to chromium and other metals. In N. Kharasch, Ed., *Trace Metals in Health and Disease.* Raven Press, New York, pp. 1–25.

Fishbein, L. (1980). *Proceedings of a Workshop/Conference on the Role of Metals in Carcinogenesis.* S. Belman, Ed., Atlanta, Georgia, March 24-28, 1980.

Fishbein, L. (1981). "Sources, transport and alteration of metal compounds: An overview. 1. Arsenic, Berylium, Cadmium, Chromium and Nickel." *Environ. Health Perspect.* **40**, 43–64.

Forbes, R.M., and Erdman, J.W. (1983). "Bioavailability of trace mineral elements." *Ann. Rev. Nutr.* **3**, 213–231.

Forstner, V., and Wittmann, G.T.W. (1981). *Metal Pollution in the Aquatic Environment.* Springer-Verlag, Berlin, Heidelberg, New York.

Furman, C.R., Owusu, V.I., and Tsang, J.C. (1984). "Interlaboratory effects of some transition metal ions on growth and pigment formation of *Serratia marcescens.*" *Microbios* **40**, 45–51.

Giordano, P.M., and Mortvedt, J.J. (1976). "Nitrogen effects on mobility and plant uptake of heavy metals in sewage sludge applied to soil columns." *J. Environ. Qual.* **5**, 165–168.

Goatcher, L.J., Qureshi, A.A., and Gaudet, I.D. (1984). "Evaluation and refinement of the *Spirillum volutans* test for use in toxicity screening." In D. Liu and B.J. Dutka, Eds., *Toxicity Screening Procedures Using Bacterial Systems.* Marcel Dekker, New York, pp. 89–108.

Gray, N.F., and Clarke, J. (1984). "Heavy metals in heterotrophic slimes in Irish rivers." *Environ. Tech. Lett.* **5**, 201–206.

Guthrie, R.K., Singleton, F.L., and Cherry, D.S. (1977). "Aquatic bacterial populations and heavy metals—II. Influence of chemical content of aquatic environments on bacterial uptake of chemical elements." *Water Res.* **11**, 643–646.

James, B.R., and Bartlett, R.J. (1984). "Nitrification in soil suspension treated with chromium (III, VI) salts or tannery wastes." *Soil Biol. Biochem.* **16**, 293–295.

Jan, T.K., and Young, D.R. (1978). "Chromium speciation in municipal wastewaters and seawater." *J. Water Poll. Cont. Fed.* **50**, 2327–2336.

Jaworski, J.F., Ed. (1976). *Effects of Chromium in the Canadian Environment.* Publication No. 15017. Natural Research Council of Canada.

Johnson, J., Flower, N., and Loutit, M.W. (1981). "Contribution of periphytic bacteria to the contribution of chromium in the Crab *Helice crassa.*" *Microb. Ecol.* **7**, 245–252.

Kemp, A.L.W., Williams, J.D.H., Thomas, R.L., and Gregory, M.C. (1978). "Impact of man's activities on the chemical composition of the sediments of Lakes Superior and Huron." *Water Air Soil Pollut.* **10**, 381–402.

Kumpulainen, J., and Koivistoinen, P. (1978). "Effects of glucose and chromium(III) concentrations in the medium on the uptake of ^{51}Cr by brewer's yeast." *Bioinorgan. Chem.* **8**, 431–438.

Loveless, L.E., Spoerl, E., and Weisman, T.H. (1954). "A survey of effects of chemicals on division and growth of yeast and *Escherichia coli*." *J. Bacteriol.* **68**, 637–644.

Luli, G.W., Talnagi, J.W., Strohl, W.R., and Pfister, R.M. (1983). "Hexavalent chromium-resistant bacteria isolated from river sediments." *Appl. Environ. Microbiol.* **46**, 846–854.

MacNaughton, M.G. (1977). "Adsorption of chromium(VI) at the oxide-water interface." In Drucker, H. and Wildung, R.E., Eds., *Biological Implications of Metals in the Environments*. Technical Information centre, ERDA, United States.

Marques, A.M., Espuny Thomas, M.J., Congregado, F., and Simon-Pujol, M.D. (1982). "Accumulation of chromium by *Pseudomonas aeruginosa*." *Microbios Lett.* **21**, 143–147.

Matthews, J.E., Kerr, R.S., and Hasting, L. (1985). "Interlaboratory evaluation of a toxicity reduction test procedure for use in land treatability screening." Proceedings of the *Second International Symposium on Toxicity Testing Using Bacteria*, May 6-10, 1985, Banff, Alberta, Canada.

Naguib, M.I., Haikal, N.Z., and Gouda, S. (1984). "Effect of chromium ions on the growth of *Fusarium oxysporum* f. sp. *lycopersici* and *Cunninghamella echinulata*." *Arab Gulf J. Scient. Res.* **2**, 149–157.

Norseth, T. (1981). "The carcinogenicity of chromium." *Environ. Health Perspect.* **40**, 121–130.

Nugent, C.E., Atchison, G.J., Nelson, D.W., and McIntosh, A.W. (1980). "The effects of heavy metals on microbial biomass in sediments of Palestine Lake." *Hydrobiolgia* **70**, 69–73.

Petrilli, F.L., and DeFlora, S. (1977). "Toxicity and mutagenicity of hexavalent chromium on *Salmonella typhimurium*." *Appl. Pollut.* (Series B). **1**, 117–126.

Pfeiffer, W.C., Fiszman, M., and Carbonell, N. (1980). "Fate of chromium in a tributary of the Iraj'a River, Rio de Janiero. *Environ. Pollut.* (Series B). **1**, 117–126.

Qureshi, A.A., Coleman, R.N., and Paran, J.H. (1984). "Evaluation and refinement of the Microtox test for use in toxicity screening." In D. Liu and B.J. Dutka, Eds., *Toxicity Screening Procedures Using Bacterial Systems*. Marcel Dekker, New York, Basel pp. 1–22.

Rogers, J.E., and Li, S.W. (1985). "Effects of metals and other inorganic ions on soil microbial activity: Soil dehydrogenase assay as a simple toxicity test." *Bull. Environ. Contam. Toxicol.* **34**, 858–865.

Romanenko, V.I., and Koren'kov, V.N. (1977). "A pure culture of bacteria utilizing chromates and bichromates as hydrogen acceptors in growth under anaerobic conditions." *Microbiology* **46**, 329–332.

Ross, D.S., Sjögren, R.E., and Bartlett, R.J. (1981). "Behaviour of chromium in soils: IV. Toxicity to microorganisms." *J. Environ. Qual.* **10**, 145–148.

Rundle, H.L., Calcroft, M., and Holt, C. (1984). "An assessment of accumulation of Cd, Cr, Cu, Ni, and Zn in the tissues of British Friesian steers fed on the products of land which has received heavy application of sewage sludge." *J. Agric. Sci. Camb.* **102**, 1–6.

Shimada, K. (1979). "Effects of sixvalent chromium on growth and enzyme production of chromium resistant bacteria." Proceedings *Annual Meeting of the American Society of Microbiology*.

Simon-Pujol, M.D., Marques, A.M., Ribera, M., and Congregado, F. (1978). "Drug resistance of chromium tolerant Gram-negative bacteria isolated from a river." *Microbios Lett.* **7**, 139–144.

Sittig, M. (1985). "Handbook of toxic and hazardous chemicals and carcinogens." *Noyes Publication*, NJ, pp. 243–248.

Smillie, R.H., Hunter, K., and Loutit, M. (1981). "Reduction of chromium(VI) by bacterially produced hydrogen sulphide in a marine environment." *Water Res.* **15**, 1351–1354.

Taylor, M.C., Reeder, S.W., and Demayo, A. (1979). "Vol. 1. Inorganic chemical substances. Chromium." In *Guidelines for Surface Water Quality*. Environment Canada, Ottawa.

Theodotou, A., Stretton, R.J., Norbury, A.H., and Massey, A.G. (1976). "Morphological effects of chromium and cobalt complexes on bacteria." *Bioinorgan. Chem.* **5**, 235–239.

Thompson, G.A., and Watling, R.J. (1984). "Comparative study of toxic metal compounds to heterotrophic bacteria." *Bull. Environ. Contam. Toxicol.* **33**, 114–120.

Thorstensen, T.C., and Shah, M. (1979). "Technical and economic aspects of tannery sludge as a fertilizer." *J. Am. Leather Chemicals Assoc.* **74**, 14–23.

Tso, W.W., and Fung, W.P. (1981). "Mutagenicity of metallic cations." *Toxicol. Lett.* **8**, 195–200.

Williams, J.W., and Silver, S. (1984). "Bacterial resistance and detoxification of heavy metals." *Enzyme Microbial Tech.* **6**, 530–537.

Zemansky, G.M. (1974). "Removal of trace metals during conventional water treatment." *J. Am. Water Works.* **October.** 606–609.

Zibilske, L.M., and Wagner, G.H. (1982). "Bacterial growth and fungal general distribution in soil amended with sewage sludge containing cadmium, chromium and copper." *Soil Sci.* **134**, 364–370.

14

GENETIC APPROACHES IN THE STUDY OF CHROMIUM TOXICITY AND RESISTANCE IN YEAST AND BACTERIA

B.-I. Ono

Laboratory of Environmental Hygiene Chemistry
Faculty of Pharmaceutical Sciences
Okayama University
Okayama, Japan

1.   INTRODUCTION

1.1   Essentiality and Toxicity of Chromium

Chromium has been recognized as an essential trace element for animals, including humans. Though molecular events have not been well understood, its importance in glucose metabolism seems established (see Chapter 2). At the same time, chromium has been recognized for a long time as an environmental pollutant. Epidemiological studies have established that chromate workers have suffered from skin, nasal, renal and hepatic lesions (Chapter 19), and have a high incidence of respiratory cancer (see Chapter 17).

1.2   Scope of Review

Genetics is a science dealing with genetic systems. Classic genetics deals with "conceptual" genes; genes and their expression are defined by biological functions. In contrast, molecular genetics deals with physical or material genes; genes and their expression are defined by physical and chemical terms. Conceptual genes are traced by examination of transmission of traits from one cell to another, or from one generation to the next. According to the present knowledge of molecular genetics, identification of a conceptual gene for a certain characteristic indicates existence of a physical gene and a gene product (protein or RNA) which eventually determines the corresponding characteristic. Thus, identification of a "gene" by any genetic means encourages biochemists to examine the process biochemically. Availability of mutants is essential in this approach. In fact, mutants have facilitated construction of various metabolic pathways and elucidation of their controls. In addition, molecular genetics has paved a way for studying "genes" as a chemical substance, namely, DNA. Not surprisingly, therefore, genetics can be used as a useful tool for elucidating various biological phenomena, including toxicity and the biological defense against toxins.

The aim of this article is two-fold: to illustrate applications of genetic approaches in the toxicology of chromium; and to discuss cellular defense mechanisms against chromium. Basic information is presented in Sections 2, 3, and 4. In Section 5, studies on available strains or mutants showing altered responses to chromate are reviewed (Section 5.1.) and chromate mutagenicity is also briefly examined (Section 5.2.). Finally, in Section 6, mechanisms of chromium toxicity and cellular defense mechanisms against chromium are discussed.

2. PHYSICOCHEMICAL ASPECTS OF CHROMIUM

A detailed description of the chemistry of chromium may be found elsewhere in this volume (see Chapter 2). However, it is unquestionable that biological effects are the consequence of chemical reactions which are governed by physical and chemical laws. Thus, the most important physical and chemical characteristics of chromium are briefly described here.

Although various valency states of chromium, ranging from Cr^{2+} to Cr^{6+}, are known, Cr^{3+} [Cr(III)] and Cr^{6+} [Cr(VI)] are prominent in nature. Chromium(VI) most often is in the forms of CrO_4^{2-} (chromate ion) and $Cr_2O_7^{2-}$ (dichromate ion), especially in solution. Since the magnitude of the reduction potential of Cr(VI) is so large, the oxidation of Cr(III) to Cr(VI) scarcely takes place at physiological conditions. Chromate and dichromate are powerful oxidizing agents. They are readily reduced to Cr(III) in acidic conditions or in the presence of organic compounds. Thus, biological effects of chromium have been intuitively attributed to the $+3$ oxidation state of chromium. However, Cr(VI) is considerably more toxic than Cr(III).

3. "RESISTANT" OR "SENSITIVE" VERSUS "NORMAL" STRAINS

"Resistant" and "sensitive" are relative terms, and they are usually used as antonyms. However, this usage of the words very often results in confusion and inconsistency. A strain is said to be sensitive when resistant mutants are obtained from it. However, the same strain would be said to be resistant if sensitive mutants were obtained from it. To avoid such ambiguity, the use of the word normal is recommended; standard, or wild type, strains are said to have a normal level of resistance. The resistance or sensitivity level of any strain should be referred to that of the wild type. In a case where the standard strain is undefined or unavailable, it is suggested that resistance level of a majority of the population be referred to as normal. This convention originates from the thought that standard or wild type strains may possess resistance mechanisms.

4. CELLULAR DEFENSE MECHANISMS AGAINST TOXINS

Before we go on to an examination of cellular defense mechanisms operating specifically against chromium, it is useful to provide some general comments.

Probable cellular defense mechanisms are listed in Table 1. They may be divided into five functional categories: (1) inhibition of entry of toxins into the cell, (2) minimizing concentration of toxins in the cell, (3) cellular target tolerance, and (4) bypassing or (5) repair of damage. Each organism adopts one or more of these mechanisms against a certain toxin. Mutations may cause an increase or decrease in the effectiveness of these mechanisms.

5.   GENETIC APPROACHES IN THE STUDY OF CHROMIUM TOXICITY

5.1   Studies on Chromate-Resistant or -Sensitive Strains (Table 2)

5.1.1   Naturally Occurring Bacterial Strains Resistant to Chromate

It has been widely known that bacterial resistance against antibiotics is often determined by genes on plasmids. Novic and Roth (1968) found that penicillinase plasmids in *Staphylococcus aureus* mediate resistance to inorganic ions such as arsenate and arsenite, and those of lead, cadmium, mercury, bismuth, antimony, and zinc. Efstathiou and McKay (1977) have found that lactose-fermenting plasmids in *Streptococcus lactis* mediate resistance to arsenate, arsenite, and chromate, and sensitivity to copper; altered responses to these inorganic ions are lost by induction of lactose nonfermentation. Summers and Jacoby (1978) found that a plasmid, pMG6, of *Pseudomonas aeruginosa* is responsible for resistance to chromate. Bopp et al. (1983) have found that chromate resistance of *Pseudomonas fluorescens* LB300 isolated from a chromium-contaminated environment is mediated by a plasmid, pLHB1, and that the plasmid effectively confers chromate resistance to *Escherichia coli*.

Mechanism(s) operating in the plasmid-mediated resistance to chromate are

Table 1   Cellular Defense Mechanisms Against Toxins

Inhibition of entry to the cell
    Membrane barrier
    Extracellular conversion to impermeable forms
Lowering of effective concentration of toxin
    Excretion
    Compartmentalization
    Binding to innert entities (conjugation)
    conversion to nontoxic forms
Cellular target tolerance
Bypassing of the incurred damage
Repair of incurred damage

not understood. Although it has been claimed that reduction of toxic Cr(VI) to less toxic Cr(III) is enhanced in *P. fluorescens* LB300 (Bopp and Ehrlich, 1980, cited by Silver 1983), no complete account of this work has been published.

Horitsu et al. (1978) have isolated, from activated sludge, a chromate-resistant strain (G-1) of *Pseudomonas ambigua*. It is shown that a derivative of the strain that is not resistant to chromate takes up about six times more chromate than the parental strain, indicating that transport barrier is the mechanism of resistance. However, it is also claimed that the G-1 chromate-resistant strain contains a Cr(VI)-reducing enzyme (Horitsu et al., 1983). Thus, it is possible that two or more mechanisms are operating in the strain. Whether this resistance, or part of it, is mediated by a plasmid is unknown.

5.1.2 Induced Bacterial Mutants Resistant to Chromate

Pardee et al. (1966) have discovered that sulfate transport is effectively inhibited by sulfite and chromate and ineffectively by selenate in *Salmonella typhimurium*. From this observation, they proceeded to select mutants resistant to chromate and found that they had a defective sulfate transport. All of the mutants characterized are dependent on cysteine for growth, and the responsible mutations fall into a single cistron, *cys*A. Since the sulfate transport system is repressed by cysteine, the parent strains are resistant to chromate if cysteine is supplied to the growth medium; thus, djenkolic acid is utilized as a sulfur source in the selection of the mutants. Similar mutants resistant to chromate are also obtained in *E. coli* (Karbonowska et al., 1977). These results clearly indicate that chromate is transported specifically via the sulfate transport system in these bacterial species.

5.1.3 Induced Fungal Mutants Resistant to Chromate

Extrachromosomal, or plasmid-mediated, chromate resistance has not been reported in eucaryotic organisms including fungi. However, chromosomal mutants resistant to chromate are obtained and studied in several fungi. They are reviewed as follows.

In *Aspergillus nidulans*, Arst (1968) has shown that the *s3* (*sB*) mutation confers resistance to both chromate and selenate, and that this mutation causes a defect in the sulfate transport system. It is suggested that the sulfate transport system of this fungus is analogous to that of bacteria described previously. Paszewski (1976) has shown that the sulfate transport system of this organism is repressed by methionine.

In *Neurospora crassa*, chromate-resistant mutations arise in the *CYS13* locus (Marzluf, 1970a, b). However, the *cys13* mutants are only partially defective in transport of inorganic sulfate, especially during the conidial stage. There is another locus involved in sulfate transport, namely *CYS14*; mutations in this locus reduce sulfate uptake, to about 25% of the wild type strains, during the mycelial stage but do not confer resistance to chromate. Strains containing both the *cys13* and *cys14* mutations have completely defective sulfate transport in all stages of the life cycle. Chromate-resistant mutations arise in another locus, *CYS3*, which is a regulatory gene of the *CYS13* and *CYS14* genes (Marzluf,

Table 2 Mutants Showing Altered Responses to Chromium(VI)

Reference	Organism	Response[a]	Plasmid or Mutation	Remarks
Efstathiou and McKay (1977)	*Streptococcus lactis*	R	pLM3001	Lactose-fermentable, resistant to arsenate and arsenite, sensitive to copper
		R	pLM2201	Lactose-fermentable, resistant to arsenate and arsenite
		R	pLM2103	Same as pLM2201
		R	pLM2001	Same as pLM2201
		R	pLM2102	Same as pLM3001
		R	pLM1801	Same as pLM3001
Summers and Jacoby (1978)	*Pseudomonas aeruginosa*	R	pMG6	Resistant to chloramphenicol, gentamicin, kanamycin, streptomycin, sulfonamide, tobramycin, mercuric chloride, and potassium tellurite
Bopp et al. (1983)	*Pseudomonas fluorescens*	R	pLHB1	Conjugally transferred from *P. Fluorescens* (pLHB1)
	Escherichia coli	R	pLHB1	
Horitsu et al. (1978)	*Pseudomonas ambigua* G-1	R	?	Resistant to copper and cadmium
Pardee et al. (1977)	*Salmonella typhimurium*	R	cysA	Dependent on cysteine due to defect in the sulfate transport system
Karbonowska et al. (1977)	*Escherichia coli*	R	cysA	Same as cysA of *Salmonella typhimurium*
Arst (1968)	*Aspergillus nidulans*	R	s3 (sB)	Deficient in sulfate transport
Marzluf (1970a, b)	*Neurospora crassa*	R	cys13	Deficient in one of the two transport systems

Reference	Organism		Gene	Description
Marzluf (1970a, b)	*Neurospora crassa*	R	*cys3*	Deficient in the two sulfate transport systems (regulatory mutation)
Breton and Surdin-Kerjan (1977)	*Saccharomyces cerevisiae*	R	*CHR*	Deficient in sulfate transport
Ono and Weng (1982b)	*Saccharomyces cerevisiae*	R	*CHR1*	Partially deficient in chromate transport
		R	*CHR2*	Same as *CHR1*
Ono and Weng (1982a)	*Saccharomyces cerevisiae*	S	*chs1* (*lys7*)	Dependent on lysine due to defect in homocitric dehydrase
		S	*chs2*	
		S	*chs3*	
		S	*chs4*	
		S	*chs5*	
		S	*chs6*	
Ono et al. (1986)	*Saccharomyces cerevisiae*	S	*cys1 cys3*	Dependent on cysteine due to defect in serine acetyltransferase and γ-cystathionase
		S	*cys2*	Dependent on cysteine due to defect in serine acetyltransferase and presumably in cystathionine β-synthase
		S	*met17*	Dependent on cysteine or methionine due to defect in O-acetylserine and O-acetylhomoserine sulfhydrylase
Christie et al. (1984)	Chinese hamster ovary cell line EM9	S	?	Sensitive to x-ray and UV light due to deficiency in DNA-repair

[a]R and S denote resistant and sensitive, respectively.

357

1970a, b). Roberts and Marzluf (1971) have indicated that sulfate transport is competitively inhibited by chromate.

In *Saccharomyces cerevisiae*, Breton and Surdin-Kerjan (1977) have obtained a mutant, G10, that is able to grow in the presence of both chromate and selenate. Genetic analysis of the mutant has indicated that it contains two mutations conferring resistance to chromate (*CHR*) and selenate (*sel*), respectively; the former is dominant and the latter is recessive against the respective wild type alleles. It should be mentioned that the double mutant of *CHR* and *sel* is resistant to higher concentrations of chromate or selenate than mutants containing each mutation, indicating interaction of these mutations. Ono and Weng (1982b) have independently obtained mutants resistant to chromate. All of the mutations are dominant, and so far as representative mutants are examined, they do not, or do only rarely, recombine with each other. Thus, it is concluded that these mutations have arisen in a narrow region in a certain chromosome, presumably in one or two loci. These chromate resistant mutants, unexpectedly, take up chromate even at a rate of about two thirds of that of wild type strains. The result indicates that chromate resistance of these mutants is hardly attributable to deficiency of chromate transport. Dominant or semidominant nature of the chromate resistant mutations is in accord with this view.

5.1.4 Induced Yeast Mutants Sensitive to Chromate

Strains showing increased sensitivity to chromate have not been reported except for *S. cerevisiae*. Studies of such mutants must complement those of resistant mutants. Moreover, existence of sensitive mutants is evidence for wild type strains possessing defense mechanisms against chromate; it is important to realize that wild type strains are not merely sensitive to chromate.

Ono and Weng (1982a) have obtained mutants sensitive to chromate after mutagenesis with ultraviolet light. Of 10 mutants obtained, six contained single mutations, each of which falls into a distinctive complementation group (*chs1* through *chs6*), indicating that a number of loci give rise to chromate-sensitive mutations. The remaining four mutants contain two or more mutations that cooperatively cause chromate sensitivity. It is clear that the "normal" level of chromate resistance is a product of "normal" functioning of many genes that act independently or cooperatively with others. It has been shown that *chs1* mutation is allelic to the previously known *lys7* mutation. An authentic *lys7* mutation is shown to cause sensitivity to chromate. Although the *LYS7* locus is known to code for homocitric dehydrase which is thought to be identical to homoaconitate hydratase (EC4.2.1.36) (Bhattacharjee et al., 1968), how this enzyme is involved in rendering resistance to chromate is not known.

Ono and Weng (1982b) have examined responses against chromate of strains containing both sensitive and resistant mutations and found that one resistant mutation (*CHR1*) results in the sensitive phenotype in the presence of either *chs1* or *chs5*, whereas another mutation (*CHR2*) results in the sensitive phenotype with *chs5* but the normal phenotype with *chs1*. Whether this difference in the resulted phenotypes is due to simple additivity or to synergistic interaction of the mutations in question is not known.

5.1.5 Chromate Sensitivity of Yeast Strains Dependent on Cysteine

In *S. cerevisiae*, the *cys1 cys3* mutant is deficient in serine acetyltransferase (EC2.3.1.30) and γ-cystathionase (EC4.4.1.1) (Halos, 1976; Ono et al., 1984); the *met17* mutant is deficient in O-acetylserine and O-acetylhomoserine sulfhydrylase (EC4.2.99.10) (Yamagata et al., 1975); and the *cys2* mutant is deficient in serine acetyltransferase (EC2.3.1.30) and presumably of cystathionine β-synthase (EC4.2.1.22) (Halos, 1976; Ono, in preparation). Despite different enzymic deficiencies, all these strains require cysteine for growth; the *met17* strain can grow if either cysteine or methionine is supplied. These strains are more sensitive to chromate than wild type strains (Ono et al., 1986). Unlike the case of the *lys7* mutant, chromate sensitivity of these strains is due to deficiency of cysteine, but not of enzymes. In fact, addition of cysteine to growth medium has caused reduction of chromate sensitivity not only of the cysteine-dependent strains but also of strains of various other responses against chromate.

This antagonistic effect of cysteine against chromate is attributed to conversion of Cr(VI) to Cr(III) at the expense of cysteine. Studies on interactions of SH-compounds with metals including chromium are reviewed previously (Gergely and Sóvágó, 1979; Rabenstein et al., 1979) (see Chapter 2). Susa (1984) has shown that DL-penicillamine, glutathione (reduced form), L-cysteine and 2,3-dimercapto-1-propanol antagonize toxicity of dichromate. These effects are explained by reduction of Cr(VI) to Cr(III). Detailed analyses of reduction of Cr(IV) to Cr(III) with D-penicillamine have indicated that resulted Cr(III) exists in a complex containing penicillamine (Sugiura et al., 1972; Hojo et al., 1977). McAuley and Olatunji (1977a, b) have suggested reduction of Cr(VI) to Cr(III) takes place with formation of a transitional complex of chromium and glutathione. A similar process may take place in interactions with various SH-compounds including cysteine. Aaseth et al. (1982) have indicated that Cr(VI) is readily taken up by red blood cells followed by rapid reduction to Cr(III) in the cells, presumably due to reduction by glutathione, and that intracellular Cr(III) cannot be removed as long as the cell membrane is intact.

Here, it may be worthwhile to mention a biologically active substance containing chromium (Hopkins et al., 1968; see Chapter 2). The substance is known as "glucose tolerance factor." It has been purified from brewer's yeast and shown to contain chromium, nicotinic acid, glycine, glutamic acid, and cysteine (Mertz et al., 1974); note that the amino acid composition of the substance resembles that of glutathione. Recently, Wada et al. (1983) have purified a compound of similar composition from the livers of dogs injected with dichromate. The compound contains Cr(III) and is presumed to play a role in chromate detoxication in the animal.

5.1.6 Chromate Sensitivity of a Mammalian Cell Line Sensitive to Radiation

Christie et al. (1984) have compared a radiation sensitive mutant cell line (EM9) with the original Chinese hamster cell line and found that the mutant cell line is much more sensitive to Cr(VI) for growth inhibition than the original cell line. This study will be discussed further in Section 5.2.2.

5.2 Mutagenicity of Chromate (Table 3)

5.2.1 Induction of Mutants by Chromate

Mutagenic effects of chromate in bacteria (Venitt and Levy, 1974; Nishioka, 1975; Petrilli and De Flora, 1977; Bianchi et al., 1983; Beyersmann et al., 1984) and a yeast (Bonatti et al., 1976) have been described. Paschin et al. (1983) have compared Chinese hamster cell lines for mutagenic sensitivity of the locus coding for hypoxanthine-guanine phosphoribosyltransferase (EC2.4.2.8) and found that the V-79 cell line is more sensitive than the CHO-AT3-2 cell line. The difference is attributed to different nucleotide rearrangements at the X chromosome in these cell lines.

5.2.2 DNA Damage Induced by Chromate

It is widely accepted that metal chromates are highly carcinogenic compounds (NRC, 1974; Sunderman, 1978; Flessel, 1978; Flessel et al., 1980; Issaq, 1980) (see Chapter 17). Fradkin et al. (1975) have observed that chromate induces irreversible morphological alterations in a cultured mammalian cell line, BHK21. Tsuda and Kato (1976) have observed that Cr(VI), but not Cr(II) and Cr(III), induces morphological transformation as well as chromosomal aberrations in hamster embryo cells. Incidence of chromosomal aberrations is decreased by a reducing agent, Na_2SO_3. Casto et al. (1979) have shown that Cr(VI) enhances transformation of Syrian hamster embryo cells by a simian adenovirus, SA7. Since chromate is shown to be both carcinogenic and mutagenic, many investigators have turned their attention to the effects of chromate on DNA. Following, some of the current developments are reviewed (see also Chapter 16).

Sirover and Loeb (1976) examined the effects of various metal compounds on the fidelity of DNA-replication and found that Cr(II) and Cr(VI) cause increased infidelity. Bianchi et al. (1982) examined effects of Cr(VI) on an endogenous adenylate pool of hamster fibroblasts and found that ATP decreases while ADP and AMP increase. It is reasonable to assume that mutagenicity of Cr(VI) may, at least in part, be a consequence of these effects.

Tsapakos et al. (1983) have shown that DNA-strand breaks, and DNA-DNA and DNA-protein crosslinks are induced in organs of rats injected with Cr(VI); the former lesions are repaired in kidney and liver but not in lung, whereas the latter lesions are repaired only in liver. It has been shown that Cr(III) is not cytotoxic or mutagenic (Tsapakos et al., 1983, Bianchi et al., 1984) (see Chapter 16). Robison et al. (1984) indicated that Cr(VI) causes DNA-strand breaks when added to cultured mammalian cells but not when added to isolated nucleoids, and that it induces substantial DNA-repair activity at concentrations and exposure times where DNA lesions are hardly detected. Christie et al. (1984) have shown that Cr(VI) induces DNA-strand breaks and DNA crosslinks; the former lesions are rapidly repaired in both radiation-sensitive and wild type Chinese hamster cell lines, whereas the latter lesions are repaired completely, if not rapidly, in the original cell line but only partially in the mutant cell line. Cantoni and Costa (1984) have claimed that chromate induces "alkali labile sites," which result in strand breaks in an alkali environment used in the experiments. Since repair systems are different between strand breaks and alkali labile sites, clarification is needed concerning their biological significance.

Many investigators agree that Cr(III), which causes DNA-strand breaks when treated directly with DNA, does not induce DNA damage when intact cells are treated (Tsuda and Kato, 1976; Tsapakos et al., 1983; De Flora et al., 1984; Bianchi et al., 1984). This observation indicates that Cr(III) is highly impermeable. On the other hand, Cr(VI) causes only slight damage to purified DNA, suggesting that metabolic conversion of Cr(VI) to Cr(III) is essential to the induction of DNA damages.

6. CHROMATE TOXICITY AND CELLULAR DEFENSE MECHANISMS

Chromium toxication mechanisms deduced from the above mentioned genetic approaches are illustrated in Figure 1.

Chromium(VI) appears to enter bacterial cells via the sulfate transport system. Thus, mutations that impair sulfate uptake also reduce uptake of chromate and eventually lead to increased resistance to chromate in bacteria. By contrast, sulfate transport systems of fungi are different from one species to another. *A. nidulans* has a single system similar to bacteria. However, *N. crassa* has two systems whose activities change in different stages of the life cycle. One of them is similar to bacteria, but the other does not mediate transport of chromate. Similarly, *S. cerevisiae* appears to have multiple systems for sulfate transport, one of which mediates chromate transport while the other does not. The sulfate transport systems are subject to repression by sulfur-containing metabolites such as cysteine and methionine. Thus, chromate resistance is greatly affected by ingredients in the growth medium and by sulfur metabolism in the cell. From the evolutionary point of view, it is reasonable to suggest that bacteria, which have an autotrophic metabolism of sulfur, and fungi, which have dual (autotrophic and heterotrophic) metabolisms of sulfur, contain specific sulfate transport systems. In this context, it is of interest to see whether animals that do not depend on sulfate for growth have specific sulfate transport systems or

Table 3 Selected Mutagenic and Other Genotoxic Effects of Cr(VI)[a]

Reference	Test System	Observed Effects
	Microorganism	
Venitt and Levy (1974)	*Escherichia coli*	Mutagenicity
Nishioka (1975)	*Bacillus subtilis*	Mutagenicity
Bonatti et al. (1976)	*Schizosaccharomyces pombe*	Mutagenicity
Petrilli and De Flora (1977)	*Salmonella typhimurium*	Cytotoxicity
		Mutagenicity
Bianchi et al. (1983)	*S. typhimurium*	Mutagenicity
Beyersmann et al. (1984)	*S. typhimurium*	Mutagenicity
	Mammal	
Tsapakos et al. (1983)	Rat	DNA damage
		Strand breaks
		DNA–DNA crosslinks
		DNA–protein crosslink

Cultured Mammalian Cell

Reference	Cell	Effect
Fradkin et al. (1975)	Hamster BHK-21	Morphological alterations
Tsuda and Kato (1977)	Hamster embryo	Chromosomal aberrations
Casto et al. (1979)	Hamster embryo	Enhanced transformation with a simian adenovirus, SA7
Bianchi et al. (1982)	Hamster BHK-21	Unbalance of the endogenous adenylate pool
Paschin et al. (1983)	Hamster V-79	Higher mutagenic sensitivity of the HGPRT locus than AT3-2
Robison et al. (1984)	Hamster CHO	DNA strand breaks
Beyersmann et al. (1984)	Hamster CHO	Cytotoxicity
Christie et al. (1984)	Hamster EM9	Cytotoxicity

Cantoni and Costa (1984)	Hamster CHO	DNA damage
		  Strand breaks
		  DNA-DNA crosslinks
		  DNA-protein crosslinks
		Alkali sensitive sites

In vitro DNA Synthesis

Reference		Effect
Sirover and Loeb (1976)		Increased infidelity
Bianchi et al. (1983)		Increased infidelity

[a] A more detailed compilation is given elsewhere in this volume (see Chapter 16).

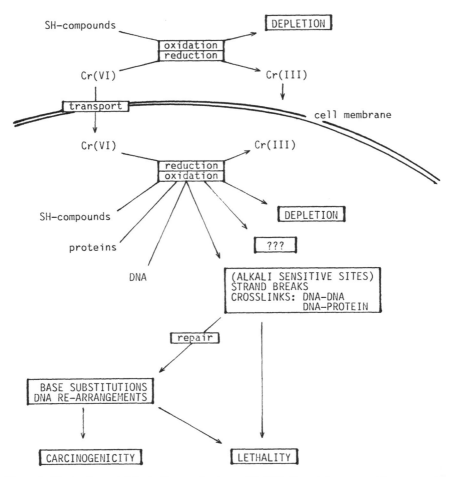

Figure 1. Toxication and detoxication mechanisms of Cr(VI). Items in boxes are either conversion processes (small letters) or consequences of them (capital letters). (Editors' note: a complementary model emphasizing the direct involvement of Cr(III) in the injurious biochemical processes is provided in Chapter 16.)

not. In these organisms, chromate may be transported via inefficient and nonspecific systems. Chromate transport systems in eucaryotes are not well characterized yet.

Extracellular reduction of Cr(VI) to Cr(III) by reducing agents such as cysteine and glutathione may be considered as a cellular defense mechanism against chromate, because Cr(III) appears to be excluded by various cell types. However, it should be stressed that this mechanism may bring about depletion of the reducing agents. If some of them are growth-limiting factors, cells have to stop growing.

Once Cr(VI) enters the cell, it interacts with various cellular components. Here again, the strong oxidizing power of Cr(VI) is the source of its biological effects. Reducing agents, including SH-compounds, act as detoxicants but increase the likelihood of depletion of the agents. It is also possible that depletion of certain metabolites causes secondary effects, so that unbalance in cellular metabolism occurs. Chromate resistance mediated by bacterial plasmids appears to be due to conversion of Cr(VI) to Cr(III), although the reducing agent and the responsible enzyme in this reduction system is not identified. Likewise, enzymes effective in metabolic detoxication in eucaryotic cells are unknown. It is of interest to determine how homocitric dehydrase of yeast acts in chromate detoxication.

Although it is expected that Cr(VI) interacts with protein, not many specific proteins or enzymes that are inactivated by chromate have been characterized. In this context, it is interesting that chromium, despite its plausible interaction with cysteine, does not bind to metallothioneins, which are small peptides rich in cysteine that bind various metals (Waalkes et al., 1984). It is also indicated that chromium does not induce metallothioneins (Onosaka et al., 1983).

It has become evident that Cr(VI) causes DNA-strand breaks, DNA-DNA, and DNA-protein crosslinks. There is also a claim that Cr(VI) induces "alkali sensitive sites" in addition to strand breaks. Nevertheless, it is certain that Cr(VI) interacts with DNA and causes damage that results in blocking of DNA synthesis leading to cell lethality. Even if the damage is subjected to repair, base substitutions and DNA rearrangements such as translocation will be inevitable. Our present knowledge suggests that at least some cancers are triggered by point mutations and/or translocations (Tabin et al., 1982; Reddy et al., 1982; Taparowsky et al., 1982; Leder et al., 1983) (but see Chapter 16). As shown in Figure 1, all aspects of chromium toxicity, cytotoxicity, mutagenicity, and carcinogenicity are explained by similar pathways. This scheme also suggests the importance of DNA-repair systems in cellular defense against chromate. In fact, a repair-deficient derivative of CHO cell line is more sensitive to chromate than the original cell line (Christie et al., 1984). In other words, repair of incurred damage appears to be an important cellular defense mechanism in chromate toxicity.

7. CONCLUDING REMARKS

It is evident that living organisms have evolved defense mechanisms against harsh environmental conditions. Thus, it is important to examine defense mechanisms of normal strains as well as resistant strains. In the case of chromium, an overall view of its effects on living organisms has not been obtained yet. However, it has become evident that it causes damages on DNA which lead to cytotoxicity, mutagenicity, and, presumably, carcinogenicity. In this respect, the roles of DNA-repair systems have to be investigated further. Of course, a better understanding of systems for chromate detoxication is also urgent.

Although genetics alone does not solve all questions about the biological effects of chromium, it provides experimental materials (mutants) and an alternative approach. Although genetic approaches are so far limited to microorganisms and cultured mammalian cells, its extension to other organisms will be needed for a better understanding of environmental and ecological effects of metals including chromium. In this effort, interdisciplinary approaches are indispensable.

REFERENCES

Aaseth, J., Alexander, J., and Norseth, T. (1982). "Uptake of ^{51}Cr-chromate by human erythrocytes—A role of glutathione." *Acta Pharmacol. Toxicol.* **50**, 310–315.

Arst, H.N., Jr., (1968). "Genetic analysis of the first steps of sulfate metabolism in *Aspergillus nidulans*." *Nature* **219**, 268–270.

Beyersmann, D., Koester, A., Buttner, B., and Flessel, P. (1984). "Model reactions of chromium compounds with mammalian and bacterial cells." *Toxicol. Environ. Chem.*, **8**, 279–286.

Bhattacharjee, N.K., Tucci, A.F., and Strasasman, M. (1968). "Accumulation of α-ketoglutaric acid in yeast mutants requiring lysine." *Arch. Biochem. Biophys.* **123**, 235–239.

Bianchi, V., Celotti, L., Lanfranchi, G., et al. (1983). "Genetic effects of chromium compounds." *Mutat. Res.* **117**, 279–300.

Bianchi, V., Debetto, P., and Zantedeschi, A. (1982). "Effects of hexavalent chromium on the adenocylate pool of hamster fibroblasts." *Toxicology* **25**, 19–30.

Bianchi, V., Zantedeschi, A., Montaldi, A., and Majone, F. (1984). "Trivalent chromium is neither cytotoxic nor mutagenic in permealized hamster fibroblasts." *Toxicol. Lett.* **23**, 51–59.

Bonatti, S., Meini, M., and Abbondandolo, A. (1976). "Genetic effects of potassium chromate in *Schizosaccharomyces pombe*." *Mutat. Res.* **38**, 147–149.

Bopp, L.H., Chakrabarty, A.N., and Ehrlich, H.L. (1983). "Chromate resistance plasmid in *Pseudomonas fluorescens*." *J. Bacteriol.* **155**, 1105–1109.

Breton, A., and Surdin-Kerjan, Y. (1977). "Sulfate uptake in *Saccharomyces cerevisiae*: Biochemical and genetic study." *J. Bacteriol.* **132**, 224–232.

Cantoni, O., and Costa, M. (1984). "Analysis of the induction of alkali sensitive sites in the DNA by chromate and other agents that induce single strand breaks." *Carcinogenesis* **5**, 1207–1209.

Casto, B.C., Meyers, J., and DiPaolo, J.A. (1979). "Enhancement of viral transformation for evaluation of the carcinogenic or mutagenic potential of inorganic metal salts." *Cancer Res.* **39**, 193–198.

Christie, N.T., Cantoni, O., Evans, et al. (1984). "Use of mammalian DNA repair-deficient mutants to assess the effects of toxic metal compounds on DNA." *Biochem. Pharm.* **33**, 1661–1670.

De Flora, S., Bianchi, V., and Levis, A.G. (1984). "Distinctive mechanisms for interaction of hexavalent chromium and trivalent chromium with DNA?" *Toxicol. Environ. Chem.* **8**, 287–294.

Efstathiou, J.D., and McKay, L.L. (1977). "Inorganic salt resistance associated with a lactose-fermenting plasmid in *Streptococcus* lactis." *J. Bacteriol.* **130**, 257–265.

Flessel, C.P. (1978). "Metals as mutagens." *Adv. Exp. Medi. Biol.* **91**, 117–128.

Flessel, C.P., Furst, A., and Radding, S.B. (1980). "A comparison of carcinogenic metals." *Metal Ions in Biological Systems.* **10**, 23–54.

Fradkin, A., Janoff, A., Lane, B.P., and Kuschner, M. (1975). "In vitro transformation of BHK21 cells grown in the presence of calcium chromate." *Cancer Res.* **35**, 1058–1063.

Gergely, A., and Sóvágó, I. (1979). "The coordination chemistry of L-cysteine and D-penicillamine." *Metal Ions in Biological Systems*. **9**, 77–102.

Halos, S. (1976) "Cysteine dependent mutants of *Saccharomyces cerevisiae*." Ph.D. thesis, University of California, Berkeley, CA.

Hojo, Y., Sugiura, Y., and Tanaka, H. (1977). "Chromium(III)-penicillamine complex: some properties and rapid chelation from chromium(VI) and penicillamine." *J. Inorg. Nucl. Chem.* **39**, 1859–1863.

Hopkins, L.L., Jr., Ransome-Kuti, O., and Majaj, A.S. (1968). "Improvement of impaired carbohydrate metabolism by chromium(III) in malnourished infants." *Am. J. Clin. Nutr.* **21**, 202–211.

Horitsu, H., Nishida, H., Kato, H., and Tomoyeda, M. (1978). "Isolation of potassium chromate-tolerant bacterium and chromate uptake by the bacterium." *Agric. Biol. Chem.* **42**, 2037–2043.

Horitsu, H., Futo, S., Ozawa, K., and Kawai, K. (1983). "Comparison of characteristics of hexavalent chromium-tolerant bacterium, *Pseudomonas ambigua* G-1, and its hexavalent chromium-sensitive mutant." *Agric. Biol. Chem.* **47**, 2907–2908.

Issaq, H.J. (1980). "The role of metals in tumor development and inhibition." *Metal Ions in Biological Systems* **10**, 55–93.

Karbonowska, H., Wiater, A., and Hulanicka, D. (1977). "Sulfate permease of *Escherichia coli* K12." *Acta Biochim. Pol.* **24**, 329–334.

Leder, P., Battey, J., Lenoir, G., et al. (1983). "Translocations among antibody genes in human cancer." *Science* **222**, 765–771.

Marzluf, G.A. (1970a). "Genetic and metabolic controls for sulfate metabolism in *Neurospora crassa:* Isolation and study of chromate-resistant and sulfate transport-negative mutants." *J. Bacteriol.* **102**, 716–721.

Marzluf, G.A. (1970b). "Genetic and biochemical studies of distinct sulfate permease species in different developmental stages of *Neurospora crassa*." *Arch. Biochem. Biophys.* **138**, 254–263.

McAuley, A., and Olatunji, M.A. (1977a). "Metal-ion oxidations in solution. Part XVIII. characterization, rates, and mechanism of formation of the intermediates in the oxidation of thiols by chromium(VI)." *Can. J. Chem.* **55**, 3328–3334.

McAuley, A., and Olatunji, M.A. (1977b). "Metal-ion oxidations in solution. Part XIX. Redox pathways in the oxidation of penicillamine and glutathione by chromate." *Can. J. Chem.* **55**, 3335–3340.

Mertz, W., Toepfer, E.W., Roginski, E.E., and Polansky, M.M. (1974). "Present knowledge of the role of chromium." *Fed. Proc.* **33**, 2275–2280.

Nishioka, H. (1975). "Mutagenic activities of metal compounds in bacteria." *Mutat. Res.* **31**, 185–189.

NRC (National Research Council, Committee on Medical and Biologic Effects of Environmental Pollutants) (1974). *Medical and Biologic Effects of Environmental Pollutants: Chromium*. National Academy of Sciences, Washington, D.C.

Novic, R.P., and Roth, C. (1968). "Plasmid-linked resistance to inorganic salts in *Staphylococcus aureus*." *J. Bacteriol.* **95**, 1335–1342.

Ono, B., and Weng, M. (1982a). "Chromium sensitive mutants of the yeast *Saccharomyces cerevisiae*." *Curr. Genet.* **5**, 215–220.

Ono, B., and Weng, M. (1982b). "Chromium resistant mutants of the yeast *Saccharomyces cerevisiae*." *Curr. Genet.* **6**, 71–77.

Ono, B., Andou, N., Weng, M., and Tong, K. (1986). "Effect of cysteine on chromate resistance in the yeast *Saccharomyces cerevisiae*." *Chem. Pharm. Bull.*, **34**, 229–234.

Ono, B., Suruga, T., Yamamoto, M., et al. (1984). "Cystathionine accumulation in *Saccharomyces cerevisiae*." *J. Bacteriol.* **158**, 860–865.

Onosaka, S., Yoshiya, S., Min, K.-S., et al. (1983). "The induced synthesis of metallothionein in various tissues of rat after injection of various metals." *Eisei Kagaku* **29**, 221–225 (in Japanese).

Pardee, A.B., Prestidge, L.S., Whipple, M.B., and Dreyfuss, J. (1966). "A binding site for sulfate and its relation to sulfate transport into *Salmonella typhimurium.*" *J. Biol. Chem.* **241**, 3962–3969.

Paschin, Y.V., Kozachenko, V.I., and Sal'nikova, L.E. (1983). "Differential mutagenic response at the HGPRT locus in V-79 and CHO cells after treatment with chromate." *Mutat. Res.* **122**, 361–365.

Paszewski, A. (1976). "Hyper-repressible operator-type mutant in sulphate permease gene of *Aspergillus nidulans.*" *Nature* **259**, 337–338.

Petrilli, F.L., and DeFlora, S. (1977). "Toxicity and mutagenicity of hexavalent chromium on *Salmonella typhimurium.*" *Appl. Environ. Microbiol.* **33**, 805–809.

Rabenstein, D.L., Guevremont, R., and Evans, C.A. (1979). "Glutathione and its metal complexes." *Metal Ions in Biological Systems* **9**, 103–141.

Reddy, E.P., Reynolds, R.K., Santos, E., and Barbacid, M. (1982). "A point mutation is responsible for the acquisition of transforming properties by the T24 human bladder carcinoma oncogene." *Nature* **300**, 149–152.

Roberts, K.R., and Marzluf, G.A. (1971). "The specific interaction of chromate with the dual sulfate permease systems of *Neurospora crassa.*" *Arch. Biochem. Biophys.* **142**, 651–659.

Robison, S.H., Cantoni, O., and Costa, M. (1984). "Analysis of metal-induced DNA-repair replication in mammalian cells." *Mut. Res.* **131**, 173–181.

Silver, S. (1983). "Bacterial interactions with mineral cations and anions: good ions and bad." In P. Webroek, and E.W. de Jong, Eds., *Biomineralization and Biological Metal Accumulation.* D. Reidel, Netherlands, pp. 439–457.

Sirover, M.A., and Loeb, L.A. (1976). "Infidelity of DNA synthesis in vitro: Screening for potential metal mutagenesis or carcinogens." *Science* **194**, 1434–1436.

Sugiura, Y., Hojo, Y., and Tanaka, H. (1972). "Studies on the sulfate-containing chelating agents. XXXIII. Interaction of penicillamine and its related compounds with chromium ion and hemoglobin-bound chromium." *Chem. Pharm. Bull.* **20**, 1362–1367.

Summers, A.O., Jacoby, G.A. (1978). "Plasmid-mediated resistance to boron and chromium compounds in *Pseudomonas aeruginosa.*" *Antimicrob. Agents Chemother.* **13**, 637–640.

Sunderman, F.W., Jr., (1978). "Carcinogenic effects of metals." *Fed. Proc.* **37**, 40–46.

Susa, N. (1984). "Antagonistic effect of DL-penicillamine of chromium toxicity in HeLa cells." *Jpn. J. Vet. Sci.* **46**, 89–98.

Tabin, C.J., Bradley, S.M., Bergmann, C.I., et al. (1982). "Mechanism of activation of a human oncogene." *Nature* **300**, 143–149.

Taparowsky, E., Suard, Y., Fasano, O., et al. (1982). "Activation of the T24 bladder carcinoma transforming gene is linked to a single amino acid change." *Nature* **300**, 762–763.

Tsapakos, M.J., Hampton, T.H., and Werrerhahn, K.E. (1983). "Chromium(VI)-induced DNA letions and chromium distribution in rat kidney, liver, and lung." *Cancer Res.* **43**, 5662–5667.

Tsuda, H., and Kato, K. (1976). "Chromosomal aberrations and morphological transformation in hamster embryonic cells treated with potassium dichromate in vivo." *Mutt. Res.* **46**, 87–94.

Venitt, S., and Levy, L.S. (1974). "Mutagenicity of chromates in bacteria and its relevance to chromate carcinogenesis." *Nature* **250**, 493–495.

Waalkes, M.P., Harvey, M.J., and Klaassen, C.D. (1984). "Relative in vitro affinity of hepatic metallothionein for metals." *Toxicol. Lett.* **20**, 33–39.

Wada, O., Wu, G.Y., Yamamoto, A., et al. (1983). "Purification and chromium-excretory function of low-molecular-weight, chromium-binding substances from dog liver." *Environ. Res.* **32**, 228–239.

Yamagata, S., Takeshima, K., and Naiki, N. (1975). "O-acetylserine and O-acetylhomoserine sulfhydrylase of yeast; studies with methionine auxotrophs." *J. Biochem.* **77**, 1029–1036.

15

THE TOXICITY
OF CHROMIUM TO FISH

D.A. Holdway

Office of the Supervising Scientist
Alligator Rivers Region Research Institute
Jabiru, Northern Territory, Australia

1. INTRODUCTION: BIOCHEMICAL MODE OF ACTION

Of all the metals, chromium has truly unique characteristics as a toxicant. Chromium apparently must be in the hexavalent and not the trivalent form in order to cross biological membranes readily (Gray and Sterling, 1950). However, hexavalent chromium is a strong oxidizing agent and once inside a biological system and upon contact with organic particulate matter or various enzymes, it is generally reduced to the trivalent form (Grogan, 1958; Mertz, 1969). Thus, toxicological impact can result both from the action of Cr(VI) itself as an oxidizing agent and by Cr(III) which is capable of complexing with various organic compounds (Mertz, 1969; NRCC, 1976) and inhibiting several metallo-enzyme systems (Arillo et al., 1982).

Chromium is often grouped with other metals in toxicity studies, implying that chromium behaves toxicologically in a manner similar to most metals. In fact, this is not the case and Doudoroff and Katz (1953) placed Cr(VI) in a class of compounds that differed both chemically and toxicologically from most metal salts.

When fish are exposed to hexavalent chromium dissolved in water, the gills function as the main route for its uptake (Knoll and Fromm, 1960), a finding similar for other metals. However, for most metals including, copper, zinc, cadmium, and mercury, the gill tissue readily accumulates them so that gill metal residue levels far exceed those of other tissues (Sellers et al., 1975; Sangalang and Freeman, 1979; Bradley and Sprague, 1985). This rapid accumulation of metal in gill tissue is usually associated with structural damage to the gill and impaired respiration and osmoregulatory function, effects that have often been cited as the acute mechanism of heavy metal toxicity (Lewis and Lewis, 1971; Burton et al., 1972).

Chromium(VI), however, appears to pass readily through the gill membrane and to accumulate rapidly in various tissues and organs at higher levels than in the gills, including opercular bone, spleen, kidney, gall bladder, gastrointestinal tract and brain (Buhler et al., 1977). Consequently, chromium can elicit its toxic effects internally well away from its site of entry at the gills.

Histopathological studies have provided direct evidence of physical injury as a consequence of acute Cr(VI) exposure, including damage to the gills (Strik et al., 1975; Van der Putte et al., 1981), stomach (Van der Putte et al., 1981), kidney (Fromm and Schiffman, 1958; Van der Putte et al., 1981), and intestine (Fromm and Schiffman, 1958).

High levels of chromium have caused increased hematocrit and hemoglobin levels in fish (Schiffman and Fromm, 1959; Strik et al., 1975; Van der Putte et al., 1982) and Strik et al. (1975) speculated that these biochemical alterations were a consequence of the observed histological damage to the gills. High concentrations of Cr(VI) have been observed to impair respiratory and osmo-regulation function (Van der Putte et al., 1981) and Arillo et al. (1982) found that chronic exposure of fish to 0.2 mg/L^{-1} of Cr(VI) for six months altered gill mucus. Thus, gill injury and the resultant respiratory and osmoregulatory impairment appear to be one aspect of chromium toxicity, one that is nonspecific and similar to the toxicities of a number of heavy metals.

Although, at present, the complete mode of action of chromium toxicity to fish is unknown, one known effect specific to chromium is on glucose metabolism. Under normal conditions, Cr(III) is thought to be an essential trace element in mammals, being a portion of the glucose tolerance factor molecule that is a cofactor of insulin and involved in the control of glucidic metabolism. Since chromium (added as VI to water but presumably converted to III inside the fish) has been shown to impair glucose transport in trout intestine (Stokes and Fromm, 1965), increase glucose levels in blood (Strik et al., 1975; Van der Putte et al., 1982), and decrease glucose levels in the liver (Arillo et al., 1982), it is likely that chromium(III) deleteriously alters the complex biochemical pathways connecting proteases (which being metallo-enzymes are inhibited by high concentrations of metals) and gluconeogenesis. Thus, for chromium, as for most other seemingly essential trace elements, a little is beneficial, while too much is deleterious. Given that chromium's normal biological requirement revolves around glucose metabolism, it seems logical that its toxic mechanisms involve the same systems, a hypothesis seemingly supported by the limited data available.

2. ACUTE TOXICITY OF CHROMIUM TO FISH

Tables 2 and 3 summarize the levels of Cr(VI) that were found in the literature to have had acute toxicity to fish; Table 2 dealing with hexavalent chromium and Table 3 with trivalent chromium.

The tabular data is arranged, wherever possible, by family, with the coldwater fishes appearing before the warmwater ones, and freshwater fish before marine fish. Chromium compounds used in the acute and chronic toxicity tables are given in Table 1.

Table 1   Chromium Compounds Tested for Toxicity to Fish

Valence (Oxidation State)	Chemical Formula	Compound Name
Cr^{+3}	$CrCl_3$	Chromic chloride
Cr^{+3}	$Cr(NO_3)_3$	Chromic nitrate
Cr^{+3}	$CrK(SO_4)_2$	Chromic potassium sulfate
Cr^{+3}	$Cr_2(SO_4)_3$	Chromic sulfate
Cr^{+6}	K_2CrO_4	Potassium chromate
Cr^{+6}	$K_2Cr_2O_7$	Potassium dichromate
Cr^{+6}	Na_2CrO_4	Sodium chromate
Cr^{+6}	$Na_2Cr_2O_7$	Sodium dichromate
Cr^{+6}	CrO_3	Chromium trioxide

Table 2 Acute Toxicity of Hexavalent Chromium to Fish

Species	Life Stage or Size	Test Conditions[a]	Effective Concentration (mg/L Chromium)	Measured Effect[b]	pH	Temp. (°C)	Hardness (mg/L)	Reference	Comments
			Freshwater Fish						
SALMONIDS									
Rainbow trout (*Salmo gairdneri*)	Fry	SR,N	14	96 hours—no effect	7.0	10	20	Garton, 1973	
Rainbow trout	10–17 cm	S,M	100	24-hour LC50	8.5–8.8	14–15	334	Schiffman and Fromm, 1959	
Rainbow trout	14 months	F,M	69	96-hour LC50	7.0–8.0	12	45	Benoit, 1976	
Brook trout (*Salvelinus fontinalis*)	5 months	F,M	59	96-hour LC50	7.0–8.0	12	45	Benoit, 1976	
MINNOWS AND CARPS									
Fathead minnow (*Pimephales promelas*)	19 mm	F,M	33.2	96-hour LC50	7.8	25	220	Broderius and Smith, 1979	
Fathead minnow	2 in	S,N	17.6	96-hour LC50	7.5	25	20	Pickering and Henderson, 1966	[c]
Fathead minnow	2 in	S,N	27.3	96-hour LC50	8.2	25	360	Pickering and Henderson, 1966	[c]
Fathead minnow	2 in	S,N	45.6	96-hour LC50	7.5	25	20	Pickering and Henderson, 1966	[d]
Fathead minnow	40–68 mm	F,M	52	96-hour LC50	7.6–8.2	15	Hard well water	Ruesink and Smith, 1975	
Fathead minnow	40–68 mm	F,M	37	96-hour LC50	8.0–8.3	25	Hard well water	Ruesink and Smith, 1975	
Fathead minnow	40–68 mm	F,M	151 hours at 38	LT50	7.6–8.7	15	Hard well water	Ruesink and Smith, 1975	
			118 hours at 30	LT50	8.0–8.3	25	Hard well water	Ruesink and Smith, 1975	
Fathead minnow	Juvenile	F,M	26	96-hour LC50	7.5–7.8	24–26	200–230	Adelman and Smith, 1976	[e]

Species	Size	Method	Value	Test	pH	Temp.	Hardness	Reference	
Fathead minnow	Juvenile	F,M	48	96-hour LC50	7.5–7.8	24–26	200–230	Adelman and Smith, 1976	[f]
Fathead minnow	1 g	F,M	36.9	96-hour LC50	7.5–8.2	25	209	Pickering, 1980a	
Fathead minnow	1 g	S,N	36.2	96-hour LC50	7.5–8.2	25	209	Pickering, 1980a	
Goldfish (*Carassius auratus*)	Juvenile	F,M	120	96-hour LC50	7.5–7.8	24–26	200–222	Adelman and Smith, 1976	
Goldfish	2 in	S,N	37.5	96-hour LC50	7.5	25	20	Pickering and Henderson, 1966	
Common carp (*Cyprinus carpio*)	—	F,M	108	96-hour LC50	5.9–6.5	20–24	4.2–5.1	Nishihara et al., 1985	
Channelfish (*Nurio denricus*)	5 cm	S,M	28.9	96-hour LC50	6.1–6.3	—	4–5	Abbasi and Soni, 1984	
			1.7	480-hour LC50					
SUNFISHES									
Bluegill sunfish (*Leopomis macrochirus*)	2 in	S,N	118	96-hour LC50	7.5	25	20	Pickering and Henderson, 1966	
Bluegill sunfish	2 in	S,N	133	96-hour LC50	8.2	25	360	Pickering and Henderson, 1966	
Bluegill sunfish	5–9 cm	S,N	170	96-hour LC50	7.6–8.7	20	10	Trama and Benoit, 1960	
Bluegill sunfish	—	S,N	113	96-hour LC50	6.5	18	10	Cairns and Scheier, 1959	[g]
Bluegill sunfish	—	S,N	135	96-hour LC50	6.5	18	110	Cairns and Scheier, 1959	[g]
Bluegill sunfish	—	S,N	113	96-hour LC50	6.5	30	10	Cairns and Scheier, 1959	[g]
Bluegill sunfish	—	S,N	130	96-hour LC50	6.5	30	110	Cairns and Scheier, 1959	[g]
Largemouth bass (*Micropterus salmoides*)	10.6 cm	S,M	195	48-hour LC50	8.5–8.8	20–21	334	Fromm and Schiffman, 1958	
Largemouth bass	Embryo—larvae	SR,N	1.17	Eight-day LC50	7.2–7.8	19–22	99.5	Birge et al., 1978	[h]

Continued

373

Table 2 (*Continued*)

Species	Life Stage or Size	Test Conditions[a]	Effective Concentration (mg/L) Chromium	Measured Effect[b]	pH	Temp. (°C)	Hardness (mg/L)	Reference	Comments
TEMPERATE BASSES									
Striped bass (*Morone saxatilis*)	Larvae (One week)	S,N	100	96-hour LC50	—	21	—	Hughes, 1971	
Striped bass	Juvenile (One month)	S,N	75	96-hour LC50	—	21	—	Hughes, 1971	
LIVEBEARERS									
Mosquito fish (*Gambusia affinis*)	—	S,N	400	96-hour LC50	7.6–8.1	17–21	Medium (alkalinity <100 ppm)	Wallen et al., 1957	i
Mosquito fish	—	S,N	280	96-hour LC50	5.4–6.7	21–23	Medium (alkalinity <100 ppm)	Wallen et al., 1957	j
Mosquito fish	—	S,N	420	96-hour LC50	7.7–8.6	20–22	Medium (alkalinity <100 ppm)	Wallen et al., 1957	k
Mosquito fish	—	S,N	264	96-hour LC50	7.7–8.6	20–22	Medium (alkalinity <100 ppm)	Wallen et al., 1957	l
Saltwater Fish									
MULLETS							Salinity (ppt)		
Yellow-eye mullet (*Aldrichetta forsteri*)	Juveniles	S,M	24	96-hour LC50	7.9	19.5	34.5	Negilski, 1976	
Yellow-eye mullet	Juveniles	F,M	31.2	96-hour LC50	8.0	18.5–18.7	34.5	Negilski, 1976	

SILVERSIDES								
Small-mouth hardyhead (Atherinasoma microstoma)	Juveniles	F,M	168-hour LC50	31.6–40.2	7.9	19.5	34.5	Negilski, 1976
KILLIFISHES								
Mummichug (Fundulus heteroclitus)	4.9 cm	S,N	96-hour LC50	91	7.8	20	20	Eisler and Hennekey, 1977
			168-hour LC50	44				
LEFT-EYE FLOUNDERS								
Speckled sanddab (Citharichthys stigmaeus)	Juvenile	S,M	96-hour LC50	30	—	13	33.5	Mearns et al, 1976
	Juvenile	F,M	21-day LC50	5.0	—	13	33.5	Mearns et al, 1976

[a] F = flow-through bioassay; S = static bioassay; SR = static-replacement bioassay; M = measured concentrations during the bioassay; N = nominal concentrations (not measured); [b] ILC50 (incipient LC50) = concentration of toxicant lethal to 50% of test organisms after an exposure period sufficiently long for acute lethal action to have essentially ceased. The asymptote of the toxicity curve approximates the value of the ILC50; LC50 = concentration of a substance which is lethal to 50% of the test organisms exposed for a defined time period; NOEC (no observed effect concentration) = the highest tested concentration of a substance that has no statistically significant adverse effect on the test population relative to controls in a toxicity test. When derived from a life cycle or partial life cycle test, it is numerically equivalent to the lower limit of the MATC; LOEC (lowest observed effect concentration) = the lowest tested concentration of a substance that has a significantly adverse effect on the exposed test population relative to controls in a toxicity test. When derived from a life cycle or partial life cycle test, it is numerically equivalent to the upper limit of the MATC; MATC (maximum acceptable toxicant concentration) = a hypothetical concentration representing a threshold of toxicity below which no adverse effects occur and above which toxic action begins. The MATC theoretically lies within a range bounded by the NOEC at the lower limit and the LOEC at the upper limit, assuming that the NOEC and LOEC were both derived from a life cycle (full chronic) or partial life cycle (partial chronic) test involving quantitative survival, growth, and reproductive toxicity data. This may be symbolically represented as NOEC < MATC < LOEC; LT50 = time at which 50% of the organisms have died at a given concentration; [c] Exposed to potassium dichromate, $K_2Cr_2O_7$; [d] Exposed to potassium chromate, K_2CrO_4; [e] Desired chromium concentrations were obtained immediately. [f] Desired chromium concentrations were obtained after three to 4 hours; [g] Hardness estimated. [h] Mean hardness over range 94–105; [i] Chromium added as K_2CrO_4; [j] Chromium added as $K_2Cr_2O_7$; [k] Chromium added as Na_2CrO_4; [l] Chromium added as $Na_2Cr_2O_7$.

Table 3 Acute Toxicity of Trivalent Chromium to Fish

Species	Life Stage or Size	Test Conditions[a]	Effective Concentration (mg/L Chromium)	Measured Effect[b]	pH	Temp. (°C)	Hardness (mg/L)	Reference	Comments
				Freshwater Fish					
SALMONIDS									
Steelhead trout (*Salmo gairdneri*)	Juvenile (Two months)	F,M	4.4	96-hour LC50	6.6–7.4	12.5	25	Stevens and Chapman, 1984	
Rainbow trout (*Salmo gairdneri*)	Juvenile (Two months)	F,M	24.1	96-hour LC50	8.1–8.3	—	—	Hale, 1977	
Rainbow trout	"Young"	S	11.2	96-hour LC50	—	—	—	Bills et al., 1977	c
MINNOWS AND CARPS									
Fathead minnow (*Pimephales promelas*)	1 g	F,M	28	96-hour LC50	7.6	20–25	203	Pickering, 1980b	
Fathead minnow	2 in	S,N	5.1	96-hour LC50	7.5	25	20	Pickering and Henderson, 1966	
Fathead minnow	2 in	S,N	67.4	96-hour LC50	8.2	25	360	Pickering and Henderson, 1966	
Goldfish (*Carassius auratus*)	2 in	S,N	4.1	96-hour LC50	7.5	25	20	Pickering and Henderson, 1966	
Common carp (*Cyprinus carpio*)	Adult	S,M	14.3	96-hour LC50	8.0	28	55	Rehwoldt et al., 1972	d
SUNFISHES									
Bluegill sunfish (*Lepomis macrochirus*)	2 in	S,N	7.5	96-hour LC50	7.5	25	20	Pickering and Henderson, 1966	
Bluegill sunfish	2 in	S,N	71.9	96-hour LC50	8.2	25	360	Pickering and Henderson, 1966	

							Salinity (ppt)		
Pumpkinseed sunfish (*Lepomis gibbosus*)	Adult	S,M	17.0	96-hour LC50	8.0	28	55	Rehwoldt et al., 1972	[d]
TEMPERATE BASSES									
Striped bass (*Morone saxatilis*)	Adult	S,M	17.7	96-hour LC50	8.0	28	55	Rehwoldt et al., 1972	[d]
White perch (*Morone americanus*)	Adult	S,M	14.4	96-hour LC50	8.0	28	55	Rehwoldt et al., 1972	[d]
LIVEBEARERS									
Guppy (*Poecilia reticulata*)	Adults	S,N	3.3	96-hour LC50	7.5	25	20	Pickering and Henderson, 1966	
FRESHWATER EELS									
American eel (*Anguilla rostrata*)	Juveniles	S,M	13.9	96-hour LC50	8.0	28	55	Rehwoldt et al., 1972	[d]
KILLIFISHES									
Banded killifish (*Fundulus diaphanus*)	Adults	S,M	16.9	96-hour LC50	8.0	28	55	Rehwoldt et al., 1972	[d]
Saltwater Fish									
MULLETS									
Yellow-eye mullet (*Aldrichetta forsteri*)	Juveniles	S,M	53.0	96-hour LC50	7.9	19.5	34.5	Negilski, 1976	[e]

[a] F = flow-through bioassay; S = static bioassay; SR = static-replacement bioassay; M = measured concentrations during the bioassay; N = nominal concentrations (not measured); [b] ILC50 (incipient LC50) = concentration of toxicant lethal to 50% of test organisms after an exposure period sufficiently long for acute lethal action to have essentially ceased. The asymptote of the toxicity curve approximates the value of the ILC50; LC50 = concentration of a substance which is lethal to 50% of the test organisms exposed for a defined time period; NOEC (no observed effect concentration) = the highest tested concentration of a substance that has no statistically significant adverse effect on the test population relative to controls in a toxicity test. When derived from a life cycle or partial life cycle test, it is numerically equivalent to the lower limit of the MATC; LOEC (lowest observed effect concentration) = the lowest tested concentration of a substance that has a significantly adverse effect on the exposed test poulation relative to controls in a toxicity test. When derived from a life cycle or partial life cycle test, it is numerically equivalent to the upper limit of the MATC; MATC (maximum acceptable toxicant concentration) = a hypothetical concentration representing a threshold of toxicity below which no adverse effects occur and above which toxic action begins. The MATC theoretically lies within a range bounded by the NOEC at the lower limit and the LOEC at the upper limit, assuming that the NOEC and LOEC were both derived from a life cycle (full chronic) or partial life cycle (partial chronic) test involving quantitative survival, growth, and reproductive toxicity data. This may be symbolically represented as NOEC < MATC < LOEC; LT50 = time at which 50% of the organisms have died at a given concentration; [c] No details of methods available; [d] Fish obtained directly from Hudson River; [e] Chromic nitrate concentrations questionable due to precipitation.

Acute toxicity was greatly affected by only a few important modifying factors. The most important appear to be pH, species, life stage, and body size. Temperature, hardness, and salinity are also modifying factors, but to a much lesser extent (see Section 4).

Generally, trivalent chromium appears to be more acutely toxic to fish than hexavalent chromium. The mean 96-hour LC50 for trivalent chromium, pooling data for all species and water conditions, was 22.0 mg/L (S.E. = 5.21; n = 17), significantly lower than 100.7 mg/L (S.E. = 17.36; n = 34) for hexavalent chromium as determined by standard error of the difference (Sprague and Fogels, 1977). Thus, although hexavalent chromium is the more likely species to be present in natural waters in dissolved form, trivalent chromium is more acutely toxic by about a factor of five times and, owing to its stability, should not be ignored with respect to establishing water quality criteria even if its presence in dissolved form is only transitory in nature.

3. CHRONIC AND SUBLETHAL TOXICITY OF CHROMIUM TO FISH

The chronic and/or sublethal effects of hexavalent and trivalent chromium to fish include histopathological damage, altered blood parameters such as hematocrit, serum protein levels, and blood glucose levels, decreased enzyme activity, and impaired respiratory and locomotory activities, all of which were previously described in Section 1. Levels of no observed effect (NOEC) and lowest observed effect (LOEC) concentrations for hexavalent and trivalent chromium are given in Tables 4 and 5. Once again, trivalent chromium appears to be more toxic to fish than hexavalent chromium. The mean LOEC for trivalent chromium calculated from the reviewed literature was 0.119 mg/L (S.E. = 0.071; n = 5), significantly lower by a factor of more than 15 times than the mean LOEC calculated for hexavalent chromium of 1.828 mg/L (S.E. = 0.662; n = 11) as determined by the standard error of the difference.

The recommended "safe" levels or maximum acceptable toxicant concentrations (MATC) for hexavalent and trivalent chromium are given in Tables 6 and 7. Depending on the species and study type, "safe" concentration ranges for hexavalent chromium varied from 0.05 to 3.95 mg/L chromium, while "safe" ranges for trivalent chromium varied from 0.000 to 0.380 mg/L chromium, a difference of over one order of magnitude.

Generally, reproduction and larval survival appear to be the most sensitive indicators of chronic chromium toxicity, a finding that supports Woltering's (1984) contention that these are the most sensitive indicators of chronic toxicity in fish. As well, the coldwater salmonids appear to be the most sensitive family of fish to chromium toxicity, while the warmwater minnows, carps, and livebearers appears to be the most tolerant families of fish to chromium toxicity, both acute and chronic.

4. MODIFYING FACTORS OF TOXICITY

Both abiotic and biotic factors have been found to significantly modify the toxicity of chromium or chromium complexes to fish. The major abiotic factors appear to be, in order of relative importance, pH, hardness, temperature, and salinity in marine systems, while the major biotic factors are fish species, developmental life stage, and body size. Other factors include interaction of chromium with other toxicants and preexposure to low levels of chromium resulting in acclimation and increased tolerance to chromium.

4.1   Abiotic Factors

4.1.1   pH

Van der Putte et al. (1981) showed that decreasing the pH of exposure water from 7.8 to 6.5 increased the toxicity of hexavalent chromium by a factor of 3.6 times for 0.2 g trout (96-hour LC50s of 12.2 and 3.4 mg/L, respectively), and by a factor of 3.2 times for 25 g fish (respective 96-hour LC50s of 65.5 and 20.2 mg/L of chromium). At intermediate fish sizes and pH, chromium toxicity was also intermediate with overall toxicity at pH 6.5, some 3.5 times greater than at pH 7.8.

In another study of the effects of pH on chromium toxicity to young rainbow trout, Hogendoorn-Roozemond et al. (1977) found that acute hexavalent chromium was 50 to 200 times greater at pH 6.4–7.2 than at pH 7.8–8.0 (48-, 72-, 96-, and 120-hour LC50s ranging from 0.22 to 0.57 mg/L at the lower pH values compared to 28 to 58 mg/L at the higher pH values).

Pickering and Henderson (1966) ran a series of experiments using several warmwater fish species that studied the toxicity of trivalent chromium (added as chromium potassium sulphate) and hexavalent chromium (added as potassium dichromate or as potassium chromate) in two water types, one soft with a lower pH (total hardness 20 mg/L as $CaCO_3$; pH 7.5) and one hard with a higher pH (total hardness 360 mg/L as $CaCO_3$; pH 8.2). The chromium salts generally reduced the pH of the higher concentration static tests, although the actual changes were not given. Trivalent chromium toxicity was affected by the different water types to a much greater degree than hexavalent chromium. Acute toxicity of trivalent chromium was increased 13.2 times for fathead minnows and 11.4 times for bluegill sunfish exposed to soft, lower pH water compared to hard, higher pH water (96-hour LC50s were 5.07 to 67.4 mg/L of chromium for fathead minnows and 7.46 to 71.9 mg/L of chromium for bluegill sunfish, respectively). Comparable increases in toxicity were much reduced for hexavalent chromium, being 1.6 and 1.1 times, respectively (96-hour LC50s of 17.6 to 27.3 mg/L of chromium for fathead minnows and 118 to 133 mg/L of chromium for bluegill sunfish). However, these differences are a combination of hardness and pH modifying factors, and thus not necessarily indicative of either factor's effect in isolation.

Table 4 Chronic or Sublethal Toxicity of Hexavalent Chromium to Fish

Species	Life Stage or Size	Test Conditions[a]	Effective Concentration (mg/L Chromium)	Measured Effect[b]	pH	Temp. (°C)	Hardness (mg/L)	Reference	Comments
				Freshwater Fish					
SALMONIDS									
Rainbow trout (*Salmo gairdneri*)	Eggs	F,M	6.1	Reduced hatchability	6.7–7.0	10	34	Sauter et al., 1976	
Rainbow trout	Eggs and fry	F,M	0.05	NOEC—survival, growth over 60 days	6.7–7.0	10	33.4	Sauter et al., 1976	
Rainbow trout	Fry	F,M	0.11	LOEC—60-day growth reduced	6.7–7.0	10	33.4	Sauter et al., 1976	
Rainbow trout	Fry	F,M	0.17	Mortality	7.6–8.2	13–19	70	Olson and Foster, 1956	c
Rainbow trout	Fingerling	F,M	0.08	Mortality	7.6–8.2	13–19	70	Olson and Foster, 1956	c
Rainbow trout	Fingerling	F,M	0.01	Reduced growth	7.6–8.2	13–19	70	Olson and Foster, 1956	c
Rainbow trout	Embryo—larvae	SR,N	0.18	28-day LC50	7.2–7.8	12–13	93–105	Birge et al., 1978	
Rainbow trout	Alevin—juvenile	F,M	0.34	Three months—100% mortality	7.0–8.0	7–15	45	Benoit, 1976	
Rainbow trout	Alevin—juvenile	F,M	0.20	Eight months—reduced growth	7.0–8.0	7–15	45	Benoit, 1976	
Brook trout (*Salvelinus fontinalis*)	Alevin—juvenile	F,M	0.76	100% mortality—three months	7.0–8.0	7–15	45	Benoit, 1976	
Brook trout	Alevin—juvenile	F,M	0.02	NOEC—eight months	7.0–8.0	7–15	45	Benoit, 1976	
Brook trout	Second generation fingerling	F,M	0.35	LOEC—22% mortality—three months	7.0–8.0	9	45	Benoit, 1976	
Lake trout (*Salvelinus namaycush*)	Eggs	F,M	24.4	Reduced hatchability	6.8–7.1	10	33.4	Sauter et al., 1976	d

Species	Life stage		Conc.	Effect	pH	Temp.		Reference	
Lake trout	Fry	F,M	0.11	NOEC—two-month growth	6.8–7.1	10	33.4	Sauter et al., 1976	
Lake trout	Fry	F,M	0.19	LOEC—reduced weight—2 months	6.8–7.1	10	33.4	Sauter et al., 1976	
Chinook salmon (*Oncorhynchus tshawytscha*)	Fry	F,N	0.2	Growth reduction—two weeks	7.7–8.0	8.3–16.1	70	Olson, 1957	[e]
Chinook salmon	Fry	F,N	0.2	12-week LC50	7.7–8.0	8.3–16.1	70	Olson, 1957	[e]
Chinook salmon	Fingerling	F,M	0.02	Reduced growth	7.6–8.2	3.5–13.5	70	Olson and Foster, 1956	[e]
Chinook salmon	Fingerling	F,M	0.08	Increased mortality	7.6–8.2	3.5–13.5	70	Olson and Foster, 1956	[e]
Chinook salmon	Fry	F,M	0.18	Increased mortality	7.6–8.2	3.5–13.5	70	Olson and Foster, 1956	
Coho salmon (*Oncorhynchus kisutch*)	Alevins—9–16 cm	F,M	0.48	Mortality after two weeks in freshwater followed by salinity shock	6.6	13	60	Sugatt, 1980	[f]
Coho salmon	Alevins—9–16 cm	F,M	0.23	Mortality after four weeks in freshwater followed by salinity shock	6.6	13	60	Sugatt, 1980	[f]
PIKES									
Northern pike (*Esox lucius*)	Eggs	F,M	1.98	NOEC hatchability	6.7–7.0	17	37.8	Sauter et al., 1976	
Northern pike	Fry	F,M	0.54	NOEC survival, growth	6.7–7.0	17	37.8	Sauter et al., 1976	
Northern pike	Fry	F,M	0.96	LOEC increased mortality—20 days	6.7–7.0	17	37.8	Sauter et al., 1976	
PERCHES									
Walleye (*Stizostedion vitreum*)	Eggs	F,M	2.17	NOEC hatchability	6.8–7.2	15	38.5	Sauter et al., 1976	

Continued

Table 4 (*Continued*)

Species	Life Stage or Size	Test Conditions[a]	Effective Concentration (mg/L Chromium)	Measured Effect[b]	pH	Temp. (°C)	Hardness (mg/L)	Reference	Comments
Walleye	Fry	F,M	2.17	NOEC survival, growth—30 days	6.8–7.2	15	38.5	Sauter et al., 1976	
MINNOWS AND CARPS									
Fathead minnow (*Pimephales promelas*)	Larvae	F,M	1.25	NOEC	8.2	25	Hard	Barron and Adelman, 1984	
Fathead minnow	Larvae	F,M	1.86	LOEC—28-day growth	8.2	25	Hard	Barron and Adelman, 1984	
Fathead minnow	Larvae	F,M	2.98	NOEC	8.2	25	Hard	Barron and Adelman, 1984	
Fathead minnow	Larvae	F,M	5.13	LOEC RNA/DNA ratio RNA/protein ratio	8.2	25	Hard	Barron and Adelman, 1984	
Fathead minnow	Larvae	F,M	3.00	NOEC	8.2	25	Hard	Barron and Adelman, 1984	
Fathead minnow	Larvae	F,M	5.59	LOEC—96-hour growth	8.2	25	Hard	Barron and Adelman, 1984	
Fathead minnow	Full life cycle	F,M	1.0	NOEC	7.5–8.2	16–24	209	Pickering, 1980a	g
			3.9	LOEC nine-week mortality					
Fathead minnow	1.9 cm	F,M	5.99	20-day LC50	7.8	25	220	Broderius and Smith, 1979	
Fathead minnow	1.9 cm	F,M	4.36	30-day LC50	7.8	25	220	Broderius and Smith, 1979	
SUNFISHES									
Bluegill sunfish (*Lepomis macrochirus*)	Eggs	F,M	1.12	NOEC hatchability	6.7–7.1	25	38.3	Sauter et al., 1976	

Organism	Life stage	Bioassay	Value	Effect	pH	Temp	Value	Reference
Bluegill sunfish	Fry	F,M	0.52	NOEC survival, growth	6.7–7.1	25	38.3	Sauter et al., 1976
Bluegill sunfish	Fry	F,M	1.12	LOEC weight reduced, 60 days	6.7–7.1	25	38.3	Sauter et al., 1976
Bluegill sunfish	Juveniles	S,N	2.4	Increased locomotor activity after two weeks	6.5	22	105	Ellgard et al., 1978
FRESHWATER CATFISH								
Channel catfish (*Ictalurus punctatus*)	Eggs	F,M	1.29	NOEC hatchability	7.0–7.4	22	36.2	Sauter et al., 1976
Channel catfish	Fry	F,M	0.15	NOEC survival, growth, 60 days	7.0–7.4	22	36.2	Sauter et al., 1976
Channel catfish	Fry	F,M	0.31	LOEC survival, growth, 30 days	7.0–7.4	22	36.2	Sauter et al., 1976
SUCKERS								
White sucker (*Catostomus commersoni*)	Eggs	F,M	1.975	NOEC hatchability	6.9–7.2	17	38.8	Sauter et al., 1976
White sucker	Larvae	F,M	0.290	NOEC growth, 60 days	6.9–7.2	17	38.8	Sauter et al., 1976
White sucker	Larvae	F,M	0.538	LOEC growth, 60 days	6.9–7.2	17	38.8	Sauter et al., 1976

[a]F = flow-through bioassay; S = static bioassay; SR = static-replacement bioassay; N = nominal concentrations (not measured); M = measured concentrations during the bioassay; N = nominal concentrations (not measured); [b]ILC50 (incipient LC50) = concentration of toxicant lethal to 50% of test organisms after an exposure period sufficiently long for acute lethal action to have essentially ceased. The asymptote of the toxicity curve approximates the value of the ILC50; LC50 = concentration of a substance which is lethal to 50% of the test organisms exposed for a defined time period; NOEC (no observed effect concentration) = the highest tested concentration of a substance that has no statistically significant adverse effect on the test population relative to controls in a toxicity test. When derived from a life cycle or partial life cycle test, it is numerically equivalent to the lower limit of the MATC; LOEC (lowest observed effect concentration) = the lowest tested concentration of a substance that has a significantly adverse effect on the exposed test poulation relative to controls in a toxicity test. When derived from a life cycle or partial life cycle test, it is numerically equivalent to the upper limit of the MATC; MATC (maximum acceptable toxicant concentration) = a hypothetical concentration representing a threshold of toxicity below which no adverse effects occur and above which toxic action begins. The MATC theoretically lies within a range bounded by the NOEC at the lower limit and the LOEC at the upper limit, assuming that the NOEC and LOEC were both derived from a life cycle (full chronic) or partial life cycle (partial chronic) test involving quantitative survival, growth, and reproductive toxicity data. This may be symbolically represented as NOEC <MATC <LOEC; LT50 = time at which 50% of the organisms have died at a given concentration; [c]Seasonal variation in temperature; [d]Growth was marginally reduced some 20% from controls; [e]Seasonal temperature fluctuation; [f]Effective concentration varied with salinity regime; [g]Survival most sensitive indicator; no significant reproductive effects but low control production.

383

Table 5 Chronic or Sublethal Toxicity of Hexavalent Chromium to Fish

Species	Life Stage or Size	Test Conditions[a]	Effective Concentration (mg/L Chromium)	Measured Effect[b]	pH	Temp. (°C)	Hardness (mg/L)	Reference	Comments
Freshwater Fish									
SALMONIDS									
Rainbow trout (*Salmo gairdneri*)	Ova	S,M	1.0	NOEC 40-minute gamete survival test	9.0	10	—	Billard and Roubaud, 1985	c
Rainbow trout	Sperm	S,M	0.005	LOEC 40-minute gamete survival test	9.0	10	—	Billard and Roubaud, 1985	d
Rainbow trout	Insemination	S,M	0.005	LOEC fertilization success after exposure during fertilization	9.0	10	—	Billard and Roubaud, 1985	e
Steelhead trout (*Salmo gairdneri*)	Eggs	F,M	0.089	NOEC fertilization	6.6-7.4	12.5	25	Stevens and Chapman, 1984	
Steelhead trout	Eggs	F,M	0.157	LOEC 50% reduction in fertilization	6.6-7.4	12.5	25	Stevens and Chapman, 1984	
Steelhead trout	Embryos	F,M	0.030	NOEC survival, growth	6.6-7.4	12.5	25	Stevens and Chapman, 1984	
Steelhead trout	Embryos	F,M	0.048	LOEC, embryo growth reduced	6.6-7.4	12.5	25	Stevens and Chapman, 1984	f

Chinook salmon (*Oncorhynchus tshawytscha*)	Fingerlings	F,N	0.2	NOEC, growth, survival	7.7–8.0	8.3–16.1	70	Olson, 1957
MINNOWS AND CARPS								
Fathead minnow (*Pimephales promelas*)	Eggs	F,M	1.4	NOEC hatchability	7.6	20–25	203	Pickering, 1980b
Fathead minnow	Larvae	F,M	0.18	NOEC survival, growth	7.6	20–25	203	Pickering, 1980b
Fathead minnow	Larvae	F,M	0.38	LOEC, second generation survival reduced at 30 days	7.6	20–25	203	Pickering, 1980b

[a]F = flow-through bioassay; S = static bioassay; SR = static-replacement bioassay; N = nominal concentrations during the bioassay; N = nominal concentrations (not measured); M = measured concentrations. [b]ILC50 (incipient LC50) = concentration of toxicant lethal to 50% of test organisms after an exposure period sufficiently long for acute lethal action to have essentially ceased. The asymptote of the toxicity curve approximates the value of the ILC50; LC50 = concentration of a substance which is lethal to 50% of the test organisms exposed for a defined time period; NOEC (no observed effect concentration) = the highest tested concentration of a substance that has no statistically significant adverse effect on the test population relative to controls in a toxicity test. When derived from a life cycle or partial life cycle test, it is numerically equivalent to the lower limit of the MATC; LOEC (lowest observed effect concentration) = the lowest tested concentration of a substance that has a significantly adverse effect on the exposed test polulation relative to controls in a toxicity test. When derived from a life cycle or partial life cycle test, it is numerically equivalent to the upper limit of the MATC; MATC (maximum acceptable toxicant concentration) = a hypothetical concentration representing a threshold of toxicity below which no adverse effects occur and above which toxic action begins. The MATC theoretically lies within a range bounded by the NOEC at the lower limit and the LOEC at the upper limit, assuming that the NOEC and LOEC were both derived from a life cycle (full chronic) or partial life cycle (partial chronic) test involving quantitative survival, growth, and reproductive toxicity data. This may be symbolically represented as NOEC < MATC < LOEC; LT50 = time at which 50% of the organisms have died at a given concentration; [c]Exposure done in solution of NaCl dissolved in distilled water with a Tris buffer; [d]Exposure employed a solution containing KCl to maintain spermatozoa immotility; [e]Dilutions of spermatozoa were varied from 10^{-1} to 10^{-3}; [f]Authors determined LOEC to be 0.156 mg/L based on "unacceptable toxic effect".

385

Table 6 Maximum Acceptable Toxicant Concentrations ("Safe") Determined from the Literature

Species	Study Type	MATC[a] (mg/L Chromium)	pH	Temp. (°C)	Hardness (mg/L)	Reference	Comment
Hexavalent Chromium							
Rainbow trout (*Salmo gairdneri*)	Egg-embryo	0.05–0.11	6.7–7.0	10	33.4	Sauter et al., 1976	
Brook trout (*Salveninus fontinalis*)	Full life cycle	0.02–0.35	7.0–8.0	7–15	45	Benoit, 1976	
Lake trout (*Salvelinus namaycush*)	Embryo-larval	0.11–0.19	6.8–7.1	10	33.4	Sauter et al., 1976	
Northern pike (*Esox lucius*)	Partial embryo-larval	0.54–0.96	6.7–7.0	17	37.8	Sauter et al., 1976	
Fathead minnow (*Pimephales promelas*)	Early life cycle	1.25–1.86	8.2	25	Hard	Barron and Adelman, 1984	b
Fathead minnow	Full life cycle	1.00–3.95	7.5–8.2	16–24	209	Pickering, 1980a	
Bluegill sunfish (*Lepomis macrochirus*)	Embryo-larval	0.52–1.12	6.7–7.1	25	38.3	Sauter et al., 1976	
Channel catfish (*Ictalurus punctatus*)	Embryo-larval	0.15–0.31	7.0–7.4	22	36.2	Sauter et al., 1976	

White sucker (*Catostomus commersoni*)	Embryo-larval	0.290–0.538	6.9–7.2	17	38.8	Sauter et al., 1976
Trivalent Chromium						
Rainbow trout (*Salmo gairdneri*)	Simultaneous gamete exposure during insemination, survival after 10 days postfertilization	0.000–0.005	9.0	10	—	Billard and Roubaud, 1985
Steelhead trout (*Salmo gairdneri*)	Egg-embryo	0.030–0.048	6.6–7.4	12.5	25	Stevens and Chapman, 1984
Fathead minnow (*Pimphales promelas*)	Full life cycle	0.180–0.380	7.6	20–25	203	Pickering, 1980b [c]

[a]MATC = maximum acceptable toxicant concentration (see Tables 1 to 5 for further explanation); [b]MATC questionable owing to cannibalism in controls and treatments preventing surviving and growth estimates beyond 20 days of age; [c]The author judged MATC to be between 0.75 and 1.4 mg/L chromium because survival of 2nd generation 30-day-old larvae exposed to 0.75 mg/L not significantly different from controls.

Table 7   Application Factors* Determined from Data Provided by the Studies in Table 6

Species	Applicator Factor	Reference
Hexavalent Chromium		
Fathead minnow (*Pimephales promelas*)	0.03–0.11	Pickering, 1980a
Brook trout (*Salvelinus fontinalis*)	0.003–0.006	Benoit, 1976
Trivalent Chromium		
Fathead minnow (*Pimephales promelas*)	0.006–0.014	Pickering, 1980b

*Application Factor = MATC/(96-hour LC50); see definition of MATC in Table 5.

The earliest study that suggested that pH was an important modifier of chromium toxicity to fish was that of Trama and Benoit (1960). They found that potassium dichromate solutions were more toxic to bluegill sunfish in static tests (96-hour LC50 of 113 mg/L) than were potassium chromate solutions (96-hour LC50 of 170 mg/L) and they hypothesized that the difference in toxicity between the two salts was due to the lower pH that resulted from the more acidic dichromate salt (pH 6.3–6.4 compared to pH 7.6–8.7 for chromate salt).

A chromate or dichromate salt readily forms either chromate (CrO_4^{-2}), hydrochromate ($HCrO_4^-$), or dichromate ($Cr_2O_7^{-2}$) ions in aqueous solution. Within the physiological pH range, chromate and hydrochromate ions predominate with hydrochromate ion formation favored at lower pH and chromate ion formation favored at higher pH. Trama and Benoit (1960) were among the first researchers to identify these species of chromium as the cause of toxicity in fish and theorize that the monovalent hydrochromate ion crossed biological membranes more easily than divalent chromium ions. Thus, they suggested that the higher toxicity of chromium solutions at lower pH is due to the higher concentration of hydrochromate ions, a theory that is now generally accepted.

4.1.2   Hardness

Very little research has been done to investigate the effect of hardness alone on chromium toxicity. Cairns and Scheier (1959) tested the effects of very soft (10.0 mg/L as $CaCO_3$) and medium hard water (108.9 mg/L as $CaCO_3$) on hexavalent chromium toxicity to bluegill sunfish. They found 96-hour LC50s of 326 and 378 mg/L of chromium at 18°C, 313 and 358 mg/L of chromium at 30°C in soft and hard water, respectively. However, although the trend was consistent, there was no significant difference between the toxicities in hard or soft water at either temperature at the 95% confidence level.

As previously described under the effects of pH, Pickering and Henderson (1966) studied the acute toxicity of trivalent and hexavalent chromium in soft and hard water to four warmwater species of fish. For both the fathead minnow and the bluegill sunfish, trivalent chromium was significantly more toxic in soft water than in hard water. Hexavalent chromium was not found to be significantly more toxic in soft water than in hard water, although after 96 hours there was a nonsignificant trend in this direction. This study, however, was confounded by different pH levels as well as different hardnesses, as was previously discussed.

Inspection of other data from the literature (Tables 2 to 5), using studies at single hardness levels, indicates that chromium toxicity (both hexavalent and trivalent) decreases with increasing hardness up to one order of magnitude.

4.1.3 Temperature

There are only a few investigations that have studied the effects of temperature on chromium toxicity to fish. From the few data available, temperature appears to be of minor importance as a modifying factor of chromium toxicity to fish, with most studies being inconclusive or showing a slight increase in toxicity in warmer water.

Cairns and Scheier (1959) failed to detect any temperature effect on the toxicity of hexavalent chromium to bluegill sunfish. For fish exposed to chromium in soft water, the 96-hour LC50s were identical at 18 and 30°C (113 mg/L chromium), while in hard water, the 96-hour LC50 was 135 mg/L chromium at 18°C and 130 mg/L chromium at 30°C.

In another study of the toxicity of hexavalent chromium to adult fathead minnows, Ruesink and Smith (1975) found the 96-hour LC50s decreased from 52 to 37 mg/L chromium when temperature increased from 15 to 25°C.

Differences between species in the effect of temperature on the acute toxicity of chromium to fish have been shown by Smith and Heath (1979). Goldfish showed the greatest sensitivity to chromium in warmer water with 24-hour LC50s at 5, 15, and 30°C being 309, 220, and 111 mg/L chromium, respectively. Corresponding values for golden shiner (*Notemigonus crysoleucus*) were 150, 111, and 102 mg/L chromium. Other species demonstrated no clear patterns of temperature effect with bluegill sunfish, channel catfish and rainbow trout showing slightly higher LC50s at higher temperatures.

4.1.4 Salinity

Hexavalent chromium was shown to be more toxic in seawater than in freshwater to coho salmon within the first four days of exposure, but by 11 days, there was no difference in toxicity with the 11-day LC50 ranging from 17.8 to 31.8 mg/L for both fresh and saltwater (Washington, State of, 1960). Trivalent chromium was found to be more toxic in freshwater than in saltwater, although this effect may have been due to a large shift in pH which occurred in the freshwater (pH fell from 7.9 to 5.6).

Juvenile coho salmon exposed to hexavalent chromium were found to have their ability to adapt to saline conditions impaired (Sugatt, 1980). Fish exposed in freshwater to 0.23 mg/L chromium for four weeks or to 0.5 mg/L chromium for two weeks had significantly reduced survival rates when subsequently transferred to 20 and 30°/oo saline waters. However, if salmon were only exposed to chromium in freshwater for one week prior to transferral to saline conditions, survival rate was much less affected.

In summary, hexavalent chromium toxicity is very similar in both fresh and saline water, while trivalent chromium may be more toxic in freshwater, although this may actually be a pH-dependent effect. Chronic chromium exposure appears to impair the ability of fish to adapt to saline waters.

4.2   Biotic Factors

The effects of species, developmental stage, and body size are all interrelated. Generally, smaller species and early life stages were the most sensitive to chromium toxicity, as shown in Tables 2 to 5. Within the same family of fish, large differences in toxicity were apparent. For example, whereas the livebearer mosquito fish was the least sensitive species to acute hexavalent chromium toxicity (96-hour LC50 400 mg/L chromium), another livebearer, the guppy, was among the species most acutely sensitive to trivalent chromium (96-hour LC50 3.3 mg/L chromium). While in acute toxicity tests, the coldwater salmonids did not appear any more sensitive to chromium toxicity than warmwater species, they were by far more sensitive to chromium in the chronic and sublethal studies.

Within life cycles, early embryo-larval survival generally appears to be one of the most sensitive life stages to chromium. However, male gamete survival and insemination were by far the most sensitive parameters measured (Billard and Roubaud, 1985), with measured detrimental effects to short-term (40-minute) trivalent chromium exposure being at least one order of magnitude lower than any other life stage measured; thus, the lowest observed effect concentration was only 0.005 mg/L chromium.

In a study of the effects of pH on hexavalent chromium toxicity to rainbow trout, Van der Putte et al. (1981) studied the effects of chromium in 0.2 and 25 g fish. They found a decrease in toxicity of at least a factor of five times between the two sizes of trout, with larger trout being more resistant to chromium toxicity.

Thus, in summary, large interspecific differences in apparent lethal susceptibilil/ to chromium exist even with the same families of fish. These differences cannot be predicted based on a coldwater or warmwater preference. Coldwater species appear, on the other hand, to be more sensitive to chronic exposure to chromium, particularly the gametes and embryo-larval stage. Larger fish are generally less sensitive to acute chromium exposure than smaller fish.

5.   BIOCONCENTRATION

A large number of accumulation studies have been performed on rainbow trout. Fromm and Stokes (1962) demonstrated that chromium uptake in trout exposed to concentrations of 0.01 and 0.0013 mg/L was a passive process with resulting tissue concentrations directly proportional to exposure concentrations. An asymptotic level appeared to have been reached after 10 days and bioconcentration factors of 1.27 and 1.34 were calculated.

Van der Putte and Part (1982) demonstrated, using isolated trout gills, that chromium and oxygen transfer are positively correlated, inferring that both cross the gill epithelium with the same mechanism by passive diffusion. This being the case, then chromium transfer would be 1.6 times more efficient at pH 6.5 than at 8.1.

Tenholder et al. (1978) and Van der Putte et al. (1981) both found that chromium was accumulated rapidly and depurated slowly. Tenholder et al. (1978) found that trout liver, kidney, and gills contained 13, 4, and 3%, respectively, of the absorbed hexavalent chromium independent of exposure concentration and that 34% of the total chromium was retained for one day after exposure ceased, while the remainder had a half-life of 26 days. They suggested a two-phase excretion process, a hypothesis supported by Buhler et al. (1977), in which there are rapid and slow turnover pools of chromium in the tissues, the former made up of the hexavalent form and the latter of the trivalent form.

Overall, fish muscle has been found to have very little capacity for accumulating chromium with a bioconcentration factor of about 3 being the highest found, giving a muscle content of 0.3 mg/kg (Calamari et al., 1982) and the lowest value giving a "background" bioconcentration value of 40 but a muscle content of only 0.01 mg/kg (Buhler et al., 1977). Gill, liver, and kidney tissues generally have been found to contain the highest chromium content, but even they seldom have bioconcentration factors exceeding 10 (Knoll and Fromm, 1960; Fromm and Stokes, 1962; Mathis and Cummings, 1973; McDermott et al., 1976; Buhler et al., 1977). In the study of Mathis and Cummings (1973), 10 freshwater fish species from the Illinois River were collected and they showed that omnivorous species had significantly higher muscle chromium concentrations than carnivorous species, implying that diet was responsible for observed differences.

The only exceptions to very low bioconcentration factors in fish appear to occur in bottom-dwelling fish, which feed from the sediments. Brown bullhead (*Ictalurus nebulosus*) from the Schuylkill River concentrated chromium in muscle and liver tissues 1,822 to 2,622 times the concentrations in the river water, but below the sediment concentrations (Reisinger, 1979). He suggested the source of chromium in the bullhead was through ingested macroinvertebrates. Common carp from the Danube River also had very high chromium levels in the gills (53 μg/g) relative to water concentrations (Rehwoldt et al., 1976), but once again lower than the chromium content of the river suspended solids (67 μg/g), suggesting gill contamination with solids.

In summary, bioconcentration of chromium in fish does not appear to be a serious problem with fish having a reasonable ability to depurate the chromium, and maximum or asymptotic levels seldom ever reach a bioconcentration factor of 10 in waters with elevated chromium levels. The only exceptions to the general rule appear to be direct chromium uptake via the diet or gills from contaminated sediments.

6.   BIOMAGNIFICATION

All available evidence suggests that chromium is not biomagnified. Whittle et al. (1977) stated that higher trophic levels contain reduced chromium levels, a statement supported by Mathis and Cummings(1973), who found that chromium concentrations in the Illinois River were highest in the sediments and in order of decreasing concentration: tubificial worms, clams, omnivorous fish, carnivorous fish, and river water. Similarly, in a study of several trophic levels in Narrangansett Bay, Phelps et al. (1975) found chromium concentrations again highest in sediments and in order of decreasing concentration: seston, plankton, worms, molluscs, fish, and bay water. These findings have been confirmed in other river systems (Reisinger, 1979; Friant, 1979) and constitute very strong evidence that chromium is not biomagnified.

7.   WATER QUALITY CRITERIA

More recent EPA criteria for protection of freshwater aquatic life for chromium (U.S. EPA, 1980) are much lower than the previous standards of 100 and 50 μg/L total chromium used for the previous decade (NAS, 1973; U.S. EPA, 1976). They are two-phased for hexavalent chromium, with a maximum 24-hour average of 0.00029 mg/L not to be exceeded, and a maximum concentration of 0.021 mg/L not to be exceeded at any one time, both measured as total recoverable hexavalent chromium. For trivalent chromium, the EPA's maximum criterion varies depending on water hardness and is expressed as the equation:

Maximum Trivalent Chromium (μg/L) $= \exp(108\,[\ln\,(\text{hardness})] + 3.48)$

Thus, at sample hardnesses of 50, 100, and 300 mg/L as $CaCO_3$, maximum total recoverable trivalent chromium should not exceed 2,219, 4,692, and 15,369 μg/L or approximately 2.2, 4.7, and 15.4 mg/L at any given time. No corresponding 24-hour average criterion is supplied.

8.   SUMMARY

From the literature reviewed, it would appear that hexavalent chromium is the more soluble form of the two chromium forms prevalent in nature and thus generally the more toxic compared to the relatively insoluble trivalent form. Toxic mechanisms for chromium appear to involve both nonspecific lesions to gills, liver, and kidneys and specific lesions to digestive tracts involving and interfering with gluconeogenesis, a process in which chromium is required as a trace element to form a portion of the glucose tolerance factor molecule.

Chromium is modified in toxicity by a number of abiotic and biotic factors and is not extremely toxic to fish at an acute level. It is, however, chronically toxic to fish at levels nearing those found in ambient waters near industrial centers, particularly with respect to reproduction and embryo-larval survival.

Water quality criteria for hexavalent chromium based on the 1980 EPA report appear to be more than adequate in protecting all species of fish from detrimental toxic effects, but the criteria recommended for trivalent chromium would appear to be grossly inadequate based on the recent literature discussed in this review. In fact, the safe levels recommended by the salmonid reproduction work of Billard and Roubaud (1985) would place the EPA water quality criteria over three orders of magnitude too high, a very disquieting observation. This definitely requires more research and a reevaluation of the water quality criteria for trivalent chromium.

Finally, there is a glaring absence of any work whatsoever investigating the possible carcinogenesis of chromium to fish. This is surprising given that chromium has been found to be mutagenic in all 15 short-term assays generally used involving prokaryotes, eukaryotic microorganisms, mammalian cells, viral transformation, *Drosophila*, mammalian in vivo assays, and even assays using higher plants (Hansen and Stern, 1984). It is thought to be responsible for tumor induction in mammals (Schlatter, 1984) and might very well explain the high incidence of papillomas found in suckers taken from the heavily industrialized Hamilton Harbour, Ontario. This incidence appears to be increasing since 29.6% were reported to have this tumor in 1977 (Sonstegard, 1977) and some 39% were reported to possess papillomas in 1983 (Cairns, 1983). Chromium makes up one of many metal contaminants in the Burlington Harbour. Nriagu et al. (1983) found total chromium levels as high as 564 mg/L in the top 4 cm of bottom cores from Hamilton Harbour, while Portt et al. (1986) in a series of some 55 sediment grab samples found chromium levels ranging from 33 to 400 mg/L and averaging at greater than 200 mg/L total chromium. Other metals found at large concentrations in the sediments included iron, manganese, polonium, cadmium, and cobalt. Bottom-dwelling fish like suckers spend much of their time at or near the sediment-water interface as well as digging into the sediment, and there is a great deal of bioturbation and mixing of the sediments in the harbour (Cairns, personal communication). Since Cr(VI) is a known and potent carcinogen in mammals, it seems only logical that it should be regarded as a prime candidate involved in inducing the observed tumors. It is strongly recommended that research be undertaken to assess the potential carcinogenic and mutagenic effects of both trivalent and hexavalent chromium on fish.

REFERENCES

Abbasi, S.A., and Soni, R. (1984). "Toxicity of lower than permissible levels of chromium(VI) to the freshwater teleost *Nuria denricus*." *Environ. Poll. A*, **36**, 75–82.

Adelman, I.R., and Smith, L.L. (1976). "Standard Test Fish Development. Part I. Fathead Minnows (*Pimephales promelas*) and Goldfish (*Carassius auratus*) as Standard Fish in Bioassays and their Reaction to Potential Reference Toxicants." U.S. Environmental Protection Agency EPA 600/3-76-061a, Duluth, MN.

Arillo, A., Margiocco, C., Melodia, F., and Mensi, P. (1982). "Biochemical effects of long term exposure to Cr, Cd, Ni on rainbow trout (*Salmo gairdneri* Rich.): Influence of sex and season." *Chemosphere* **11**, 47–57.

Barron, M.G., and Adelman, I.R. (1984). "Nucleic acid, protein content, and growth of larval fish sublethally exposed to various toxicants." *Can. J. Fish. Aquat. Sci.* **41**, 141–150.

Benoit, D.A. (1976). "Toxic effects of hexavalent chromium on brook trout (*Salvelinus fontinalis*) and rainbow trout (*Salmo gairdneri*)." *Water Res.* **10**, 497–500.

Billard, R., and Roubaud, P. (1985). "The effect of metals and cyanide on fertilization in rainbow trout (*Salmo gairdneri*)." *Water Res.* **19**, 209–214.

Bills, T.D., Marking, L.L. and Olson, L.E. (1977). "Effects of residues of the polychlorinated biphenyl Aroclor 1254 on the sensitivity of rainbow trout to selected environmental contaminants." *Prog. Fish Cult.* **39**, 150.

Birge, W.J., Hudson, J.E., Black, J.A., and Westerman, A.G. (1978). "Embryo-larval bioassays on inorganic coal elements and *in situ* biomonitoring of coal-waste effluents." In *Proceedings Surface Mining and Fish/Wildlife Needs in the Eastern United States*, pp. 97–104, FLOS/ OBS 78/81.

Bradley, R.W., and Sprague, J.B. (1985). "Accumulation of zinc by rainbow trout as influenced by pH, water hardness and fish size." *Env. Toxicol. Chem.* **4**, 685–694.

Broderius, S.J., and Smith, L.L., Jr. (1979). "Lethal and sublethal effects of binary mixtures of cyanide and hexavalent chromium, zinc, or ammonia to the fathead minnow (*Pimephales promelas*) and rainbow trout (*Salmo gairdneri*)." *J. Fish. Res. Bd. Can.* **36**, 164–172.

Buhler, D.R., Stokes, R.M., and Caldwell, R.S. (1977). "Tissue accumulation and enzymatic effects of hexavalent chromium in rainbow trout (*Salmo gairdneri*)." *J. Fish. Res. Bd. Can.* **34**, 9–18.

Burton, D.T., Jones, A.H., and Cairns, J., Jr. (1972). "Acute zinc toxicity to rainbow trout (*Salmo gairdneri*): Confirmation of the hypothesis that death is related to tissue hypoxia." *J. Fish. Res. Bd. Can.* **29**, 1463–1466.

Cairns, J., Jr., and Scheier, A. (1959). "The effects of temperature and hardness of water upon the toxicity of potassium dichromate to the common bluegill sunfish." In *Trans. N.E. Wildlife Conference Tenth Annual Meeting*, Montreal, Quebec, Canada, pp. 86–98.

Cairns, V.W. (1983). "Tumors in Great Lakes fish: A surveillance tool for bioeffects monitoring." Abstracts of papers presented at the 113th annual meeting of the American Fisheries Society, Milwaukee, WI, August 17, 1983.

Calamari, D., Gaggino, G.F., and Pacchetti, G. (1982). "Toxicokinetics of low levels of Cd, Cr, Ni and their mixture in long-term treatment on *Salmo gairdneri* Rich." *Chemosphere* **11**, 59–70.

Doudoroff, P., and Katz, M. (1953). "Critical review of literature on the toxicity of industrial wastes and their components to fish. II. The metals, as salts." *Sewage Ind. Wastes,* **25**, 802–839.

Eisler, R., and Hennekey, R.J. (1977). "Acute toxicities of Cd^{2+}, Cr^{+6}, Hg^{+2}, Ni^{2+}, and Zn^{2+} to estuarine macrofauna." *Arch. Environ. Contam. Toxicol.* **6**, 315–323.

Ellgaard, E.G., Tusa, J.E., and Malizia, A.A., Jr. (1978). "Locomotor activity of the bluegill *Lepomis macrochirus*: Hyperactivity induced by sublethal concentrations of cadmium, chromium and zinc." *J. Fish. Biol.* **12**, 19–23.

Friant, S.L. (1979). "Trace metal concentrations in selected biological, sediment, and water column samples in a northern New England river." *Water, Air, Soil Pollut.* **11**, 455–465.

Fromm, P.O., and Schiffman, R.H. (1958). "Toxic action of hexavalent chromium on largemouth bass." *J. Wildl. Man.* **22**, 40–44.

Fromm, P.O., and Stokes, R.M. (1962). "Assimilation and metabolism of chromium by trout." *J. Water Pollut. Control Fed.* **34**, 1151–1155.

Garton, R.B. (1973). "Biological effects of cooling tower blowdown." *Am. Inst. Chem. Eng. J.* **69**, 284–292.

Gray, S.J., and Sterling, K. (1950). "The tagging of red cells and plasma proteins with radioactive chromium." *J. Clin. Invest.* **29**, 1604–1613.

Grogan, C.H. (1958). "Experimental studies in metal cancerigenesis. II. On the penetration of Cr into the cell nucleus." *Cancer* **11**, 1195–1203.

Hale, J.G. (1977). "Toxicity of metal mining wastes." *Bull. Environ. Contam. Toxicol.* **17**, 66–73.

Hansen, K., and Stern, R.M. (1984). "A survey of metal-induced mutagenicity in vitro and in vivo." *Toxicol. Env. Chem.* **9**, 87–91.

Hogendoorn-Roozemond, A.S., Tenholder, J.J.H.M., Strik, J.J.T.W.A., et al. (1977). "The influence of pH on the toxicity of hexavalent chromium to rainbow trout (*Salmo gairdneri*). In *Aquatic Pollutants—Transformation and Biological Effects*. Proceedings of the Second International Symposium on Aquatic Pollutants, Pergamon, 1978, pp. 477–478.

Hughes, J.S. (1971). "Tolerance of striped bass, *Morone saxatilis* (Walbaum), larvae and fingerlings to nine chemicals used in pond culture. *Proc. Annu. Conf. Southeast. Assoc. Game Fish Comm.* **24**, 431–438.

Knoll, J., and Fromm, P.O. (1960). "Accumulation and elimination of hexavalent chromium in rainbow trout." *Physiol. Zool.* **33**, 1–8.

Lewis, S.D., and Lewis, W.M. (1971). "The effect of zinc and copper on the osmolality of blood serum of the channel catfish *Ictalurus punctatus* Rafinesque, and the golden shiner *Notemigonus crysolencas* Mitchill." *Trans. Am. Fish. Soc.* **100**, 639–643.

Mathis, B.J., and Cummings, T.F. (1973). "Selected metals in sediments, water, and biota in the Illinois River." *J. Water Pollut. Control Fed.* **45**, 1573–1583.

McDermott, D.J., Alexander, D.V., Young, D.R., and Mearns, A.J. (1976). "Metal contamination of flatfish around a large submarine outfall." *J. Water Pollut. Control Fed.* **48**, 1913–1918.

Mearns, A.J., Oshida, P.S., Sherwood, M.J., et al. (1976). "Chromium effects on coastal organisms." *J. Water Pollut. Control Fed.* **48**, 1929–1939.

Mertz, W. (1969). "Chromium occurrence and function in biological systems." *Physiol. Rev.* **49**, 163–239.

National Academy of Sciences (NAS), National Academy of Engineering. (1973). "Water Quality Criteria 1972." *EPA Ecol. Res. Ser. EPA-R3-73-033.* U.S. Environmental Protection Agency, Washington, D.C.

National Research Council of Canada. (1976). "Effects of chromium in the Canadian environment." Ottawa, Associate Committee on Scientific Criteria for Environmental Quality, Publication No. 15017 of the Environmental Secretariat.

Negilski, D.S. (1976). "Acute toxicity of zinc, cadmium and chromium to the marine fishes, yellow-eye mullet (*Aldrichetta forsteri* C. and V.) and small-mouthed hardyhead (*Atherinasoma microstoma* Whitley)." *Aust. J. Mar. Freshwater Res.* **27**, 137–149.

Nishihara, T., Shimamoto, T., Wen, K.C., and Kondo, M. (1985). "Accumulation of lead, cadmium and chromium in several organs and tissues of carp." *Eisei Kaguku* **31**, 119–123.

Nriagu, J.O., Wong, H.K.T., and Snodgrass, W.J. (1983). "Historical records of metal pollution in sediments of Toronto and Hamilton Harbours." *J. Great Lakes Res.* **9**, 365–373.

Olson, K.R., and Harrel, R.C. (1973). "Effect of salinity on acute toxicity of mercury, copper, and chromium for *Rangia cuneata* (Pelecypoda, Mactridae)." *Contrib. Mar. Sci.* **17**, 9–13.

Olson, P.A., and Foster, R.F. (1956). "Effect of chronic exposure to sodium dichromate on young chinook salmon and rainbow trout." In *Hanford Biol. Res. Ann. Rep. for 1955.* **HW-41500**, pp. 35–47.

Olson, P.A. (1957). "Comparative toxicity of CR (VI) and CR (III) in salmon." In *Hanford Biol. Res. Ann. Rep. for 1957*, pp. 215–218.

Phelps, D.K., Telek, G., and Lapan, R.L., Jr. (1975). "Assessment of heavy metal distribution within the food web." In E.A. Pearson, and E.D.F. Frangipani, Eds., *Marine Pollution and Marine Waste Disposal*, Pergamon Press, New York, pp. 341–348.

Pickering, Q.H., and Henderson, C. (1966). "The acute toxicity of some heavy metals to different species warm water fishes." *Air Water Pollut. Int. J.* 10, 453–463.

Pickering, Q.H. (1980a). "Chronic toxicity of hexavalent chromium to the fathead minnow (*Pimephales promelas*)." *Arch. Environ. Contam. Toxicol.* 9, 405–413.

Pickering, Q.H. (1980b). "Chronic toxicity of trivalent chromium to the fathead minnow (*Pimephales promelas*) in hard water." Draft. U.S. EPA, Environmental Research Laboratory, Duluth. Newtown Fish Toxicology Station, Cincinnati.

Portt, C., Minns, K., and Cairns, V.W. (1986). "The changing status of benthic invertebrates in Hamilton Harbour." (In preparation).

Rehwoldt, R., Menapace, L.W., Nerrie, B., and Alessandrello, D. (1972). "The effect of increased temperature upon the acute toxicity of some heavy metal ions." *Bull. Environ. Contam. Toxicol.* 8, 91–96.

Rehwoldt, R., Karimian-Teherani, D., and Altmann, H. (1976). "Distribution of selected metals in tissue samples of carp, *Cyrpinus carpio*." *Bull. Environ. Contam. Toxicol.* 15, 374–377.

Reisinger, H.J., II. (1979). *Heavy Metals in a Small Schuylkill River, Pennsylvania Impoundment*. Master's thesis. Millersville State College, PA.

Ruesink, R.G., and Smith, L.L., Jr. (1975). "The relationship of the 96 hr. LC50 to the lethal threshold concentration of hexavalent chromium, phenol, and sodium pentachlorophenate for fathead minnows (*Pimephales promelas* Rafinesque)." *Trans. Am. Fish. Soc.* 3, 567.

Sangalang, G.B., and Freeman, H.C. (1979). "Tissue uptake of cadmium in brook trout during chronic sub-lethal exposure." *Arch. Environ. Contam. Toxicol.* 8, 77–84.

Sauter, S., Buxton, K.S., Macek, K.J., and Petrocelli, S.R. (1976). *Effects of Exposure to Heavy Metals on Selected Freshwater Fish: Toxicity of Copper, Cadmium, Chromium and Lead to Eggs and Fry of Seven Fish Species*. EP-600/3-76-105.

Schiffman, R.H., and Fromm, P.O. (1959). "Chromium-induced changes in the blood of rainbow trout, *Salmo gairdneri*." *Sewage Ind. Wastes* 31, 205–211.

Schlatter, C. (1984). "Speculations on mechanisms of metal carcinogenesis." *Toxicol. Env. Chem.* 9, 127–137.

Sellers, C.M., Heath, A.G., and Bass, M.L. (1975). "The effect of sublethal concentrations of copper and zinc on ventilatory activity, blood oxygen and pH in rainbow trout (*Salmo gairdneri*)." *Water Res.* 9, 401–408.

Smith, M.J., and Heath, A.G. (1979). "Acute toxicity of copper, chromate, zinc and cyanide to freshwater fish: effect of different temperatures." *Bull. Environ. Contam. Toxicol.* 22, 13–119.

Sonstegard, R.A. (1977). "Environmental carcinogenesis studies in fishes of the Great Lakes of North America." *Ann. NY Acad. Sci.* 298, 261–269.

Sprague, J.B., and Fogels, A. (1977). "Watch the Y in bioassay." Proceedings 3rd Aquatic Toxicity Workshop, Halifax, Nova Scotia, Nov. 2-3, 1976. *Env. Prot. Service Tech. Rep. No. EPS-5-AR-77-1*, Halifax, Canada, pp. 107–118.

Stevens, D.G., and Chapman, G.A. (1984). "Toxicity of trivalent chromium to early life stages of steelhead trout." *Env. Toxicol. Chem.* 3, 125–133.

Stokes, R.M., and Fromm, P.O. (1965). "Effect of chromate on glucose transport by the gut of rainbow trout." *Physiol. Zool.* 38, 202–205.

Strik, J.J.T.W.A., de Iongh, H.H., van Rijn van Alkemade, J.W.A., and Wuite, T.P. (1975). "Toxicity of chromium(VI) in fish, with special reference to organoweights, liver and plasma enzyme activities, blood parameters and histological alterations." In *Sublethal Effects of Toxic Chemicals on Aquatic Animals*, Elsevier, Amsterdam, pp. 31–41.

Sugatt, R.H. (1980). "Effects of sublethal sodium dichromate exposure in freshwater on the salinity tolerance and serum osmolality of juvenile coho salmon, *Oncorhynchus kisutch*, in seawater." *Arch. Environ. Contam. Toxicol.* **9**, 41–52.

Tenholder, V.J.H.M., Roozemond, A.S.H., Kolar, Z., et al. (1978). "The uptake, tissue distribution and retention of hexavalent chromium by young rainbow trout *(Salmo gairdneri)*." In O. Hutzinger, I.H. Van Lelyveld and B.C.J. Zoitman, Eds., *Aquatic Pollutants: Transformation and Biological Effects*, Pergamon Press, Elmsford, NY, pp. 475–476.

Train, R.E. (1979). "Quality criteria for water." *Tunbridge Wells*, Kent, Castle House Publications Ltd.

Trama, F.B., and Benoit, R.J. (1960). "Toxicity of hexavalent chromium to bluegills." *J. Water Pollut. Control Fed.* **32**, 868–877.

U.S. EPA. (1976). *Quality Criteria for Water.* Office of Water and Hazardous Materials, U.S. EPA, Washington, D.C., EPA-440/9-76-023.

U.S. EPA. (1980). *Ambient Water Quality Criteria for Chromium.* Office of Water Regulations and Standards, Criteria and Standards Division, Washington, D.C., EPA/440/5-80-035.

Van der Putte, I., and Part, P. (1982). "Oxygen and chromium transfer in perfused gills of rainbow trout *(Salmo gairdneri)* exposed to hexavalent chromium at two different pH levels." *Aquat. Toxicol.* **2**, 31–45.

Van der Putte, I., Brinkhorst, M.A., and Koeman, J.H. (1981). "Effect of pH on the acute toxicity of hexavalent chromium to rainbow trout *(Salmo gairdneri)*." *Aquat. Toxicol.* **1**, 129–142.

Van der Putte, I., Lubbers, J., and Kolar, Z. (1981). "Effect of pH on uptake, tissue distribution and retention of hexavalent chromium in rainbow trout *(Salmo gairdneri)*." *Aquat. Toxicol.* **1**, 3–18.

Van der Putte, I., Laurier, M.B.H.M., and Van Eijk, G.J.M. (1982). "Respiration and osmoregulation in rainbow trout *(Salmo gairdneri)* exposed to hexavalent chromium at different pH values." *Aquat. Toxicol.* **2**, 99–112.

Wallen, I.E., Greer, W.C., and Lasater, R. (1957). "Toxicity to *Gambusia affinis* of certain pure chemicals in turbid waters." *Sewage Ind. Wastes* **29**, 695–701.

Whittle, K.J., Hardy, R., Holden, A.V., et al. (1977). "Occurrence and fate of organic and inorganic contaminants in marine animals." In H.F. Kraybell, C.J. Dawe, J.C. Harshbanger, and R.G. Tardiff, Eds., *Aquatic Pollutants and Biological Effects with Emphasis on Neoplasia. Ann. N.Y. Acad. Sci.* **298**, 47–79.

Woltering, D.M. (1984). "The growth response in fish chronic and early life stage toxicity tests: a critical review." *Aquat. Toxicol.* **15**, 1–21.

16

MUTAGENIC AND OTHER GENOTOXIC EFFECTS OF CHROMIUM COMPOUNDS

E. Nieboer and S. L. Shaw

Department of Biochemistry and
Occupational Health Program
McMaster University
Hamilton, Ontario, Canada

1.   INTRODUCTION

1.1   Preliminary Remarks

Since occupational exposure to Cr(VI) compounds has been linked to an increased risk of respiratory cancer in workers (Chapter 17), a summary of their mutagenic and other genotoxic effects is pertinent and is the subject matter of this chapter. Before defining the scope of our review, the case for the etiological role of DNA damage and chromosome abnormalities in chemical carcinogenesis is examined.

1.2   DNA Damage and the Somatic Mutation Model of Chemical Carcinogenesis

1.2.1   Somatic Mutation Theory

The preneoplastic and neoplastic states of cells are heritable. For this reason it is believed that cancer pathogenesis involves permanent alterations in gene expression and in gene structure (i.e., a mutation has taken place). The somatic mutation model of cancer is succinctly summarized in the following quotations from Friedberg (1985).

> Agents that initiate neoplastic transformation (carcinogens) do so by interacting with the DNA of cells, causing damage. If the cellular response(s) to this damage involves the misrepair of lesions, or if DNA repair fails to occur and the predominant response is an error-prone mechanism, mutations may arise in the descendants of the affected cells. These mutations, affecting specific genes (or sets of genes) that regulate critical aspects of cellular growth, may express themselves phenotypically as neoplastic transformation.

Rigorous proof of the somatic mutation theory of cancer obviously requires the demonstration in a single experimental system of at least the following:

1. that exposure of target calls to known carcinogens results in defined DNA damage;
2. that the cellular response(s) to such damage results in mutations;
3. that these mutations are expressed in the descendants of the damaged cells and that their localization to specific genes is always correlated with cellular transformation.

In cancer studies, experimental approaches of the type described involving man are unsuitable and thus " evidence for the somatic mutation theory is mainly indirect and comes from correlative studies that have examined selected aspects of the postulated relation between DNA damage, DNA repair, mutagenesis and neoplastic transformation in both prokaryotic and eukaryotic systems" (Friedberg, 1985). For example, such indirect evidence has been derived from: (1) neoplastic transformation in tissue culture of mammalian cells after treatment with ionizing radiation, chemicals, or ultraviolet (UV) radiation—agents that are known to damage DNA; (2) correlation between pyrimidine dimer lesions in DNA and tumors in the fish *Poecillia formosa* when exposed to UV radiation and maintained under non-DNA repair conditions; (3) in vitro mutagen screening tests that demonstrate that many known carcinogens (e.g., as determined by epidemiologic data and/or animal studies) are mutagens; and (4) tests with known mutagens and carcinogens that produce sister chromatid exchanges (intrachromosomal rearrangement) in many instances (Kohn, 1983; Friedberg, 1985).

Corroborating evidence for the somatic mutation hypothesis, although again indirect, comes from the observation that malignant cells of most cancers in humans have a chromosomal defect (Yunis, 1983; Friedberg, 1985). In addition, it is well established that individuals with gene or constitutional chromosomal imbalances are susceptible to increased risk of certain cancers. This occurs for DNA-repair defects in xeroderma pigmentosum patients who are light-sensitive and have a very high incidence of skin cancer, and in Fanconi's anemia patients who suffer an excess risk of leukemia (Sanger, 1983; Yunis, 1983). Similarly, diseases with chromosome deletions have been linked to a higher incidence of carcinomas.

1.2.2 Oncogenes in Chemical Carcinogenesis

Chemicals and radiation are not the only agents implicated in DNA damage and carcinogenesis. In animals, viruses can induce cancers, and specific viral genes are said to act as oncogenes. It has been proposed that oncogenes similar in base sequence (composition) to viral oncogenes, called proto-onc genes, exist in all cells (Scarpelli, 1983). It is hypothesized that proto-onc genes can be "turned on" by mutations induced by chemical mutagens, ionizing radiation, or even oncogenic viruses (Brodeur, 1986). Alternatively, neoplastic transformation could involve the translocation of a cellular oncogene or of a nearby regulatory gene that normally represses the oncogene (Sunderman, 1984). Gene products, such as specific enzymes, are believed to regulate cell transformation. However, in a recent review, Duesberg (1985) points out that human proto-onc genes are not consistently found in tumors and that no activated proto-onc gene with transforming function has been isolated from a tumor. Tumor suppressors or antioncogenes have also been postulated. Loss or inactivation of both copies of such genes (e.g., by mutation) is believed to be necessary for tumor formation (Knudson, 1985; Sager, 1986).

1.2.3 Limitations of the Somatic Mutation Theory

"Despite considerable efforts to demonstrate clear relationships between the sites of adduct formation on DNA bases, mutagenesis, and subsequent carcinogenesis, these have not been unequivocally established" (Scarpelli, 1983). The precise role of DNA damage (such as strand breaks, sister chromatid exchanges, chromosome deletions and translocations, interstrand DNA crosslinks, and protein-DNA crosslinks) and the cellular responses to DNA damage in the pathogenesis of cancer are unknown.

> However, it seems reasonably well established that many forms of neoplastic transformation are a consequence of DNA damage. It is not at all unlikely that, depending on the precise nature of the damage, its location with respect to active replicons, the nature of the cellular response to the damage and perhaps other variables of DNA structure and function, permanent alterations in gene expression may arise by mutation, by one of several modes of gene rearrangement or by other mechanisms, thereby leading ultimately to the phenotypic states of the cell that are currently defined as neoplastic transformation (Friedberg, 1985).

Strictly speaking, the somatic mutation point of view remains a postulate. As pointed out by Farber (1981, 1984), cancer is a multistage process. From the biological perspective, animal studies have identified the following processes: (1) initiation (a single or brief exposure to a carcinogen induces a permanent change in tissues recognized after promotion as focal proliferations, such as papillomas, nodules or polyps); (2) promotion (the process involving cell proliferation whereby tumor formation is accelerated or encouraged in a tissue that has been initiated); and (3) progression (focal proliferative lesions resulting from a promoting environment become precancerous lesions before malignant behavior is expressed) (Farber, 1981). Stages 2 and 3 are often reversible. The developmental phases beyond the early biochemical events of initiation are poorly studied. Recent evidence suggests that chemical promoters of cancers share an ability to produce radicals derived from molecular oxygen (O_2), namely the superoxide anion (O_2^-), the hydroxyl radical ($\cdot OH$), and the peroxy radical ($RO_2 \cdot$) (Marx, 1983). In vitro, dioxygen radicals can damage DNA. It is known that tumor promoters induce the formation of peroxide hormones derived from arachidonic acid. These hormones are intimately involved in cell division, differentiation, and apparently also in tumor growth (Ames, 1983).

From this brief summary, it is obvious that there are a lot of unresolved questions about cancer initiation, especially concerning the subsequent stages in tumor development. Nevertheless, even though the exact roles of chromosomal and DNA damage in chemical carcinogenesis remain to be defined, the conventional wisdom is that permanent alterations in gene expression are critical and most likely arise by mutation or gene rearrangement. It is also obvious that mutagenicity, DNA damage, and clastogenicity (chromosomal breakage) are not sufficient evidence for carcinogenicity. Corroborative evidence must be sought from in vitro short-term assays, animal data, and from human epidemiological surveys when available.

1.3   Objectives of the Review

Copious experimental data are available on the in vitro mutagenic and DNA-damaging effects of chromium compounds, especially for Cr(VI) compounds. Nor surprisingly, numerous reviews have appeared on this topic (Léonard and Lauwreys, 1980; Norseth, 1981; Petrilli and De Flora, 1982; Levis and Bianchi, 1982; Heck and Costa, 1982a,b). Of these surveys, that by Levis and Bianchi (1982) is the most comprehensive and informative. It is not our intention to reproduce the efforts of previous reviewers, but to provide a comprehensive compilation of the test results reported to date, to highlight trends, and to emphasize relevant aspects of the biochemistry and solution chemistry of chromium. Data is presented to illustrate that Cr(VI) compounds are considerably more genotoxic than Cr(III) compounds. Although mutagenic, water-insoluble Cr(VI) compounds have not been examined as comprehensively as water-soluble forms. A number of reports have a mechanistic focus and allow the formulation of a model consonant with the somatic-mutation view of chromium carcinogenesis. It links chromium speciation and oxidation state, chromium metabolism, and DNA damage with mutational consequences. Cytogenetic screening for clastogenic effects employing peripheral lymphocytes of workers exposed to chromium compounds have been conducted and are also summarized.

2.   MUTAGENICITY AND OTHER GENOTOXIC EFFECTS OF CHROMIUM COMPOUNDS

2.1   Overview of Experimental Results

A summary of the bacterial mutagenicity assays reported is presented in Table 1. The material in the table identifies the type of mutation, the bacterial cell system, the test compound (hexavalent or trivalent chromium), and the doses employed. Gene mutation changes were assessed predominantly by the Ames test which employs mutants reverting to prototrophy (histidine independence, his+) in a population of auxotrophic (histidine requiring, his−) cells of the bacterium *Salmonella typhimurium*. Strains of *Escherichia coli* have also been designed for comparable tests. Induction of gene mutations by Cr(VI) compounds has likewise been demonstrated in eukaryotic systems, namely, yeasts and cultured

Table 1 Summary of Bacterial Tests for Mutagenicity and Other Genotoxic Effects of Chromium Compounds

Bacterial Assay System	Trivalent Chromium Test Compound(s)[a]	Dose Range[b]	Result[c]	Hexavalent Chromium Test Compound(s)[a]	Dose Range[b]	Result[c]	Reference
Mutagenicity Tests							
Reverse mutation: E. coli Hs30R ar⁻ → → trp⁺	None tested			Na_2CrO_4; K_2CrO_4; $CaCrO_4$	0.05–0.2 μmol	+	Venitt and Levy, 1974
	None tested			$K_2Cr_2O_7$	1.2–4 mM	+	Nishioka, 1975
	None tested			K_2CrO_4	0.5–2.5 μg/ml	+	Green et al., 1976
	None tested			$PbCrO_4$	4–40 μM	+*	Nestmann et al., 1979
				CrO_3 (*positive in fluctuation test; negative in plate test)	2–100 μM	+*	
	None tested			CrO_3; K_2CrO_4	Not indicated	−	Kanematsu et al., 1980
				$K_2Cr_2O_7$	Not indicated	+	
	$Cr_2(SO_4)_3 \cdot 8H_2O$	Not indicated	−	K_2CrO_4 (*mutagenic using both standard and modified medium)	6.25–25 μM	+*	Arlauskas et al., 1985
Reverse mutation: E. coli Hs30R arg⁻ → → arg⁺	$Cr(CH_3COO)_3$	16–130 mM	+	$K_2Cr_2O_7$; K_2CrO_4	1.6–13 mM	+	Nakamuro et al., 1978
Reverse mutation: S. typhimurium various strains his⁻ → → his⁺	$CrCl_3 \cdot 6H_2O$	Not indicated	−	$K_2Cr_2O_7$; K_2CrO_4	Not indicated	+	Tamaro et al, 1975
	$CrK(SO_4)_2 \cdot 12H_2O$ $CrCl_3 \cdot 6H_2O$	3.125–800 μg not indicated	−	$Na_2Cr_2O_7$; CrO_3; $CaCrO_4$; K_2CrO_4	0.125–8 μmol (as Chromium)	+	Petrilli and De Flora, 1977
	None tested			Chromate and dichromate	0.1–0.2 μmol (as Chromium)	+	Löfroth, 1978
	None tested			K_2CrO_4	10–80 μg	+*	Petrilli and De Flora, 1978a
				$Na_2Cr_2O_7$	5–160 μg	+*	

Cr(III) compound	Amount	Result	Cr(VI) compound	Amount	Result	Reference
			$ZnCrO_4 \cdot Zn(OH)_2$ + CrO_3 (as Zn yellow)	Not indicated	+*	Petrilli and De Flora, 1978b
			$PbCrO_4 \cdot PbO$ (as Cr orange) (*mutagenicity suppressed by addition of rat liver microsomes, erythrocyte lysates, ascorbic acid, and sodium sulfite)	Not indicated	+*	
$CrK(SO_4)_2 \cdot 12H_2O$	10–8000 µg	–*	None tested			
$Cr(NO_3)_3 \cdot 9H_2O$	10–8000 µg	–*				
$CrCl_3 \cdot 6H_2O$	10–8000 µg	–*				
$Cr(OH)SO_4$ (as neochromium)	10–8000 µg	–*				
$Cr_2(SO_4)_3$ (as Cr alum) (*addition of $KMnO_4$ resulted in mutagenicity)	10–8000 µg	–*				
Cr_2O_3 (as chromite) (*contaminated by Cr(VI))	2000 µg	+*				
None tested			$PbCrO_4$	50–400 µg	+	Nestmann et al., 1979
			CrO_3	7–29 µg	+	
None tested			$Na_2Cr_2O_7$ (*mutagenicity decreased by preincubation with human gastric juice)	40 µg	+*	De Flora and Boido, 1980
None tested			$K_2Cr_2O_7$ (spot test)	Not indicated	–	Kanematsu et al., 1980
$CrCl_3 \cdot 6H_2O$	30 µmol	–	$Na_2Cr_2O_7$	0.05–0.25 µmol	+	De Flora, 1981
$Cr(NO_3)_3 \cdot 9H_2O$	20 µmol	–	K_2CrO_4	0.08–0.41 µmol	+	
$CrK(SO_4)_2 \cdot 12H_2O$	16 µmol	–	$CaCrO_4$	0.06–0.29 µmol	+	
$Cr(CH_3COO)_3$	70 µmol	–	$(NH_4)_2CrO_4$	0.05–0.32 µmol	+	

Continued

405

Table 1 (*Continued*)

Bacterial Assay System	Trivalent Chromium Test Compound(s)[a]	Dose Range[b]	Result[c]	Hexavalent Chromium Test Compound(s)[a]	Dose Range[b]	Result[c]	Reference
	$Cr(OH)SO_4$ (as neochromium)	50 µmol	−	CrO_3	0.04–0.22 µmol	+	Tso and Fung, 1981
				$ZnCrO_4 \cdot Zn(OH)_2$ (as Zn yellow)	0.09–0.59 µmol	+	
	Cr_2SO_4 (as Cr alum)	43 µmol	−	Cl_2CrO_2	0.10–0.43 µmol	+	Warren et al., 1981
	Cr_2O_3 (as chromite; spot test)	2000 µg	+	$PbCrO_4 \cdot PbO$ (spot test)	2000 µg	+	
				$PbCrO_4$ (as Mo orange)	2000 µg	+	
				$PbCrO_4$ (as Cr yellow)	Not indicated	−	
	$CrCl_3 \cdot 6H_2O$	Up to 10^4 µM	−	CrO_3	Up to 10^4 µM	+	
	cis-[Cr(bipy)$_2$ox]|[Cr(urea)$_6$]Cl$_3$	Up to 1.2 µmol	+	None tested			
	cis-[Cr(bipy)$_2$Cl$_2$]Cl	Up to 1.5 µmol	+				
	cis-[Cr(phen)$_2$Cl$_2$]Cl (only compounds scoring positive are indicated)	Up to 0.125 µmol	+				
		Up to 1.4 µmol	+				
	None tested		−	$Na_2Cr_2O_7$ (*mutagenicity decreased by addition of rat liver S9 and trout liver S9 mix)	20 µg	+*	De Flora et al., 1982
	$Cr(CH_3COO)_3$	2.478–9.912 µmol (as chromium)	−	$Na_2Cr_2O_7$	0.030–0.122 µmol (as chromium)	+*	Bennicelli et al., 1983
	$Cr(NO_3)_3 \cdot 9H_2O$	1.625–6.5 µmol (as chromium)	−	$CaCrO_4$	0.042–0.169 µmol (as chromium)	+*	
	$CrK(SO_4)_2 \cdot 12H_2O$	1.042–4.168 µmol (as chromium)	−	CrO_3 (*mutagenicity suppressed by addition of rat liver S9 and human lung S12 mix)	0.052–0.208 µmol (as chromium)	+*	
	$CrCl_3 \cdot 6H_2O$	18.1–145 µg	−	$K_2Cr_2O_7$	2.5–40 µg	+	Bianchi et al., 1983

Compound	Dose	Result	Compound	Dose	Result	Reference
$Cr(NO_3)_3 \cdot 9H_2O$	No indicated	+*				
Cr_2O_3 (as chromite) (*contaminated by Cr(VI))	Not indicated	+*				
cis-$[Cr(bipy)_2Cl_2]Cl$	20–200 μmol	+	Chromate	20–200 μmol	+	Beyersmann et al., 1984
cis-$[Cr(phen)_2Cl_2]Cl$	20–200 μmol	+				
$CrCl_3 \cdot 6H_2O$	Not indicated	−	$Na_2Cr_2O_7$; K_2CrO_4; $CaCrO_4$; $(NH_4)_2CrO_4$; CrO_3 (*mutagenicity decreased by addition of S9 mix)	Not indicated	+*	De Flora et al., 1984a
$Cr(NO_3)_3 \cdot 9H_2O$		−				
$CrK(SO_4)_2 \cdot 12H_2O$		−				
$Cr(CH_3COO)_3$		−				
$CrCl_3 \cdot 6H_2O$	Not indicated	−	$Na_2Cr_2O_7$	Not indicated	+	De Flora et al., 1984b
$Cr(NO_3)_3 \cdot 9H_2O$		−				
$CrK(SO_4)_2 \cdot 12H_2O$		−				
$Cr(CH_3COO)_3$		−				
None tested			K_2CrO_4 (*weakly mutagenic; mutagenicity enhanced in the presence of NaN_3)	5.7–45.7 μM	+*	LaVelle and Witmer, 1984
$Cr_2(SO_4)_3 \cdot 8H_2O$	Not indicated	−	K_2CrO_4	2.314–9.259 μM (plate assay); 6.25–25 μM (fluctuation assay)	+; +	Arlauskas et al., 1985
None tested			$Na_2Cr_2O_7$; $PbCrO_4$ (*addition of an oil dispersant caused no change)	20–40 μg; 40–640 μg	+*; −*	De Flora et al., 1985a

Continued

Table 1 (Continued)

Bacterial Assay System	Trivalent Chromium Test Compound(s)[a]	Dose Range[b]	Result[c]	Hexavalent Chromium Test Compound(s)[a]	Dose Range[b]	Result[c]	Reference
	None tested			$Na_2Cr_2O_7$ (*mutagenicity decreased by addition of rat liver S12 mix; mutagenicity restored by addition of S12 + dicumarol)	30 µg	+*	De Flora et al., 1985b
	$CrCl_3 \cdot 6H_2O$ (*weakly mutagenic)	0.02–2 µmol	+*	K_2CrO_4	0.02–0.5 µmol	+	Langerwerf et al., 1985
	$CrK(SO_4)_2 \cdot 12H_2O$	0.02–50 µmol	−				
	soluble Cr(III)-amino acid complexes	0.02–50 µmol	−				
	$Cr_2(SO_4)_3 \cdot nH_2O$	0.5–5000 µg	−	Na_2CrO_4	3.3–321 µg	+	Loprieno et al., 1985
				$K_2Cr_2O_7$	2.5–40 µg	+	
				$PbCrO_4$	20–160 µg	+*	
				$PbCrO_4 \cdot PbO$ (as Cr orange) (*mutagenic activity elicited only upon addition of NTA)	50–800 µg	+*	
	$CrCl_3 \cdot 6H_2O$ (*toxic at high dose)	0.005–3 µM	−*	$K_2Cr_2O_7$; Na_2CrO_4 (*toxic at high dose)	0.005–0.3 µM	+*	Marzin and Phi, 1985
	Tannins (mainly Cr(III) sulfates)	Up to 10 mg	−	None tested			Venier et al., 1985
Forward mutation: E. coli K-12/343/113(λ) gal+	None tested			$PbCrO_4$	2.5–100 µg	−	Nestmann et al., 1979
Forward mutation: E. coli KMBL 3835 lacI+	$CrCl_3 \cdot 6H_2O$ (*small increase in mutation frequency if any)	Not indicated	Not clear*	$K_2Cr_2O_7$	Not indicated	+	Zakour and Glickman, 1984

DNA-Repair Tests

Differential killing: B. subtilis rec+/rec− (rec assay)

Compound	Concentration	Result	Compound	Concentration	Result	Reference
CrCl₃·6H₂O	2.5 μmol (spot test)	−	K₂CrO₄; K₂Cr₂O₇	2.5 μmol (spot test)	+	Nishioka, 1975
			K₂Cr₂O₇ (*rec-effect suppressed by addition of Na₂SO₃)	5–200 mM	+*	
Cr(CH₃COO)₃	160–1300 mM	+	K₂Cr₂O₇	16–130 mM	+	Nakamuro et al., 1978
CrNO₃)₃·9H₂O	160–650 mM	+	K₂CrO₄	20–320 mM	+	
CrCl₃·6H₂O	320–1300 mM	−	CrO₃	16–260 mM	+	
K₂Cr₂(SO₄)₄	Not indicated		K₂Cr₂O₇; K₂CrO₄	5 mM	+	Kanematsu et al., 1980
Cr₂(SO₄)₃ (*high toxic)		−*	CrO₃	10 mM	+	
CrCl₃·6H₂O (*positive rec-effect in the presence of salicylate or citrate)	50 mM	−*	K₂Cr₂O₇ (*reduced rec-effect in the presence of complexing/chelating agents)	50 mM	+*	Gentile et al., 1981
CrK(SO₄)₂·12H₂O	50 mM	−	CrO₃; (NH₄)₂Cr₂O₇; Na₂CrO₄; Na₂Cr₂O₇	50 mM	+	
Cr₂(SO₄)₃	50 mM	−				

Differential killing: S. typhimurium rec+/rec− (rec assay)

Compound	Concentration	Result	Compound	Concentration	Result	Reference
CrCl₃·6H₂O	50 mM	−	K₂Cr₂O₇; CrO₃; (NH₄)₂Cr₂O₇; Na₂Cr₂O₇; Na₂CrO₄	50 mM	+	Gentile et al., 1981
CrK(SO₄)₂·12H₂O	50 mM	−				
Cr₂(SO₄)₃	50 mM	−				

Differential killing: E. coli polA⁺/polA⁻

Compound	Concentration	Result	Compound	Concentration	Result	Reference
None tested			PbCrO₄	200–300 μg	−	Nestmann et al., 1979

Differential killing: E. coli

Strain	Compound	Concentration	Result	Compound	Result	Reference
uvrA⁺/uvrA⁻	cis-[Cr(bipy)₂ox]l	Not indicated	+	None tested		Warren et al., 1981
recA⁺/recA⁻	cis-[Cr(bipy)₂Cl₂]Cl		+			
lexA⁺/lexA⁻	cis-[Cr(phen)₂Cl₂]Cl		+			
recA⁺ uvrA⁺/	[Cr(urea)₆]Cl₃		+*			
recA⁻ uvrA⁻	[Cr(en)₃](SCN)₃		+*			
	[Cr(pm)₃]Cl₃					

Continued

Table 1 (Continued)

Bacterial Assay System	Trivalent Chromium Test Compound(s)[a]	Trivalent Chromium Dose Range[b]	Result[c]	Hexavalent Chromium Test Compound(s)[a]	Hexavalent Chromium Dose Range[b]	Result[c]	Reference
recA+lexA+/	trans-[Cr(en)2(SCN)2] SCN		+*	Na2Cr2O7	Up to 50 µg	+	De Flora et al., 1984a
recA−lexA−	[Cr(en)3]Cl3 (*less active than first four compounds; only compounds scoring positive are indicated)		+*	K2CrO4	Up to 75 µg	+	
				CaCrO4	Up to 100 µg	+*	
				(NH4)2CrO4	Up to 75 µg	+*	
				CrO3 (*weakly genotoxic)	Up to 37.5 µg	+*	
Differential killing:							
E. coli							
uvrA+polA+	CrCl3·6H2O	Up to 2000 µg	+*				
uvrA−polA−	Cr(NO3)3·9H2O	Up to 2500 µg	+*				
uvrA+recA+	CrK(SO4)2·12H2O	Up to 2000 µg	+*				
uvrA+/uvrA−	Cr(CH3COO)3	Up to 2500 µg	+*				
lexA+/uvrA−	(*genotoxicity diminished in the presence of S9 mix)						
recA−lexA−							
SOS chromotest:							
E. coli PQ37 sfi expression	CrK(SO4)2·12H2O (*toxic at high dose)	0.4–50 µg	−*	Na2Cr2O7 (*toxic at high dose)	0.4–50 µg	+*	De Flora et al., 1985a

[a]Abbreviations: bipy = bipyridine, en = ethylenediamine, NTA = nitrilotriacetic acid, ox = oxalate, phen = 1,10-phenanthroline, pn = 1,2-propanediamine. [b]Unless specified to the contrary, concentrations expressed by weight or in weight/volume units refer to the actual test compound and not specifically to its chromium content. [c]For explanation of superscripts (*) refer to definition or statement in the corresponding column labelled Test Compound(s); + positive; − negative result.

Chinese hamster cells. (For a critical review of these test systems, the reader is referred to ICPEMC, 1982; Hoffman, 1982; Maron and Ames, 1983.) The mutagenic and other genotoxic in vitro effects in mammalian cells of Cr(III) and Cr(VI) compounds are summarized in Table 2. The genotoxic effects observed include both chromosomal and DNA damage (assessed directly or indirectly): chromosomal aberrations, sister-chromatid exchanges, DNA strand breaks, alkali lability of DNA, DNA interstrand and DNA-protein crosslinking, DNA fragmentation, inhibition of DNA replication, unscheduled DNA synthesis, mitotic cycle alterations, and inhibition of cell growth. Results for cell types other than mammalian or bacterial (e.g., yeast, and chick embryo hepatocytes) are presented in Table 3.

In Figures 1 to 4, the data compiled in Tables 1 and 2 are depicted graphically for individual compounds. The bibliographic data in Table 4 identify the original references to which the data in the histograms refer. It is obvious from the histograms in Figures 1 and 2 that Cr(VI) compounds have been studied extensively and routinely score positively in mutagenicity and chromosome aberration tests. Of the eight water-soluble hexavalent chromium compounds, potassium chromate (K_2CrO_4) and potassium dichromate ($K_2Cr_2O_7$) have been most intensely studied. By contrast, water-insoluble hexavalent chromium compounds have not been examined to the same extent. Again, most tests were positive. As discussed in Section 2.2, the equivocal response of lead chromate in the bacterial tests may be related to its severe insolubility.

An examination of the data in Figures 3 and 4 reveals that water-soluble trivalent chromium compounds most often score negatively. Positive studies occur most frequently for Cr(III) chloride and Cr(III) acetate. Inactivity is also found for Cr(III) sulfates as depicted in Figure 3 [anhydrous Cr(III) sulfate is of low solubility, while hydrated derivatives are generally soluble in water] (CRC, 1985).

It is obvious from Tables 5 and 6 that hexavalent chromium compounds are considerably more mutagenic and genotoxic than trivalent chromium compounds. For a comparable number of studies, 87% of the hexavalent chromium mutagenicity tests were positive, compared to 25% for trivalent compounds. A similar trend occurs for genotoxicity assessments in mammalian cell cultures (96% of hexavalent chromium compounds and 26% of trivalent compounds tested positively, although it must be noted that considerably more tests were carried out with hexavalent compounds).

2.2 Differential Responses of Chromium(VI) Compounds

2.2.1 Properties and Cellular Uptake of the Chromate Ion

As indicated by Nieboer and Jusys (Chapter 2), chromium may be designated a "borderline" metal. This means that it has a catholic affinity for ligands containing either oxygen, nitrogen, or the sulfur atom as the point of attachment. This has been well established for Cr(III) complex formation. Chromium(VI)

Table 2 Summary of Genotoxic Effects of Chromium Compounds (Cultured Mammalian Cell Systems)

Mammalian In Vitro Test System	Trivalent Chromium Test Compound(s)[a]	Dose Range[b]	Result[c]	Hexavalent Chromium Test Compounds[a]	Dose Range[b]	Result[c]	Assay Type[d]	Reference
Mutagenicity Tests								
Chinese hamster V79 cell line	$Cr(CH_3COO)_3$	50–200 μg/ml	−	$K_2Cr_2O_7$ $ZnZrO_4$ $PbCrO_4$	0.35–0.78 μg/ml 1–4 μg/ml 5–10 μg/ml	+ + −	HGPRT mutants (8-azaguanine resistance)	Newbold et al., 1979
	None tested			$K_2Cr_2O_7$ (*mutation frequency increases in the presence of triethylene phosphoramide)	0.1–0.5 μg/ml	+*	HGPRT mutants (8-azaguanine resistance)	Paschin and Kozachenko, 1982
L5178 mouse lymphoma cells	$Cr(CH_3COO)_3$	200–800 μM	−	$K_2Cr_2O_7$	7–41 μM	+	6-thioguanine resistance	Bianchi et al., 1983
	None tested			K_2CrO_4 $K_2Cr_2O_7$	1–8 μg/ml 1–8 μg/ml	+ +	Forward mutation at the thymidine locus	Oberly et al., 1982
Chromosome Damage Tests								
Chinese hamster V79 cell line	None tested			$K_2Cr_2O_7$	0.35–0.80 μg/ml	+	SCE, aberrations, fragmentation	Newbold et al., 1979
	None tested			K_2CrO_4	0.5 μg/ml	+	SCE aneuploid/polyploid	Price-Jones et al., 1980
	$CrCl_3$ (*whole cell exposure; +DNA exposure)	1000–5000 μM 10–100 μM	−* ++	$K_2Cr_2O_7$ (*whole cell exposure; +DNA exposure)	50–5000 μM 10–100 μM	−* ++	Fragmentation, SB, ISCL	Bianchi et al., 1983
	$CrCl_3 \cdot 6H_2O$ (*low SCE frequency)	9.76–39.04 μg/ml (as chromium)	+*	Na_2CrO_4	0.032–0.128 μg/ml (as chromium)	+	SCE	Elias et al., 1983

Cell line	Compound	Concentration	Result	Compound	Concentration	Result	Endpoint	Reference
	Cr_2O_3	34.21–136.84 µg/ml (as chromium)	+	$Na_2Cr_2O_7$	0.035–0.140 µg/ml (as chromium)	+	Aberrations, SCE	Levis and Majone, 1979
				K_2CrO_4	0.026–0.106 µg/ml (as chromium)	+		
Chinese hamster CHO cell line	$CrCl_3 \cdot 6H_2O$	5–50 µg/ml (as chromium)	+	$K_2Cr_2O_7$; $Na_2Cr_2O_7$	0.1–1 µg/ml (as chromium)	+		
	$CrK(SO_4)_2 \cdot 12H_2O$	150 µg/ml (as chromium	+	K_2CrO_4	0.25 µg/ml (as chromium)	+		
	$Cr(CH_3COO)_3$	5–20 µg/ml (as chromium)	+	Na_2CrO_4	0.25–0.5 µg/ml (as chromium)	+		
	$Cr(NO_3)_3 \cdot 9H_2O$	50–150 µg/ml (as chromium)	+	CrO_3	0.1–0.25 µg/ml (as chromium)	+		
				$CaCrO_4$	0.5 µg/ml (as chromium)	+		
	$CrCl_3 \cdot 6H_2O$	0.6–0.9 µM	−	K_2CrO_4	0.08–3 µM	+	SCE	MacRae et al., 1979
	$CrCl_3 \cdot 6H_2O$; $Cr(NO_3)_3 \cdot 9H_2O$; $CrK(SO_4)_2 \cdot 12H_2O$; $Cr(CH_3COO)_3$	20–150 µg/ml (as chromium)	−	$Na_2Cr_2O_7$; Na_2CrO_4; $K_2Cr_2O_7$; K_2CrO_4; CrO_3	0.25–1 µg/ml (as chromium)	+	Aberrations, SCE	Bianchi et al., 1980
	$Cr(OH)SO_4$ (as neochromium)	5–150 µg/ml	+*	$PbCrO_4$ (as Cr yellow)	5–150 µg/ml	+	Aberrations, SCE	Levis and Majone, 1981
	$Cr_2(SO_4)_3$ (as Cr alum)	5–150 µg/ml	+*	$PbCrO_4 \cdot PbO$ (as Cr orange)	5–150 µg/ml	+		
	$CrCl_3 \cdot 6H_2O$	5–150 µg/ml	+*	$PbCrO_4$ (as Mo orange)	5–150 µg/ml	+		
	$Cr_2(SO_4)_3$ (*induced aberrations but not SCE)	5–150 µg/ml	+*	$ZnCrO_4 \cdot Zn(OH)_2$ + CrO_3 (as Zn yellow)	5–150 µg/ml	+		
	Cr_2O_3 (as chromite)	5–150 µg/ml	+	K_2CrO_4	0.3 µg/ml	+		

Continued

Table 2 (Continued)

Mammalian In Vitro Test System	Trivalent Chromium Test Compound(s)[a]	Dose Range[b]	Result[c]	Hexavalent Chromium Test Compounds[a]	Dose Range[b]	Result[c]	Assay Type[d]	Reference
	$CrCl_3 \cdot 6H_2O$ None tested	1000 μM	−	$K_2Cr_2O_7$ $CaCrO_4$	1 μM 10–50 μM	+ +	SCE Alkali labile sites	Bianchi et al., 1983 Cantoni and Costa, 1984
	None tested			$CaCrO_4$ (*no detectable damage when added directly to nucleoids)	50–1000 μM	+*	SB	Robison et al., 1984
	$Cr_2(SO_4)_3$	2.5 μM	−	$K_2Cr_2O_7$ $PbCrO_4$ (*increase in SCE after addition of NTA)	0.3 μM 0.3 μM	+ +*	SCE	Loprieno et al., 1985
	Tannins (mainly Cr(III) sulfates) (*contaminated with Cr(VI) or other impurities)	10 μg/ml	+*	$PbCrO_4 \cdot PbO$ None tested	1.8 μM	+	SCE	Venier et al., 1985
Chinese hamster ovary cells	$CrCl_3 \cdot 6H_2O$ None tested	25–50 μg/ml	+	$PbCrO_4$	10–100 μM	−	Aberrations, fragmentation	Douglas et al., 1980
Don Chinese hamster cells	$CrCl_3 \cdot 6H_2O$ $Cr_2(SO_4)_3$	32 μg/ml 6 μg/ml	+ −	CrO_3 K_2CrO_4; $K_2Cr_2O_7$	0.32 μg/ml 0.8 μg/ml	+ +	SCE	Ohno et al., 1982
Syrian hamster fibroblasts BHK cell line	$CrCl_3 \cdot 6H_2O$	10–1000 μM	−	$K_2Cr_2O_7$	10–1000 μM	+	SCE	Bianchi et al., 1984
Primary Syrian hamster embryo cells	$CrCl_3 \cdot 6H_2O$	0.35–3.5 μg/ml (as chromium)	−	$K_2Cr_2O_7$	0.035–0.175 μg/ml (as chromium)	+	Aberrations, SCE	Tsuda and Kato, 1977

Primary mouse fetal cells	Cr$_2$(SO$_4$)$_3$ · 4H$_2$O	0.35–3.5 μg/ml (as chromium)	−	CrO$_3$	0.035–0.175 μg/ml (as chromium)	+	Aberrations	Raffetto et al., 1977
	CrCl$_3$ · 6H$_2$O	0.04–0.4 μg/ml (as chromium)	+	K$_2$Cr$_2$O$_7$	0.015–0.1 μg/ml (as chromium)	+	Aberrations	
Frm3A (mouse mammary carcinoma)	Cr$_2$(SO$_4$)$_3$	32–1000 μM	−	K$_2$Cr$_2$O$_7$	0.6–3.2 μM	+	Aberrations, SCE	Umeda and Nishimura, 1979
				K$_2$CrO$_4$	3.2–10 μM	+		
				CrO$_3$	1–10 μM	+		
Mouse L1210 leukemia cells	CrCl$_3$ · 6H$_2$O (*increase in crosslinking when isolated nuclei exposed)	200 μM	−*	K$_2$CrO$_4$	25–200 μM	+	SB, DPCL	Fornace et al., 1981
				K$_2$Cr$_2$O$_7$	200 μM	+		
BALB/c mouse primary lymphocytes	CrCl$_3$ · 6H$_2$O	1000 μM	−	K$_2$Cr$_2$O$_7$	1 μM	+	SCE	Bianchi et al., 1983
BALB/Mo mouse primary lymphocytes	CrCl$_3$ · 6H$_2$O	1000 μM	−	K$_2$Cr$_2$O$_7$	1 μM	+	SCE	Bianchi et al., 1983
LSTRA cell line mouse lymphocytes	CrCl$_3$ · 6H$_2$O	1000 μM	−	K$_2$Cr$_2$O$_7$	1 μM	+	SCE	Bianchi et al., 1983
Human peripheral blood lymphocytes	CrCl$_3$ · 6H$_2$O	1500–2255 μM	+	CrO$_3$	20–60 μM	+	Aberrations	Kaneko, 1976
	Cr(CH$_3$COO)$_3$	16–32 μM	+	K$_2$Cr$_2$O$_7$	0.5–4 μM	+	Aberrations, SCE	Nakamuro et al., 1978
	CrCl$_3$ · 6H$_2$O	32 μM	−	K$_2$CrO$_4$	4–8 μM	+	SCE	Douglas et al., 1980
	None tested			PbCrO$_4$	4.3–77 μM	+		
				K$_2$CrO$_4$	20–100 μM	+		
	CrCl$_3$ · 6H$_2$O	10–7500 μM	−	Na$_2$Cr$_2$O$_7$	0.1–50 μM	+	Aberrations	Sarto et al., 1980
	None tested			K$_2$Cr$_2$O$_7$; CrO$_3$	0.025–0.1 μg/ml	+	SCE	Gomez-Arroyo et al., 1981

Continued

Table 2 (*Continued*)

Mammalian In Vitro Test System	Trivalent Chromium Test Compound(s)[a]	Dose Range[b]	Result[c]	Hexavalent Chromium Test Compounds[a]	Dose Range[b]	Result[c]	Assay Type[d]	Reference
	CrCl₃·6H₂O (*only aberrations induced at high dose)	1–1000 μM	+*	CaCrO₄	0.01–0.02 μg/ml	+	Aberrations, SCE	Stella et al., 1982
				K₂Cr₂O₇	0.01–10 μM	+		Andersen, 1983
Human bronchial epithelial cells	Cr(CH₃COO)₃	10–1000 μM	+	K₂Cr₂O₇	0.1–10 μM	+	SCE	Andersen, 1983
	CrCl₃·6H₂O	200 μM	−	K₂CrO₄	25–200 μM	+	SB, DPCL	Fornace et al., 1981
				K₂Cr₂O₇	200 μM	+		
Human IMR-90 cells	CrCl₃·6H₂O	200 μM	−	K₂CrO₄	25–200 μM	+	SB, DPCL	Fornace et al., 1981
				K₂Cr₂O₇	200 μM	+		
Human embryonic fibroblasts	CrCl₃·6H₂O	200 μM	−	K₂CrO₄	25–200 μM	+	SB, DPCL	Fornace et al., 1981
				K₂Cr₂O₇	200 μM	+		
Human skin fibroblasts	None tested			K₂CrO₄	0.01–3 μM	+	Aberrations, SCE	MacRae et al., 1979
				K₂Cr₂O₇	0.01–1 μM	+		
Human fibroblast cell lines (normal and XP)	None tested			K₂CrO₄	100–5000 μM	+	Fragmentation	Whiting et al., 1979
	None tested			K₂CrO₄ (*normal and XP cells)	2–50 μM	+*	SB	Fornace, 1982
P388D, macrophage cell line	Cr(CH₃COO)₃	1–100 μM	+	K₂Cr₂O₇	0.01–1 μM	+	SCE	Andersen, 1983
	Cr₂O₃	10–1000 μM	−					
Tests for Other Genetic Damage								
Chinese hamster CHO cell line	None tested			Chromate	>20 μM	+	Inhibition of cell growth	Beyersmann et al., 1984
Syrian hamster fibroblasts BHK cell line	CrCl₃·6H₂O	30–300 μg/ml (as chromium)	+*	K₂Cr₂O₇	0.1–2 μg/ml (as chromium)	+	Inhibition of cell growth	Levis and Majone, 1979
	Cr(NO₃)₃·9H₂O	30–300 μg/ml (as chromium)	+*	Na₂Cr₂O₇	0.1–2 μg/ml (as chromium)	+		

Compound	Concentration		Compound	Concentration		Endpoint	Reference
CrK(SO$_4$)$_2$ · 12H$_2$O	30–300 µg/ml (as chromium)	+*	K$_2$CrO$_4$	0.1–2 µg/ml (as chromium)	+	Inhibition of cell growth	Levis and Majone, 1981
Cr(CH$_3$COO)$_3$ (*chronic exposure; for acute exposure Cr(III) compounds were almost inactive)	20–300 µg/ml (as chromium)	+*	Na$_2$CrO$_4$	0.1–2 µg/ml (as chromium)	+		
			CrO$_3$	0.1–2 µg/ml (as chromium)	+		
			CaCrO$_4$	0.1–2 µg/ml (as chromium)	+		
Cr(OH)SO$_4$ (as neochromium)	500 µg/ml	+	PbCrO$_4$ (as Cr yellow)	50–500 µg/ml	+		
Cr$_2$(SO$_4$)$_4$ (as Cr alum)	500 µg/ml	+	PbCrO$_4$ (as Mo orange)	50–500 µg/ml	+		
Cr$_2$O$_3$ (as chromite)	500 µg/ml	+	PbCrO$_4$ · PbO (as Cr orange)	50–500 µg/ml	+		
Cr$_2$(SO$_4$)$_3$	500 µg/ml	+	ZnCrO$_4$ · Zn(OH)$_2$ +CrO$_3$ (as Zn yellow)	50–500 µg/ml	+		
CrCl$_3$ · 6H$_2$O	500 µg/ml	−	K$_2$Cr$_2$O$_7$	50–500 µg/ml	+	Morphological transformation	Bianchi et al., 1983
CrCl$_3$ · 6H$_2$O	75–300 µg/ml	−	K$_2$Cr$_2$O$_7$	2.5–20 µg/ml	+	Morphological transformation	
			CaCrO$_4$	3.3–25 µg/ml	+	Inhibition of DNA replication	Bianchi et al., 1984
CrCl$_3$ · 6H$_2$O	10–1000 µg/ml	−	K$_2$Cr$_2$O$_7$	10–1000 µM	+		
None tested			K$_2$Cr$_2$O$_7$	0.035–0.175 µg/ml (as chromium)	+	Morphological transformation	Tsuda and Kato, 1977
Primary Syrian hamster embryo cells							
CrCl$_3$ · 6H$_2$O	11.3 µM	−	K$_2$CrO$_4$ (*synergistic enhancement when combined with B(a)P, N-OH-AAF of NQO)	2.6 µM	+*	Morphological transformation	Rivedal and Sanner, 1981

Continued

Table 2 (*Continued*)

Mammalian In Vitro Test System	Trivalent Chromium Test Compound(s)[a]	Dose Range[b]	Result[c]	Hexavalent Chromium Test Compounds[a]	Dose Range[b]	Result[c]	Assay Type[d]	Reference
Primary mouse fetal cells	None tested			$CaCrO_4$	5–100 μM	+	UDS	Robison et al., 1984
	$CrCl_3 \cdot 6H_2O$	0.04–0.4 μg/ml (as chromium)	+	$K_2Cr_2O_7$	0.015–0.1 μg/ml (as chromium)	+	Morphological transformation	Raffetto et al., 1977
Human skin fibroblasts	Cr(III)-glycine complex	$1–10^4$ μM (as chromium)	–	K_2CrO_4	$0.1–10^4$ μM (as chromium)	+	UDS	Whiting et al., 1979
Human EUE heteroploid cell line	None tested			$K_2Cr_2O_7$	0.001–0.1 μM	–	UDS	Bianchi et al., 1982
Heteroploid HEp-2 human epithelial-like line	$CrCl_3 \cdot 6H_2O$	Not indicated	–	$K_2Cr_2O_7$	10–1000 μM	–	UDS	Bianchi et al., 1983
	$Cr(OH)SO_4$ (as neochromium)	500 μg/ml	+*	$PbCrO_4$ (as Cr yellow)	500 μg/ml	+*	Mitotic cycle alterations	Levis and Majone, 1981
	$Cr_2(SO_4)_3$ (as Cr alum)	500 μg/ml	+*	$PbCrO_4$ (as Mo orange)	500 μg/ml	+		
	$Cr_2(SO_4)_3$	500 μg/ml	+*	$PbCrO_4 \cdot PbO$ (as Cr orange)	500 μg/ml	+		
	$CrCl_3 \cdot 6H_2O$ (*weak)	500 μg/ml	+*	$ZnCrO_4 \cdot Zn(OH)_2 + CrO_3$ (as Zn yellow)	500 μg/ml	+		
	Cr_2O_3 (as chromite)	500 μg/ml	+	$K_2Cr_2O_7$	500 μg/ml	+		

[a]Abbreviations: B(a)P = benzo(a)pyrene, N-OH-AAF = N-hydroxy-2-acetylaminofluorene, NQO = 4-nitroquinoline-1-oxide, XP = xeroderma pigmentosum. [b]Unless specified to the contrary, concentrations expressed in weight/volume units refer to the actual test compound and not specifically to its chromium content. [c]For explanation of superscripts (*, [+]) refer to definition or statement in the corresponding column labelled Test Compound(s); + positive, – negative result. [d]Abbreviations: DPCL = DNA-protein crosslinks, ISCL = inter-strand crosslinks, SCE = sister chromatid exchange, SB = strand breaks, UDS = unscheduled DNA synthesis, aberrations and fragmentation refer to chromosomal damage and DNA damage, respectively.

Table 3 Summary of Genotoxic Effects of Chromium Compounds (Miscellaneous Cultured Cell Systems)

Miscellaneous In Vitro Test System	Trivalent Chromium			Hexavalent Chromium			Assay Type[c]	Reference
	Test Compound(s)	Dose Range[a]	Result[b]	Test Compound(s)	Dose Range[a]	Result[b]		
Schizosaccharomyces pombe	None tested			$K_2Cr_2O_7$	10^2–10^5 μM	+	Forward mutation	Bonatti et al., 1976
Saccharomyces cerevisiae D5	None tested			$PbCrO_4$ (*reduced recombination in the presence of S9 mix)	31.2–500 μg/ml	+*	Mitotic recombination	Nestmann et al., 1979
Chicken embryo hepatocytes	None tested			Na_2CrO_4 (*SB repaired three hours after exposure; low frequency of ISCL)	5 μM	+*	SB, ISCL, DPCL	Cupo and Wetterhahn, 1984
	None tested			Na_2CrO_4	5 μM	+	SB, ISCL, DPCL	Cupo and Wetterhahn, 1985b

[a] Unless specified to the contrary, concentrations expressed in weight/volume units refer to the actual test compound and not specifically to its chromium content. [b] For explanation of superscripts (*) refer to definition or statement in the corresponding column labelled Test Compound(s); + positive, – negative result. [c] Abbreviations: DPCL = DNA-protein crosslinks, ISCL = interstrand crosslinks, SB = strand breaks.

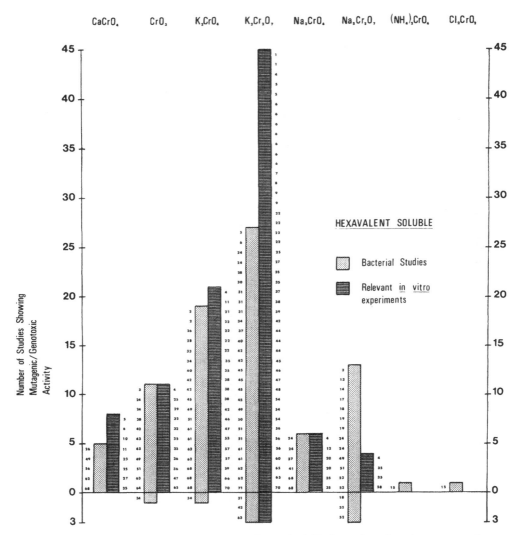

Figure 1. Mutagenicity and other genotoxic effects of soluble hexavalent chromium compounds (includes both moderately and highly water-soluble chromates). Numbers identify the original reference to which the data refer. References and details of the test systems employed are summarized in Table 4. An individual summary for most studies is provided in Table 1 for the bacterial studies and Tables 2 and 3 for relevant in vitro experiments. Entries below the line y = o denote negative results.

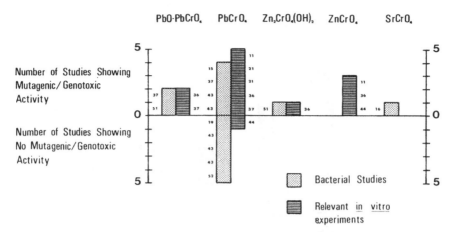

Figure 2. Mutagenicity and other genotoxic effects of insoluble hexavalent chromium compounds (includes water-insoluble to sparingly water-soluble chromates). Numbers identify the original reference to which the data refer. References and details of the test systems employed are summarized in Table 4. An individual summary for most studies is provided in Table 1 for the bacterial studies and Table 2 and 3 for relevant in vitro experiments.

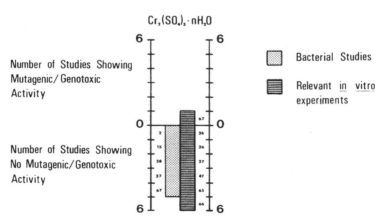

Figure 3. Mutagenicity and other genotoxic effects of trivalent chromium sulphates of unspecified solubilities. Numbers identify the original reference to which the data refer. References and details of the test systems employed are summarized in Table 4. An individual summary for most studies is provided in Table 1 for the bacterial studies and Tables 2 and 3 for the relevant in vitro experiments.

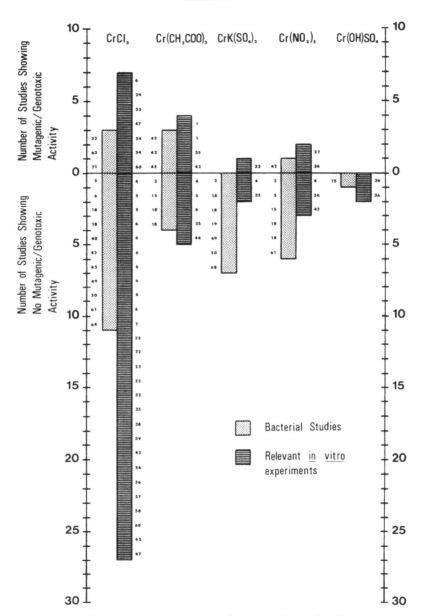

Figure 4. Mutagenicity and other genotoxic effects of water-soluble trivalent chromium compounds. Numbers identify the original reference to which the data refer. References and details of the test systems employed are summarized in Table 4. An individual summary for most studies is provided in Table 1 for the bacterial studies and Tables 2 and 3 for relevant in vitro experiments.

Table 4 References for Mutagenicity and Genotoxicity Data Summarized in Figures 1 to 4[a,b]

Reference Number	Reference	Reference Number	Reference
1	Andersen, 1983	37	Loprieno et al., 1985
2	Arlauskas et al., 1985	38	MacRae et al., 1979
3	Bennicelli et al., 1983	39	Majone et al., 1982
4	Bianchi et al., 1980	40	Marzin and Phi, 1985
5	Bianchi et al., 1982	41	Mohn and Ellenberger, 1977
6	Bianchi et al., 1983	42	Nakamuro et al., 1978
7	Bianchi et al., 1984	43	Nestmann et al., 1979
8	Bigaliev et al., 1977	44	Newbold et al., 1979
9	Bonatti et al., 1976	45	Nishioka, 1975
10	Cantoni and Costa, 1984	46	Oberly et al., 1982
11	Casto et al., 1979	47	Ohno et al., 1982
12	Cupo and Wetterhahn, 1984	48	Paschin and Kozachenko, 1982
13	De Flora et al., 1980	49	Petrilli and De Flora, 1977
14	De Flora and Boido, 1980	50	Petrilli and De Flora, 1978a
15	De Flora, 1981	51	Petrilli and De Flora, 1978b
16	De Flora, 1982	52	Petrilli and De Flora, 1982
17	De Flora et al., 1982	53	Price-Jones et al., 1980
18	De Flora et al., 1984b	54	Raffetto et al., 1977
19	De Flora et al., 1985a	55	Rainaldi et al., 1980
20	DiPaolo and Casto, 1979	56	Rainaldi et al., 1982
21	Douglas et al., 1980	57	Rivedal and Sanner, 1981
22	Fornace et al., 1981	58	Sarto et al., 1980
23	Fradkin et al., 1975	59	Siegfried and Nesnow, 1984
24	Gentile et al., 1981	60	Stella et al., 1982
25	Gomez-Arroyo et al., 1981	61	Tamaro et al., 1975
26	Green et al., 1976	62	Tindall et al., 1978
27	Jacquet and Draye, 1982	63	Tkeshelashvili et al., 1980
28	Kada et al., 1980	64	Tso and Fung, 1981
29	Kaneko, 1976	65	Tsuda and Kato, 1977
30	Kanematsu and Kada 1978	66	Umeda and Nishimura, 1979
31	Kanematsu et al., 1980	67	Venier et al., 1985
32	Koshi, 1979	68	Venitt and Levy, 1974
33	Langerwerf et al., 1985	69	Warren et al., 1981
34	LaVelle and Witmer, 1984	70	Whiting et al., 1979
35	Levis and Majone, 1979	71	Zakour and Glickman, 1984
36	Levis and Majone, 1981		

[a]Bacterial mutagenicity studies include: mutation changes in *Salmonella typhimurium* and *Escherichia coli;* and DNA repair in repair-deficient strains of *Bacillus subtilis, E. coli* and *S. typhimurium.* [b]Relevant *in vitro* cell studies include experiments testing for: chromosomal aberrations, sister chromatid exchanges, DNA fragmentation, DNA strand breaks, DNA interstrand and DNA-protein cross-links, alkali lability of DNA, unscheduled DNA synthesis, inhibition of DNA replication, inhibition of cell growth, mitotic cycle alterations, morphological transformation and gene mutations.

Table 5 A Summary of Bacterial Mutagenicity Studies of Chromium Compounds[a,b]

	Number of Studies		
	+	−	Total
Hexavalent			
Soluble	78	8	86
Insoluble	8	5	13
Total	86	13	99
Trivalent			
Soluble	21	54	75
Insoluble	0	8	8
Total	21	62	83

[a]Bacterial mutagenicity studies include mutation changes in *Salmonella typhimurium* and *Escherichia coli,* and DNA repair in repair-deficient strains of *Bacillus subtilis, E. coli,* and *S. typhimurium.* [b]Soluble includes moderately and highly water-soluble; insoluble includes sparingly water-soluble and water-insoluble.

Table 6 A Summary of Relevant In Vitro Genotoxicity Studies of Chromium Compounds[a,b]

	Number of Studies		
	+	−	*Total*
Hexavalent			
Soluble	95	3	98
Insoluble	11	1	12
Total	106	4	110
Trivalent			
Soluble	15	38	53
Insoluble	1	7	8
Total	16	45	61

[a]Relevant *in vitro* cell studies include experiments testing for chromosomal aberrations, sister chromatid exchanges, DNA fragmentation, DNA strand breaks, DNA interstrand and DNA-protein crosslinks, alkali lability of DNA, unscheduled DNA synthesis, inhibition of cell growth, inhibition of DNA replication, mitotic cycle alterations, morphological transformation, and gene mutations. [b]Soluble includes moderately and highly water-soluble; insoluble includes sparingly water-soluble and water-insoluble.

would also appear to exhibit borderline character since it has affinity for both oxygen (e.g., as in the chromate ion) and sulfur donor groups (e.g., the formation of chromate-thioesters; Connett and Wetterhahn, 1983). What distinguishes chromium from many other metals is its existence as oxyanions in its higher oxidation states. Generally speaking, cations of metals (including the transition metals) are excluded from cells, although specific transport mechanisms exist for endogenous metal ions (e.g., involving metallothionein for Cu^{2+} and Zn^{2+}; transferrin for Fe^{3+}; and ion pumps for Na^+, K^+, Ca^{2+}). Such mechanisms are highly specific and circumvent general access to cells. By contrast, anions usually are not excluded as readily as cations. Presumably this occurs because recognized transport pathways for similar but endogenous anions exist (e.g., phosphate and sulphate). There is good evidence that the chromate ion can utilize such existing anion channels in eukaryotic and prokaryotic cells (Campbell et al., 1981; Kitagawa et al., 1982) (see Chapter 14). Furthermore, toxic anions such as chromate, arsenate, and sulphite have the ability to alter membranes chemically. Numerous experiments have indeed confirmed that chromate freely diffuses across cell membranes, while the Cr(III) ion does not.

Cell-culture studies have shown that cellular uptake in vitro of chromate is at least 10 times greater than that of Cr(III) from equimolar solutions (Gray and Sterling, 1950, for erythrocytes; Levis et al., 1978b for Syrian hamster BHK-21 cells; see Cupo and Wetterhahn, 1984 for mention of other cell types). (Discordant results for the BHK-21 cell line are reported by Levis and Majone, 1981.) Speciation studies have revealed that intracellular reduction of chromate to Cr(III) occurs (Levis et al., 1978a,b; Levis and Majone, 1981; Kitagawa et al., 1982; Levis and Bianchi, 1982). At least in the case of erythrocytes, the cell-associated chromium is primarily cytosolic and is predominantly bound to hemoglobin (Kitagawa et al., 1982). Since the resultant Cr(III) has a high affinity for intracellular binding sites (see Chapter 2), it is not surprising that little of the chromate accumulated by bovine erythrocytes can be removed by washing with isotonic buffer (Kitagawa et al., 1982).

It appears that chromate uptake is not a prerequisite for all cytotoxic effects. Studies with intact cultured hamster fibroblasts have demonstrated that chromate, on initial exposure, enhances the cell membrane permeability to nucleosides. This is followed by a second phase in which inhibition of uptake occurs. Structural modification of cell membranes, presumably by the oxidative action of chromate, is believed to stimulate simple inward diffusion of the nucleosides (Bianchi et al., 1979). Subsequently, when Cr(III) accumulates at the membrane surface, nucleoside permeases are thought to be inactivated and perhaps even denatured. Separate administration of Cr(III) salts do indeed inhibit nucleoside uptake (Levis et al., 1978b).

2.2.2 Experimental Responses and Observations

As discussed in Chapter 2, chromates have a range of water solubilities. The positive results indicated in Figure 2 for hexavalent "insoluble" chromium compounds imply that some dissolution in the test medium occurred. Because biological media contain ligands (e.g., proteins, amino acids) that have a high affinity for metal ions (e.g., Pb^{2+} and Zn^{2+}), the solubility of relatively water-insoluble metal chromates may be expected to be augmented. Such enhancement of solubility and of the concomitant mutagenic potency has recently been demonstrated for basic lead chromate ($PbCrO_4 \cdot PbO$) by adding the chelating agent nitrilotriacetic acid to the test medium (Loprieno et al., 1985). The response in in vitro tests of water-soluble forms depends on the effective dissolved concentration of the chromate anion attained. In many of the studies, the "insoluble" Cr(VI) compounds were dissolved in base (e.g., 0.5 M sodium hydroxide) or acid (e.g., 0.5M hydrogen chloride). In test systems employing cells with phagocytic properties, direct uptake of particulates of micron size may also occur (see Section 2.3.2).

The mutagenicity of Cr(VI) compounds in the Ames test varies depending upon the bacterial stain used (Bennicelli et al., 1983). Of the *Salmonella* strains tested, TA102 is most sensitive. This is not surprising since this bacterium scores strongly with mutagens possessing oxidation/reduction properties. The reduction of hexavalent chromium to the trivalent state involves a number of intermediate oxidation states of high reactivity (see Sections 2.4.2 and 2.4.3). The interaction of these intermediates with molecular oxygen to generate toxic radicals (e.g., the superoxide anion and the hydroxyl radical) may be anticipated. Studies with a number of other *Salmonella* strains implicate the involvement of frameshift mutations and basepair substitutions (Petrilli and De Flora, 1977; Maron and Ames, 1983; Bianchi and Levis, 1984). And finally, it is of interest to mention the work of Baker et al. (1984), who showed that "false or phenotypic reversion plays an insignificant role in Cr(VI) mutagenesis." In fact, for *Salmonella* TA100, they were able to identify unequivocally that the majority of nonreplicating colonies were his[+] and that the cytotoxicity of chromate inhibited their growth.

As discussed in Section 2.4.1, microsomal fractions employed to activate organic mutagens suppress mutagenicity of chromate.

2.3 Differential Responses of Chromium(III) Compounds

2.3.1 Experimental Responses and Observations

Trivalent chromium salts, as indicated by the data in Figures 3 and 4 and Tables 5 and 6, have yielded some positive outcomes in in vitro test systems. To produce equivalent effects, usually about 10- to 100-fold less chromate is needed than of Cr(III) (e.g., Nakamuro et al., 1978; Levis and Majone, 1979; Elias et al., 1983). As already pointed out, this difference in response appears to be a matter of intracellular availability.

A number of determinants of the cellular uptake of Cr(III) compounds have been identified. The work of Warren et al. (1981) suggests that if the right ligand environment is provided, trivalent chromium can cross the cell membrane. They found that four of the 17 trivalent chromium compounds tested were active in both an *E. coli* differential repair assay and in the Ames test; four others were positive only in the repair assay. The doubly active complexes of greatest potency (three of four) contained aromatic amine ligands, namely, 2,2'-bipyridine and 1,10-phenanthroline, and bore a single positive charge. Unfortunately, controls for the inherent mutagenicity of these ligands were not determined and thus confirmation of these observations is required. All but one of the complexes active only in the repair assay had a $3+$ charge and had saturated amines as ligands (e.g., ethylenediamine). It is evident that of the trivalent compounds referenced in Figure 4, the Cr(III) chloride hexahydrate has scored positively most often relative to the other Cr(III) compounds. Interestingly, the commercially available green-colored hexahydrate of Cr(III) chloride, $[CrCl_2(H_2O)_4]Cl \cdot 2H_2O$, yields the singly charged ion $CrCl_2(H_2O)_4^+$ on initial dissolution, while the violet-colored salt $[Cr(H_2O)_6]Cl_3$ gives $Cr(H_2O)_6^{3+}$. Perhaps the reduced charge of the former species facilitates its uptake. Equilibrium considerations predict the presence of similar low-charged species for Cr(III) acetate [e.g., $Cr(CH_3COO)^{2+}$, $Cr(CH_3COO)_2^+$. Salicylate and citrate complexes of Cr(III) have also been found to be genotoxic by the *Bacillus subtilis* rec-assay, while the chloride and sulfate salts were not positive (Gentile et al., 1981). Since Cr(III) salts are prone to hydrolysis, including polymerization, biological responses may be expected to show a strong dependence on the nature of the salt tested and on the composition of the test medium and its pH (see Chapter 2). It is noteworthy that Langerwerf et al. (1985) observed that amino acid complexes of Cr(III) were not mutagenic in *Salmonella typhimurium* assays.

2.3.2 Endocytosis and Genotoxicity

There is some evidence that particulates and dissolved forms of Cr(III) compounds are taken up by endocytosis. Sister-chromatid exchanges (SCE) has been the outcome measurement employed to assess this. In a continuous murine macrophage cell-line, the uptake of Cr(III) oxide (Cr_2O_3) was confirmed by phase-contrast microscopy (Andersen, 1983). Similarly, when Chinese hamster V79 cells were treated with the hexahydrate of Cr(III) chloride (water-soluble) and Cr_2O_3 (water-insoluble), cell-cycle delay and a statistically significant ($p <$ 0.001) dose- and time-dependent increase in SCE frequency over the spontaneous level were observed (Elias et al., 1983). The phagocytosed Cr_2O_3 particulates were usually seen in the cytoplasm, often near the nucleus. The time-dependence of the SCE damage incurred by Cr(III) chloride suggested uptake by pinocytosis (i.e., the internalization of a dissolved substance). Uptake by this route has also been inferred for SCE induction in Chinese hamster fibroblasts by Cr(III) chloride and tannins derived from Cr(III) sulfates (Venier et al., 1985). Although these compounds were genotoxic, they failed to score positively in the Ames test. More recently, Elias et al. (1986) have demonstrated cellular uptake and intracellular localization of particulates ($< 1~\mu m$ in size) of crystalline Cr_2O_3 and observed the induction of a statistically significant ($p < 0.001$) increase in the mutation frequency of up to 10-fold over the controls in V79 Chinese hamster cells.

2.3.3   Contamination of Trivalent Chromium with Hexavalent Chromium Compounds

A word of caution is warranted concerning in vitro tests with Cr(III) compounds. A number of reports have demonstrated that Cr(III) compounds, especially those of industrial origins such as tannins, may be contaminated with hexavalent chromium. In such instances, Cr(VI) contamination may account for all or part of the observed genotoxicity (Bianchi et al., 1983; Elias et al., 1983; Venier et al., 1985).

2.4   Mechanistic Considerations

2.4.1   Metabolic Reduction of Hexavalent Chromium

The reduction in mutagenic response observed when S9 microsomal fractions of mammalian liver are added to in vitro assays finds a ready explanation in the differential bioavailability discussed previously (Section 2.2.1) for the Cr(VI) and Cr(III) oxidation states. The microsomal fraction has been shown to reduce hexavalent chromium to trivalent chromium (De Flora, 1978; Löfroth, 1978; Gruber and Jennette, 1978). In fact, many reducing agents (e.g., ascorbic acid, sodium sulphite, glutathione, NADPH, and NADH) are known to deactivate hexavalent chromium directly (Petrilli and De Flora 1978a; Connett and Wetterhahn, 1983; De Flora et al., 1985b). Even human gastric juice and saliva, as well as human erythrocyte lysate, reduced the mutagenic capacity of sodium dichromate (De Flora and Boido, 1980; Petrilli and De Flora, 1982). Conversely, prior oxidation of Cr(III) to Cr(VI) restores the mutagenic response (Petrilli and De Flora, 1978b). Thus, it is reasonable to conclude that the metabolic deactivation of chromate mutagenicity involves the conversion of chromate to Cr(III), which as reviewed in Section 2.3 is nonmutagenic or possesses low mutagenic potency in most assay systems.

Evidence is mounting that chromate is reduced by microsomal electron-transport cytochrome P-450 systems. The relatively slow rate of reduction in control samples (microsomes without NADH or NADPH; NADH or NADPH in the absence of microsomes), and the inhibition by P-450–specific inhibitors confirm that this reaction is enzymatic (Gruber and Jennette, 1978; Garcia and Wetterhahn Jennette, 1981). It is also known that mitochondrial preparations from rat liver possess Cr(VI) reducing capability (Ryberg and Alexander, 1984). Sodium chromate in the micromolar range caused the fast inhibition of mitochondrial respiration, which was specific for NAD-linked substrates. Ryberg and Alexander (1984) suggest that Cr(VI) reduction is therefore probably coupled to the electron flow in Complex I (the NADH-Q reductase component) of the electron-transport chain. Further, microsomal fractions from rat liver and lung have been shown to involve NADPH- or NADH-requiring mechanisms (De Flora et al., 1985b). These authors demonstrated that the major chromate-reducing activity resided in the cytosolic fraction and experimental indications support a role for the enzyme DT-diaphorase. This enzyme has been implicated in the detoxification of some organic compounds.

2.4.2 Chromate-Induced DNA Lesions

Chromate-induced DNA strand breaks, DNA interstrand crosslinks and DNA-protein crosslinks have been observed in in vitro and in vivo systems as outlined later. Calcium chromate has also been suspected of promoting alkali-labile sites in Chinese hamster ovary cells (Cantoni and Costa, 1984).

In vitro, chromate-induced strand breakage is often accompanied by stimulation of DNA repair as measured by unscheduled DNA synthesis (Whiting et al., 1979; Robison et al., 1984). In rats treated with sodium dichromate, it has been demonstrated that DNA interstrand crosslinks are repaired more quickly than DNA-protein crosslinks (Tsapakos et al., 1981, 1983). It is interesting that in rat kidney and lung, DNA-protein crosslinks persisted longer than in the liver. Differential repair tendencies have also been reported for chromate-induced DNA damage in cultures of chick embryo hepatocytes (Cupo and Wetterhahn, 1984).

Recently, Cupo and Wetterhahn (1985a) showed that chromium was bound to whole chromatin, polynucleosomes (DNA-histone octamer complex), DNA, nuclear proteins, and a cytoplasmic RNA fraction in liver and kidney tissues of rat after intraperitoneal injection of sodium dichromate or Cr(III) chloride. Initially, Cr(VI) treatment resulted in a greater fraction of chromatin-bound chromium than the Cr(III) treatment. At 40 hours after the injections, the levels of chromium bound to DNA were similar in both treatments. By contrast to chromate exposure, the in vitro Cr(III) complexes associated with chromatin and resulting from Cr(III) chloride injection did not result in consistently detectable DNA-protein crosslinks, DNA interstrand crosslinks, or DNA strand breaks. This suggests that Cr(III) binding to chromatin is not sufficient to cause lesions. Recent reports help to unravel these divergent observations and warrant closer examination.

In rats treated intravenously "with ^{51}Cr-labelled sodium chromate a large portion of the liver radioactivity was detected in the supernatant, whereas in the animals given trivalent chromium the majority of the liver ^{51}Cr was localized in cell organelles" including nucleus, mitochondria, lysosomes, and microsomes (Manzo et al., 1983). Biliary excretion of a Cr(VI) administered dose is considerably higher than for the corresponding Cr(III) dose (Norseth et al., 1982; Manzo et al., 1983). Plasma-to-liver chromium concentration ratios were similar for both hexavalent and trivalent doses (Manzo et al., 1983). The significantly higher plasma-to-bile and liver-to-bile ratios for Cr(III) administration is consistent with lower excretion via the bile and the smaller fraction of chromium associated with low-molecular weight substances. Excretion into the bile appears to depend on the latter species. Yamamoto et al. (1984) and Wada et al. (1983) have identified a low-molecular-weight complex of relative molecular mass of 1500 with glutamic acid, glycine, and cysteine as the major constituent amino acids. This ligand binds Cr(III) avidly and significant exchange (25%) was observed to transferrin but not to albumin (0.5%). It is significant that transferrin, the iron-transport protein, is taken up by cells in tissues (e.g., liver) by receptor-mediated internalization (Dautry-Varsat et al., 1983). It is also the major protein in the plasma binding Cr(III) (see Chapter 2). The apo-form of the above low-molecular-weight complex incorporates trivalent chromium but not chromate. Presumably, a reducing agent is required for the latter. Indeed, Norseth et al. (1982) have demonstrated in rat that biliary excretion is reduced when livers are depleted of glutathione with cyclohexene treatment.

The differential interaction with DNA in vivo observed by Cupo and Wetterhahn (1985a) may well depend on speciation factors similar to those described for biliary excretion. Because Cr(III) is substitutionally inert, Cr(III) complexes formed in situ during reduction of Cr(VI) may be quite different from those derived from Cr(III) by direct interaction. By contrast, the intermediate IV and V oxidation states are labile and may facilitate the attachment of a ligand during the reduction of the hexavalent to the trivalent state (Connett and Wetterhahn, 1983). In studies with isolated DNA, the interaction with Cr(III) was minimal and Cr(VI) only formed complexes in the presence of a complete microsomal reducing system (Tsapakos and Wetterhahn, 1983). Recently, for chicken embryo hepatocytes in vitro, Cupo and Wetterhahn (1985b) have proposed that under glutathione-rich conditions in the cell, chromate is reduced to Cr(IV) which would appear to promote the formation of DNA strand breaks and DNA-protein crosslinks. However, upon depletion of glutathione with concurrent elevated levels of cytochrome P-450, the formation of Cr(V) is postulated, which is purported to promote DNA lesions less effectively (see later).

2.4.3 Oxidative Damage of Chromosomes and DNA

An explanation for the differential effects of hexavalent and trivalent chromium not considered widely in published reports is that the intermediate chromium oxidation states can promote the formation of active oxygen radicals and/or potentiate hydrogen peroxide. Oxygen radicals are known to damage chromosomes and DNA (Halliwell and Gutteridge, 1984). Reactive oxygen species have been implicated as mediators of the mutagenic effects of gamma radiation and of tumor promotion (Section 1.2.3). The strong scoring of chromate in TA102 *Salmonella* strain mutagenicity testing supports this perspective (Section 2.2.2). So does the co-requirement of glutathione as a reducing agent in the test systems described earlier.

Considerable evidence has accumulated that both Cr(V) and Cr(IV) are formed as transient intermediates in the reduction of Cr(VI) (i.e., chromates) to Cr(III). Of these intermediates, those of Cr(V) are somewhat more stable. The magnitude of the half-lives amounts to minutes, allowing their characterization by spectrometric techniques (Wiberg, 1965; Beattie and Haight, 1972).

Chromium(V) has also been characterized as a stable intermediate (lifetimes of minutes to hours) by electron spin resonance when Cr(VI) is reduced by rat liver microsomes (Polnaszek, 1981; Wetterhahn Jennette, 1982) or by reduced glutathione (Goodgame and Joy, 1986). Goodgame et al. (1982) were also able to generate Cr(V) when chromate was mixed with ribonucleotides at pH 6 to 7. Interestingly, the reduction did not occur with the corresponding deoxy-ribonucleotides, suggesting that the oxidation of the ribose ring is involved.

2.4.4   Infidelity of DNA Replication In Vitro

DNA lesions (e.g., DNA strand breaks, DNA interstrand crosslinks, and DNA-protein crosslinks) were considered in the rationalization of chromium muta-genicity in the previous sections. An alternative perspective of chromium mutagenicity involves the direct effect of chromium compounds on the fidelity of DNA replication. Three modes of metal-ion binding influencing the fidelity of DNA synthesis have been identified: (1) attachment to the incoming nucleoside triphosphate that alters substrate conformation; (2) binding to the enzyme altering its conformation and function; and (3) complex formation with the DNA template altering the base specificity (Zakour et al., 1981).

Initially, Cr(II) chloride ($CrCl_2$) and Cr(VI) oxide (CrO_3) were observed to decrease the fidelity of DNA synthesis in vitro (Sirover and Loeb, 1976). Bianchi et al. (1983) later showed that potassium dichromate also scored positively. Chromium(II) is unstable in an aerobic environment and is converted to Cr(III) by air oxidation (see Chapter 2). As already discussed (Section 2.4), Cr(VI) would be reduced to Cr(III) in the presence of reducing agents (perhaps the nucleotides themselves or other additives to the test medium). Presumably, it is the Cr(III) form that interferes with the replication process. Support for this interpretation is derived from the results of Tkeshelashvili et al. (1980), who found that both Cr(VI) oxide (CrO_3) and Cr(III) chloride (hexahydrate) induced the same error rates employing both a synthetic polynucleotide template and a natural template (DNA from a mutant of bacteriophage $\Phi_x 174$). They avoided the often-mentioned Cr(III) precipitation problem by removing the precipitate before isolating the newly synthesized product. The formation of the precipitate, which likely is a nucleotide complex complicated by hydrolysis at the basic pH of 8 employed, appears to be avoided when starting with Cr(II) or Cr(VI). As discussed in Section 2.4.2, such differential interaction with the template, substrate, or enzyme by direct and indirect sources of Cr(III) may be expected.

3.   CYTOGENETIC STUDIES

In vitro studies with human peripheral lymphocytes have shown that dissolved Cr(VI) compounds induce a dose-related increase in sister-chromatid exchanges (SCE) (Douglas et al., 1980; Gomez-Arroyo et al., 1981; Stella et al., 1982). Only in one incident has a similar effect been attributed to Cr(III) compounds.

Andersen (1983) found a statistically significant increase with Cr(III) acetate. Chromium compounds of both oxidation states have been shown in vitro to produce chromosome aberrations, chromosome and chromatid gaps, and breaks (Douglas et al., 1980; Stella et al., 1982). In a comparison of the hexahydrate of Cr(III) chloride and potassium dichromate, considerably higher concentrations (100-fold) were required to increase the frequency of chromosome aberrations (Stella et al., 1982). Statistically significant increases ($p < 0.001$) of chromosome-type aberrations and of SCE have been reported in peripheral lymphocytes of workers exposed to chromic acid fumes, CrO_3 (Sarto et al., 1982; Stella et al., 1982). It is noteworthy that the SCE frequencies increased most significantly in younger workers. In a preliminary study (Imreh and Radulescu, 1982), workers exposed to potassium dichromate were found to exhibit significant diffferences in chromosome-type aberration frequency, in chromatid-type aberration frequency, and in micronucleus scoring of lymphocytes. Collectively, these data support the view that Cr(VI) compounds are genotoxic in vivo.

4.   MODEL OF CHROMIUM MUTAGENICITY AND CARCINOGENICITY

The ability of the chromate anion to diffuse readily through cell membranes and its reasonable longevity at physiological pH, even though it is relatively strongly oxidizing (see Chapter 2), are crucial to toxicity model depicted in Figure 5. In part, the uptake seems to involve existing anion pathways (e.g., of sulphate). Chromate is not only reduced to Cr(III) directly by extracellular and intracellular low-molecular-weight molecules (e.g., glutathione, ascorbic acid, lactic acid) and proteins, but also enzymatically within the cell by microsomal, mitochondrial, and cytosolic enzyme systems. Furthermore, because of its avid nonlabile complex-formation ability, Cr(III) accumulates in storage compartments (such as cell nuclei) from which it is released slowly. Providing Cr(III) is in the correct ligand environment or form, it too can be transported across biological membranes, by diffusion as low-molecular-weight complexes (e.g., of amino acids and oxalate), when bound to protein by pinocytosis (e.g., transferrin), or as particulates (phagocytosis). All these transport and distribution mechanisms are depicted in Figure 5.

Inside the nucleus, Cr(III) is believed to be the critical form. Because the Cr(III) cation interacts strongly with nuclear enzymes, proteins, nucleotides, and DNA, Cr(III)-induced structural alterations may result in DNA replication and repair errors (Bianchi and Levis, 1984). Although not indicated in Figure 5, direct oxidative damage of Cr(VI) to genetic material (see Chapter 14), as well as indirect oxidative damage via the generation of oxygen radicals, are also potential mechanisms.

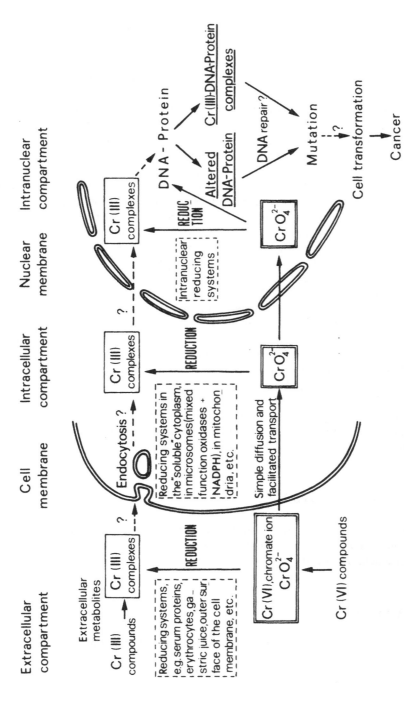

Figure 5. Cellular model of chromium mutagenicity and carcinogenicity. Reproduced with permission from Levis and Bianchi (1982). A complementary model emphasizing the importance of direct oxidative damage is provided by Ono (Chapter 14).

433

5.   SUMMARY AND CONCLUDING REMARKS

Nearly all Cr(VI) compounds have been shown to be potent mutagens, in contrast to Cr(III) compounds. The majority of mutagenicity tests employed the Ames assay. Other genotoxicity tests in vitro and in vivo reproduced this differential response for these two common chromium oxidation states. Chromate has ready access to cells, while entry by Cr(III) is much more restricted. In vivo, the metabolism, distribution, and transport of hexavalent chromium and trivalent chromium are quite different. This may account for the lower genotoxic effect of Cr(III) observed in rodents. Both oxidation states were equally effective in reducing the fidelity of DNA synthesis in vitro with synthetic and natural templates. Collectively, the experimental data concur with the interpretation that the Cr(III) form is the major putative agent responsible for the molecular events leading to mutagenicity. A contribution from oxidative damage at the chromosomal and molecular levels can, however, not be discounted.

The proven DNA damaging effects, clastogenicity, and mutagenicity of most Cr(VI) compounds appear to be predictive of carcinogenicity in man. There is evidence that all forms of Cr(VI), both water-soluble and water-insoluble compounds, are respiratory carcinogens in humans. There is no corresponding evidence that Cr(III) compounds increase the risk of respiratory cancer (see Chapter 17). This finding is consistent with the much weaker response of Cr(III) in the in vitro test systems reviewed. As pointed out in Sections 1.1 and 1.2, the exact relationship of DNA damage, clastogenicity, and mutagenicity in in vitro tests to cancer initiation and development in humans are not known (also see ICPEMC, 1982). Neither is their exact relevance to teratogenicity and developmental toxicity understood, although it is estimated that some 20% of human developmental deficiencies can be attributed solely or primarily to mutation in a prior germ line (Wilson, 1977).

ACKNOWLEDGMENTS

Financial support for the preparation of this review was received from the Occupational Health and Safety Division of the Ontario Ministry of Labour, and is gratefully acknowledged.

REFERENCES

Ames, B.N. (1983). "Dietary carcinogens and anticarcinogens. Oxygen radicals and degenerative diseases." *Science* **221**, 1256–1264.

Andersen, O. (1983). "Effects of coal combustion products and metal compounds on sister chromatid exchange (SCE) in a macrophage like cell line." *Environ. Health Perspect.* **47**, 239–253.

Arlauskas, A., Baker, R.S.U., Bonin, A.M., et al. (1985). "Mutagenicity of metal ions in bacteria." *Environ. Res.* **36**, 379–388.

Baker, R.S.U., Bonin, A.M., Arlauskas, A., et al. (1984). "Chromium(VI) and apparent phenotypic reversion in *Salmonella* TA100." *Mutat. Res.* **138**, 127–132.

Beattie, J.K., and Haight, G.P., Jr. (1972). "Chromium(VI) oxidations of inorganic substrates." In J.O. Edwards, Ed., *Inorganic Chemistry Series, Inorganic Reaction Mechanisms, Part II.* Interscience Publishers (John Wiley & Sons), New York, pp. 93–145.

Bennicelli, C., Camoirano, A., Petruzelli, S., et al. (1983). "High sensitivity of Salmonella TA102 in detecting hexavalent chromium mutagenicity and its reversal by liver and lung preparations." *Mutat. Res.* **122**, 1–5.

Beyersmann, D., Köster, A., and Buttner, B. (1984). "Model reactions of chromium compounds with mammalian and bacterial cells." *Toxicol. Environ. Chem.* **8**, 279–286.

Bianchi, V., and Levis, A.G. (1984). "Mechanisms of chromium genotoxicity." *Toxicol. Environ. Chem.* **9**, 1–25.

Bianchi, V., Levis, A.G., and Saggioro, D. (1979). "Differential cytotoxic activity of potassium dichromate on nucleoside uptake in BHK fibroblasts." *Chem.-Biol. Interactions* **24**, 137–151.

Bianchi, V., Dal Toso, R., Debetto, P., et al. (1980). "Mechanisms of chromium toxicity in mammalian cell cultures." *Toxicology* **17**, 219–224.

Bianchi, V., Nuzzo, F., Abbondandolo, A., et al. (1982). "Scintillometric determination of DNA repair in human cell lines: A critical appraisal." *Mutat. Res.* **93**, 447–463.

Bianchi, V., Celotti, L., Lanfranchi, G., et al. (1983). "Genetic effects of chromium compounds." *Mutat. Res.* **117**, 279–300.

Bianchi, V., Zantedeschi, A., Montaldi, A., and Majone, F. (1984). "Trivalent chromium is neither cytotoxic nor mutagenic in permeabilized hamster fibroblasts." *Toxicol. Lett.* **23**, 51–59.

Bigaliev, A.B., Elemesova, M.S., and Turebaev, M.N. (1977). "Evaluation of the mutagenous activity of chromium compounds." *Gig. Tr. Prof. Zabol.* **6**, 37–40.

Bonatti, S., Meini, M., and Abbondandolo, A. (1976). "Genetic effects of potassium dichromate in *Schizosaccharomyces pombe.*" *Mutat. Res.* **38**, 147–150.

Brodeur, G.M. (1986). "Molecular correlates of cytogenetic abnormalities in human cancer cells: Implication for oncogene activation." *Prog. Hematol.* **14**, 229–256.

Campbell, C.E., Gravel, R.A., and Worton, R.G. (1981). "Isolation and characterization of Chinese hamster cell mutants resistant to the cytotoxic effects of chromate." *Somat. Cell Genet.* **7**, 535–546.

Cantoni, O., and Costa, M. (1984). "Analysis of the induction of alkali sensitive sites in the DNA by chromate and other agents that induce single strand breaks." *Carcinogenesis* **5**, 1207–1209.

Casto, B.C., Meyers, J., and DiPaolo, J.A. (1979). "Enhancement of viral transformation for evaluation of the carcinogenic or mutagenic potential of inorganic metal salts." *Cancer Res.* **39**, 193–198.

Connett, P.H., and Wetterhahn, K.E. (1983). "Metabolism of the carcinogen chromate by cellular constituents." *Structure and Bonding (Berlin)* **54**, 93–124.

CRC (1985). *Handbook of Chemistry and Physics.* 66th ed. CRC Press, Florida, pp. 13–80.

Cupo, D.Y., and Wetterhahn, K.E. (1984). "Repair of chromate-induced DNA damage in chick embryo hepatocytes." *Carcinogenesis* **5**, 1705–1708.

Cupo, D.Y., and Wetterhahn, K.E. (1985a). "Binding of chromium to chromatin and DNA from liver and kidney of rats treated with sodium dichromate and chromium(III) chloride *in vivo.*" *Cancer Res.* **45**, 1146–1151.

Cupo, D.Y., and Wetterhahn, K.E. (1985b). "Modification of chromium(VI)-induced DNA damage by glutathione and cytochromes P-450 in chicken embryo hepatocytes." *Proc. Natl. Acad. Sci. USA* **82**, 6755–6759.

Dautry-Varsat, A., Ciechanover, A., and Lodish, H.F. (1983). "pH and the recycling of transferrin during receptor-mediated endocytosis." *Proc. Natl. Acad. Sci. USA* **80**, 2258–2262.

De Flora, S. (1978). "Metabolic deactivation of mutagens in the *Salmonella*-microsome test." *Nature* **271**, 455–456.

De Flora, S. (1981). "Study of 106 organic and inorganic compounds in the *Salmonella* microsome test." *Carcinogenesis* **2**, 283–298.

De Flora, S. (1982). Cited in Levis and Bianchi, 1982.

De Flora, S., and Boido, V. (1980). "Effect of human gastric juice on the mutagenicity of chemicals." *Mutat. Res.* **77**, 307–315.

De Flora, S., Boido, V. and Picciotto, A. (1980). "Metabolism of mutagens in the gastric environment." *Mutat. Res.* **74**, 187–188.

De Flora, S., Zanacchi, P., Bennicelli, C., and Arillo, A. (1982). "Influence of liver S9 preparations from rats and rainbow trout on the activity of four mutagens." *Toxicol. Lett.* **10**, 345–349.

De Flora, S., Zanacchi, P., Camoirano, A., et al. (1984a). "Genotoxic activity and potency of 135 compounds in the Ames reversion test and in a bacterial DNA-repair test." *Mutat. Res.* **133**, 161–198.

De Flora, S., Camoirano, A., Zanacchi, P., and Bennicelli, C. (1984b). "Mutagenicity testing with TA97 and TA102 of 30 DNA-damaging compounds, negative with other *Salmonella* strains." *Mutat. Res.* **134**, 159–165.

De Flora, S., DeRenzi, G.P., Camoirano, A., et al. (1985a). "Genotoxicity assay of oil dispersants in bacteria (mutation, differential lethality, SOS DNA-repair) and yeast (mitotic crossing-over)." *Mutat. Res.* **158**, 19–30.

De Flora, S., Morelli, A., Basso, C., et al. (1985b). "Prominent role of DT-diaphorase as a cellular mechanism reducing chromium(VI) and reverting its mutagenicity." *Cancer Res.* **45**, 3188–3196.

DiPaolo, J.A., and Casto, B.C. (1979). "Quantitative studies of *in vitro* morphological transformation of Syrian hamster cells by inorganic metal salts." *Cancer Res.* **39**, 1008–1013.

Douglas, G.R., Bell, R.D.L., Grant, C.E., et al. (1980). "Effect of lead chromate on chromosome aberration, sister-chromatid exchange and DNA damage in mammalian cells *in vitro*." *Mutat. Res.* **77**, 157–163.

Duesberg, P.H. (1985). "Activated proto-onc genes: sufficient or necessary for cancer?" *Science* **228**, 669–677.

Elias, Z., Poirot, O., Schneider, O., et al. (1986). "Cellular uptake, cytotoxic and mutagenic effects of insoluble chromic oxide in V79 Chinese hamster cells." *Mutat. Res.* **169**, 159–170.

Elias, Z., Schneider, O., Aubry, F., et al. (1983). "Sister chromatid exchanges in Chinese hamster V79 cells treated with the trivalent chromium compounds chromic chloride and chromic oxide." *Carcinogenesis* **4**, 605–611.

Farber, E. (1981). "Chemical carcinogenesis." *New Engl. J. Med.* **305**, 1379–1389.

Farber, E. (1984). "Chemical carcinogenesis: A current biological perspective." *Carcinogenesis* **5**, 1–5.

Fornace, A.J., Jr. (1982). "Detection of DNA single-strand breaks produced during the repair of damage by DNA-protein cross-linking agents." *Cancer Res.* **42**, 145–149.

Fornace, A.J., Jr., Seres, D.S., Lechner, J.F., and Harris, C.C. (1981). "DNA-protein cross-linking by chromium salts." *Chem.-Biol. Interact.* **36**, 345–354.

Fradkin, A., Janoff, A., Lane, B.P., and Kuschner, M. (1975). "*In vitro* transformation of BHK21 cells grown in the presence of calcium chromate." *Cancer Res.* **35**, 1058–1063.

Friedberg, E.C. (1985). *DNA Repair*. W.H. Freeman, New York, pp. 505–574.

Garcia, J.D., and Wetterhahn Jennette, K. (1981). "Electron-transport cytochrome P-450 system is involved in the microsomal metabolism of the carcinogen chromate." *J. Inorg. Biochem.* **14**, 281–295.

Gentile, J.M., Hyde, K., and Schubert, J. (1981). "Chromium genotoxicity as influenced by complexation and rate effects." *Toxicol. Lett.* **7**, 439–448.

Gomez-Arroyo, S., Altamirano, M., and Villalobos-Pietrini, R. (1981). "Sister-chromatid exchanges induced by some chromium compounds in human lymphocytes *in vitro*." *Mutat. Res.* **90**, 425–431.

Goodgame, D.M.L., and Joy, A.M. (1986). "Relatively long-lived chromium(V) species are produced by the action of glutathione on carcinogenic chromium(VI)." *J. Inorg. Biochem.* **26**, 219–224.

Goodgame, D.M.L., Hayman, P.B., and Hathway, D.E. (1982). "Carcinogenic chromium(VI) forms chromium(V) with ribonucleotides but not with deoxyribonucleotides." *Polyhedron* **1**, 497–499.

Gray, S.J., and Sterling, K. (1950). "The tagging of red cells and plasma proteins with radioactive chromium." *J. Clin. Invest.* **29**, 1604–1613.

Green, M.H.L., Muriel, W.J., and Bridges, B.A. (1976). "Use of a simplified fluctuation test to detect low levels of mutagens." *Mutat. Res.* **38**, 33–42.

Gruber, J.E., and Jennette, K.W. (1978). "Metabolism of the carcinogen chromate by rat liver microsomes." *Biochem. Biophys. Res. Commun.* **82**, 700–706.

Halliwell, B., and Gutteridge, J.M.C. (1984). "Oxygen toxicity, oxygen radicals, transition metals and disease." *Biochem. J.* **219**, 1–14.

Heck, J.D., and Costa, M. (1982a). "*In vitro* assessment of the toxicity of metal compounds. I. Mammalian cell transformation." *Biol. Trace Element Res.* **4**, 71–82.

Heck, J.D., and Costa, M. (1982b). "*In vitro* assessment of the toxicity to metal compounds. II. Mutagenesis." *Biol. Trace Element Res.* **4**, 319–330.

Hoffmann, G.A. (1982). "Mutagenicity testing in environmental toxicology." *Environ. Sci. Technol.* **16**, 560A-574A.

ICPEMC (International Commission for Protection Against Environmental Mutagens and Carcinogens) (1982). "Mutagenesis testing as an approach to carcinogenesis." *Mutat. Res.* **99**, 73–91.

Imreh, St., and Radulescu, D. (1982). "Cytogenetic effects of chromium *in vivo* and *in vitro*." *Mutat. Res.* **97**, 192–193.

Jacquet, P., and Draye, J.P. (1982). "Toxicity of chromium salts to cultured mouse embryos." *Toxicol. Lett.* **12**, 53–57.

Kada, T., Hirano, K., and Shirasu, Y. (1980). "Screening of environmental chemical mutagens by the rec-assay system with *Bacillus subtilis*." In F.J. de Serres and A. Hollaender, Eds., *Chemical Mutagens: Principles and Methods for their Detection*. Vol. 6. Plenum Press, New York, pp. 149–173.

Kaneko, T. (1976). "Chromosome damage in cultured human leukocytes induced by chromium chloride and chromium trioxide." *Jpn. J. Ind. Health* **18**, 136–137.

Kanematsu, N., and Kada, T. (1978). "Mutagenicity of metal compounds." *Mutat. Res.* **54**, 215–216.

Kanematsu, N., Hara, M., and Kada, T. (1980). "Rec assay and mutagenicity studies on metal compounds." *Mutat. Res.* **77**, 109–116.

Kitagawa, S., Seki, H., Kametani, F., and Sakurai, H. (1982). "Uptake of hexavalent chromium by bovine erythrocytes and its interaction with cytoplasmic components; the role of glutathione." *Chem.-Biol. Interact.* **40**, 265–274.

Knudson, A.G. (1985). Hereditary cancer, oncogenes and antioncogenes. *Cancer Res.* **45**, 1437–1443.

Kohn, B.H. (1983). "Sister chromatid exchange in cancer." *Ann. Clin. Lab. Sci.* 13, 267–274.

Koshi, K. (1979). "Effects of fume particles from stainless steel welding on sister chromatid exchanges and chromosome aberrations in cultured Chinese hamster cells." *Industrial Health* 17, 39–49.

Langerwerf, J.S.A., Bakkeren, H.A., and Jongen, W.M.J. (1985). "A comparison of the mutagenicity of soluble trivalent chromium compounds with that of potassium chromate." *Ecotoxicol. Environ. Safety* 9, 92–100.

LaVelle, J.M., and Witmer, C.M. (1984). "Chromium(VI) potentiates mutagenesis by sodium azide but not ethylmethanesulfonate." *Environ. Mutagen.* 6, 311–320.

Léonard, A., and Lauwreys, R.R. (1980). "Carcinogenicity and mutagenicity of chromium." *Mutat. Res.* 76, 227–239.

Levis, A. G., and Bianchi, V. (1982). "Mutagenic and cytogenetic effects of chromium compounds." In S. Langård, Ed., *Topics in Environmental Health, Biological and Environmental Aspects of Chromium,* Vol. 5. Elsevier Biomedical Press, Amsterdam, pp. 171–208.

Levis, A.G., and Majone, F. (1979). "Cytotoxic and clastogenic effects of soluble chromium compounds on mammalian cell cultures." *Br. J. Cancer* 40, 523–533.

Levis, A.G., and Majone, F. (1981). "Cytotoxic and clastogenic effects of soluble and insoluble compounds containing hexavalent and trivalent chromium." *Br. J. Cancer* 44, 219–235.

Levis, A.G., Bianchi, V., Tamino, G., and Pegoraro, B. (1978b). "Cytotoxic effects of hexavalent and trivalent chromium on mammalian cells *in vitro.*" *Br. J. Cancer* 37, 386–396.

Levis, A.G., Buttignol, M., Bianchi, V., and Sponza, G. (1978a). "Effects of potassium dichromate on nucleic acid and protein synthesis and on precursor uptake in BHK fibroblasts." *Cancer Res.* 38, 110–116.

Löfroth, G. (1978). "The mutagenicity of hexavalent chromium is decreased by microsomal metabolism." *Naturwissenschaften* 65, 207–208.

Loprieno, N., Boncristiani, G., Venier, P., et al. (1985). "Increased mutagenicity of chromium compounds by nitrilotriacetic acid." *Environ. Mutagen.* 7, 185–200.

MacRae, W.D., Whiting, R.F., and Stich, H.F. (1979). "Sister chromatid exchanges induced in cultured mammalian cells by chromate." *Chem.-Biol. Interact.* 26, 281–286.

Majone, F., Montaldi, A., and Ronchese, F. (1982). Cited in Levis and Bianchi, 1982.

Manzo, L., DiNucci, A., Edel, J., et al. (1983). "Biliary and gastrointestinal excretion of chromium after administration of Cr-III and Cr-VI in rats." *Res. Comm. Chem. Pathol. Pharmacol.* 42, 113–125.

Maron, D.M., and Ames, B.N. (1983). "Revised methods for the *Salmonella* mutagenicity test." *Mutat. Res.* 113, 173–215.

Marx, J.L. (1983). "Do tumor promoters affect DNA after all? Tumor promoters may contribute to cancer development by generating activated oxygen compounds that damage DNA." *Science* 219, 158–159.

Marzin, D.R., and Phi, H.V. (1985). "Study of the mutagenicity of metal derivatives with *Salmonella typhimurium* TA102." *Mutat. Res.* 155, 49–51.

Mohn, G.R., and Ellenberger, J. (1977). "The use of *Escherichia coli* K12/343/113 (λ) as a multi-purpose indicator strain in various mutagenicity testing procedures." In B.J. Kilbey, M. Legator, W. Nichols, and C. Ramel, Eds., *Handbook of Muta-genicity Test Procedures.* Elsevier/North-Holland Biomedical Press, Amsterdam, pp. 95–118.

Nakamuro, K., Yoshikawa, K., Sayato, Y., and Kurata, H. (1978). "Comparative studies of chromosomal aberration and mutagenicity of trivalent and hexavalent chromium." *Mutat. Res.* 58, 175–181.

Nestmann, E.R., Matula, T.I., Douglas, G.R., et al. (1979). "Detection of the mutagenic activity of lead chromate using a battery of microbial tests." *Mutat. Res.* 66, 357–365.

Newbold, R.F., Amos, J., and Connell, J.R. (1979). "The cytotoxic, mutagenic and clastogenic effects of chromium-containing compounds on mammalian cells in culture." *Mutat. Res.* **67**, 55–63.

Nishioka, H. (1975). "Mutagenic activities of metal compounds in bacteria." *Mutat. Res.* **31**, 185–189.

Norseth, T. (1981). "The carcinogenicity of chromium." *Environ. Health Perspect.* **40**, 121–130.

Norseth, T., Alexander, J., Aaseth, J., and Langård, S. (1982). "Biliary excretion of chromium in the rat: a role of glutathione." *Acta Pharmacol. et Toxicol.* **51**, 450–455.

Oberly, J.T., Piper, C.E., and McDonald, D.S. (1982). "Mutagenicity of metal salts in the L5178Y mouse lymphoma assay." *J. Toxicol. Environ. Health* **9**, 367–376.

Ohno, H., Hanaoka, F., and Yamada, M. (1982). "Inducibility of sister-chromatid exchanges by heavy-metal ions." *Mutat. Res.* **104**, 141–145.

Paschin, Y.V., and Kozachenko, V.I. (1982). "The modifying effects of hexavalent chromate on the mutagenic activity of thio-TEPA." *Mutat. Res.* **103**, 367–370.

Petrilli, F.L., and De Flora, S. (1977). "Toxicity and mutagenicity of hexavalent chromium on *Salmonella typhimurium*." *Appl. Environ. Microbiol.* **33**, 805–809.

Petrilli, F.L., and De Flora, S. (1978a). "Metabolic deactivation of hexavalent chromium mutagenicity." *Mutat. Res.* **54**, 139–147.

Petrilli, F.L., and De Flora, S. (1978b). "Oxidation of inactive trivalent chromium to the mutagenic hexavalent form." *Mutat. Res.* **58**, 167–173.

Petrilli, F.L., and De Flora, S. (1982). "Interpretations on chromium mutagenicity and carcinogenicity." In M. Sorsa, and H. Vainio, Eds., *Mutagens in Our Environment*. Alan R. Liss, New York, pp. 453–464.

Polnaszek, C.F. (1981). "Stable chromium(V) free radical species formed by the enzymatic reduction of chromate." *Fed. Proc.* **40**, 715.

Price-Jones, M.J., Gubbings, G., and Chamberlain, M. (1980). "The genetic effects of crocidolite asbestos; comparison of chromosome abnormalities and sister-chromatic exchanges." *Mutat. Res.* **79**, 331–336.

Raffetto, G., Parodi, S., Parodi, C., et al. (1977). "Direct interaction with cellular targets as the mechanism for chromium carcinogenesis." *Tumori* **63**, 503–512.

Raindaldi, G., Colella, C., and Piras, A. (1980). "Mutagenicity of $K_2Cr_2O_7$ in Chinese hamster cells in culture." *Mutat. Res.* **74**, 221.

Rainaldi, G., Colella, C.M., Piras, A., and Mariani, T. (1982). "Thioguanine resistance, ouabain resistance and sister chromatid exchanges in V79/AP4 Chinese hamster cells treated with potassium dichromate." *Chem.-Biol. Interact.* **42**, 45–51.

Rivedal, E., and Sanner, T. (1981). "Metal salts as promoters of *in vitro* morphological transformation of hamster embryo cells initiated by benzo(a)pyrene." *Cancer Res.* **41**, 2950–2953.

Robison, S.H., Cantoni, O., and Costa, M. (1984). "Analysis of metal-induced DNA lesions and DNA-repair replication in mammalian cells." *Mutat. Res.* **131**, 173–181.

Ryberg, D., and Alexander, J. (1984). "Inhibitory action of hexavalent chromium (Cr(VI)) on the mitochondrial respiration and a possible coupling to the reduction of Cr(VI)." *Biochem. Pharmacol.* **33**, 2461–2466.

Sager, R. (1986). Genetic suppression of tumor formation: A new frontier in cancer research. *Cancer Res.* **46**, 1573–1580.

Sanger, W.G. (1983). "Chromosomes and neoplasia." *Ann. Clin. Lab. Sci.* **13**, 366–370.

Sarto, F., Levis, A.G., and Paulon, C. (1980). "Clastogenic activity of hexavalent and trivalent chromium in cultured human lymphocytes." *Caryologia* **33**, 239–250.

Sarto, F., Cominato, I., Bianchi, V., and Levis, A.G. (1982). "Increased incidence of chromosomal aberrations and sister chromatid exchanges in workers exposed to chromic acid (CrO_3) in electroplating factories." *Carcinogenesis* **3**, 1011–1016.

Scarpelli, D.G. (1983). "Recent developments toward a unifying concept of carcinogenesis." *Ann. Lab. Clin. Sci.* **13**, 249–259.

Siegfried, J.M., and Nesnow, S. (1984). "Cytotoxicity of chemical carcinogens towards human bronchial epithelial cells evaluated in a clonal assay." *Carcinogenesis* **5**, 1317–1322.

Sirover, M.A., and Loeb, L.A. (1976). "Infidelity of DNA synthesis *in vitro*: Screening for potential metal mutagens or carcinogens." *Science* **194**, 1434–1436.

Stella, M., Montaldi, A., Rossi, R., et al. (1982). "Clastogenic effects of chromium on human lymphocytes *in vitro* and *in vivo*." *Mutat. Res.* **101**, 151–164.

Sunderman, F.W., Jr. (1984). "Recent advances in metal carcinogenesis." *Ann. Clin. Lab. Sci.* **14**, 93–122.

Tamaro, M., Banfi, E., Venturini, S., and Monti-Bragadin, C. (1975). "Hexavalent chromium compounds are mutagenic for bacteria." In Proceedings of the 17th *National Congress of the Italian Society of Microbiology*, Padua, pp. 411–415.

Tindall, K.R., Warren, G.R., and Skaar, P.D. (1978). "Metal ion effects in microbial screening systems." *Mutat. Res.* **53**, 90–91.

Tkeshelashvili, L.K., Shearman, C.W., Zakour, R.A., et al. (1980). "Effects of arsenic, selenium, and chromium on the fidelity of DNA synthesis." *Cancer Res.* **40**, 2455–2460.

Tsapakos, M.J., and Wetterhahn, K.E. (1983). "The interaction of chromium with nucleic acids." *Chem.-Biol. Interact.* **46**, 265–277.

Tsapakos, M.J., Hampton, T.H., and Wetterhahn Jennette, K. (1981). "The carcinogen chromate induces DNA cross-links in rat liver and kidney." *J. Biol. Chem.* **256**, 3623–3626.

Tsapakos, M.J., Hampton, T.H., and Wetterhahn, K.E. (1983). "Chromium(VI)-induced lesions and chromium distribution in rat kidney, liver and lung." *Cancer Res.* **43**, 5662–5667.

Tso, W., and Fung, W. (1981). "Mutagenicity of metallic cations." *Toxicol. Lett.* **8**, 195–200.

Tsuda, H., and Kato, K. (1977). "Chromosomal aberrations and morphological transformation in hamster embryonic cells treated with potassium dichromate *in vitro*." *Mutat. Res.* **46**, 87–94.

Umeda, M., and Nishimura, M. (1979). "Inducibility of chromosomal aberrations by metal compounds in cultured mammalian cells." *Mutat. Res.* **67**, 221–229.

Venier, P., Montaldi, A., Busi, L., et al. (1985). "Genetic effects of chromium tannins." *Carcinogenesis* **6**, 1327–1335.

Venitt, S., and Levy, S.L. (1974). "Mutagenicity of chromates in bacteria and its relevance to chromate carcinogenesis." *Nature* **250**, 493–495.

Wada, O., Wu, G.Y., Yamamoto, A., et al. (1983). "Purification and chromium-excretory function of low-molecular-weight, chromium-binding substances from dog liver." *Environ. Res.* **32**, 228–239.

Warren, G., Shultz, P., Bancroft, D., et al. (1981). "Mutagenicity of a series of hexacoordinate chromium(III) compounds." *Mutat. Res.* **90**, 111–118.

Wetterhahn Jennette, K. (1982). "Microsomal reduction of the carcinogen chromate produces chromium(V)." *J. Am. Chem. Soc.* **104**, 874–875.

Whiting, R.F., Stich, H.F., and Koropatnick, D.J. (1979). "DNA damage and DNA repair in cultured human cells exposed to chromate." *Chem.-Biol. Interact.* **26**, 267–280.

Wiberg, K.B. (1965). "Oxidation by chromic acid and chromyl compounds." In K.B. Wiberg, Ed., *Organic Chemistry, Part A*. Academic Press, New York, pp. 69–184.

Wilson, J.G. (1977). "Current status of teratology." In J.G. Wilson, and F.C. Fraser, Eds., *Handbook of Teratology*. Vol. 1. Plenum Press, New York, pp. 47–74.

Yamamoto, A., Wada, O., and Ono, T. (1984). "Distribution and chromium-binding capacity of a low-molecular-weight, chromium-binding substance in mice." *J. Inorg. Biochem.* **22**, 91–102.

Yunis, J.J. (1983). "The chromosomal basis of human neoplasia." *Science* **221**, 227–236.

Zakour, R.A., and Glickman, B.W. (1984). "Metal-induced mutagenesis in the *lac I* gene of *Escherichia coli*." *Mutat. Res.* **126**, 9–18.

Zakour, R.A., Kunkel, T.A., and Loeb, L.A. (1981). "Metal-induced infidelity of DNA synthesis." *Environ. Health Perspect.* **40**, 197–205.

17

CARCINOGENICITY OF CHROMIUM COMPOUNDS

A. Yassi

Occupational Health Program
Department of Community Health Sciences
University of Manitoba
Winnipeg, Manitoba, Canada

E. Nieboer

Department of Biochemistry and
Occupational Health Program
Faculty of Health Sciences
McMaster University
Hamilton, Ontario, Canada

1. OVERVIEW AND SCOPE

It has been over a half century since a link was first suggested between occupational exposure to chromates and increased risk of lung cancer (Lehmann, 1932), and almost a century has passed since Newman (1890) reported a case of nasal cancer in a chrome pigment worker. Several further reports (e.g., Pfeil, 1935; Alwens et al., 1936) prompted the German health authorities in 1936 to accept lung cancer in chromate-producing workers as an occupational disease (Langård, 1983). Ten years later in the United States, lung cancer was officially accepted as being caused by chromate exposure (Hueper, 1966). Epidemiological studies conducted in numerous countries, including the United States (e.g., Machle and Gregorius, 1948; Mancuso and Hueper, 1951; Taylor, 1966; Mancuso, 1975; Hayes et al., 1979), Great Britain (e.g., Bidstrup and Case, 1956; Royle, 1975a, b), Scandinavian countries (e.g., Langård et al., 1980), the Netherlands and West Germany (e.g., Frentzel-Beyme and Claude, 1980; Frentzel-Beyme, 1983), Italy (e.g., Franchini et al., 1983), the U.S.S.R. (e.g., Pokrovskaya and Shabynina, 1973), and Japan (e.g., Ohsaki et al., 1978; Satoh et al., 1981) have since confirmed the existence of an increased cancer risk in association with exposure to chromium-containing compounds. It is evident from these references and others (e.g., Langård and Vigander, 1983 and Davies, 1984a) that studies on the high incidence of lung cancer in workers exposed to chromates are still appearing and still generate considerable debate (Norseth, 1986).

Extensive reviews on the carcinogenicity of chromium have been published by government bodies, international organizations, and academic institutions (e.g., IARC, 1973, 1980; NAS, 1974; NIOSH, 1973, 1975) and by individuals (e.g., Sunderman, 1976, 1984; Langård and Norseth, 1979; Norseth, 1979, 1981, 1986; Hayes, 1980, 1982; Langård, 1983); the U.S. EPA has also prepared a reappraisal of the literature (EPA, 1984). A critical analysis of the early epidemiological evidence was conducted by Enterline (1974).

It is evident from the epidemiological data reviewed in this chapter that exposure to chromium in a variety of different forms and occupational settings has been linked to increased respiratory cancer risk. Although risk assessment studies assign high carcinogenic potency to exposures to chromium compounds in pigment production and in chromate manufacturing, attempts to identify the specific causative agent(s) of chromium-associated lung cancer by biostatistical methods alone have generally not been successful. The reasons for this include the fact that workplaces are often contaminated with a variety of trivalent and hexavalent chromium compounds resulting in mixed exposures, as explained in Section 4 and elsewhere in this volume (Chapter 2). Consequently, the laboratory animal studies reviewed later play a major role in evaluating the carcinogenic potential of various chromium compounds. As outlined in Chapter 16, important advances in the understanding of chromium carcinogenicity have also emerged from recent studies of mutagenic activity in bacteria and from related genotoxicity data employing other in vitro cell systems. These three independent sources of information are tapped to formulate a number of concluding remarks concerning the feasibility of differentiating between "carcinogenic" and "noncarcinogenic" chromium compounds.

2.   CARCINOGENICITY STUDIES IN ANIMALS

Studies of chromium carcinogenesis in animals have been reviewed and comprehensively tabulated in recent documents (IARC, 1980; EPA, 1984). The tables produced by IARC (1980) are included as an Appendix to this chapter.

Even though observations in humans of a link between exposure to chromium compounds and increased risk of cancer had been reported since 1932, early attempts to induce cancer in experimental animals (including mice, rats, guinea pigs, cats, and rabbits) by inhalation or parenteral administration of metallic chromium, chromite ore, and various chromium compounds were largely unsuccessful (Lukanin, 1930; Akatsuka and Fairhall, 1934; Shimkin and Leiter, 1940; Schinz, 1942; Schinz and Uehlinger, 1942; Hueper, 1955; Baetjer et al., 1959a).

Hueper and his colleagues (Hueper, 1958; Hueper and Payne, 1959) reported the development of squamous cell carcinomas and sarcomas of the lung and fibrosarcomas of the thighs in rats who had received intrapleural or intramuscular implants of chromite ore. He showed that implantation of calcium chromate also produced sarcomas. Hueper concluded that chromium is carcinogenic when present in a form that provides for its adequate biological availability, which, in turn, appears to depend on the degree of solubility of the chromium compound in the particular biological medium. Through a series of experiments with a variety of chromium compounds possessing a range of solubilities in water and Ringer's solution (including calcium chromate, zinc chromate, strontium chromate, lead chromate, and barium chromate), Hueper and Payne (1959) confirmed the validity of this solubility hypothesis. Subsequent studies (e.g., by Payne, 1960a; Dvizhkov and Fedorova, 1967; Roe and Carter, 1969) confirmed that intraosseous, intramuscular, subcutaneous, intrapleural, and intraperitoneal injections of chromium-metal powder and hexavalent chromium compounds can produce local sarcomas in rodents. Steffee and Baetjer (1965), however, were still unable to produce malignant tumors of the respiratory tract in rabbits, guinea pigs, rats, or mice by administering various chromate materials by inhalation and/or intratracheal injection.

Figures 1 to 6 illustrate the negative and positive findings reported to date for groups of chromium compounds tested by implantation, ingestion, or injection (also see the Appendix). Negative studies in which the follow-up period was felt to be too short (i.e., less than two years) and positive studies in which the strain had a high incidence of spontaneous tumors were excluded from these figures. The literature references for the data in these figures are provided in Table 1. A perusal of the data indicates that very convincing evidence exists that calcium chromate (Fig. 1) and lead chromates (Fig. 2) are carcinogenic. While fewer studies have been carried out on zinc, strontium, and chromium chromates, these compounds also appear to be carcinogenic (Fig. 3). Clear trends are absent for other insoluble chromium compounds (Fig. 4) and chromium oxides (Fig. 5), and data for soluble forms of chromium (mostly hexavalent chromates) are essentially negative (Fig. 6). The only positive response occurred for repeated application by intratracheal administration of sodium dichromate.

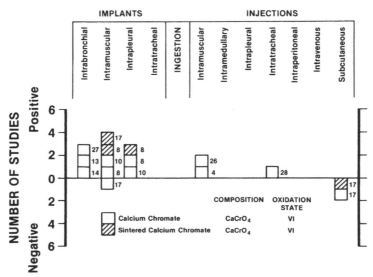

Figure 1. Carcinogenicity of calcium chromate in animals. Numbers identify the original reference to which the data refer (see appended index in Table 1). An individual summary for most studies is provided in the Appendix.

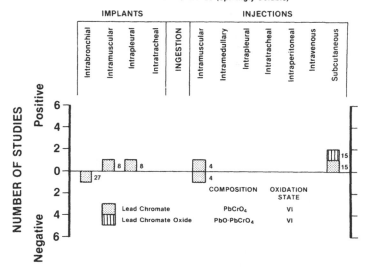

Figure 2. Carcinogenicity of lead chromates in animals. See legend to Figure 1 for explanatory notes.

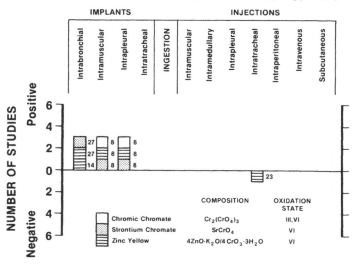

Figure 3. Carcinogenicity of Cr(III), strontium, and zinc chromates in animals. See legend to Figure 1 for explanatory notes.

447

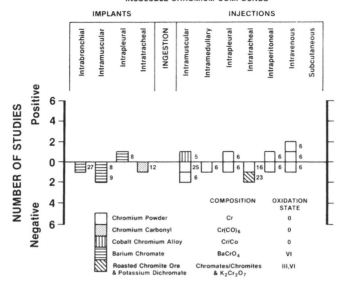

Figure 4. Carcinogenicity of water-insoluble chromium compounds in animals. See legend to Figure 1 for explanatory notes.

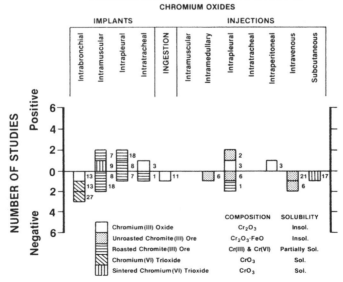

Figure 5. Carcinogenicity of chromium oxides in animals. See legend to Figure 1 for explanatory notes.

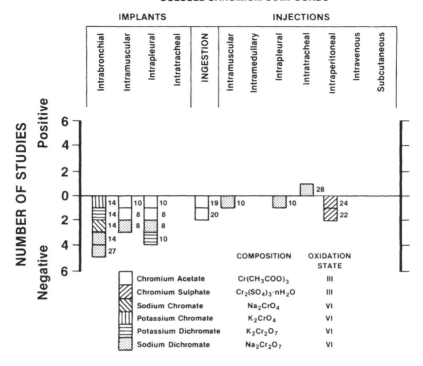

Figure 6. Carcinogenicity of water-soluble Cr(III) and Cr(VI) compounds in animals. The only positive response occurred for weekly administration for two years and eight months of sodium dichromate by intratracheal instillation with a cumulative dose of 56 mg Cr/kg. See legend to Figure 1 for explanatory notes.

As clearly shown in Figure 7, which depicts the inhalation studies, it has proven very difficult to demonstrate a carcinogenic response in the lungs of experimental animals. Thus, the question of which compounds are the most potent respiratory carcinogens is still not resolved. In fact, the only study reported to demonstrate tumorigenic activity of a chromate with inhalation techniques was that of Nettesheim et al. (1971) in which adenocarcinomas and alveologenic adenomas (but no bronchogenic tumors) developed in mice exposed to chronic inhalation of calcium chromate (Fig. 7 and Appendix). Tests in progress have suggested positive results for unspecified intermediates from chromate production as well (as cited by Norseth, 1981). As noted by Langård (1983), inhalation exposure in animals, while constituting the experimental route of administration most representative of human exposure, is fraught with

[a]See Appendix for a summary of the experimental design and results reported for most of these studies.

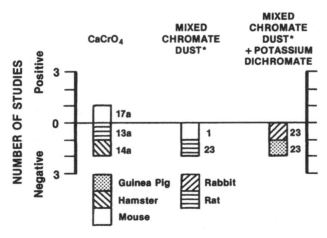

ANIMAL CARCINOGENICITY STUDIES
Administration by Inhalation

*Mixed chromate dust: Roasted chromite ore (III + VI)

Figure 7. Carcinogenicity in animals of selected chromium compounds administered by inhalation. See legend to Figure 1 for explanatory notes.

difficulties, both in the design of the experiments and in the interpretation of the results (see also Phalen, 1976; Campbell, 1976; Langård and Nordhagen, 1980). Therefore, Kuschner and Laskin (1971) developed a different method that consisted of intrabronchial implantation of stainless steel wire-mesh pellets as a carrier of different chromates. Using this method they succeeded in inducing squamous cell and adenocarcinomas in lungs of rats exposed to calcium chromate. By comparing the carcinogenicities of calcium chromate, chromic chromate, Cr(III), and Cr(VI) oxides they concluded that the moderately soluble hexavalent chromate (calcium chromate) was the most potent respiratory carcinogen. Levy and Venitt (1975; 1986) and Levy et al. (1986) have also tested a large number of chromium-containing materials by the intrabronchial pellet technique. Positive results have been reported for calcium, strontium, and zinc chromates and negative responses for chromite ore, Cr(III) oxide, Cr(VI) oxide, sodium chromate and dichromate, barium chromate, and lead chromate. Unfortunately, only a single low dose was employed in these studies: ~ 2 mg of test material by Levy and Venitt (1975, 1986) and an unspecified dose (but likely 2 mg) by Levy et al. (1986).

Little can be concluded from animal studies regarding the potential of trivalent chromium compounds to cause respiratory cancer in humans. Hueper (1961) did report a low incidence (3%) following Cr(III) acetate implantation (which was considered as negative for the compilation in Figure 6), compared to a high incidence of respiratory tumors following intrapleural implantation of, for example, calcium chromate, strontium chromate, and chromic chromate (57, 49, and 74%, respectively in 35 rats tested). A similarly low incidence for Cr(III) acetate was later confirmed by Hueper and Payne (1962) (also reported as negative in Figure 6; see Appendix). While the small number of animals used by Hueper and Payne (1962) and the lack of detail in the report by Hueper (1961) make it difficult to evaluate the carcinogenicity of the compounds tested, these studies do suggest that of the two valences, hexavalent chromium seems to be more potent. Furthermore, ingestion of trivalent chromium acetate failed to induce tumors (Fig. 6; Schroeder et al., 1964, 1965; Ivankovic and Preussmann, 1975), although only low doses were tested. Hexavalent chromium has not been tested by this route. Although it can be said that not a single animal study has unequivocally demonstrated that metallic chromium or trivalent chromium is carcinogenic, as can be seen from Figures 4 to 6, it is also true that these compounds have been studied less extensively than many of the hexavalent chromium compounds.

One may conclude that there is certainly strong evidence that at least some Cr(VI) compounds are carcinogenic to animals when administered by certain routes. Generally, as illustrated in Figures 8 and 9, low-dose experiments tend to have negative results and high-dose tests produce positive results. A threshold value from 2 to 4 mg chromium is suggested. The data in Figure 9 suggest that the tumor yield increases with the dose of metal chromate administered (expressed in milligrams of chromium), independent of the metal cation (Cr^{3+}, Pb^{2+}, Sr^{2+}, or Zn^{2+}) and mode of administration. Jones and Walsh (1985) have also concluded for a number of experimental carcinogens, including chromium compounds, that dose is the most important variable when considering mode of administration, number of treatments, and pathological classification of the induced neoplasia in rats. The saturation phenomena suggested by the data in this figure also appear to be a common feature. The apparent dose-response relationship in Figure 9 must, nevertheless, be viewed with caution as the number of data points for each metal chromate and mode of administration is few. Conceivably, each mode of injection and each type of metal chromate might generate individual perturbations which would then secondarily affect the sensitivity of the neoplastic response.

The animal data presented in this section have confirmed the carcinogenicity of sparingly soluble and moderately soluble hexavalent chromium compounds and intimate that soluble chromates, when administered repeatedly, may have carcinogenic potential. Data for Cr(III) compounds are inconclusive. Thus, animal studies have contributed to our knowledge on the potential role of solubility and valency. As we see later, they also provide supportive evidence with respect to quantitative epidemiological trends. The epidemiological data, which usually form the cornerstone for dose-response assessments, are reviewed in the next section. An overview of the carcinogenicity of chromium compounds that considers all the evidence—namely, from bioassays, animal studies, and epidemiological reports—is subsequently presented.

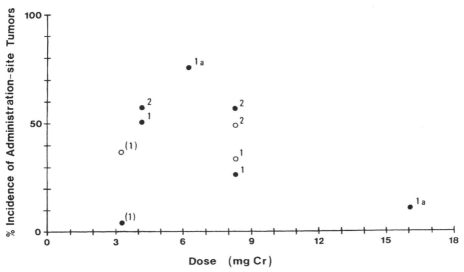

Figure 8. Percentage incidence of administration-site tumors *versus* dose for calcium chromates. The data correspond to the results of pertinent studies summarized in Figure 1 and refer to experiments with rats or mice (given in parentheses). ● = $CaCrO_4$. O = $CaCrO_4$, sintered. 1. = intramuscular injection or implantation. 2. = intrapleural implantation. a. = Cumulative dose, rather than a single application.

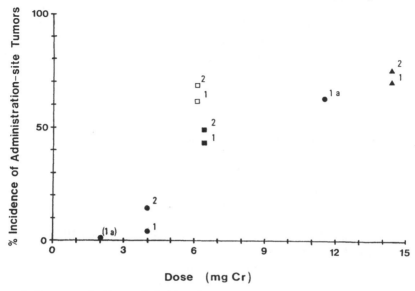

Figure 9. Percentage incidence of administration-site tumors *versus* dose for metal chromates (of chromium, lead, strontium, and zinc). The data correspond to the results of pertinent studies summarized in Figures 2 and 3, and refer to experiments with rats or mice (given in parentheses). ● = $PbCrO_4$. ■ = $SrCrO_4$. ▲ = $Cr_2(CrO_4)_3$. □ = zinc yellow   ($K_2CrO_4 \cdot 3ZnCrO_4 \cdot Zn(OH)_2 \cdot 2H_2O$). 1. = intramuscular injection or implantation. 2. = intrapleural implantation. a. = Cumulative dose rather than a single application.

3. EPIDEMIOLOGICAL STUDIES ON THE CARCINOGENICITY OF CHROMIUM

3.1 Introductory Remarks

As already mentioned, epidemiological studies on the carcinogenicity of chromium have been carried out in numerous countries and have clearly documented the link between chromium and cancer. These studies are summarized in Tables 2 to 7. The most extensive investigations have focused on the chromate production industry (Tables 2 to 4), on chromate pigment production (Table 5), and on chromium plating (Table 7). Most of these studies provide evidence for an excess risk of respiratory cancer. Studies on ferrochromium production (Table 6) and other secondary users of chromates are more equivocal. They nevertheless suggest that these workers may also be at increased risk of developing cancer. Furthermore, the possibility that exposure to chromium compounds might cause cancer at sites other than the respiratory tract has been raised in a number of reports (Tables 4 and 7). The most important issue, however, is the characterization of level and duration of exposure associated with cancer development in these studies, so that preventive measures may be rationally designed.

In order to appraise properly the results of 35 years of epidemiological investigation, certain methodological factors must be taken into account. The Advisory Council on Occupational Health and Occupational Safety in Ontario, Canada, published a detailed memorandum to the Ontario Minister of Labour (ACOHOS, 1983) stipulating the principles and procedures for the interpretation of epidemiological studies. These criteria for evaluating historical prospective and proportional mortality studies have been adopted in the construction of the tables and in our discussion of the data.

3.2   Studies in Chromium-Producing Industries

3.2.1   Production of Monochromates and Dichromates

Clinical observations from Germany in the 1930s strengthened earlier suspicions that workers in chromate production plants had an increased risk of developing lung cancer. (Baetjer in 1950 reviewed 122 cases of respiratory cancer among U.S. and German chromate workers that were reported before 1950.) Alwens and Jonas (1938), for example, reported 20 cases of lung cancer in dichromate-producing workers. The workers in the chromate industry were exposed to trivalent chromite ore as well as hexavalent chromium in the form of sodium chromate, sodium and potassium bichromates, calcium chromate, and Cr(VI) oxide. No quantitative exposure data were provided, but analysis of the lungs of some of the cancer cases revealed high water/acid soluble and insoluble chromium contents (Baetjer et al., 1959b). As analytical methods were quite different then, no direct comparison can be made with more recent results (Langård, 1983; see Chapter 2).

These early reports provided the impetus for a number of epidemiological studies in the chromate-producing industry in the United States, Great Britain, and Japan. A list of the plants studied by the various investigators is provided in Table 2, and the available information on exposure conditions in these plants is outlined in Table 3. The reader is referred to Chapter 2 for a discussion of the chemical speciation of chromium encountered in industry. The methodological details and epidemiological findings by the various investigators are outlined in Table 4.

U.S. Studies. Machle and Gregorius (1948) studied seven chromate plants in the United States, for which adequate employment records existed in six (see Tables 2 and 3). Their data showed that a similar high incidence of respiratory cancer existed in five of the plants assessed individually. Their overall lung cancer death figures represented over 16 times the expected ratio. Some authors (e.g., Langård, 1983) express the opinion that this is an underestimate, as minimum initiation dose and latent period were not taken into consideration, retired workers were not included, and the reference population was not comparable. A slight increased risk of digestive cancer was also reported. The

Table 2 The Location of Chromate Manufacturing Plants From Which Vital Statistics Were Obtained in Epidemiological Studies[a,c]

Location of Plant	Machle and Gregorius, 1948	Brinton et al., 1952	Gafafer, 1953	Taylor, 1966	Enterline, 1974	Mancuso and Hueper, 1951	Mancuso, 1975	Baetjer, 1950a	Hayes et al., 1979	Hill and Ferguson, 1979	Bidstrup, 1951	Bidstrup and Case, 1956	Alderson et al., 1981	Ohsaki et al., 1978	Satoh et al., 1981
Painesville, OH	+	+	+	+	+	+	+	−	−	−	−	−	−	−	−
Baltimore, MD	+	+	+	+[b]	+[b]	−	−	+	+	+	−	−	−	−	−
Jersey City, NJ (two plants)	+	+	+	+[b]	+[b]	−	−	−	−	−	−	−	−	−	−
Newark, NJ	+	+	+	−	−	−	−	−	−	−	−	−	−	−	−
Kearny, NJ	+	+	+	−	−	−	−	−	−	−	−	−	−	−	−
Glens Falls, NY	+	−	+	−	−	−	−	−	−	−	−	−	−	−	−
Bolton, England	−	−	−	−	−	−	−	−	−	−	+	+	+	−	−
Eaglescliffe, UK	−	−	−	−	−	−	−	−	−	−	+	+	+	−	−
Rutherglen, Scotland	−	−	−	−	−	−	−	−	−	−	+	+	+	−	−
Hokkaido Island, Japan	−	−	−	−	−	−	−	−	−	−	−	−	−	+	−
Tokyo, Japan	−	−	−	−	−	−	−	−	−	−	−	−	−	−	+

[a] +, Participating chromate manufacturing plant; −, Nonparticipating chromate manufacturing plant. [b]Only the larger of the two plants was studied. [c]The available environmental data are reported in Table 3.

Table 3A Environmental Data for Chromate Manufacturing Plants (in mg/m³). Chromate Production in the United States and Great Britain

Plant	Great Britain[a]			Painesville[b]			E(Painesville)[c]	C(Glens Falls)[c]	D1[c]	A1[c]
Exposure to Chromium	Alkali Insoluble	Alkali Soluble	Total	Water Insoluble	Water Soluble	Total	Total	Total	Total	Total
Ore prep	2.14			4.53	0.03	1.56				
Mixing of reactants	0.17	0.005	2.15	1.20	0.10	1.30	0.005–0.70	0.15–1.40	0.40–2.30	0.03–0.50
Roasting	0.04	0.03	0.20	0.39	0.23	0.62				
Extraction of monochromate		0.52	0.56	0.10	0.10	0.20				
Neutralization				0.03	0.06	0.09				
Formation of dichromate	0.01	0.06	0.07	0.09	0.10	0.19				
Separation of dichromate/liquor				0.13	0.27	0.40				
Processing sodium sulfate				0.88	0.24	1.12				
Concentration	0.005	0.88	0.89				0.07–0.14	0.01	1.09	0.01–0.10
Drying				0.30			0.13	0.02	0.17	0.10
Granulation	0.009	0.47	0.48		0.18	0.48	0.06–0.07	0.005–0.37		0.02–0.29
Packing							2.5–10.5			0.10
General air							0.002		0.02	0.01–0.50

[a] The reported values are the result of a survey performed in 1949–1950 when 96 air samples from the three plants were taken and analyzed (Buckell and Harvey, 1951). [b] The environmental hygiene survey was conducted in early 1949 and the levels reported do not reflect the greater, though steadily decreasing, atmospheric chromium contamination of the plant during the time since it began operations in 1932. A total of 121 air samples were analyzed (Bourne and Yee, 1950). [c] Air sampling and analysis was conducted at some time during 1941–1948 (Machle and Gregorius, 1948). It was reported that the magnitude of the chromium exposure had progressively lessened during the 15–20 years prior to 1948.

Table 3B Detailed Survey of the Six United States Plants (Gafafer, 1953) (1,600 Samples Collected Circa 1950)

Functional Group	Total Number of Samples	0.00 0.05	0.06 0.10	0.11 0.15	0.16 0.20	0.21 0.25	0.26 0.30	0.31 0.35	0.36 0.40	0.41 0.50	0.51 0.60	0.61 0.70	0.71 0.80	0.81 0.90	0.91 1.00	1.01 1.50	1.51 2.00	2.01 3.00	Over 3.00
Frequency Distribution of Air Concentrations of Total Chromium for Various Plant Operations																			
Ore processing	28	14	7	18	4	0	7	3	0	4	0	4	0	4	0	14	11	7	3
Residue processing	16	6	6	12	6	6	13	13	7	19	0	0	6	0	0	6	3	0	3
Roasting	32	6	13	13	16	3	9	6	0	12	4	6	3	0	0	3	3	0	3
Leaching	15	6	13	13	7	0	7	7	0	13	0	7	0	3	3	0	3	7	6
Neutralizing and treating	33	9	15	12	9	9	9	7	6	6	6	0	0	3	3	0	3	3	6
Concentrating and granulating	10	30	10	20	30	0	0	10											
Filtering, drying and packing	15	0	0	27	7	13	7	7	0	7	13	7	0	0	0	6	0	0	6
Other products	30	27	13	17	13	4	4	0	0	0	3	0	3	0	3	3	0	3	10
Frequency Distribution of Air Concentrations of Water Soluble Chromium (Hexavalent) for Various Plant Operations																			
Ore processing	76	92	7	1															
Residue processing	90	51	18	10	2	6	2	4	4	2	0	0	0	0	0	0	1		
Roasting	116	53	14	9	4	3	2	2	2	2	2	1	1	1	0	2	0	2	1
Leaching	98	38	22	7	11	6	5	1	2	2	1	1	1	1	0	0	0	1	
Neutralizing and treating	206	51	15	14	5	5	1	2	1	2	1	1	½	1	½				
Concentrating and granulating	84	63	25	7	2	1	0	2	1										
Filtering, drying and packing	101	46	11	5	6	9	4	1	2	4	2	4	4	0	1	0	0	0	1
Other products	169	68	16	5	2	3	1	1	1	½	0	½	0	½	0	0	0	0	½

Percent of Samples Within Indicated Range (Milligrams Chromium per Cubic Meter of Air)

Continued

Table 3B (Continued)

	Total Number of Samples	*Percent of Samples Within Indicated Range (Milligrams Chromium per Cubic Meter of Air)*																	
Functional Group		0.00 0.05	0.06 0.10	0.11 0.15	0.16 0.20	0.21 0.25	0.26 0.30	0.31 0.35	0.36 0.40	0.41 0.50	0.51 0.60	0.61 0.70	0.71 0.80	0.81 0.90	0.91 1.00	1.01 1.50	1.51 2.00	2.01 3.00	Over 3.00
Frequency Distribution of Air Concentrations of Acid Soluble–Water Insoluble Chromium (Mostly Trivalent, > 96%) for Various Plant Operations																			
Ore processing	28	54	18	11	10	0	4	0	3										
Residue processing	16	25	38	7	12	0	6	0	6	6									
Roasting	32	69	13	3	9	0	3	0	3	3									
Leaching	21	62	0	0	14	14	0	0	0	10									
Neutralizing and treating	32	69	9	4	6	3	0	0	0	6	0	0	0	0	0	3			
Concentrating and granulating	10	80	20																
Filtering, drying and packing	15	67	7[a]	7[a]	0	7[a]	0	6[a]	0	0	0	0	6[a]						
Other products	30	53	7	10[a]	4[a]	3	0	0	0	0	0	0	0	0	0	3[a]	3[a]	4[a]	7[a]
Frequency Distribution of Air Concentrations of Acid-Insoluble Chromium (Chromite Ore) for Various Plant Operations																			
Ore processing	28	25	11	7	4	0	4	0	3	7	0	0	3	0		18	7	7	4
Residue processing	16	50	25	13	0	0	0	0	12	0	0	0	3	3					
Roasting	22	59	19	4	0	6	3	0	3	0	0	3	0	3					
Leaching	21	70	0	0	10	5	0	0	0	5	5	0	0	0		5			
Neutralizing and treating	33	76	15	3	3	3	0	0	0										
Concentrating and granulating	10	100																	
Filtering, drying and packing	15	100																	
Other products	31	97	3																

[a]Primarily tanning compound; Cr as basic chromium sulfate.

Table 4 Epidemiological Studies of Cancer in Workers in Chromate-Producing Industries

Reference	Study Design	Definition of Cohort, Follow-up (FU), and Analysis	Comparison Population	Respiratory Cancer[a]		Other Cancers		
				No. of Cases	Estimated Relative Risk	Site	No. of Cases	Estimated Relative Risk
Machle and Gregorius, 1948[b]	Historical prospective cohort	Active workers employed for 4–17 years prior to 1948; mortality based on life insurance claims; excluded those who left the industry	Oil-refinery workers, cancer mortality 1933–1938	42	25[c] (Ranged from 18–50 in five of the six plants) 5.4	Digestive	13	2[c]
Baetjer 1950b[b]	Case control	Lung cancer U.S. chromium plant	Random sample of hospital admissions	Oral: 3 11	Zero in controls[c]		Not reported	
Brinton, et al., 1952; Gafafer 1953[b]	Proportional mortality study	Members of sick-benefit associations who died within one year of becoming disabled during 1940–1948	U.S. male population 1940–1948 —white males —black males	26 10 16	29[c] 14 80	All other	6	0.95
Mancuso and Hueper, 1951[b]	Proportional mortality study	Workers employed one year or more 1931–1949; mortality based on death certificates, included those who left	Proportional mortality for county, 1937–1947	6	15[c]		Not reported	
Mancuso, 1975[b]	Proportional mortality study	Workers employed for one year or more in 1931–1937; FU until 1974	None	41			Not reported	
Taylor, 1966[b]	Historical prospective cohort	Workers employed three or more months between 1937–1940 and have earnings reported to OASDI[d]; mortality based on death certificate until Dec. 1960 and who had death claim filed with OASDI, included workers who left the industry	Age-specific mortality rates for the U.S. civilian male population	71	8.5[c]	All other	32	1.3
Enterline, 1974[b]	Historical prospective cohort	As above—based on recalculation of data for the years 1941–1960 originally reported by Taylor, 1966	As above	69	9.4	Digestive	16	1.5

Continued

459

Table 4 (*Continued*)

Reference	Study Design	Definition of Cohort, Follow-up (FU), and Analysis	Comparison Population	Respiratory Cancer[a]		Other Cancers		
				No. of Cases	Estimated Relative Risk	Site	No. of Cases	Estimated Relative Risk
Hayes et al., 1979[b]	Historical prospective cohort	Workers employed three or more months hired 1945–1974; 88% follow-up by mid-1977	Baltimore City mortality rates	59	2[c]	All other	86	1.02
	Case-control analysis w.r.t. work history	Analysis for cohort employed three years or more exclusively in the new plant with improved hygiene standards, hired 1950–1959	Internal comparison by worksite	3	4[c]	Digestive	13	0.6
Bidstrup and Case, 1956[b]	Historical prospective cohort	Numbers employed in 1949 derived from x-ray survey; mortality based death certificate entry to 1955; workers who left the industry (30%) were not followed	Males from England and Wales 1951–1953	12	3.6[c]	All other	9	1.1
Alderson et al., 1981[b]	Historical prospective cohort	Workers employed at least one year between 1948–1977	National mortality rates for England, Wales, and Scotland	116 (lung)	2.4[c]	All other	80	1.2
				2 (nasal)	7.1			
Satoh et al., 1981	Historical prospective cohort	Workers employed for ≥ one year between 1918–1975; FU until 1978	Japanese males 1950–1978	26	9.5[c]	Other plus sinus (5), nasal (1)	18	0.8
Ohsaki et al., 1978	Case series	Cases diagnosed during 1972–1976 and found in death certificates; insufficient detail	Japanese population 1975	14	50		Not reported	
Tsuneta, 1982		1982 update of Ohsaki study		24	16			
Korallus et al., 1982	Historical prospective cohort	Workers of two plants employed ≥ one year, starting 1948; FU until 1979	Regional mortality rates	21	1.9[c]	All other	12	0.55[c]
				30	2.2[c]		29	1.1

[a] Here and in subsequent tables, respiratory cancer usually includes that of the trachea, bronchus, lung, and larynx (ICD-161/2). Laryngeal cancer (ICD-161) is also reported separately if statistically significant on its own. Nasal cancer (ICD-160) statistics are shown separately. [b] Environmental data for these plants are summarized in Table 3. [c] Statistical significance with p < 0.05. [d] OASDI, Old Age and Survivors Disability Insurance; ICD, International Classification of Diseases.

Table 5 Summary of Epidemiological Studies of Cancer in Workers Employed in the Chromate Pigment Industry

Reference	Study Design	Definition of Cohort, Follow-up (FU), and Analysis	Comparison Population	Respiratory Cancer		Other Cancers			Exposure Data
				No. of Cases	Estimate of Relative Risk	Site	No. of Cases	Estimate of Relative Risk	
Langård and Norseth, 1975	Historical prospective cohort	24 Workers with more than three years employment in a Norwegian chromate pigment plant; FU Norwegian Cancer Registry 1951–1972	Cancer incidence rates for Norwegian males 1955–1971	3	38	G.I. Nasal	1[a] 1[a]	n/a n/a	Yes[e]
American Dry Color Manufacturers, 1976 (EEH, 1976)	Historical prospective cohort	Workers hired before 1960, employed 10 years or more; mortality based upon death certification as reported to OASDI.	Cancer mortality rates for U.S. males and for counties where plants were located						No[b]
		–exposed to lead chromate only (plants 1 & 2)		2	2.4	G.I.	0	n/a	
		–exposed to lead and zinc chromates (plant 3) (but see 1983 report)		6	2.4[g]	Stomach	5	7.8	
Davies, 1978, 1979, 1984a	Historical prospective cohort	Males initially employed ≥ one year prior to 1967 at three chromate pigment plants; deaths reported to the N.H.S. cancer registry to Dec. 31, 1981	Quinary quinquennial national death rates covering 1931–1935 onwards; not adjusted for regional or social class factors						No[c]
		–exposed to lead chromate only		7	1.1				
		–zinc and lead chromates							
		–high and medium exposure[c]		32	2.7[g]	Other resp.	3	2.7	
		–low exposure[c]							
Frentzel-Beyne, 1983	Historical prospective cohort	Workers in five factories in two countries with average year of entry between 1958–1963; FU until 1976	Regional mortality rates	7 5 2 3 2 7	1.0 2.3 2.5 3.9[g] 1.6 1.6		Not reported		No
Sheffet et al., 1982	Historical prospective cohort	Males employed for greater than one month during 1940–1969; 324 of 1946 lost to FU	Cancer mortality rates for U.S. males adjusted for race	Various expected rates given according to assumptions about those lost to FU, relative risk ranging from 1.3–1.6[g]		Stomach Pancreas	ranging 1.4–2.0 ranging 1.5–1.7		Yes[d]
	Proportional mortality study			Ranging from 1.5–2.0[g]		Stomach Pancreas	ranging 1.6–2.8[g] ranging 1.6–2.2		

Continued

Table 5 (*Continued*)

Reference	Study Design	Definition of Cohort, Follow-up (FU), and Analysis	Comparison Population	Respiratory Cancer — No. of Cases	Respiratory Cancer — Estimate of Relative Risk	Other Cancers — Site	Other Cancers — No. of Cases	Other Cancers — Estimate of Relative Risk	Exposure Data
Langård and Vigander, 1983	Historical prospective cohort	FU of cohort in study by Langård and Norseth, 1975		6	44	G.I.	4[a]	n/a	Yes[e]
						Nasal	1[a]	n/a	Yes[b]
American Dry Color Manufacturers, 1983 (Cooper, 1983)	Historical prospective cohort	Six months exposure prior to Jan. 1975:	Cancer mortality rates for U.S. white males						
		–Plant 1 (lead chromate)		4	1.6	Stomach	0	n/a	
		–Plant 2 (mixed: Pb, Zn, Sr, Ba)		2	2.0	Stomach	0	n/a	
		–Plant 3 (mixed: Pb & Zn)		9	2.2[g]	Stomach	5	7.9[g]	

[a] These deaths refer to the cohort of all men who were ever employed at the pigment plants between 1948 and 1972, which numbered 133. [b] Reported in the 1983 report as follows: Lead levels of 11.0–51.6 mgPb/m³ in 1957 in one plant. In a second plant the ranges and averages for lead were 0.01–10 mg/m³ and 0.58 mgPb/m³, respectively; in 1965–1968; for 1968–1972, 0.05 – 24.3 and 2.47 mgPb/m³, respectively. Most ratios of lead to chromium were between 4:1 and 6:1 in 1975. [c] Exposure categories were defined as follows: high = in "dry" departments where pigments were ground, blended, and packaged; medium = in "wet" departments where precipitates were washed, pressed, and store-dried; low = most lab jobs, boiler stoking, painting, bricklaying. [d] Exposure estimates during the study period, based on air samples collected in later years, were as follows (in mgCr/m³): high, >2; medium, 0.5–2; low, <0.1. The proportion of $PbCrO_4$ to $ZnCrO_4$ was approximately nine to one in most areas of the plant. [e] Exposure to airborne chromium (in mg/m³) in three chromate pigment plants in Norway (Langård and Norseth, 1975):

	Plant		
Operation	A	B	C
Sackfilling	0.43	1.35	0.08
Mixing raw materials	0.35	0.33	0.01
Foreman	0.19	0.04	

The hygiene survey was conducted circa 1972 with portable personal samplers, where N = 4 or 5 for each job class. Past exposures were at the same level as at the time of the survey, as ascertained by interviews with older employees. Plant C was constructed in 1972. [f] n/a = not available. [g] Statistical significance at p < 0.05.

Table 6   Summary of Epidemiological Studies of Cancer in Workers Employed in the Ferrochromium Industry

Reference	Study Design	Definition of Cohort, Follow-up (FU), and Analysis	Comparison Population	Respiratory Cancer		Other Cancers			Exposure Data
				No. of Cases	Estimate of Relative Risk	Site	No. of Cases	Estimate of Relative Risk	
Pokrovskaya and Shabynina, 1973	Historical prospective cohort	Workers employed in a Russian chromium ferroalloy smelting plant between 1955 and 1969; FU of death certificates from city archives; results given for males age 50–59	Cancer mortality rates for the city in which the plant was located	n/a	6.6^d	Stomach Esophagus All sites	n/a n/a n/a	3.2 2.0^d 3.3^d	Yes
Langård et al., 1980	Historical prospective cohort	Males employed for at least one year during 1928–1959 in a Norwegian ferroalloy plant; FU of deaths reported during 1953–1977	Cancer mortality rates for: (a) Norway (b) Hordaland county where the plant is located (c) internal plant group (all workers excluding ferrochromium workers)	7 7 7	2.3 3.9 7.3^d	All other sites	16	0.8	Yes^a
Axelsson et al., 1980	Historical prospective cohort	Males employed for at least one year during 1930–1975 in a Swedish ferroalloy plant; FU of deaths reported since 1951 to the National Bureau of Statistics	Cancer mortality rates for Älvsborg county in which the plant was located	5	0.7	All other sites	65	0.9	Yes^b
Axelsson and Rylander, 1980	Proportional mortality study	Population of Trollhättan and Vänersborg, communities in Älvsborg county which have ferrochromium-producing plants, for the period 1961–1975	Population of Älvsborg county and communities therein, grouped according to size for the period 1961–1975	No increase in mortality rate was found when comparisons were made with communities of the same size			n/a		Yes^c

[a] Environmental chromium levels (in mg/m^3)—Norwegian plant

Operation	Total Chromium		
	Avg.	Range	N
Ferrochromium department (except packing)	0.05	0.01–0.37	61
Packing	0.23	0.01–1.34	28

Air sampling was conducted in 1975; 11% to 33% of the chromium was water soluble.

[b] Environmental chromium levels (in mg/m^3)—Swedish plant

Operation	Est. levels of exposure	
	Cr°+Cr(III) (insol.)	Cr(VI) (sol.)
Arc furnace	2.5	0.25
Transport, metal grinder, sampling	0.5–2.5	0.01–0.05
Maintenance	2.5	0.05

Approximate calculations on exposure levels were made based on recent measurements and discussions with retired workers and foremen employed in the 1930s. General working conditions are known to have improved considerably over the 1970s. The authors state that the estimated exposure data should not be used to construct general dose-response relationships or to define threshold values. Exposure to Cr(VI) also occurred in chrome alum production from 1950–1956.

[c] Environmental chromium levels (in mg/m^3)—Swedish communities near plants.

Community	Total Chromium
Most polluted areas in Trollhättan and Vänersborg	$(0.1–0.4)\times10^{-3}$
Rural areas of Älvsborg	$(1–8)\times10^{-6}$

Monthly readings were taken in 1976–1979.

[d] Statistical significance with $p < 0.05$.

Table 7 Summary of Epidemiological Studies of Cancer in Workers Employed in the Chromium Plating Industries

Reference	Study Design	Definition of Cohort, Follow-up (FU), and Analysis	Comparison Population	Respiratory Cancer		Other Cancers			Exposure Data
				No. of Cases	Estimate of Relative Risk	Site	No. of Cases	Estimate of Relative Risk	
Royle, 1975a	Historical prospective cohort	Current and former workers employed for more than three months in 54 British plants	Nonexposed workers individually matched on age, sex, and when last known to be alive	24	1.8	G.I.	11	2.8	Yes[a]
Waterhouse, 1975	Historical prospective cohort	About 5000 workers, just over half being male, employed since 1946 in a British plant; FU incomplete	Not specified	49 (males)	1.4[c]	Other sites	16 / Not reported	2.3	No
Okubu and Tsuchiya, 1977	Historical prospective cohort	Workers with six months of employment in the Tokyo chromium plating industry and registered with a health insurance society between 1970 and 1976; FU through reports from plant managements (70% response) and FU of retirees	Tokyo mortality rates. Also compared with nonexposed chemical and unskilled workers	0	—	Other sites	5	0.5	No
Franchini et al., 1983	Historical prospective cohort	Males employed for more than one year prior to Jan. 1972 in nine plants in Italy; FU for the years 1951–1981: —overall —"hard" platers[b]	Cancer mortality rates in Italian males	3 3	3.8 5[c]	All sites	7 6	2 2.7[c]	No[b]
Waterhouse, 1980	Historical prospective cohort	2,297 workers who commenced employment in 1946–1970, worked for six months or more, and left the company by Dec. 1970; 13.6% not traced. FU for 1946–1980 for subcohort of nonbath workers: —overall —high exposure (three to eight years)	Cancer mortality in England and Wales	35 12	1.7[c] 3.6[c]	Stomach Larynx All sites	12 5 20	2.0[c] 10.0[c] 2.6[c]	No

[a] Air sampling was conducted during 1969–1970. Of the 54 plants in the study, 42 were involved in the hygiene survey, with the others either having ceased operations or refusing permission.

No. of Plants (vat type)		Chromium Air Levels (mg/m^3)
3	(automatic)	<0.005 in one; not detectable in the other 2
2	(manual)	>0.05
37	(mostly manual)	<0.02
12	(mostly manual)	not surveyed

[b] "Hard" platers were more exposed to chromic acid mist than were "bright" platers; [c] Statistical significance with p < 0.05.

U.S. Public Health Service (Brinton et al., 1952; Gafafer, 1953) studied the same chromate plants. In this study, the cohort was limited to men who were members of a sick benefit association, membership of which was voluntary before 1949. Only those workers who died within one year after becoming disabled were included in this study. They reported a 4.5-fold increase in cancer at all sites and a nearly 29-fold increase in respiratory cancer. They found no excess risk for cancer at other sites. As shown in Table 4, the respiratory cancer risk was greatest for black males. Additional evidence supporting these observations comes from a case-control study of lung cancer in two Baltimore hospitals (Baetjer, 1950a,b), the details of which are reproduced in Table 4.

Mancuso and Hueper (1951) studied chromium-induced lung cancer in the Ohio plant, using proportionate mortality in their analyses. They found six respiratory cancer deaths in exposed workers, and an observed/expected ratio for lung cancer of 15. The available exposure information for this plant, as provided in Table 3, showed that workers were exposed to three distinct types of chromium compounds: (1) the insoluble crude or chromite, which contains trivalent chromium [Cr(III)]; (2) the water-soluble hexavalent sodium monochromate (the intermediate formed after the roasting treatment); and (3), the water-soluble sodium dichromate end-product which again has chromium in the hexavalent state [Cr(VI)] (see Figure 1, Chapter 2). Average frequency distributions of these fractions in air samples for various plant operations in six of the United States plants are also reproduced in Table 3. Based on the hygiene assessments conducted by Bourne and Yee (1950), Mancuso and Hueper (1951) indicated that approximately 96% of the workers were predominantly exposed to water-insoluble Cr(III) (chromite), with no workers exposed exclusively to hexavalent chromium. Mixed exposures thus occurred, which is typical of chromate production facilities. Mancuso and Hueper's weighted average exposure estimates for the men who died of lung cancer range (in mg chromium m^{-3}): 0.01–0.15 water-soluble chromium, 0.10–0.58 for insoluble, and 0.11–0.66 for total chromium. These air levels fall within the ranges reported by Gafafer (Table 3). On analysis of tissue from one worker who died of lung cancer and one who died of bladder cancer, significant amounts of chromium were found in the lungs (26–46 $\mu g\ g^{-1}$ compared to 0.3 $\mu g\ g^{-1}$ for controls; wet weight). Mancuso and Hueper believed this to be consistent with the high cancer rate observed. Thus, while postulations concerning the etiology of lung cancer had centered on hexavalent chromium, this study emphasized the latent importance of the insoluble trivalent chromium as potentially carcinogenic. Mancuso (1975) presented a further report, detailing findings from the 1931–1937 part of the cohort, followed up until 1974. Again using hygiene data collected in 1949, the time-weighted average of exposure to insoluble, soluble, and total chromium (per m^3) was calculated for each occupation and for each worker in the plant. From detailed work-history records, duration of exposure in each job was established, thereby allowing the calculation of exposure years to each type of chromium for each worker. His data suggest that the age-adjusted lung cancer death rates per

100,000 increased with both insoluble and total chromium exposures. Mancuso, however, also noted that for a given exposure to insoluble chromium, an increase in exposure to total chromium resulted in an increased death rate from lung cancer, whereas for a given exposure to total chromium, increasing exposure to insoluble chromium led to a decreasing lung cancer death rate. He concludes that the data are consistent with the lung cancer risk being a function of both the water-soluble [mostly Cr(VI)] and water-insoluble chromium [mostly Cr(III)], namely, the total chromium rather than to one class of compounds. On this basis, Mancuso (1975) articulated the controversial conclusion that "carcinogenic potential extends to all forms of chromium." IARC (1980) agreed that the data demonstrated a dose-response effect for lung cancer death with increasing chromium exposure indices, and that both soluble and insoluble chromium appeared to be implicated in carcinogenesis. However, in their view, Mancuso's generalized conclusion that all chromium compounds must be regarded as suspect was not justified by his data.

In 1966, Taylor presented an extensive follow-up study of chromate workers who had been employed in one of three United States chromate-producing plants between 1937 and 1940. Information on deaths was derived from old-age and survivor disability insurance, and latency period was not considered when calculating the expected death figures. Nonetheless, chromate workers were found to have an excess of respiratory cancer of 8.5 times the expected rate. The only measure of exposure to chromate dust used in this study was duration of employment. Taylor (1966) did indeed demonstrate a dose-response relationship with this parameter. The observed-to-expected ratio varied from 4 to 18 with increasing duration of work. Deaths due to other cancers were also slightly increased. Enterline (1974) recalculated the data obtained by Taylor (1966), using different values for expected incidence, and he reported a 9.4-fold increase in respiratory cancer for the entire 20-year observation period. Enterline also specifically referred to the great excess in neoplasms of the mucocilliary sinuses and to the slight excess in digestive cancers. Moreover, he noted that the greatest risk was in the period immediately after the cohort was identified, and concluded that the rather short latency period observed in this and previous studies probably reflected exposure to a very potent carcinogen.

Hayes et al. (1979) conducted a well-designed study of dichromate-producing workers from the Baltimore plant to determine if improvements in the hygienic standards introduced in the United States chromate-producing industry in the early 1950s had eliminated the increased cancer risk. The cohort was divided into a "high exposure" group consisting of workers who had been employed in the old plant or whose exposure was undetermined, and a "low exposure" group consisting of workers employed only in the new plant since 1950. The workers were further subdivided into those employed from 1945 to 1949, from 1950 to 1959, and from 1960 onward. Although there was no information in the report on actual exposure levels, there was an apparent dose-relationship as associated with length of employment in the high exposure group. No lung cancers were observed in the most recently hired group, as would be expected, since they had not been under observation long enough to develop chromate-induced cancer. In the 1950–1959 subgroup, the observed/expected ratio for high exposure workers varied from 1.8 to 3.4 ($p < 0.05$) with no significant differences between them. Hayes et al. (1979) concluded that the risk of cancer was not eliminated in the new plant, as the relative risk for low exposure long-term employees (three years plus) was 4.0. They also pointed out that even short-term workers, employed three months to two years and working exclusively in the old plant, had a higher lung cancer mortality than expected. Further, these authors noted that employees at the "wet end" of the production process, whose exposure was largely to highly soluble chromate compounds, had the highest cancer risk. As pointed out, and as already alluded to, many of these workers had mixed work exposures.

Hill and Ferguson (1979) investigated the same Baltimore plant. In contrast to the comprehensive study by Hayes et al. (1979), they employed the statistical method of "probability window analysis" to demonstrate trends in lung cancer risk in association with the modernization of the plant. A statistically significant decline was found from 23 to 7 lung cancers in 24-year time periods covering the intervals 1927–1951 and 1952–1976, respectively. IARC (1980) stresses that no conclusions on improved safety in the plant can be drawn, as the method of analysis compares only lung cancer cases, not rates of lung cancer, and does not make allowance for the latency period.

British Studies. Bidstrup (1951) and Bidstrup and Case (1956) studied the risk of lung cancer in chromate-producing plants in England. In the early survey (Bidstrup, 1951), only one cancer was detected from chest x-rays performed on 724 workers in three chromate plants. In the follow up of these workers (Bidstrup and Case, 1956), 12 lung cancer cases were observed, while only 3.3 were expected for a statistically significant increase of 3.6-fold as shown in Table 4. Langård (1983) has noted that if all of the lung cancer cases occurred in the group of 237 workers who had been employed for more than 15 years, only this group should have served as the base for estimation of expected cases and a larger increased risk would have resulted. From the exposure data (collected in 1949) included in Table 3, it may be seen that mixed exposure to soluble Cr(VI) and insoluble Cr(III) occurred for all operations in the British plants, a situation similar to that described earlier for the American counterparts.

Recently, Alderson et al. (1981) followed up this British chromate-producing population, and included workers employed in subsequent years to assess the effect of improvements in plant hygiene. They found 116 lung cancer cases while 48 were expected, for an overall relative risk of 2.4 ($p < 0.05$). They also reported two cases of nasal cancer. The relative risk for developing lung cancer decreased from 3.0 before plant modification to about 1.9 afterward, which the authors noted failed to reach statistical significance but indicated an appreciable reduction of the lung cancer risk. No quantitative industrial hygiene data were provided.

German Studies. The early German studies have already been mentioned. Recently, Korallus et al. (1982) found statistically significant lung cancer risks for two West German chromate-producing plants during the period 1948–1979, as indicated in Table 4. Although they detected a decrease in the lung cancer occurrence subsequent to a change in production to "no-lime" processing of the chrome ore, this reduction failed to reach statistical significance. By contrast, a lowering in nasal perforations, symptomatic of chromium exposure (see Chapter 19), did so ($p < 0.05$).

Japanese Studies. Ohsaki et al. (1978) studied the occurrence of lung cancer in a small cohort of active and retired dichromate and chromic acid production workers. As shown in Table 4, they found 14 lung cancer cases, which was calculated to be more than 50 times the expected figures. Although no quantitative exposure data were given, the large number of workers with nasal perforations, combined with the fact that the plant was old and there were few industrial hygiene measures enforced, suggested that chromate exposure was likely very high. In an update, Tsuneta (1982) reported the incidence of lung cancer in this plant as 16 times that of the general population. The histological tumour types observed in these case-series studies were mostly squamous and small cell carcinomas (see also Nishiyama et al., 1985; Kim et al., 1985). There were many heavy smokers among the cases. However, according to Ohsaki et al. (1978), differences in smoking histories could not explain the very high risks observed (see also Abe et al., 1982).

Sano and Mitohara (1978) reported that 19 of the 36 deceased chromate workers in Tokyo died of cancer of the respiratory tract, and this was interpreted as evidence of a high risk of cancer associated with the chromate-production industry. Furthermore, it was reported that cancers of other organs might be caused by engaging in chromate work (Sano and Ebihara, 1979). By contrast, Satoh et al. (1981) who found a 9.5-fold increase ($p < 0.05$) in respiratory cancer among workers in a Tokyo chromate manufacturing plant, failed to note increases in cancers of other sites.

In summary, the studies of the chromate manufacturing industry have consistently indicated a strong association between exposure in this industry and increased risk of respiratory cancer. Often the relative risk has been found to be extremely high. Lack of exposure data precludes the establishment of a rigorous generalizable dose-response relationship. Because mixed exposure occurred in all plants studied, it is unclear whether trivalent chromium or hexavalent or both have an etiological role (but see later, Section 4). Neither can the relative potencies of soluble versus insoluble hexavalent chromium be determined from these data, although both have been implicated.

3.2.2 Chromate Pigment Production

Although the chromate chemical industry has been the most extensively studied, the earliest reports of cancer related to chromium exposure stemmed from chromium pigment exposure (Newman, 1890; Gross and Kolsch, 1943; Baetjer, 1950a,b).

Langård and Norseth (1975) published the first epidemiological study linking chromium pigment manufacturing to lung cancer. They investigated a small work force exposed to lead chromate, zinc chromate, sodium dichromate, and other substances in three Norwegian pigment plants. Three lung cancers were identified in the 24 workers who had been exposed for more than three years before 1973, for a relative risk of 38. As indicated in Table 5, chromium exposure at the time of the epidemiological study showed that workers were exposed to significant amounts of chromates (of sodium and zinc). Levels in the past were estimated to be similar, but exposure was to both zinc and lead chromates. A slight excess in cancer of the gastrointestinal tract was also observed in these workers (Langård and Norseth, 1979). The follow-up study (Langård and Vigander, 1983) reported three additional lung cancer cases for a relative risk of 44. As high as this is, Langård and colleagues (Langård and Anderson, 1980; Langård, 1983) felt that the reported relative risks are an underestimate. Langård (1983) noted that except for one of the cases, these cancer patients had been exclusively exposed to zinc chromate, implicating this pigment as the most likely causative agent in this study.

In 1976, the American Dry Color Manufacturers' Association commissioned a study of three United States plants that mainly produced lead chromate (EEH, 1976). The results, as shown in Table 5, indicate a considerably lower cancer risk than was seen in the Norwegian population. The authors felt that their findings remained consistent with the hypothesis that lead chromate is a carcinogen. The excess was found generally in long-term workers, albeit when excluding those exposed also to zinc chromate, the excess was based on only three deaths. In any case, they noted that due to the small number of deaths, the relative excess of respiratory cancer could not be determined precisely. The authors also found a Standardized Mortality Ratio (SMR) of 778 for stomach cancer in the plant with exposure to both lead and zinc chromate, but felt that no causative link was yet warranted.

Davies (1978, 1979, 1984a,b) studied three chromate pigment plants in Great Britain. As shown in Table 5, an elevated risk of lung cancer was found only in the plants that produced both zinc and lead chromate, and was absent in workers who were employed only in the manufacture of lead chromate. Davies concludes: "the results provide no indication that lead chromate induces lung cancer in man" and "exposure which is relatively mild or lasts less than a year may not constitute an effective risk."

A study including five factories in two European countries was conducted in order to assess further the possibility of a cancer risk in workers exposed to lead chromate (Frentzel-Beyme and Claude, 1980; Frentzel-Beyme, 1983). A moderately but consistently increased risk of lung cancer was found, but only reached statistical significance in one of the factories. Unfortunately, almost all of the workers had been exposed to both lead and zinc chromates (Davies and

Kirsch, 1984). Similarly, in the recent report by Sheffet et al. (1982) of increased cancer risk at a pigment plant (Table 5), exposure was primarily to lead chromate but zinc chromate was also present.

In 1983 the American Dry Color Manufacturers' Association (Cooper, 1983) presented the findings of a further five-year follow up of their previous cohort and gave an indication of the exposure levels (see Table 5, footnote b). No evidence was found to support their earlier hypothesis that lead chromate was carcinogenic as no additional lung cancer deaths were observed in this subgroup (plant 1 workers). In contrast, for workers with mixed exposure (plants 2 and 3), a statistically significant elevated risk of lung cancer was again found, supporting the hypothesis that zinc chromate is carcinogenic. Note that exposure in plant 2 was not to lead chromate alone as believed in the 1976 study (see Table 5). An increased risk of stomach cancer was also again noted.

It thus seems that exposure to chrome pigments definitely increases the risk of respiratory cancer and possibly other cancers, such as stomach cancer. While it is not yet possible to differentiate conclusively between the cancer risk associated with zinc chromate as opposed to lead chromate or other pigments, the evidence suggests zinc chromate is the most potent carcinogen in this group.

3.2.3   Ferrochromium-Producing Industries

While exposure in the manufacturing of chromates and chrome pigments is predominantly to hexavalent chromium, workers employed in ferrochromium production are primarily exposed to trivalent chromium (chromite ore is the starting material) and chromium metal. Typically, the ambient chromium is 10 to 30% water soluble. This fraction is largely hexavalent chromium (Langård et al., 1980; see Table 6).

Pokrovskaya and Shabynina (1973) studied workers employed in the production of chromium ferroalloys in the U.S.S.R. between 1955 and 1969, as shown in Table 6. Exposure to hexavalent chromium and benzo(a)-pyrene was noted as well as to trivalent chromium. The authors reported a relative risk for lung cancer in males from 4.4 to 6.6 ($p < 0.001$), depending on age group. A statistically significant ($p < 0.001$) increase in mortality from all malignancies and from esophageal cancer also occurred in the 50 to 59 year age category. Too few details were provided in the report to permit an evaluation of the validity of these findings (EPA, 1984).

Langård et al. (1980) studied a Norwegian ferrochromium and ferrosilicon production plant. A slight increase in lung cancer was demonstrated in workers who had been employed at the ferrochromium furnaces for many years, with the excess varying from 2.3 to 7.3 ($p < 0.05$) depending on the reference population used. Current exposure data from this cohort are reported in Table 6. The authors concluded that the role of trivalent chromium in the increase in lung cancer noted cannot be evaluated in view of the concomitant exposure to hexavalent chromium. They suggested, however, that the lower relative risk found in this study compared to studies in the chromate manufacturing industry may indicate that chromic compounds are less potent carcinogens than chromates.

Axelsson et al. (1980) studied the employees of a ferrochromium plant in Sweden. No increase in cancer-related deaths was noted among arc furnace workers who likely had the highest exposure to trivalent and hexavalent chromium, as shown in Table 6. These authors also concluded that the low incidence of lung cancer in this study reflected the lack of carcinogenicity of trivalent chromium. The different results obtained relative to the study by Langård et al. (1980) may relate to differences in statistical power, in cohort definition, in the observation period, or possibly in exposures to hexavalent chromium.

3.3 Studies in Chromium-Using Industries

3.3.1 Chromium-Plating Industry

The chromium-plating industry has been in existence for a long time and has grown considerably in recent years (Hayes, 1980). While a number of case reports of lung cancer in chromium platers have been published (e.g., Barborík et al., 1958; Sehnalová and Barborík, 1965; Michel-Briand and Simonin, 1977), the fact that this work tends to be done in a large number of small workplaces has not been conducive to epidemiological studies in this industry. Exposure in this setting is to dissolved hexavalent chromium trioxide (chromic acid) mist, which essentially contains chromate (as CrO_4^{2-}, $HCrO_4^-$, H_2CrO_4, or $Cr_2O_7^{2-}$ depending on pH; all hexavalent forms). Stern (1982) explains that the exposure to Cr(VI) in plating shops depends strongly on the degree of ventilation available. With good local ventilation, ambient levels can be below 0.01 mg m^{-3} compared to values near 1 mg^{-3} in its absence (see also Table 7). Floating plastic chips that top electrolysis baths can reduce the ambient levels significantly.

In 1975, Royle conducted a historical retrospective as well as prospective mortality study of past and current chromium platers, as outlined in Table 7. The data indicate that the death rate for cancer of all sites was significantly increased in the exposed workers (p < 0.01). Rates of lung cancer and gastrointestinal cancer were also increased but failed to reach statistical significance. The results of the first two years of the prospective study demonstrated the same trend, but firm conclusions await the completion of this study.

Waterhouse (1975) presented preliminary results from a study of some 5,000 chromium platers, of whom about half were defined as "chrome bath workers." About 80% of these workers were traced, and 49 lung cancer cases were observed with 35 expected. Although these preliminary results indicate a relative risk of only 1.4, the excess was found to be significant. Langård (1983) suggests that upon subdividing this cohort into different exposure categories and observation periods, it may be possible to determine dose-response relationships. Indeed, in a follow-up study of the nonbath workers in a large chrome-plating plant, Waterhouse (1980) noted that when the cohort was subdivided into low exposure (zero to two years) and high exposure (three to eight years), there was a striking difference in cancer mortality, as shown in Table 7. For example, 12 cases of bronchial cancer were found in the low exposure male group, exactly the number expected. In contrast, 10 cases were found in the high exposure group, compared to 3.3 expected ($p < 0.01$).

Okubu and Tsuchiya (1977) studied cancer mortality in Japanese chromium platers, as indicated in Table 7. The cohort was based on worker registration with an insurance society in 1970, and vital status was ascertained by questionnaire. No increased risk was found. However, the manner of defining the cohort, the possibility that the controls used may have also been exposed, and the incomplete ascertainment of vital status has led most researchers and organizations (e.g., Hayes, 1980; IARC, 1980; Langård, 1983) to suggest that no conclusions can be drawn from this study.

Franchini et al. (1983) recently reported their findings from a cohort of Italian chromium platers in three "hard" and six "bright" chromium electroplating plants. Eight cancer deaths were found in the cohort, against 4.2 expected. Statistical significance was not achieved, even when the latency period was specified to exceed 10 years (Table 7). In the subcohort of hard platers, who were exposed to roughly 10-fold the levels of Cr(VI) compared to the bright platers (see Stern, 1982) and had a latency ≤ 10 years, seven cancer deaths occurred against 2.7 expected ($p < 0.02$). Three lung cancers were found, all in the subgroup of hard chromium platers; 0.7 were expected with $p < 0.05$ (see also data in Table 7 for latency > 10 years). These statistically significant results led the authors to conclude that exposure to water-soluble hexavalent chromium mist indeed increases cancer risk.

3.3.2 Spray-Painting and Related Uses of Chromium Pigments

Chromate pigments are widely used in paints and inks. Dalager et al. (1980) conducted a proportionate mortality study on zinc chromate spray painters from two aircraft maintenance bases. They found 21 respiratory cancers compared to 11.4 expected ($p < 0.01$). In addition, the risk appeared to increase with increasing duration of exposure. By contrast, Chiazze et al. (1980), in their study of automobile spray painters who had been exposed to chromate pigments for more than five years, failed to find an excess lung cancer risk. Differences in exposure levels, pigment composition, or personal protection could account for the discrepancies between these two studies. Definitive conclusions can therefore not be drawn about cancer risk for users of chromium pigments.

3.3.3 Stainless Steel Welding

It is now well known that welding fumes derived from stainless steel welding contain hexavalent chromium compounds (Fregert and Övrum, 1963; Naranjit et al., 1979, Thomsen and Stern, 1979). Evidence that stainless steel welders absorb chromium was presented by Nieboer and Jusys (Chapter 2) (see also

Gylseth et al., 1977; Tola et al., 1977; Mutti et al., 1979; Welinder et al., 1983). In vitro studies strongly indicate that stainless steel (SS) welding fumes generated by manual metal arc welding (MMA) contain mutagenic compounds (White et al., 1981; Hedenstedt et al., 1977; Maxild et al., 1978), and specifically that Cr(VI) compounds constitute the active component (Hedenstedt et al., 1977). Stern (1982) provides data illustrating that the SS/MMA combination yields the largest amount of water soluble Cr(VI). In his review of the mutagenicity studies of welding fumes, Stern (1980, 1981) concluded that the water-soluble hexavalent chromium, as well as the nickel, contained in welding fumes exhibited the same type of genotoxicity as is expected for pure fractions of Ni^{2+} and chromate. He continues: "Since there is no evidence to the contrary, one should assume that the actual risk for genetic damage and other delayed health effects to welders can be expected to be similar to that found in other industries with exposure to these substances."

Although welding is carried out on a large scale in the industrialized world, few epidemiological studies have focused on welders as a distinct group. Breslow et al. (1954) conducted a case-control study in which 14 welders were found among the patients with lung cancer against two of the controls ($p < 0.01$). Milham (1976) noted a proportional mortality ratio of 137 ($p < 0.05$) for lung cancer among welders and flame cutters; Menck and Henderson (1976) also demonstrated a standardized mortality ratio of 137 in their study. Beaumont and Weiss (1980) found 53 lung cancers against 40 expected in a group of welders; when taking into account a 20-year latency period the SMR increased from 131 to 169 ($p < 0.001$). Unfortunately, the type of welding performed by the workers in these studies was not identified. Similarly, Gottlieb (1980) found eight lung cancer cases among welders against two among the controls, but again insufficient information was provided to implicate chromium exposure. Sjögren (1980), however, determined the mortality of 234 welders who had stainless steel welding as their major task for more than five years. Three lung cancer cases were found against 0.7 expected. Although the numbers were not large, the finding was significant ($p<0.03$), even when adjusted for smoking, and is suggestive of a chromium-related cancer risk. (The relative risk from nickel is thought to be much lower than that for chromium). In a retrospective follow-up study among chromium- and nickel-exposed welders in 25 West German factories, Becker et al. (1985) found that, relative to an internal control group of turners, milling cutters, and drillers (not similarly exposed), a significant increase ($p < 0.05$) in all malignant neoplasms (International Classification of Diseases, ICD 140–209) occurred. Statistical significance was not maintained when the comparison was made with the general population, nor for individual neoplasms. However, contrary to trends for the internal control group, there was a trend in lung and all cancers to higher values with increasing time since first exposure. Statistical significance was only attained for all cancers with exposures ≥ 30 years ($p < 0.05$). These correlations were strongest for welders using coated electrodes, for which the release of hexavalent chromium during welding is known to be significant.

Stern (1983a,b) has recently reviewed 20 epidemiological studies of lung cancer incidence among welders. His conclusions are summarized in the following quote from his paper (Stern, 1983a):

> Analysis of epidemiologic studies of lung cancer among the general welding population indicates that the values of excess risk found can be accounted for by chance provided there is an average risk ratio of approximately 1.5, which almost completely can be accounted for by a general excess tobacco use, bystander exposure to asbestos predominantly among shipyard welders, and the presence in the total cohort of 5–10% stainless steel welders who are at the relatively high risk of a factor of 4–7 probably due to their unique exposure to Cr and Ni.

3.4   Summary of Epidemiological Evidence

The risk of lung cancer has been shown consistently to be high in the chromate manufacturing industry, where exposure is to a mixture of insoluble (largely trivalent) and both moderately water-soluble and soluble (largely hexavalent) chromium compounds. Very little water-insoluble Cr(VI) occurred in chromate production plants as demonstrated in Table 2. The data suggest that even the more recently employed workers, who presumably are exposed to lower levels of chromium than workers employed previously, may have an excess cancer risk with prolonged employment.

In the chromate-pigment production industry, where the exposure is largely to sparingly soluble hexavalent compounds, high risk was also demonstrated in several studies. Moreover, one study indicated that the cancer hazard associated with production of chrome pigments may extend to the users of these pigments, such as spray painters.

In the ferrochromium-producing industry, where workers are exposed to chromium which predominantly is metallic and trivalent, but nevertheless may contain 10 to 30% soluble hexavalent chromates, the possibility of increased risk for lung cancer has also been raised. However, the relative risks reported were considerably lower than those generally experienced in chromate production. Since there have been conflicting reports, and in view of various methodological problems in the studies, the findings should be regarded as tentative.

Similarly, an increased risk of lung cancer has also been reported for chromium platers exposed primarily to highly soluble chromates. Again, the limited number of studies and limited strength and consistency of the association leave some room to question these findings, but the suggestion of a dose-response relationship in several of these studies (e.g., Waterhouse, 1980; Franchini et al., 1983) is certainly noteworthy.

Various studies have indicated that welders in general are at increased risk of cancer. Since there is as yet little epidemiological data on selected subcohorts of welders with restricted exposures, Stern (1980) stresses that supposition of process-dependent "hot spots" or risk must be sought on the basis of indirect (i.e., nonhuman) experiments. Recent epidemiological studies (Sjögren, 1980;

Becker et al., 1985) have indeed suggested that increased cancer risk is most strongly associated with stainless steel welding, specifically implicating chromium exposure as a causative factor. Stern (1983a) estimates that the chromium-induced risk ratio for respiratory tract cancer among stainless steel welders resulting from cumulative exposures of the order of three (mg m^{-3}) years may be near 6.

An excess of cancers at sites other than the respiratory system has been reported in chromate production workers (Machle and Gregorius, 1948; Enterline, 1974), chromeplaters (Royle, 1975a,b; Waterhouse, 1980), chromate pigment producers (EEH, 1976; Langård and Norseth, 1979; Cooper, 1983; Davies, 1984a), and ferrochromium producton workers (Pokrovskaya and Shabynina, 1973). Specifically, cancers of the gastrointestinal tract, as well as nasal and laryngeal cancers, have been noted to be slightly increased in various investigations (see Tables 4 to 7). Such findings, however, have not been consistently found in all studies, and no firm conclusions are possible. Interestingly, Hernberg et al. (1983) in a case-referent investigation of living patients with nasal and sinonasal cancers found a significant correlation (p < 0.05) with occupations in which there was suspected exposure to chromium (mostly welders).

4. CONCLUSIONS

4.1 On the Differentiation Between "Carcinogenic" and "Noncarcinogenic" Chromium Compounds

In order to synthesize an overview and to formulate conclusions, the evidence from animal and biochemical studies, bioassays, epidemiological reports, and exposure data are first summarized and put into perspective.

Bioassays and related in vitro experiments indicate that all hexavalent chromates (both water-soluble and insoluble) can induce mutations and inflict chromosomal damage (Chapter 16). By contrast, positive animal carcinogenicity data are only convincing for $CaCrO_4$ (intermediate solubility) and $PbCrO_4$ (sparingly soluble), and perhaps also for the sparingly soluble compounds $ZnCrO_4$, $SrCrO_4$, and $Cr_2(CrO_4)_3$, although fewer experiments have been reported for these. The results for soluble chromates are negative, unless long-term administration is employed. For trivalent chromium, the animal studies show no consistent carcinogenic potential; and only when large concentrations are employed, when a special ligand environment is generated around Cr(III) or when phagocytosis can occur, can mutagenicity be demonstrated.

While the Cr(III) ion and its complexes are generally excluded from cells, the chromate anion (CrO_4^{2-}) is readily taken up. By analogy to nickel compounds, crystalline forms of both Cr(III) and Cr(VI) may well gain access to cells by phagocytosis. This has recently been demonstrated for Cr(III) oxide (Cr_2O_3) in macrophages (Andersen, 1983) and in V79 Chinese hamster lung cells (Elias et al., 1986). Of course, suitable intracellular dissolution must follow phagocytic accumulation for the toxicity to be expressed. Evidence is mounting that once inside a cell, the chromate anion is reduced to the Cr(III) ion. Because of its high charge ($+3$), and because it possesses almost inert (irreversible) kinetics, the Cr(III) ion has a great affinity for proteins, small chelating molecules (e.g., citrate), and nucleotides including DNA and RNA. Because of these interactions, intracellular accumulation of Cr(III) is believed to induce mutations and thus the Cr(III) ion is designated as the ultimate carcinogen (see Chapter 16). And, finally, intracellular bioavailability would appear to be a major determinant in chromium carcinogenesis. From this vantage point, hexavalent chromium compounds that generate the chromate anion have an advantage in comparison to Cr(III) compounds which yield Cr(III) on dissolution, because a well-defined cellular uptake pathway exists for CrO_4^{2-} but not for Cr(III). Furthermore, the more insoluble chromates (compared to soluble chromates) have the potential of providing long-lived pools in the target organ, and there is considerable evidence that this occurs (see Chapter 2). In contrast, exposure to soluble chromates is characterized by excretion components with relatively short biological half-times (matter of days), and on this account may not be able to exercise fully their carcinogenic potential.

The epidemiological data are only partially successful in assigning risk. As indicated, strong evidence for respiratory cancer has been established for chromate production and chromate pigment workers, with the data for chrome platers being somewhat less certain; that for ferrochromium manufacture and welders may be considered equivocal. Pigment workers were exposed mainly to the relatively insoluble chromates of zinc and lead (hexavalent chromium), and platers mainly to solutions of chromates (again hexavalent chromium). However, chemical production of chromates was associated with and characterized by mixed exposure. Workers were exposed to ambient materials containing chromite ore [a Cr(III) compound also containing Fe(II) as a major constituent], the moderately soluble $CaCrO_4$, and soluble sodium chromates (hexavalent; both mono and dichromates of sodium were handled). From this summary, it is evident that both relatively insoluble chromates (for pigment workers) and soluble chromates (for chrome platers) are implicated in chromium cancer etiology, while no clear judgment is possible about the relative carcinogenic potential or potency of the various compounds to which chromate production workers were exposed. However, the risk assessment data discussed in the next section suggest that water-soluble chromates are strongly incriminated.

Proponents of the carcinogenicity of Cr(III) compounds fail to find strong support from animal studies and bioassays. Opponents of this view suggest that the equivocal nature of the epidemiological data for ferrochrome production workers, and perhaps also for welders, is evidence for the noncarcinogenicity of Cr(III)-containing substances. However, since only the sum of chromium metal (Cr°) and Cr(III) has been determined in fumes associated with these processes, the exact Cr(III) content has not been reported. This gap in knowledge weakens this counterargument considerably. Ferrochrome exposures contain about 10–30% soluble chromium, presumably as Cr(VI), while stainless steel fumes have 0.005–4.3% water-soluble Cr(VI) (the top of the range refers to MMA/SS welding). It is obvious from this that these two groups of workers experience mixed exposure.

We conclude that the current integrated evidence does not permit the clear division of chromium compounds as carcinogenic or noncarcinogenic. It is apparent that the sparingly soluble chromates (hexavalent) have recognized carcinogenic potency, while incriminating evidence is mounting for soluble chromates (hexavalent). Although the projected risk associated with exposure to trivalent chromium does not seem to be of the same order of magnitude as hexavalent chromium, it cannot be concluded that Cr(III) compounds are totally devoid of carcinogenic potential (but see Section 4.2). Since not much is known about exposures to chromium metal, no conclusions are possible.

The issue of chromium carcinogenicity is more complex than just valence state and water solubility. Other characteristics that are likely of importance are the size distribution and other factors that influence the deposition, clearance, retention, and absorption of the inhaled particles and mists (e.g., chemical composition, crystallinity, phagocytic index, chemical transformation, and body-fluid dissolution rates). These factors will also determine the biological half-life and the intracellular bioavailability of the inhaled materials, and presumably their carcinogenicity.

4.2   Dose-Response Relationships

Throughout the discussion of the epidemiological evidence in Section 3, the available exposure data were reported and integrated into our discussions. However, epidemiological studies meet great difficulty in considering these factors. As Langård (1983) has pointed out, even when atmospheric concentrations of substance(s) in question are cited in these studies, measurements usually have only been carried out during a short time period at the end of the observation period (Machle and Gregorius, 1948; Mancuso and Hueper, 1951; Mancuso, 1951; Gafafer, 1953; Langård and Norseth, 1975; Langård et al., 1980; Axelsson et al., 1980). Such measurements do not correspond to the period of exposure that is the focus of the epidemiological study, nor do they provide information about average exposure or changes in exposure over time. Consequently, the information is not suitable for attempting to relate quantitatively exposure and risk; it does not allow extrapolation to "safe" levels, nor the determination of maximum allowable exposures in the occupational setting (Hayes, 1980, 1982; Langård, 1983). Moreover, ideally the dose should be defined as the amount of a specific carcinogenic substance taken up during a given time. This requires the identification of the carcinogen. To complicate the matter further, the numerical assessment of risk is also dependent on a variety of methodological variables related to the epidemiological method adopted. In spite of these limitations and concerns, a quantitative risk assessment has recently been attempted by Nieboer et al. (1985), and is described in some detail below. In addition, the epidemiological studies reviewed in Section 3 do provide some indirect evidence concerning dose-response relationships. Using the duration of employment as an indicator of exposure, authors were frequently able to demonstrate that workers employed longer had a higher risk of respiratory cancer. The associated average latency period estimated from the interval between initial employment to the date of cancer diagnosis is estimated to be 18 ± 5 years (Enterline, 1974; Hayes, 1982).

The approach taken by Nieboer et al. (1985) to risk assessment is similar to that employed by the U.S. Environmental Protection Agency (EPA, 1984) and is summarized in Table 8. Excess lifetime risk is equal to baseline risk x (relative risk-1), with the relative risk being derived from the appropriate epidemiological data (see Tables 3 to 7) and the baseline lifetime risk for lung cancer being taken as 0.036 (EPA, 1984). The equivalent lifetime dose of Cr(VI) is calculated as follows: ambient chromium x fraction of chromium as Cr(VI) x fraction of year worked x fraction of lifetime worked. The exposure data for the Hayes et al. (1979) risk assessment is based on the data of Gafafer (1953), is representative of the Baltimore plant studied (see Tables 2 and 3), corresponds to an overall mean concentration weighted according to job category, and overlaps the observation period of the epidemiological study. That for the Mancuso (1975) study was assessed 12 years after the actual employment period (employed 1931–1937, followed up until 1974) and was standardized according to job category and person-years of exposure (EPA, 1984). Ambient assessments similar to those selected have been independently reported recently for the Baltimore factories

Table 8   Excess Lifetime Risks for Occupational Exposure to Chromium(VI) (Adapted from Nieboer et al., 1985)

Data Base	Occupational Category	Equivalent Lifetime Dose of Chromium(VI) $(\mu g/m^3)^c$	Excess Lifetime Risk[c]
Hayes et al. (1979) and Gafafer (1953)	Chromate production	0.33	0.029[a]
Franchini et al. (1983) and Guillemin and Berode (1978)	"Hard" chrome platers	2.0	0.14
Mancuso (1975) and Bourne and Yee (1950)	Chromate production	3.3[b]	0.22

[a]Based on Hayes' "short-term" employees (90 days to two years), who were initially hired 1945–1949. [b]Based on the observation that, on the average, 21% of the ambient chromium measured for the various job categories was Cr(VI). [c]A straight-line fit of these data (not constrained to pass through the origin) gives a fitted line: excess risk $= 0.0088 + 0.064 \times$ (equivalent dose). The goodness of fit is $r^2 = 0.9996$.

(Braver et al., 1985). The environmental data employed for the Franchini et al. (1983) calculations were suggested by them to match reasonably well with the epidemiological time frame and are also in agreement with estimates by Stern (1982).

Several conclusions may be drawn from the results summarized in Table 8. First, the dose-response relationship observed is significant, and is independent of whether exposure occurred during chrome plating (solely water-soluble chromates) or chromate production (mixture of water-soluble chromates and $Cr(III)$ ore). Second, the "goodness of fit" for risks associated with these two different chromium exposures suggests that it is the water-soluble $Cr(VI)$ component of the ambient chromium in chromate production that appears to account adequately for the excess risk of respiratory cancer. Third, the dose-response curve has a small positive y-intercept implying the absence of a measurable threshold concentration below which no adverse effect occurs. And, finally, the unit risk corresponding to the slope of the line restricted to go through the origin was 0.069 $[\mu g\ Cr(VI)/m^3]^{-1}$. Thus, the excess risk of respiratory cancer resulting from a lifetime exposure to 1 $\mu g\ Cr(VI)/m^3$ is estimated as 69/1000. This indicates that, relatively speaking, $Cr(VI)$ is a potent human carcinogen (an extensive comparison with other carcinogens is provided in EPA, 1984).

ACKNOWLEDGMENTS

Financial support for the preparation of this review was received from the Occupational Health and Safety Division of the Ontario Ministry of Labour, and is gratefully acknowledged.

REFERENCES

Abe, S., Ohsaki, Y., Kimura, K., Tsuneta, Y., Mikami, H. and Murao, M. (1982). "Chromate lung cancer with special reference to its cell type and relation to the manufacturing process." Cancer, **49**, 783–787.

ACOHOS (Advisory Council on Occupational Health and Occupational Safety). (1983). "Principles and procedures for the interpretation of epidemiological studies." In Fifth Annual Report, April 1, 1982, to March 31, 1983, Vol. 1, pp. 135–191, *Advisory Memorandum 82-V to the Ontario Ministry of Labour*, Ontario Ministry of Labour, Toronto, Canada.

Akatsuka, K., and Fairhall, L.T. (1934). "The toxicology of chromium." *J. Indust. Hygiene* **16**, 1–24.

Alderson, M.R., Rattan, N.S., and Bidstrup, L. (1981). "Health of workmen in the chromate-producing industry in Britain." *Br. J. Ind. Med.* **38**, 117–124.

Alwens, W., Bauke, E.E., and Jonas, W. (1936). "Auffallende haufung von bronchialkrebs bei arbeitern derchemischen industrie." *Arch. Gewerbepathol. Gewerbehyg.* **7**, 69.

Alwens, W., and Jonas, W. (1938). "Der chromat-lungenkrebs." *Act-Unio Internationalis Contra Cancrum* **3**, 103–118.

Andersen, O. (1983). "Effects of coal combustion products and metal compounds on sister chromatid exchange (SCE) in a macrophagelike cell line." *Environ. Health Persp.* **47**, 239–253.

Axelsson, G., and Rylander, R. (1980). "Environmental chromium dust and lung cancer mortality." *Environ. Res.* **23**, 469–476.

Axelsson, G., Rylander, R., and Schmidt, A. (1980). "Mortality and incidence of tumours among ferrochromium workers." *Br. J. Ind. Med.* **37**, 121–127.

Baetjer, A.M. (1950a). "Pulmonary carcinoma in chromate workers. I. A review of the literature and report of cases." *A.M.A. Arch. Ind. Hyg. Occup. Med.* **2**, 487–504.

Baetjer, A.M. (1950b). "Pulmonary carcinoma in chromate workers. II. Incidence on basis of hospital records." *A.M.A. Arch. Ind. Hyg. Occup. Med.* **2**, 505–516.

Baetjer, A.M., Damron, C., and Budacz, V. (1959b). "The distribution and retention of chromium in men and animals." *A.M.A. Arch. Ind. Health* **20**, 136–150.

Baetjer, A.M., Lowney, J.F., Steffee, H., and Budacz, V. (1959a). "Effect of chromium on incidence of lung tumors in mice and rats." *A.M.A. Arch. Ind. Health* **20**, 124–135.

Barbořík, M., Hanslian, L., Oral, L., et al. (1958). "Carcinoma of the lungs in personnel working at electrolytic chromium plating." *Pracov. Lek.* **10**, 413–417. English abstract in *Exerpta Med. Sect.* **16**, *Cancer* **7**, 1395 (1959).

Beaumont, J.J., and Weiss, N.S. (1980). "Mortality of welders, shipfitters, and other metal trades workers in Boilermakers Local No. 104, AFL-CIO." *Am. J. Epidemiol.* **112**, 775–786.

Becker, N., Claude, J., and Frentzel-Beyme, R. (1985). "Cancer risk of arc welders exposed to fumes containing chromium and nickel." *Scand. J. Work Environ. Health* **11**, 75–82.

Bidstrup, P.L. (1951). "Carcinoma of the lung in chromate workers." *Br. J. Ind. Med.* **8**, 302–305.

Bidstrup, P.L., and Case, R.A.M. (1956). "Carcinoma of the lung in workmen in the bichromates-producing industry in Great Britain." *Br. J. Ind. Med.* **13**, 260–264.

Bourne, H.G., Jr., and Yee, H.T. (1950). "Occupational cancer in a chromate plant—an environmental appraisal." *Ind. Med. Surg.* **19**, 563–567.

Braver, E.R., Infante, P., and Chu, K. (1985). "An analysis of lung cancer risk from exposure to hexavalent chromium." *Teratogenesis Carcinog. Mutagen* **5**, 365–378.

Breslow, L., Hoaglin, L., Rasmussen, G., and Abrams, H.K. (1954). "Occupations and cigarette smoking as factors in lung cancer." *Am. J. Public Health* **44**, 171–181.

Brinton, H.P., Frasier, E.S., and Koven, A.L. (1952). "Morbidity and mortality experience among chromate workers." *Public Health Rep.* **67**, 835–847.

Buckell, M., and Harvey, D.G. (1951). "An environmental study of the chromate industry." *Br. J. Ind. Med.* **8**, 298–301.

Campbell, K.I. (1976). "Inhalation toxicology." *Clin. Toxicol.* **9**, 849–921.

Chiazze, L., Jr., Ference, L.D., and Wolf, P.H. (1980). "Mortality among automobile assembly workers. I. Spray painters." *J. Occup. Med.* **22**, 520–526.

Cooper, W.C. (1983). "Mortality in employees of three plants which produced chromate pigments." The Dry Color Manufacturers' Association, Arlington, VA.

Dalager, N.A., Mason, T.J., Fraumeni, J.F., et al. (1980). "Cancer mortality among workers exposed to zinc chromate paints." *J. Occup. Med.* **22**, 25–29.

Davies, J.M. (1978). "Lung-cancer mortality of workers making chrome pigments." *Lancet* **1**, 384.

Davies, J.M. (1979). "Lung-cancer mortality of workers in chromate pigment manufacture: An epidemiological survey." *J. Oil Colour Chem. Assoc.* **62**, 157–163.

Davies, J.M. (1984a). "Lung mortality among workers making lead chromate and zinc chromate pigments at three English factories." *Br. J. Ind. Med.* **41**, 158–169.

Davies, J.M. (1984b). "Long term mortality study of 57 chromate pigment workers who suffered lead poisoning." *Br. J. Ind. Med.* **41**, 170–178.

Davies, J.M., and Kirsch, P. (1984). "Lung cancer mortality of workers employed in chromate pigment factories." *J. Cancer Res. Clin. Oncol.* **107**, 65.

Davis, J.M.G. (1972). "The fibrogenic effects of mineral dusts injected into the pleural cavity of mice." *Br. J. Exp. Pathol.* **53**, 190–201.

Dvizhkov, P.P., and Fedorova, V.I. (1967). "On blastomogenic properties of chromic oxide." *Vop. Onkol.* **13**, 57–62.

EEH (Equitable Environmental Health, Inc.) (1976). "An epidemiological study of workers in lead chromate plants." Final report. The Dry Color Manufacturers' Association, Alexandria, VA.

Elias, Z., Poirot, O., Schneider, O., et al. (1986). "Cellular uptake, cytotoxic and mutagenic effects of insoluble chromic oxide in V79 Chinese hamster cells." *Mutat. Res.* **169**, 159–170.

Enterline, P.E. (1974). "Respiratory cancer among chromate workers." *J. Occup. Med.* **16**, 523–526.

EPA. (1984). *Health Assessment Document for Chromium.* United States Environmental Protection Agency, Environmental Criteria and Assessment Office, EPA-600/8-83-014F. Research Triangle Park, NC.

Franchini, I., Magnani, F., and Mutti, A. (1983). "Mortality experience among chromeplating workers: Initial findings." *Scand. J. Work Environ. Health* **9**, 247–252.

Fregert, S., and Övrum, P. (1963). "Chromate in welding fumes with special reference to contact dermatitis." *Acta Dermato-Vener.* **43**, 119–124.

Frentzel-Beyme, R. (1983). "Lung cancer mortality of workers employed in chromate pigment factories." *J. Cancer Res. Clin. Oncol.* **105**, 183–188.

Frentzel-Beyme, R., and Claude, J. (1980). "Lung cancer mortality in workers employed in chromate pigment factories. A multicentric Central European epidemiologic study." *Am. J. Epidemiol.* **112**, 443.

Furst, A., Schlauder, M., and Sasmore, D.P. (1976). "Tumorigenic activity of lead chromate." *Cancer Res.* **36**, 1779–1783.

Gafafer, W.M., Ed. (1953). *Health of Workers in Chromate Producing Industry: A Study.* U.S. Public Health Service, Division of Occupational Health, Pub. No. 192, Washington, D.C.

Gottlieb, M.S. (1980). "Lung cancer and the petroleum industry in Louisiana." *J. Occup. Med.* **22**, 384–388.

Gross, E., and Kolsch, F. (1943). "Uber den lungenkrebs in der chromfarbenindustrie." *Arch. Gewerbepathol. Gewerbehyg.* **12**, 164–170.

Guillemin, M.P., and Berode, M. (1978). "A study of the difference in chromium exposure in workers in two types of electroplating process." *Ann. Occup. Hyg.* **21**, 105–112.

Gylseth, B., Gundersen, N., and Langård, S. (1977). "Evaluation of chromium exposure based on a simplified method for urinary chromium determination." *Scand. J. Work Environ. Health* **3**, 28–31.

Hayes, R.B. (1980). "Cancer and occupational exposure to chromium chemicals." *Rev. Cancer Epidemiol.* **1**, 293–333.

Hayes, R.B. (1982). "Carcinogenic effects of chromium." In S. Langård, Ed., *Biological and Environmental Aspects of Chromium, Topics in Environmental Health.* Vol. 5. Elsevier Biomedical Press, Amsterdam, pp. 221–247.

Hayes, R.B., Lilienfeld, A.M., and Snell, L.M. (1979). "Mortality in chromium chemical production workers: A prospective study." *Int. J. Epidemiol.* **8**, 365–374.

Heath, J.C., Freeman, M.A.R., and Swanson, S.A.V. (1971). "Carcinogenic properties of wear particles from prostheses made in cobalt-chromium alloy." *Lancet* **1**, 564–566.

Hedenstedt, A., Jenssen, D., Lindesten, B.-M., et al. (1977). "Mutagenicity of fume particles from stainless steel welding." *Scand. J. Work Environ. Health* **3**, 203–211.

Hernberg, S., Westerholm, P., Schultz-Larsen, K., et al. (1983). "Nasal and sinonasal cancer.

Connection with occupational exposures in Denmark, Finland and Sweden." *Scand J. Work Environ. Health* **9**, 315–326.

Hill, W.J., and Ferguson, W.S. (1979). "Statistical analysis of epidemiological data from a chromium chemical manufacturing plant." *J. Occup. Med.* **21**, 103–106.

Hueper, W.C. (1955). "Experimental studies in metal cancerigenesis. VII. Tissue reactions to parenterally introduced powdered metallic chromium and chromite ore." *J. Natl. Cancer Inst.* **16**, 447–462.

Hueper, W.C. (1958). "Experimental studies in metal cancerigenesis. X. Cancerigenic effects of chromite ore roast deposited in muscle tissue and pleural cavity of rats." *Arch. Ind. Health* **18**, 284–291.

Hueper, W.C. (1961). "Environmental carcinogenesis and cancers." *Cancer Res.* **21**, 842–857.

Hueper, W.C. (1966). *Occupational and Environmental Cancers of the Respiratory System.* Springer Verlag, New York, pp. 56–85.

Hueper, W.C., and Payne, W.W. (1959). "Experimental cancers in rats produced by chromium exposures and their significance to industry and public health." *Am. Ind. Hyg. Assoc. J.* **20**, 274–280.

Hueper, W.C., and Payne, W.W. (1962). "Experimental studies in metal carcinogenesis— chromium, nickel, iron, arsenic." *Arch. Environ. Health* **5**, 51–68.

IARC. (1973). " Chromium and inorganic chromium compounds." *IARC Monogr. Eval. Carcinog. Risk Chem. Man* **2**, 100–125.

IARC. (1980). "Chromium and chromium compounds." *IARC Monogr. Eval. Carcinog. Risk Chem. Humans* **23**, 205–323.

Ivankovic, S., and Preussmann, R. (1975). "Absence of toxic and carcinogenic effects after administration of high doses of chromic oxide pigment in subacute and long-term feeding experiments in rats." *Food Cosmet. Toxicol.* **13**, 347–351.

Jones, T.D., and Walsh, P.J. (1985). " Animal studies and prediction of human tumours can be aided by graphical sorting of animal data: neoplastic risk from B(a)P, benzene, benzidine and chromium." *Am. J. Ind. Med.* **7**, 185–217.

Kim, S., Iwai, Y., Fujino, M., et al. (1985). "Chromium-induced pulmonary cancer. Report of a case and a review of the literature." *Acta. Pathol. Jpn.* **35**, 643–654.

Korallus, V., Lange, H.J., Ness, A., et al. (1982). "Relationship between precautionary measures and bronchial carcinoma mortality in the chromate-producing industry." *Arb. Soc. Preven.* **17**, 159–167.

Kuschner, M., and Laskin, S. (1971). "Experimental models in environmental carcinogenesis." *Am. J. Pathol.* **64**, 183–196.

Lane, B.P., and Mass, M.J. (1977). "Carcinogenicity and cocarcinogenicity of chromium carbonyl in heterotopic tracheal grafts." *Cancer Res.* **37**, 1476–1479.

Langård, S. (1983). "The carcinogenicity of chromium compounds in man and animals." In D. Burrows, Ed., *Chromium: Metabolism and Toxicity*, CRC Press Inc., Boca Raton, FL, pp. 13–30.

Langård, S., and Andersen, Aa. (1980). "Betydningen av bruk av latenstid, kritisk alder og valg avreferansepopulasjon ved beregning av cancerinsidens i kohort-studies." Presented at *29. Nordike Yrkeshygieniske Mote i Norge*, November 1980, Rep. HD 842, Yrkeshygienisk Institutt, Oslo.

Langård, S., Andersen, Aa., and Gylseth, B. (1980). "Incidence of cancer among ferrochromium and ferrosilicon workers." *Br. J. Ind. Med.* **37**, 114–120.

Langård, S., and Hensten-Pettersen, A. (1981). "Chromium toxicology." In D.F. Williams, Ed., *Systematic Aspects of Biocompatibility*, CRC Press, Boca Raton, FL, pp. 143–161.

Langård, S., and Nordhagen, A.-L. (1980) " Small animal inhalation chambers and the significance of dust ingestion from the contaminated coat when exposing rats to zinc chromate." *Acta Pharmacol. Toxicol.* **46**, 43–46.

Langård, S., and Norseth, T. (1975). "A cohort study of bronchial carcinomas in workers producing chromate pigments." *Br. J. Ind. Med.* **32**, 62–65.

Langård, S., and Norseth, T. (1979). "Cancer in the gastrointestinal tract in chromate pigment workers." *Arh. Hig. Rada Toksikol.* **30**, (Suppl.), 301–304.

Langård, S., and Vigander, T. (1983). "Occurrence of lung cancer in workers producing chromium pigments." *Br. J. Ind. Med.* **40**, 71–74.

Laskin, S. (1972). *Research in Environmental Sciences*, Institute of Environmental Medicine, Washington, D.C., pp. 92–97.

Laskin, S., Kuschner, M., and Drew, R.T. (1970). "Studies in pulmonary carcinogenesis." In M.G. Hanna, Jr., P. Nettesheim, and J.R. Gilbert, Eds. *Inhalation Carcinogenesis*, U.S. Atomic Energy Commission, Oak Ridge, TN, pp. 321–351.

Lehmann, K.B. (1932). "Ist grund zu einer besonderen beunruhigung wegen desauftretens von lugenkrebs bei chromatarbeitern vorhaden?" *Zentralbl. Gewerbehyg. Unfallverhuet.* **19**, 168.

Levy, L.S., Martin, P.A., and Bidstrup, P.L. (1986). "Investigation of the potential carcinogenicity of a range of chromium containing materials on rat lung." *Br. J. Ind. Med.* **43**, 243–256.

Levy, L.S., and Venitt, S. (1975). "Carcinogenic and mutagenic activity of chromium containing materials." *Br. J. Cancer.* **32**, 254–255.

Levy, L.S., and Venitt, S. (1986). "Carcinogenicity and mutagenicity of chromium compounds: The association between bronchial metaplasia and neoplasia." *Carcinogenesis* **7**, 831–835.

Lukanin, W.P. (1930). "Zur pathologie der chromat-pneumokoniose." *Archiv Fur Hygiene* **104**, 166–174.

Machle, W., and Gregorius, F. (1948). "Cancer of the respiratory system in the United States chromate producing industry." *Public Health Rep.* **63**, 1114–1127.

Maltoni, C. (1974). "Occupational carcinogenesis." In C. Maltoni, Ed., *Advances in Tumour Prevention, Detection and Characterizations*. Vol. 2. *Cancer Detection and Prevention*. Excerpta Medica, Amsterdam, pp. 19–26.

Maltoni, C. (1976). "Predictive value of carcinogenesis bioassays." *Ann. NY Acad. Sci.* **271**, 431–443.

Mancuso, T.F. (1951). "Occupational cancer and other health hazards in a chromate plant: A medical appraisal. II. Clinical and toxicologic aspects." *Ind. Med. Surg.* **20**, 393–407.

Mancuso, T.F. (1975). "Consideration of chromium as an industrial carcinogen." In *Symposium Proceedings*. Vol. III. International Conference on Heavy Metals in the Environment, Toronto, Canada, October 27–31, pp. 343–356.

Mancuso, T.F., and Hueper, W.C. (1951). "Occupational cancer and other health hazards in a chromate plant: A medical appraisal. I. Lung cancers in chromate workers." *Ind. Med. Surg.* **20**, 358–363.

Maxild, J., Andersen, M., Kiel, P., and Stern, R.M. (1978). "Mutagenicity of fume particles from metal arc welding on stainless steel in the Salmonella/microsome test." *Mutat. Res.* **56**, 235–243.

Menck, H.R., and Henderson, B.E. (1976). "Occupational differences in rates of lung cancer." *J. Occup. Med.* **18**, 797–801.

Michel-Briand, C., and Simonin, M. (1977). "Cancers bronchopulmonaires survenus chez deux salariés occupés à un poste de travail dans le même atelier de chromage electrolytique." *Arch. Mal. Prof. Med. Trav. Secur. Soc.* **38**, 1001–1013.

Milham, S., Jr. (1976). "Cancer mortality patterns associated with exposure to metals." *Ann. NY Acad. Sci.* **271**, 243–249.

Mukubo, K. (1978). "Studies on experimental lung tumor by the chemical carcinogens and inorganic substances. III. Histopathological studies on lung tumors in rats induced by pertracheal vinyl tube infusion of 20-methylcholanthrene combined with chromium and nickel powders." *J. Nara Med. Assoc.* **29**, 321–340.

Mutti, A., Cavatorta, A., Pedroni, C., et al. (1979). "The role of chromium accumulation in the relationship between airborne and urinary chromium in welders." *Int. Arch. Occup. Environ. Health* **43**, 123–133.

Naranjit, D., Thomassen, Y., and Van Loon, J.C. (1979). "Development of a procedure for studies of the chromium(III) and chromium(VI) contents of welding fumes." *Anal. Chim. Acta* **110**, 307–312.

NAS. (1974). *Chromium.* National Academy of Sciences, Washington, D.C.

Nettesheim, P., Hanna, M.G., Jr., Doherty, D.G., et al. (1971). "Effect of calcium chromate dust, influenza virus, and 100 R whole-body X-radiation on the lung tumor incidence in mice." *J. Natl. Cancer Inst.* **47**, 1129–1144.

Newman, D. (1890). "A case of odeno-carcinoma of the left inferior turbinated body, and perforation of the nasal septum, in the person of a worker in chrome pigments." *Glasgow Med. J.* **33**, 469–470.

Nieboer, E., Verma, D.K., and Easson, I. (1985). Health hazard assessment of exposures to hexavalent chromium in gas mask wear. Special Report, McMaster University, Hamilton, Ontario, Canada.

NIOSH. (1973). "Criteria for a recommended standard: Occupational exposure to chromic acid." DHEW (NIOSH), Pub. No. 73-11021, National Institute for Occupational Safety and Health, Cincinnati, OH.

NIOSH. (1975). "Criteria for a recommended standard: Occupational exposure to chromium(VI)." DHEW (NIOSH), Pub. No. 76-129, National Institute for Occupational Safety and Health, Cincinnati, OH.

Nishiyama, H., Yano, H., Nishiwaki, Y., et al. (1985). "Lung cancer in chromate workers—analysis of 11 cases." *Jpn. J. Clin. Oncol.* **15**, 489–497.

Norseth, T. (1979). "Health effects of nickel and chromium." In E. Di Ferrante, Ed., *Trace Metals Exposure and Health Effects*, Pergamon Press, Oxford, pp. 135–146.

Norseth, T. (1981). "The carcinogenicity of chromium." *Environ. Health Persp.* **40**, 121–130.

Norseth, T. (1986). "The carcinogenicity of chromium and its salts." *Br. J. Ind. Med.* **43**, 649–651.

Ohsaki, Y., Abe, S., Kimura, K., et al. (1978). "Lung cancer in Japanese chromate workers." *Thorax* **33**, 372–374.

Okubo, T., and Tsuchiya, K. (1977). "An epidemiological study on lung cancer among chromium plating workers." *Keio J. Med.* **26**, 171–177.

Payne, W.W. (1960a). "Production of cancers in mice and rats by chromium compounds." *Arch. Ind. Health* **21**, 530–535.

Payne, W.W. (1960b). "The role of roasted chromite ore in the production of cancer." *Arch. Environ. Health* **1**, 20–26.

Pfeil, E. (1935). "Lungentumoren als berufserkrankung in chromatbetrieben." *Dtsch. Med. Wochenschr.* **61**, 1197–1202.

Phalen, R.F. (1976). "Inhalation exposure to animals." *Environ. Health Persp.* **16**, 17–24.

Pokrovskaya, L.V., and Shabynina, N.K. (1973). "Carcinogenous hazards in the production of chromium ferroalloys." *Gig. Tr. Prof. Zabol.* **10**, 23–26 (in Russian, English abstract).

Rivedal, E., and Sanner, T. (1981). "Metal salts as promoters of *in vitro* morphological transformation of hamster embryo cells initiated by benzo(a)pyrene." *Cancer Res.* **41**, 2950–2953.

Roe, F.J.C., and Carter, R.L. (1969). "Chromium carcinogenesis: Calcium chromate as a potent carcinogen for the subcutaneous tissues of the rat." *Br. J. Cancer* **23**, 172–176.

Royle, H. (1975a). "Toxicity of chromic acid in the chromium plating industry (1)." *Environ. Res.* **10**, 39–53.

Royle, H. (1975b). "Toxicity of chromic acid in the chromium plating industry (2)." *Environ. Res.* **10**, 141–163.

Sano, T., and Ebihara, I. (1979). "Pathological and clinical findings of chromate and chrome plating workers." *J. Sci. Labour* **55**, 21–31.

Sano, T., and Mitohara, I. (1978). "Occupational cancer among chromium workers." *Jpn. J. Chest Disorders* **37**, 90–101.

Satoh, K., Fukuda, Y., Torii, K., and Katsuno, N. (1981). "Epidemiological study of workers engaged in the manufacture of chromium compounds." *J. Occup. Med.* **23**, 835–838.

Schinz, H.R. (1942). "Der Metallkrebs—ein neues prinzip der krebserzeugung." *Schweizerische Medizinische Wochenschrift* **39**, 1070–1074.

Schinz, H.R., and Uehlinger, E. (1942). "Der Metallkrebs. Ein neues prinzip der krebserzeugung." *Zeitschrift fur Krebsforschung* **52**, 425–437.

Schroeder, H.A., Balassa, J.J., and Vinton, W.H. Jr. (1964). "Chromium, lead, cadmium, nickel and titanium in mice: Effect on mortality, tumors and tissue levels." *J. Nutr.* **83**, 239–250.

Schroeder, H.A., Balassa, J.J., and Vinton, W.H., Jr. (1965). "Chromium, cadmium and lead in rats: Effects of life span, tumors and tissue levels." *J. Nutr.* **86** 51–66.

Sehnalová, H., and Barborík, M. (1965). "Kotazce ukladane sloucenin chromu u lide (On the problem of deposition of chromium compounds in man)." *Pracov. Lek.* **17**, 399.

Sheffet, A., Thind, I., Miller, A.M., and Louria, D.B. (1982). "Cancer mortality in a pigment plant utilizing lead and zinc chromates." *Arch. Environ. Health* **37**, 44–52.

Shimkin, M.B., and Leiter, J. (1940). "Induced pulmonary tumors in mice. III. The role of chronic irritation in the production of pulmonary tumors in strain A mice." *J. Natl. Cancer Inst.* **1**, 241–254.

Shimkin, M.B., Stoner, G.D., and Theiss, J.C. (1978). "Lung tumor response in mice to metals and metal salts." *Adv. Exp. Med. Biol.* **91**, 85–91.

Sjögren, B. (1980). "A retrospective cohort study of mortality among stainless steel welders." *Scand. J. Work Environ. Health* **6**, 197–200.

Steffee, C.H., and Baetjer, A.M. (1965). "Histopathologic effects of chromate chemicals. Report of studies in rabbits, guinea pigs, rats, and mice." *Arch. Environ. Health* **11**, 66–75.

Steinhoff, S., Gud, C., Hatfield, G.K., and Mohr, U. (1983). "Testing Sodium Dichromate and Soluble Calcium Chromate for Carcinogenicity in Rats." Bayer, A.G. Institute of Toxicology, unpublished report (data is summarized in EPA, 1984).

Stern, R.M. (1980). "A chemical, physical and biological assay of welding fume." In R.C. Brown, I.P. Gromley, M. Chamberlain, and R. Davies, Eds., *The in vitro Effects of Mineral Dusts,* Academic Press, London, pp. 203–209.

Stern, R.M. (1981). "Process-dependent risk for delayed health effects for welders." *Environ. Health Persp.* **41**, 235–253.

Stern, R.M. (1982). "Chromium compounds: Production and occupational exposure." In S. Langård, Ed., *Biological and Environmental Aspects of Chromium, Topics in Environmental Health.* Vol. 5. Elsevier Biomedical Press, Amsterdam, pp. 5–47.

Stern, R.M. (1983a). "The Assessment of Risk: Application to the Welding Industry Lung Cancer." The Working Environment Research Group, The Danish Welding Institute, Copenhagen, Denmark.

Stern, R.M. (1983b). "Assessment of risk of lung cancer for welders." *Arch. Environ. Health* **38**, 148–155.

Stoner, G.D., Shimkin, M.B., Troxell, M.C., et al. (1976). "Test for carcinogenicity of metallic compounds by the pulmonary tumor response in strain A mice." *Cancer Res.* **36**, 1744–1747.

Sunderman, F.W., Jr. (1976). "A review of the carcinogenicities of nickel, chromium and arsenic compounds in man and animals." *Prev. Med.* **5**, 279–294.

Sunderman, F.W., Jr. (1984). "Recent advances in metal carcinogenesis." *Ann. Clin. Lab. Sci.* **14**, 93–122.

Sunderman, F.W., Jr., Lau, T.J., and Cralley, L.J. (1974). "Inhibitory effect of manganese upon muscle tumorigenesis by nickel subsulfide." *Cancer Res.* **34**, 92–95.

Taylor, F.H. (1966). "The relationship of mortality and duration of employment as reflected by a cohort of chromate workers." *Am. J. Public Health* **56**, 218–229.

Thomsen, E., and Stern, R.M. (1979). "A simple analytical technique for the determination of hexavalent chromium in welding fumes and other complex matrices." *Scand. J. Work Environ. Health* **5**, 386–403.

Tola, S., Kilpio, J., Virtamo, M., and Haapa, K. (1977). "Urinary chromium as an indicator of the exposure of welders to chromium." *Scand. J. Work Environ. Health* **3**, 192–202.

Tsuneta, Y. (1982). "Investigations of the pathogenesis of lung cancer observed among chromate factory workers." *Hokkaido J. Med. Sci.* **57**, 175–187.

Waterhouse, J.A.H. (1975). "Cancer among chromium platers." *Br. J. Cancer* **32**, 262.

Waterhouse, J.A.H. (1980). "A mortality study of chromeplaters." Unpublished results.

Welinder, H., Littorin, M., Gullberg, B., and Skerfving, S. (1983). "Elimination of chromium in urine after stainless steel welding." *Scand. J. Work Environ. Health* **9**, 397–403.

White, L.R., Hunt, J., Tetley, T.D., and Richards, R.J. (1981). "Biochemical and cellular effects of welding fume particles in the rat lung." *Ann. Occup. Hyg.* **24**, 93–101.

Appendix Table Summary of Carcinogenicity Studies of Chromium and Chromium Compounds in Animals[a]

Compound	Species	Route and Dosage	Findings	Reference
Chromium powder	Mouse	Four IP injections of 0.2 ml of a 0.005% solution	One myeloid leukemia in 50 treated animals	Hueper, 1955
	Mouse	Six IV injections of 0.05 ml of a 0.005% solution	No tumors	Hueper, 1955
	Mouse	Six intrapleural injections of 0.2 ml of a 0.005% suspension	No tumors in 50 treated mice	Hueper, 1955
	Rat	One intratracheal injection of 10 mg	No squamous-cell carcinomas of the lung in 12 treated rats	Mukubo, 1978
	Rat	One IM injection of 2 mg	No local tumors in 22 surviving treated animals	Sunderman et al., 1974
	Rat	Six IP injections of 0.1 ml of a 0.05% suspension	No increase in round-cell sarcoma incidence compared with controls; two insulinomas in treated animals, none in controls	Hueper, 1955
	Rat	Six IV injections of 0.18 ml of a 0.05% suspension	Two rats with pulmonary adenomas; no increase in sarcomas compared with controls	Hueper, 1955
	Rat	Six intrapleural injections of 0.05 ml of a 33.6% (by weight) suspension or six intrapleural injections of 0.1 ml of a 0.5% suspension	Two haemangiomas and 1 angiosarcoma in 50 treated animals and 0/25 controls	Hueper, 1955
	Rat	Intramedullary injection into the femur of 46 mg	No injection-site tumors in 25 treated animals	Hueper, 1955

Continued

Appendix Table (*Continued*)

Compound	Species	Route and Dosage	Findings	Reference
Unroasted chromite(III) ore	Rabbit	18 IV injections of 0.5 ml/kg bw of a 5% suspension	One carcinoma of lymph node in three treated survivors and 0/4 controls	Hueper, 1955
	Mouse	Intrapleural injection of 10 mg in 0.5 ml distilled water	Granulomas	Davis, 1972
	Rat	Six intrapleural injections of 0.05 ml of a 73.4% (by weight) suspension	Injection-site sarcoma in 1/25 treated animals	Hueper, 1955
	Rat	Intramedullary injection into the femur of 58 mg	No injection-site tumors in 25 treated rats	Hueper, 1955
	Rabbit	12 IV injections of 5 ml of a 5% suspension	No tumors	Hueper, 1955
Roasted chromite(III) ore	Mouse	IM implantation of 10 mg (equivalent to 0.79 mg chromium)	No implantation-site tumors	Payne, 1960b
	Rat	IM implantation of 25 mg	Sarcomas at implantation site in 3/31 treated animals and zero vehicle controls	Hueper, 1958
	Rat	IM implantation	Tumors (type unspecified) at implantation site in 1/34 treated animals and 0/32 vehicle controls	Hueper, 1961
	Rat	Intrapleural implantation	Tumors (type unspecified) at implantation site in 5/32 treated animals and 0/34 controls	Hueper, 1961
	Rat	Intrapleural implantation of 25 mg	Lung squamous-cell carcinomas in two of four treated survivors and one lung adenoma in one of four controls	Hueper, 1958

488

	Species	Treatment	Result	Reference
	Rat	Intrapleural implantation of 25 mg (equivalent to 2 mg chromium)	Implantation-site tumors (type unspecified) in 3/35 treated rats and 0/35 controls	Payne, 1960b
Mixed chromate(VI) dust	Mouse	Inhalation, four hours/day, five days/week for 16–58 weeks (total dose chromium inhaled: 480–1205 mg–hrs)	No lung carcinomas; no significant increase in lung adenomas in 500 treated animals compared with controls	Baetjer et al., 1959b
	Mouse	Five to six intratracheal instillations of dust (equivalent to 0.04 mg chromium trioxide)	No more lung tumors in 506 treated mice than in controls	Baetjer et al., 1959b
	Mouse	Four intrapleural injections of 0.05 ml of a 2 or 4% suspension	No increase in lung tumor incidence in 55 treated animals compared with 41 controls	Baetjer et al., 1959b
	Rat	Inhalation, four to five hours/day, four days/week for lifespan chromic oxide concentration of 3–4 mg/m³)	No significant increase in tumor incidence in 78 treated rats compared with controls	Steffee and Baetjer, 1965
Mixed chromate dust + potassium dichromate(VI)	Rat	16 intratracheal injections of 0.1 ml of suspension of 0.5% roasted chromate + 0.6% potassium dichromate (equivalent to 0.07 mg chromium/dose)	No significant increase in tumor incidence compared with controls	Steffee and Baetjer, 1965
Mixed chromate(VI) dust + potassium dichromate(VI) + sodium chromate(VI) + pulverized residue dust	Rabbit	Inhalation, four days/week for 50 months	No increase in tumor incidence compared with controls	Steffee and Baetjer, 1965
	Guinea-pig	Inhalation, four days/week for lifespan	Pulmonary carcinomas in 3/50 treated animals	Steffee and Baetjer, 1965

Continued

489

Appendix Table *(Continued)*

Compound	Species	Route and Dosage	Findings	Reference
Barium chromate(VI)	Rat	IM implantation of 25 mg	No implantation-site tumors in 35 rats	Hueper and Payne, 1959
	Rat	IM implantation	No implantation-site tumors in 34 rats	Hueper, 1961
	Rat	Intrapleural implantation	Implantation-site tumors in 1/31 treated rats and 0/34 controls	Hueper, 1961
Calcium chromate(VI)	Mouse	Inhalation, 13 mg/m^3 five hours/day, five days/week for lifespan	Lung adenomas in 14/136 treated animals and 5/136 controls	Nettesheim et al., 1971
	Mouse	IM implantation of 10 mg	Implantation-site sarcomas in 2/50 and in 0/50 controls	Payne, 1960a
	Mouse	One SC injection of 10 mg	Injection-site sarcomas in 1/13 and in 0/52 controls	Payne, 1960a
	Rat	Bronchial implantation	Six squamous-cell carcinomas and two adenocarcinomas of the lung in 100 treated rats; 0/24 controls	Laskin et al., 1970
	Rat	Bronchial implantation	Increased incidence of bronchial squamous-cell carcinomas	Levy and Venitt, 1975
	Rat	Inhalation, 2 mg/m^3, 589 exposures of five hours over 891 days	One squamous-cell carcinoma of lung, one of larynx, one peritruncal tumor (number of treated animals unspecified)	Laskin, 1972
	Rat	IM implantation of 12.5 mg	Malignant tumors at implantation site in four of eight treated animals	Hueper and Payne, 1962
	Rat	20 injections, total dose 19 mg	Injection-site sarcomas in 18/24 and zero vehicle controls	Roe and Carter, 1969

490

Rat	12 injections of 4 mg	Injection-site sarcomas in 5/45 and 0/22 in vehicle controls	Furst et al., 1976
Rat	IM implantation of 25 mg	Injection-site sarcomas in 8/35 treated animals and 0/32 controls	Hueper and Payne, 1959
Rat	IM implantation	Tumors (type unspecified) at implantation site in 9/32 treated animals and 0/32 controls	Hueper, 1961
Rat	Intrapleural implantation of 12.5 mg	Malignant tumors (unspecified) at implantation site in 8/14 treated animals	Hueper and Payne, 1962
Rat	Intrapleural implantation	Tumors (type unspecified) at implantation site in 20/32 treated animals and 0/34 controls	Hueper, 1961
Hamster	Inhalation, 2 mg/m^3, 589 exposures	One squamous-cell carcinoma and one papilloma of larynx (number of treated animals unspecified)	Laskin et al., 1972
Mouse	IM implantation of 10 mg	Implantation-site sarcomas in 9/46 treated animals and 0/50 controls	Payne, 1960a
Mouse	SC injection of 10 mg	No injection-site sarcomas	Payne, 1960a
Rat	IM implantation of 25 mg	Implantation-site sarcomas in 8/35 treated animals and zero controls	Hueper and Payne, 1959
Sintered calcium chromate(VI) Rat	IM implantation	Tumors (type unspecified) at implantation site in 12/34 treated animals and 0/32 controls	Hueper, 1961

Continued

Appendix Table *(Continued)*

Compound	Species	Route and Dosage	Findings	Reference
	Rat	Intrapleural implantation	Tumors (type unspecified) at implantation site in 17/33 treated rats and 0/34 controls	Hueper, 1961
Chromic(III) acetate	Mouse	PO, 5 mg/L drinking water for life	No increase in tumor incidence	Schroeder et al., 1964
	Rat	PO, 5 mg/L drinking water for life	No increase in tumor incidence	Schroeder et al., 1965
	Rat	Eight IM implantations of 25 mg each, over 24 months	Implantation-site sarcoma in 1/35 treated animals	Hueper and Payne, 1962
	Rat	IM implantation	Implantation-site tumor (type unspecified) in 1/34 and in 0/32 controls	Hueper, 1961
	Rat	Eight intrapleural implantations of 25 mg each over 13 months	No implantation-site tumors after two years in 42 treated animals	Hueper and Payne, 1962
	Rat	Intrapleural implantation	Implantation-site tumor (type unspecified) in 1/34 and in 0/34 controls	Hueper, 1961
Chromic(III) oxide	Rat	PO, 1, 2, and 5% in bread on five days/week for two years	1% dose: 3/60 mammary fibroadenomas 2% dose: 1/60 mammary fibroadenomas 5% dose: 3/60 mammary fibroadenomas Controls: 1/60 mammary carcinoma; 2/60 fibroadenomas	Ivankovic and Preussmann, 1975
	Rat	Single intratracheal application of 50 or 20 mg	50 mg dose: 7/34 with tumors (four with lung sarcomas) 20 mg dose: 6/18 with tumors (five with lung sarcomas)	Dvizhkov and Fedorova, 1967

Compound	Species	Treatment	Result	Reference
	Rat	Bronchial implantation	No controls	Laskin et al., 1970
	Rat	One IP injection of 20 mg	No lung tumors in 98 animals	Dvizhkov and Fedorova, 1967
	Rat	Two intrapleural injections of 5 mg	Lung sarcomas in 4/20; no controls. Reticulum-cell sarcomas of lung in 3/17 treated animals; no controls	Dvizhkov and Fedorova, 1967
Chromium carbonyl	Rat	Injection of 2.5 mg into subcutaneously implanted tracheal rings	Squamous-cell carcinomas in 2/22 animals; none in four vehicle controls	Lane and Mass, 1977
Chromium(III) sulphate	Mouse	24 IP injections (total doses: 480, 1200, and 2400 mg/kg bw)	No significant increase of pulmonary adenoma incidence in 60 treated rats compared with 40 vehicle and untreated controls	Stoner et al., 1976
Chromium(VI) trioxide	Rat	Bronchial implantation	No increase in lung tumor incidence in 100 treated rats compared with 24 controls	Laskin et al., 1970
Sintered chromium(VI) trioxide	Mouse	One SC injection of 10 mg	No injection-site tumors in 52 treated animals	Payne, 1960a
Cobalt-chromium alloy	Rat	IM implantation of 25 mg	Implantation-site sarcomas in 15/35 treated animals and 0/35 controls	Hueper and Payne, 1959
	Rat	IM injection of 28 mg	Injection-site sarcomas in 7/74 treated rats; other tumors in 7/74	Heath et al., 1971
Lead chromate(VI)	Mouse	Four IM injections of 3 mg	Two lymphomas and three lung adenocarcinomas in 17 mice necropsied; similar incidences in controls	Furst et al., 1976

Continued

Compound	Species	Route and Dosage	Findings	Reference
	Rat	One SC injection of 30 mg	Injection-site sarcomas in 26/40 treated animals and 0/60 vehicle controls	Maltoni, 1974, 1976
	Rat	Nine IM injections of 8 mg	Injection-site sarcomas in 31/47 treated rats; three renal carcinomas; 0/22 in vehicle controls	Furst et al., 1976
	Rat	IM implantation	Tumor (type unspecified) at implantation site in 1/33 treated rats and 0/32 controls	Hueper, 1961
	Rat	Intrapleural implantation	Tumors (type unspecified) at injection site in 3/34 treated rats and 0/34 controls	Hueper, 1961
Lead chromate(VI) oxide	Rat	One SC injection of 30 mg	Injection-site sarcomas in 27/40 treated rats and 0/60 vehicle controls	Maltoni, 1974, 1976
Potassium chromate(VI)	Rat	Bronchial implantation	No increased incidence of lung tumors	Levy and Venitt, 1975
Potassium dichromate(VI)	Rat	Bronchial implantation	No increased incidence of lung tumors	Levy and Venitt, 1975
Sodium chromate(VI)	Rat	Bronchial implantation	No increased incidence of lung tumors	Levy and Venitt, 1975
Sodium dichromate(VI)	Rat	16 IM injections of 2 mg	No injection-site tumors	Hueper and Payne, 1962
	Rat	IM implantation	No implantation-site tumors	Hueper, 1961
	Rat	16 intrapleural injections of 2 mg	One lung adenocarcinoma in 39 treated animals; no injection-site tumors in 60 vehicle controls	Hueper and Payne, 1962

	Rat	Intrapleural implantation	No injection-site tumors in 26 treated animals	Hueper, 1961
Strontium chromate(VI)	Rat	Bronchial implantation	No increase in lung tumors	Levy and Venitt, 1975
	Rat	IM implantation	Implantation-site tumors in 15/33 treated animals and 0/32 controls	Hueper, 1961
Zinc potassium chromate(VI)	Rat	Bronchial implantation	Increased incidence of bronchial squamous-cell carcinomas	Levy and Venitt, 1975
Zinc yellow	Mouse	Six intratracheal injections of 0.03 ml of a 0.2% suspension	No pulmonary carcinomas; pulmonary adenomas in 31/62 treated animals and 7/18 untreated controls	Steffee and Baetjer, 1965
	Rat	IM implantation	Tumors (type unspecified) at implantation site in 16/34 treated animals and 0/32 controls	Hueper, 1961[b]
	Rat	Intrapleural implantation	Tumors (type unspecified) at implantation site in 22/33 treated animals and 0/34 controls	Hueper, 1961[b]

[a]IM, intramuscular; IP, intraperitoneal; IV, intravenous; PO, orally; SC, subcutaneous. [b]It was not specified whether this compound was zinc chromate, zinc potassium chromate, or zinc yellow.

Source: Reproduced in whole and with permission from IARC Monographs on the Evaluation of the Carcinogenic Risk of Chemicals to Humans. Some Metals and Metallic Compounds, Vol. 23, pp. 264–273 (1980). International Agency for Research on Cancer, Lyon, France.

18

CHROMIUM HYPERSENSITIVITY

A. T. Haines

*Department of Clinical Epidemiology
and Biostatistics, and Occupational
Health Program
McMaster University
Hamilton, Ontario, Canada*

E. Nieboer

*Department of Biochemistry and
Occupational Health Program
McMaster University
Hamilton, Ontario, Canada*

1. INTRODUCTION

The concept of hypersensitivity, as it is referred to here, conforms to that of Pepys, the eminent immunologist, in his definition of " allergy": " The profound biologic transformation of the intact organism resulting from adequate exposure to viable and non-viable agents, so that re-exposure, often too small, indeed often minute, and apparently innocuous amounts of extrinsic agents, elicits a different sort of reaction (than that occurring at first exposure) . . ." (Pepys, 1977). Pepys further describes the traditional criteria for regarding a reaction as hypersensitivity-mediated (with or without supporting immunologic evidence):

1. history of previous exposure (i.e., a period of sensitization);
2. a tendency for the specific sensitivity to increase with further exposure;
3. a low (usually) proportion of affected individuals in the exposed population.

The hypersensitivity phenomena associated with chromium exposure that are considered in this chapter are generally consistent with this definition and these criteria.

Most of the literature on hypersensitivity induced by chromium compounds concerns skin allergies. Thus, the following discussion of prevalence, clinical patterns, specific known causes, prognosis, prophylaxis, and pathogenesis has a dermatologic focus. The available evidence concerning respiratory hypersensitivity effects of chromium is also summarized.

2. CUTANEOUS CHROMIUM HYPERSENSITIVITY— CLINICAL CONSIDERATIONS

2.1 Prevalence

Chromium is generally accepted as being the second most common skin allergen in the general population, after nickel (Polak, 1983); in men, it is the most frequent sensitizer (Cronin, 1980). Data on the prevalence of chromium sensitivity in the general population are scarce. In a recent study conducted in Finland, 2% of men and 1.5% of women showed a positive patch test reaction to potassium dichromate (Peltonen and Fräki, 1983). Occupational exposure could be confirmed in many of the cases. Numerous reports cite the proportion of patients investigated for contact dermatitis who reacted positively to patch testing with 0.5% (0.017 M) potassium dichromate. In these studies, the prevalence of patch test sensitivity to chromate showed wide variability, perhaps because the breakdown of subjects by sex was not reported. Table 1 shows the prevalence of chromate sensitivity found in eight important investigations in which sex was taken into account. The greater frequency of chromate sensitivity among males is apparent. Both differing sex ratios of prevalence and differing overall rates by geographical setting have been attributed largely to the proportion of the local population engaged in the building industry, particularly under hot humid conditions. In North America (North American Contact Dermatitis Group, 1973), considerable variability in prevalence was observed among study sites (Table 2).

The age distribution of chromate-sensitive patients seen at St. John's Hospital for Diseases of the Skin in London, England, appears to correspond to the

Table 1   Prevalence of Positive Patch Tests to Potassium Dichromate Among Patients Investigated for Contact Dermatitis, By Sex

Setting	N	Chromate Positive (%)		Reference
		M	F	
Belgium	400	17.3	9.7	Lachapelle and Tennstedt, 1979
Brazil	536	25.5	5.7	Moriearty et al., 1978
"Europe"	4825	10.7	3.6	Fregert et al., 1969
Finland	1790	12.0	3.0	Peltonen and Fräki, 1983
Nigeria	453	11.0	4.6	Olumide, 1985
"North America"	1200	10.0	6.0	North American Contact Dermatitis Group, 1973
"Scandinavia"	5558	11.7	3.0	Magnusson et al., 1968
Scotland	1312	7.0	4.0	Husain, 1977

Table 2   Prevalence of Positive Patch Tests to Potassium Dichromate Among Patients Investigated for Contact Dermatitis, North America (Men and Women Combined)[a]

City	N	Chromate Positive (%)
Bangor	59	10
Detroit	20	20
Hanover	197	9
Marshfield	129	2
New Orleans	24	12
New York	44	9
Portland	229	10
Richmond	207	4
San Francisco	126	9
Vancouver	165	8

[a]Based on data reported by North America Contact Dermatitis Group, 1973.

working years of life (Table 3). Within given geographical settings, the prevalence of positive patch tests to chromate has remained reasonably constant over the years (Baer et al., 1964; Hammershoy, 1980). Moreover, there is no evidence of changing frequency patterns within sex categories (Cronin, 1980).

At St. John's Hospital, the existence of concomitant sensitivity to cobalt and/or nickel was also documented among patients with chromate sensitivity. As can be seen from the data in Table 4, the large majority of subjects reacted to chromate only.

Table 3   Age Ranges of Chromate-Sensitive Subjects, St. John's Hospital for Diseases of the Skin, 1967–1976, by Sex

Age Range	Men	Women
Under 21	15	13
21–30	61	35
31–40	122	23
41–50	125	29
51–60	121	41
61–70	65	15
Over 70	4	6

[a]Reproduced in modified form from Cronin (1980).

Table 4   Associated Sensitivity to Cobalt and/or Nickel Among Chromate-Sensitive Subjects, St. John's Hospital for Diseases of the Skin, 1973–1976[a,b]

	Men		Women	
Sensitivity	N	(%)	N	(%)
Cr_6^+ Co^- Ni^-	143	(70)	47	(57)
Cr_6^+ Co^+ Ni^-	38	(19)	7	(9)
Cr_6^+ Co^- Ni^+	10	(5)	9	(11)
Cr_6^+ Co^+ Ni^+	12	(6)	19	(23)
Total Cr_6^+	203		82	

[a]Reproduced from Cronin (1980); [b]Cr_6 denotes Cr(VI); superscript +, positive reaction; superscript −, negative reaction.

2.2   Clinical Patterns

The hands are involved in the majority of chromate dermatitis cases. Typically affected, as with other contact sensitizers, are fronts of wrists, backs of hands, sides and backs of fingers, and finger webs (Burrows, 1983). Not infrequently, however, other parts of the body are involved and there may be a striking resemblance to other dermatitides. For example, among 134 patients with cement dermatitis, Burrows and Calnan (1965) observed patterns resembling nummular dermatitis in 13%, seborrheic dermatitis in 10%, and stasis dermatitis in 7%. The palms were involved in 9% of cases. Fourteen individuals with positive patch test results to potassium dichromate were reported by Shanon (1965) to have a rash closely resembling atopic dermatitis; these patients improved after removal from chromate exposure. Engel and Calnan (1963) noted a number of cases of chromate dermatitis associated with primer paint exposure which were indistinguishable from seborrheic, nummular, or atopic dermatitis in distribution and clinical appearance.

3. ASSOCIATED OCCUPATIONAL AND ENVIRONMENTAL EXPOSURES

3.1 Introduction

In 1925, chromium dermatitis was first reported (Parkhurst, 1925) in a blueprint production worker with dermatitis who reacted positively to patch testing with 0.5% potassium dichromate. Since then, reports linking dermatitis with chromium exposure in diverse occupational and environmental contexts have multiplied considerably. However, the occupational or other possible sources of chromate exposure have been infrequently investigated among chromate-sensitive patients. Among 140 such subjects, Pirilä (1954) noted a predominance of cement workers (57%) with 11% being lithographers and photographers and 7% plating and metal workers. Among another series of 245 cases (Kresbach, 1967), exposure to cement in 27% was the only clear-cut category. At St. John's Hospital for Skin Diseases (Cronin, 1980), workers from building trades (46%) again figured prominently among patients with chromate sensitivity. Among female patients, about one third had domestic work as a plausible source of exposure. Significantly, no source was found in a large proportion of both male (16%) and female subjects (44%).

Most of the information that we have concerning the allergenic risks of chromium derives from case and case series reports. Table 5 lists the major sources of occupational and environmental chromium exposure that have been associated with the development of contact sensitivity; the chromium compounds considered responsible are also shown.

The important features of these occupational and environmental exposures and their observed effects are briefly examined in this section. Extensive reviews of these topics are available (Cronin, 1980; Burrows, 1983).

3.2 Cement

Cement is the most important cause of chromate allergy and its role in producing dermatitis has long been recognized. In 1700, Ramazzini noted that "lime makes the hands of brick layers wrinkle and sometimes ulcerates them" (cited in Burrows, 1983). More than 30 years ago, Jaeger and Pelloni (1950) showed that 30 of 32 cement dermatitis patients reacted positively on patch testing to potassium dichromate, whereas only 5% of other patients did so. These investigators confirmed the link between cement dermatitis and chromium allergy by demonstrating the presence of chromium in cement. Burrows (1978) has reported a declining rate of cement exposure among patients with chromate

Table 5   The Ten Most Important Examples of Eczematogenic Exposures to Chromium Compounds[a]

Chromium Containing Materials or Objects	Profession or Places of Contact	Chromium Compounds Responsible
Chromium ore	Industrial chromium production	Chromite
Chrome baths	Electroplating industry	Chromic acid Sodium dichromate
	Graphic trade	Chromates
	Metal industry	Chromates Zinc chromate
Chrome colors and dyes	Painters and decorators, graphic trades, textile, rubber, glass, and china industries	Chromic oxide green, chromic hydroxide green, chrome yellow (lead chromate)
Lubricating oils and greases	Metal industry	Chromic oxide
Anticorrosive agents in water system	Diesel locomotive workshops and sheds, central heating and air-conditioning systems	Alkali dichromates
Wood preservation (Wolman salts)	Wood impregnation, furniture industry, carpenters, miners	Alkali dichromates
Cement, cement products, quick-hardening agents for cement (e.g., Sika 1)	Cement production, manufacture of cement products, building trades	Chromates
Cleaning materials (l'eau de Javelle), washing and bleaching materials	Housewives, cleaners, laundry workers	Chromates
Textiles, furs	Textile and fur industries, everyday life	Chromates
Leather and artificial leather tanned with chromium	Leather and footwear industry, everyday life	Chromium sulfate and chromium alum

[a]Reproduced in modified form from Polak (1983).

sensitivity, and he postulated that this may reflect changing methods of material handling in the building industry.

Cement is a skin irritant because of its alkaline, abrasive, and hygroscopic properties. In the dry form, it is rarely associated with the development of dermatitis (Calnan, 1960). When cement becomes wet, by the addition of water or by absorbing it from skin, alkaline calcium hydroxide is liberated. The abrasiveness of cement increases as it absorbs water and when it is mixed with sand.

The trivalent chromium compounds present in cement precursors are oxidized to the hexavalent form during heating in kilns at temperatures exceeding 1000°C. The chromate content of cement has been measured by numerous investigators, and significant amounts have been found to be present with considerable consistency (Cronin, 1980). For example, Fregert and Gruvberger (1972) detected measurable quantities of water-soluble chromate (1–40 μg chromium g^{-1} in all of 52 cement samples tested, although only in two were concentrations in excess of 35 μg chromium g^{-1}. Perone et al. (1974) found 18 of 42 samples of Portland cement to contain measurable water-soluble chromate, in amounts ranging from 0.1 to 5.4 μg chromium g^{-1}. Possible chromate sources in finished cement include: (1) the ingredients—chalk, clay, and gypsum; (2) chromium-containing refractory bricks; (3) chromium-containing steel grinders; (4) ash. It should be noted, however, that cement may contain numerous additives, of which some may be sensitizers or irritants in their own right (Calnan et al., 1969).

Cement dermatitis is a major occupational disease. In Europe in recent years, it has consistently constituted more than 25% of occupational dermatoses (Cronin, 1980). As already indicated (Section 3.1), nearly half of chromate-sensitive men examined by Cronin (1980) were likely to have acquired their sensitivity from cement. Cross-sectional studies in Scandinavia found approximately 5% of construction workers to have cement dermatitis (Wahlberg, 1969; Høvding, 1970, cited in Cronin, 1980). Burrows (1972) has estimated that 3.5% of workers in the building industry develop a dermatitis that leads to time off work. In European studies, over 75% of men with cement dermatitis have been found to react positively upon patch testing to chromate (Høvding, 1970, cited in Cronin, 1980).

Data concerning cement dermatitis in North America are limited. Although Perone et al. (1974) found that among 95 construction workers more than one third had manifestations of mild to moderate dermatitis, only one reacted positively to patch testing with 0.25% (0.008 M) dichromate. In contrast, between 2.5 and 6% of European cement workers with *normal* skin had positive patch test results (Cronin, 1980).

3.3   Anti-Corrosion Agents

Chromium-associated dermatitis has been found in relation to exposure to anti-rust and anti-corrosion agents in a number of settings. A water-soluble chromate (usually potassium dichromate) is the active ingredient. For example, chromate-containing primer paint has been linked with the development of dermatitis in aircraft workers (Hall, 1944), and in workers who were wet-sandpapering car bodies (Engel and Calnan, 1963). Adams et al. (1976) reported that the content of water-soluble chromate in primer paints formulated with zinc potassium chromate was of the order of that found in patch test solutions, and much greater than that in cement. Chromate is currently being excluded from paints because of the carcinogenic risk during manufacturing (Burrows, 1983; Chapter 17).

The use of sodium dichromate as an anti-corrosive in coolants has been associated with the development of chromate dermatitis, particularly among diesel maintenance workers (Winston and Walsh, 1951; Calnan and Harman, 1961; Guy, 1954). The replacement of chromate by other anti-corrosives in coolant systems has been linked to a decline in chromate dermatitis in these circumstances.

The handling of chromate-treated galvanized sheets has also been related to the development of contact dermatitis by Fregert et al. (1970a). These investigators demonstrated a recovery of between 1 and 5 mg chromium per person (about 3% was hexavelent) from the hand-wash water of workers handling galvanized sheets. This amount was much greater than that found in hand washes from bricklayers using cement (0–7 μg). Earlier, Newhouse (1963) reported 47 cases of chromate sensitivity among 130 car assembly workers, who were handling chromate-treated zinc-plated parts.

3.4   Bleaches and Detergents

In some countries, chromate in detergent and bleaches contributes substantially to the prevalence of hand dermatitis among those engaged in domestic work. A well-recognized source of chromium allergy has been bleach (l'eau de Javelle) to which chromate has been added for stabilizing and coloring purposes (Rabeau and Ukrainczyk, 1939; Hilt, 1954). Chromate has been removed from l'eau de Javelle in France, but, as of 1973, not in Italy, Spain, or Belgium (García-Pérez et al., 1973). Feuerman (1969) attributed a high prevalence of chromate sensitivity among housewives with hand eczema to the use of both bleaches and detergents. Subsequently, García-Pérez et al. (1973) linked a declining frequency of chromate-positive hand dermatitis in female domestic workers to reduced quantities of chromate in detergents and bleaches. Over the same period, no change in the rate of chromium sensitivity in men was observed. Chromate is not added to bleach in North America (Burrows, 1983).

3.5   Leather

Chromium compounds used in leather tanning have been extensively associated with sensitization of both tannery workers (Morris, 1958; Fregert, 1975; Pedersen, 1982; Burrows, 1983) and the users of leather materials (Pirilä and Kilpiö, 1954; Rudzki and Kozlowska, 1980; Angelini et al., 1980; Weston and Weston, 1984). Morris (1958) reported four individuals with shoe dermatitis as reacting positively to patch testing with 0.2% basic (trivalent) chromic sulfate, used in the one-bath tanning process [involves Cr(III) only]. Samitz and Gross (1961) reported the presence of both trivalent and hexavalant chromium in extracts from leather of human sweat. However, Fregert and Gruvberger (1970) were able to identify only trivalent chromium in synthetic sweat and water extracts from shoe leather. Most of the chromium compounds used in the tanning process are Cr(III) derivatives (Stern, 1982). In the two-bath process used to prepare specialty leather, the skins are first treated with dichromate, which is followed by a reduction step that converts absorbed Cr(IV) to Cr(III). It should be noted that foot dermatitis can be associated with contact allergy to a number of other nonchromium shoe components (Cronin, 1966). The replacement of chrome-tanning by vegetable-tanning of gloves has markedly reduced their chromium content (Fregert and Gruvberger, 1979), but the impact of this change on the prevalence of hand dermatitis has not been reported.

3.6   Machine Oils

Exposure to chromate in unused (Holz et al., 1961; Calnan, 1978) and used (Weiler, 1969; Fregert and Gruvberger, 1976a) mineral oil has been associated with contact hypersensitivity manifestations. Rycroft (cited in Burrows, 1983) documented an increase in the level of chromium in soluble oil used in lathe machines over a period of five weeks; however, none of 70 operators reacted positively on patch testing to chromate.

3.7   Dyes

Chromate used in dyeing applications has been associated with the development of hypersensitivity in an upholstery factory worker (Fregert and Gruvberger, 1976b), card players in contact with green felt tables (Fregert et al., 1970b; Fisher, 1976), and the wearers of green military textiles (Fregert et al., 1978). Skin inflammation has also been reported in association with the green Cr(III) oxide pigment in tattoos (Scutt, 1966; Cairns and Calnan, 1962).

3.8   Welding

Exposure to welding fumes has been linked to contact dermatitis of the face (Fregert and Övrum, 1963). Inhalation of welding fumes has been incriminated in episodes of hand dermatitis in a chromate-sensitive worker (Shelley, 1964). Chromium is commonly present in the coating and core of welding rods as well as in the workplace. The latter is a minor source of fumes. Particularly, manual metal arc welding of stainless steel liberates considerable amounts of chromium, 90% of which is hexavalent and water soluble (Stern, 1982; Chapter 2). Gas-shielded (inert) welding of stainless steel also liberates significant quantities of chromium, but a smaller fraction ($\sim$5%) is Cr(VI).

3.9   Foundry Sand, Boiler Linings, Ashes

As in chromate dermatitis associated with cement, much of the allergenicity of chromium exposure in certain other occupational settings is attributable to conversion of trivalent chromium into hexavalent forms, during heating under alkaline conditions. Thus, hand dermatitis associated with chromate sensitivity has developed in men working with molding sand in foundries (Hellier, 1962). Investigation has revealed the source as the presence in foundry sand of ground chromium magnesite bricks, previously used as a refractory material (Fregert, 1963). Chromate dermatitis has also been reported in a fitter who was repairing the refractory lining of a boiler (Rycroft and Calnan, 1977). And, finally, Fregert (1962) observed positive patch tests in chromium-sensitive individuals elicited by water extracts of wood ashes. Unlike the ashes of coal, coke, and fuel oil, wood ashes are alkaline and contain hexavalent chromium.

3.10   Printing

Following the report of Parkhurst (1925), a risk of chromate dermatitis was recognized in the printing industry. Pirilä and Kilpiö (1954) reported chromate sensitivity in 27% of 149 cases of dermatitis among printing workers. Of 76 cases of dermatitis reported among lithographers (Levin et al., 1959), 17 patch tested positively to potassium dichromate. Spruit and Malton (1975) demonstrated high chromium levels in the materials handled by three men working in an offset printing factory who developed dermatitis and chromate sensitivity. The use of potassium dichromate in the printing industry has declined following suitable substitution by other agents (Burrows, 1983).

3.11   Chromate Production and Chromium Plating

While chromate production and chromium-plating industries are high-risk environments for chrome ulcerations (Gafafer, 1953; Anon, 1963; Henning, 1972), sound prevalence data are not available with regard to chromate sensitization in this setting. Royle (1975) found 25% of 997 electroplaters to have dermatitis. However, patch testing was not performed, and the reported prevalence was based on answers to a questionnaire. Among 897 chromate production workers examined, Gafafer (1953) noted 17 individuals with chrome dermatitis of the erythematous and papular type. They attributed this low prevalence to the existence of adequate washing facilities and the availability of clean working clothes. The possibility of the development of tolerance was not discussed. It is of interest to note that Edmundson (1951) found that only two of 56 chromate workers with chrome ulcers had positive patch tests to 0.5% (0.017 M) of potassium dichromate. In other studies, little association has been observed between the finding of chrome ulcerations and the presence of chromium contact sensitivity (Cronin, 1980; Burrows, 1983).

3.12   Matches

Chromium in matches has been associated with dermatitis both in match factory employees (Pirilä, 1954) and in match users (Fregert, 1961; Rudzki and

Kozlowska, 1980). Potassium chromate and dichromate are added to match heads as oxidizing agents.

3.13 Miscellaneous Exposures

Finally, mention is made of a number of miscellaneous sources of chromium exposure that have been associated with chromate dermatitis. Cumulatively, their impact is likely to be large.

Contact dermatitis attributed to chromium compounds has been reported in the pulp and paper industry (Fregert et al., 1972). Burrows (1983) has observed chromate dermatitis in association with exposure to timber preservatives. Chromate sensitivity (on patch testing) and dermatitis have also been demonstrated in laboratory workers involved in analyzing the fat and protein content of milk samples preserved with potassium dichromate (Huriez et al., 1975; Rogers and Burrows, 1975; Rudzki and Czerwinska-Dihnz, 1977). In an equally unusual situation, chromate sensitivity and dermatitis developed in a laboratory technician utilizing 10% potassium chromate as an indicator in determining the sodium chloride content in food (Pedersen, 1977). Contact dermatitis has also been associated with exposure to chromium-containing glue (Morris, 1955). Among 26 workers involved in making television screens in a process using ammonium dichromate, 18 were found to have chromate dermatitis (Stevenson, 1975). Krook et al. (1977) reported dermatitis in a telegraphist who regularly handled magnetic tapes containing Cr(IV) oxide; patch testing revealed sensitivity to cobalt and nickel, as well to chromate. A final instance deserving mention concerns a tire fitter who developed hand dermatitis from chromate dissolved in a solution used to facilitate fitting the tire to the rim (Burrows, 1981).

Stainless steel alloys containing significant amounts of chromium are employed in orthopedic implants (Hierholzer et al., 1984), while chromium-cobalt alloys are common in dental devices (de Melo et al., 1983). Corrosion of these protheses releases significant quantities of chromium and other metals into the surrounding tissues, especially on infection (Hierholzer et al., 1984). Some patients experience localized dermatitis over the prosthesis or a more generalized eruption and test positively for chromium sensitivity (Langård and Hensten-Pettersen, 1981; Pedersen, 1982; Fisher, 1984). The management of such cases is controversial, as well as the employment of patch tests preoperatively (Carlsson et al., 1980; Ziegler and Höhndorf, 1984; Fisher, 1984).

3.14 Comment on Dose-Response

In most of the reports cited previously that included enough subjects to allow an assessment of dose-response relationships for induction and elicitation of sensitivity, the concentrations and types of chromium compounds present were documented inadequately for this purpose. Generally speaking there is little evidence that the prevalence of contact dermatitides correlates with dose. On the other hand, separate, detailed investigations of chromium speciation have been undertaken for a number of problematic exposure circumstances, such as cement use (Johnston and Calnan, 1958). It is obvious from such studies that the hexavalent chromate ion is the most potent sensitizer. From the evidence cited, it is apparent that in numerous cases chromate dermatitis developed on contact of the skin to relatively low concentrations of chromate. It is concluded that the presence of chromium-containing compounds in the workplace, particularly in their hexavalent forms, should be cause for concern with respect to their sensitizing potential. As reviewed in Section 6, there is also considerable evidence that Cr(III) compounds are allergenic, although of inferior eliciting capacity in epicutaneous applications when compared to hexavalent chromium.

4. PROGNOSIS

The consensus is that the outlook for persons with chromium dermatitis is bleak. In follow-up postal surveys, the majority of chromium dermatitis patients report ongoing skin symptoms (Harrison, 1979; Dooms-Goossens et al., 1980). After four to seven years' follow up in 48 patients, positive patch testing persisted in 38 individuals: in 27 subjects, chronic dermatitis was present; in 16, it was recurring; and in only five cases had the dermatitis cleared. In a 10-year follow up of occupational chromium dermatitis, Fregert (1975) found only 7% of women and 10% of men to have healed (compared to 16 and 24%, respectively, for dermatitis of all causes). Burrows (1972) reported that only 8% of individuals with cement dermatitis had cleared after 10 to 14 years from initial diagnosis, compared to 25% of those with other industrial dermatoses. Healing of cement dermatitis has been reported—while the workers continued in their trade—in four of 46 building workers (Geiser and Girard, 1965) and eight of 17 bricklayers (Hunziker and Musso, 1960).

The chronicity of chromium dermatitis is attributable to a number of factors: the persistence of hypersensitivity; the ubiquity of chromium compounds in foods and in objects coming into cutaneous contact; the persistence of chromium compounds in skin; and the coexistence of other contact allergies—for example, to cobalt and nickel (Fregert and Rorsman, 1966) or rubber in rubber gloves (Fregert, 1975).

5. PROPHYLAXIS

In this section, strategies for primary prevention of chromium dermatitis are outlined (Burrows, 1984). First, maintenance of workplace chromium exposure to the lowest feasible levels, good housekeeping, and a high level of awareness among workers and management are important. And second, numerous alternatives to the use of hexavalent chromium in industry have been sought and are available. For example, nitrites and amines can replace chromates as anti-rust agents. It is also possible to employ trivalent chromium in plating processes. Furthermore, Cr(III) salts may be replaced in part by aluminum(III) salts in tanning (Covington and Sykes, 1981), and a nonchromium, vegetable-tanning process is also possible. Combinations of aluminum salts and vegetable tannins are currently employed in some countries (Sykes et al., 1980). The use of ferrous sulphate in cement (to reduce hexavalent to trivalent chromium) also holds considerable promise. In a number of other industries, similar replacements or additions have been implemented.

6. PATHOGENESIS

6.1 Introduction

While the causal role of certain chromium compounds in inducing and eliciting contact dermatitis is undisputed, the allergenicity of chromium-containing substances varies considerably. With regard to research into the elicitation of sensitivity reactions, Polak (1983) states this in the converse fashion: "Under favourable conditions such as a high degree of hypersensitivity, appropriate solvents, sufficiently long exposure time, etc., a positive skin reaction to all chromium compounds used (is) observed even if different concentrations of them are necessary."

The main determinants of the sensitizing and eliciting capacity of chromium compounds are generally considered to be valency, solubility, concentration, pH, and presence of organic matter (Fregert, 1981). Our discussion of the pathogenesis of chromium hypersensitivity begins with a consideration of how the influence of these variables is mediated, with emphasis on the part played by chemical speciation and valency. Then, the role of certain significant modulating factors is reviewed, namely, genetic influences, noncutaneous absorption, and interaction of light. Lastly, relevant aspects are reviewed with respect to the phenomena of tolerance, desensitization, and flare reactions.

6.2 Relevant Chemical Principles

A brief synopsis concerning the interaction of chromium compounds with tissues and the importance of chemical speciation is presented in this section. These topics are covered more comprehensively elsewhere in this volume (Chapter 2), including a pictorial illustration of the anatomy of the skin.

The chromate ion is stabilized in alkaline solutions and therefore is less oxidizing. It readily penetrates skin and cell membranes and eventually is reduced to the Cr(III) ion by skin and intracellular components. In contrast, the hydrated Cr(III) ion, $Cr(H_2O)_6^{3+}$, does not penetrate skin well and appears largely to be excluded from cells. The lack of skin permeability is explained by the strong binding of this cation to skin constituents, especially proteins, and by its propensity to polymerize extensively due to hydrolysis. The latter process becomes serious at neutral pH values, and increases with increasing pH. Both types of interactions may be expected to reduce the ability of Cr(III) species to diffuse through the epidermis and dermis. It is also clear that the nature of the Cr(III) salt can have considerable bearing on the extent of binding and hydrolysis, and thus penetration. For example, the commercially purchased Cr(III) chloride on dissolution yields a singly charged chloride complex, $[CrCl_2(H_2O)_4]^+$, rather than the triply charged hydrated ion obtained by dissolving Cr(III) nitrate or sulphate. Basic chromic sulphate produces a doubly charged species, $[Cr(H_2O)_5OH]^{2+}$. Such complex cations bear a lower net charge, and may be expected to be less reactive and to diffuse considerably better through skin. Furthermore, the state of hydration of Cr(III) salts often determines water solubility (e.g., anhydrous $CrCl_3$ is insoluble, while the hydrate $[CrCl_2(H_2O)_4]Cl \cdot 2H_2O$ is).

6.3   The Role of Valency: Clinical and Experimental Observations

6.3.1   Human Studies

Controversy surrounds the issue of the valence state in which chromium participates in the formation of the antigenic determinant. Chromium has at least six valence states (Chapter 2). However, only tri- and hexavalent compounds are stable enough to be considered as potential haptens (Burrows, 1978). The metal itself is considered nonsensitizing, and this lack of reactivity is attributed to a surface layer of insoluble chromium oxide (Cronin, 1980).

Early clinical experience suggested that only hexavalent compounds were associated with the induction of hypersensitivity (Jaeger and Pelloni, 1950; Calnan, 1960). However, the occurrence of contact allergy to trivalent chromium was demonstrated subsequently as well (Fregert and Rorsman, 1964; Rudzki et al., 1978). Nevertheless, while both valence states have been associated with the induction and elicitation of hypersensitivity reactions, it is generally recognized that the hexavalent form is the more potent agent in these functions; clinical experience remains consonant with this perspective. In an experiment with volunteers, Kligman (1966) found that, upon epidermal administration by the "maximization procedure," about half of a group of 23 human subjects became sensitized to trivalent chromium salts [Cr(III) oxide and Cr(III) sulphate], whereas potassium dichromate sensitized each of 23 individuals. With respect to elicitation of sensitivity, Zelger (1964) demonstrated that the threshold concentrations for the production of positive skin test results in hypersensitive patients were generally much lower for hexavalent than trivalent compounds (see Table 6). Thus, in five subjects reacting positively to 0.25% (0.008 M) potassium dichromate (Samitz and Shrager, 1966), only one showed a (slightly) positive patch test reaction to 5% (0.19 M) Cr(III) trichloride and 0.5% (0.008 M) and 1% (0.016 M) Cr(III) sulphate. None reacted to Cr(III) nitrate. Anderson (1960a) found that, of 19 patients reacting positively to patch testing

Table 6   Threshold Concentrations for Positive Skin Tests in Chromium-Hypersensitive Patients[a]

Agent	Valency	Concentration (% w/v)	Number of Positive Patients[b]
K_2CrO_4	VI	0.001	3
		0.005	9
		0.01	10
		0.05	11
H_2CrO_4	VI	0.001	1
		0.005	5
		0.01	2
		0.05	5
$K_2Cr_2O_7$	VI	0.01	2
		0.05	17
		0.1	14
$KCr(SO_4)_2 \cdot 12 H_2O$	III	0.05	3
		0.1	7
		0.5	12
		Negative	6
$Cr(NO_3)_3$	III	0.05	3
		0.1	2
		0.5	5
		Negative	18

[a] Reproduced in modified form from Polak (1983); also see Zelger (1964); [b] Total number of patients tested corresponds to the sum of the entries for each compound.

with potassium dichromate, only one was positive to testing with the trivalent chloride at comparable chromium concentrations. On the other hand, other investigators observed that 11 of 17 chromium-sensitive subjects reacted positively to epicutaneous tests with 0.5 M chromium trichloride (Fregert and Rorsman, 1964), although only four of 22 responded to this compound at a concentration of 0.07 M. Subsequently, it was found that 70 of 90 dichromate-positive subjects gave positive patch test reactions to 0.1 M Cr(III) lactate (Fregert and Rorsman, 1965), in which the active ingredient appears to be CrL_2^+ (with L denoting lactate; Morris, 1958). Forstrom et al. (1969) noted that 23 of 46 dichromate-positive patients reacted positively to Cr(III) chloride in 1 (0.04 M) to 5% (0.19 M) solutions. It has been demonstrated (Rudzki et al., 1978) that differences in positive reaction rates associated with trivalent and hexavalent chromium cannot be explained by differing release from the patch test vehicle.

6.3.2 Animal Studies

The role of valence states in chromium hypersensitivity has been extensively investigated in animal studies, which mainly have utilized guinea pigs. As the guinea pig cannot be sensitized epicutaneously by chromium compounds in aqueous solutions, and is sensitizable only inconsistently by this route by chromium in a variety of matrices, special techniques that enhance sensitization are needed in this animal model. The most successful method incorporates the hapten in Freund's complete adjuvant (a water-in-oil emulsion), and involves a combination of intramuscular, intradermal, and epicutaneous administration (Polak, 1983). Thus, in guinea pigs and applying the latter method, Polak et al. (1973) found that trivalent chromium chloride possessed a sensitizing capacity *equal* to that of hexavalent potassium dichromate, but upon subsequent epicutaneous and intradermal challenging elicited reactions of lesser intensity than the latter compound.

The findings of most investigators concerning chromium contact sensitivity in the guinea pig have been compatible with those of Polak et al. (1973). The observations of Gross et al. (1968) were very similar, and van Neer (1963) found that intradermal application of hexavalent potassium dichromate and of trivalent chromium sulphate induced contact sensitivities of the same degree. However, upon epicutaneous testing, only potassium dichromate yielded positive results; intradermal challenges were not given. Mali et al. (1963) observed that subdermal administration of Cr(III) sulphate induced contact sensitivity less frequently and less intensely than hexavalent salts; skin responses to challenges with Cr(III) sulphate were negative. Polak (1983) notes, however, that Mali et al. utilized lower amounts of elementary chromium equivalents in their experiments than did Polak et al. (1973). Shmunes et al. (1973) reported that guinea pigs sensitized to potassium dichromate reacted more intensely to challenges with this compound than with trivalent chromium chloride.

With the exception of two reports of equivalent induced *and* eliciting capacity in guinea pigs (Jansen and Berrens, 1968; Schwarz-Speck, 1968, cited in Polak, 1983), the animal findings are consistent with near equipotency of hexavalent and trivalent compounds in sensitivity induction (by the methods used in guinea pigs), but superior elicitory capacity of hexavalent chromium when applied intradermally or epicutaneously. As already pointed out (Section 6.2), permeability differences appear to account for these differential eliciting responses. The clinical and experimental evidence for this is presented in the next section.

6.4   The Relationship Between Valency and Penetrability

Hexavalent chromium readily traverses the epidermis to the dermis (Mali et al., 1964; Czernilewski et al., 1965). In contrast, the rate of skin penetration by trivalent chromium has generally been reported to be much lower than that of hexavalent chromium (Mali et al., 1964; Spruit and van Neer, 1966). From a chemical viewpoint, the nature and concentration of trivalent chromium salts strongly determine their penetrating capacity (see earlier comments, Section 6.2). Samitz et al. (1967) have shown in laboratory diffusion experiments that for equivalent concentrations in phosphate buffers in the pH range 5–7, the order of diffusion through human skin was: Cr(III) nitrate < Cr(III) sulphate < Cr(III) chloride < potassium dichromate. Ion exchange experiments suggested that the diffusivity of the Cr(III) salts declined with increasing positive charge on the active (diffusing) species. Wahlberg and Skog (1965), utilizing "isotope disappearance measurements" found that for concentrations of 0.017 to 0.239 M, Cr(III) chloride and sodium chromate readily penetrated undamaged skin. However, in higher concentrations (0.26 to 0.40 M), hexavalent sodium chromate penetrated twice as fast as trivalent chromium chloride.

Lidén and Lundberg (1979) studied the histological distribution of hexavalent chromium following penetration of intact human skin in vivo. Tangential sections of punch biopsies from patch-tested skin were analyzed for chromium content 5, 24, and 72 hours after application of potassium dichromate. Two maxima of chromium concentration were observed: at the dermal-epidermal junction and in the upper mid-dermis. A steady state of skin chromium content was attained by 5 hours, with penetration continuing at 72 hours.

The local fate of chromium has been studied when it is administered intracutaneously to human subjects (thus bypassing the skin barrier). Doses of sodium chromate and chromium trichloride were given in concentrations required for elicitation of positive test reactions. Ten minutes after administration of 0.5 μg chromium, only 50% of the sodium chromate remained at the test site, but about 15% was identifiable after two days and 5% after two months. Chromium trichloride disappeared at a much slower rate. Half of the injected chromium (5 μg) disappeared in 5 to 12 days, with a retention of about 45% of the original dose at the test site after one month (Pedersen et al., 1969; Pedersen and Naversten, 1973). A dose of 0.5 μg chromium administered as Cr(III) chloride disappeared at a rate intermediate between that described for the 0.5 μg chromate and the 5.0 μg of the chloride. Similar results were observed in both allergic and nonallergic subjects. Interestingly, in two chromium-sensitive men tested three and four years earlier by intracutaneous injection with chromium trichloride, Fregert (1971) found it necessary to excise the test sites because of relapsing "flare-up" reactions. Analysis of the excised skin revealed persistence of 21% and 25% of the original injected chromium dose.

Pedersen et al. (1970) have also quantitatively assessed the disposition of sodium chromate applied epicutaneously to two chromate-sensitive patients and two normal controls. After two days about 10% of the dose (20.8 μg chromium) was identified at the test site and approximately 90% on the patch itself. That some absorption had occurred was demonstrated by positive reactions in the sensitive subjects. At one month, between 0.6 and 1% of the labelled chromium remained in the skin of control subjects at the test site; in one allergic subject, 0.1% of the original dose persisted and in the other none was identifiable.

Evidently, upon intracutaneous administration, allergic and nonallergic subjects do not differ in their handling of chromium doses. On the other hand, sensitive subjects may have more rapid clearance of epicutaneously administered chromium, perhaps because of desquamation of cells from the inflamed test site.

Other research results are also inconsistent with respect to relative chromate absorption in the skin of sensitized and normal subjects (Czernilewski et al., 1965; Heise and Mattheus, 1968, cited in Polak, 1983).

6.5 The Relationship Between Valency and Protein-Binding Capacity

Following absorption, hexavalent and trivalent chromium behave differently, with respect to protein binding. It is generally accepted that low-molecular-weight substances, such as chromium compounds, become immunogenic only after conjugation with protein carriers (Landsteiner and Jacobs, 1935; see Section 6.14). As documented elsewhere in this volume (Chapter 2), chromate is readily reduced in vitro and in vivo, and the resultant Cr(III) ion binds strongly to proteins, polynucleotides, and other complexing biomolecules. For example, Gray and Sterling (1950) found, following intravenous injection of dogs with ^{51}Cr-labelled Cr(III) chloride, that almost the complete dose was bound strongly to plasma proteins, while labelled sodium chromate was associated mainly with erythrocytes. Intracutaneous injection in humans of labelled sodium chromate and Cr(III) chloride gave similar results (Bang Pedersen et al., 1969; Bang Pedersen and Naversten, 1973). It is known that erythrocyte chromium derived from chromate uptake is bound to hemoglobin as Cr(III).

As emphasized by Nieboer and Jusys (Chapter 2), chromate is an anion that can act as a complexing agent (ligand) toward metal ions. In contrast, Cr(III) is a metal ion with considerable binding capacity. Not surprisingly, therefore, Anderson (1960b) has demonstrated that trivalent chromium reacted more readily with skin proteins than chromate, although the valence state of ultimate binding for the latter was not studied. Similarly, Polak et al. (1973) were unable to demonstrate any protein binding of chromate following incubation of guinea pig serum with potassium dichromate. Incubation with Cr(III) chloride resulted in considerable binding to serum proteins. Neither could Magnus (1958) demonstrate any stable binding of chromate to a number of proteins by paper electrophoresis. Samitz and Katz (1964) have demonstrated the reduction of hexavalent potassium dichromate to trivalent chromium by skin proteins and other constituents such as ascorbic acid and lactic acid. Chromate has demonstrated potential for oxidizing most biological reducing agents at physiological pH values (both low- and high-molecular-weight molecules). Anderson's work (1960b) illustrated the considerable influence of pH on in vitro protein binding of potassium dichromate, with more binding occurring at lower pH levels. Undoubtedly, this conjugation followed reduction to trivalent chromium, especially since the oxidizing power of chromates increases with decreasing pH (See Chapter 2).

6.6 The Role of Concentration

The influence of concentration of chromium compounds on sensitizing and eliciting potential is not as clear-cut as might be supposed. Wahlberg and Skog

(1965) in isotope disappearance studies (see Section 6.3) demonstrated the occurrence of a peak for relative absorption (at 0.26 M) and a plateau for absolute absorption (0.40 M) of sodium chromate. While total absorption certainly does not decrease with increasing concentration, it has been postulated (Jansen and Berrens, 1968) on the basis of clinical and experimental studies that high concentrations of chromium compounds are less allergenic than low concentrations. The low prevalence of chromium sensitivity among chromate production workers is a case in point (Gafafer, 1953). The high frequency of contact sensitivity in the building industry—where the chromium content of materials is generally low, relative to lower prevalences in certain other industrial settings with ostensibly greater chromium exposure—has been cited in support of this postulate (Polak, 1983). It has been suggested that the induction of immunological tolerance and desensitization (which can be produced in animals) could underlie such a phenomenon (Polak, 1983; see Section 6.13). However, at variance with this hypothesis are examples of workplaces where high levels of exposure are associated with a high rate of contact sensitivity, such as the apparent high prevalence reported for car assembly workers (Section 3.3) and chrome platers (Section 3.11). Further, the independent effect of concentration has not been determined in such observations, after accounting for that of valency, solubility, pH, and presence of organic material. Nonetheless, the possible tolerogenic impact of chromium in the workplace could complicate attempts to establish dose-response relationships.

6.7   The Roles of Solubility, pH, and Reducing Agents

It might be expected that given chromium compounds of a particular valency and at a certain level of concentration, their sensitizing and eliciting potential would correlate with water solubility. Consistent with this clinically observed relationship are the results of investigations in which various chromium compounds were administered to hypersensitive individuals (Zelger, 1964; Bockendahl, 1954, cited in Polak, 1983). For example, the water-insoluble lead chromate induced skin reactions only in the most sensitive patients.

Wahlberg (1968) has demonstrated a pH-dependence of percutaneous absorption of sodium chromate, with skin penetration in the guinea pig increasing with increasing pH. This effect of pH is mediated largely through its influence on valence state (Burrows, 1978). A consideration of electrode potentials suggests that the chromate ion is stabilized in basic solution because of a loss in oxidizing power (Chapter 2).

The remaining significant determinant of sensitizing and eliciting potential is the presence of organic matter. It exerts its effect by acting as a reducing agent toward hexavalent chromium (e.g., due to the availability of sulfhydryl groups), or through an ability to tie up Cr(III) by complex formation, perhaps even inducing its precipitation.

6.8   In Vitro Indicators of Hypersensitivity

In addition to the patch test (Fischer and Rystedt, 1985), there are a number of in vitro indicators of chromium hypersensitivity, which to date have been of limited clinical applicability. These are briefly summarized in this section.

Polak (1983) has demonstrated that, upon stimulation with Cr(III) chloride guinea-pig serum conjugates, peritoneal or lymph node lymphocytes from hexavalent chromium-sensitive guinea pigs release an inflammatory factor called the "skin reactive factor." Injection of supernatants containing this factor induces local inflammatory reactions in normal guinea pigs. T-effector cells also release a lymphokine on stimulation with antigen called the "macrophage migration inhibition factor." Hexavalent and trivalent chromium salts have been found in in vitro studies to inhibit macrophage migration with peritoneal exudates from chromate-sensitive guinea pigs (Polak, 1983) and leukocyte migration with white blood cells from chromium-allergic human subjects (Tio, 1976; Nordlind and Sandberg, 1983). Increased DNA synthesis upon in vitro stimulation by lymphocytes from sensitized individuals comprises the "lymphocyte transformation test." Until recently, its use appears to have been limited to contact sensitivity studies in animals (Polak, 1983; Siegenthaler et al., 1983). Al-Tawil et al. (1983) have shown that for 60% of patients with a positive potassium dichromate patch test lymphocyte stimulation could be induced. They conclude that the DNA synthesis test is a reliable in vitro aid in the diagnosis of chromium sensitivity.

6.9   The Role of Genetic Factors

Relevant to consideration of the pathogenesis of chromium hypersensitivity is reference to the possible role of genetic factors. Polak (1983) demonstrated, in guinea pigs, strain differences in the likelihood of becoming sensitive to potassium dichromate. No evidence was identified concerning the frequency of chromium sensitivity among the relatives of affected human subjects. However, the frequency of histocompatability antigens has been investigated in chromium-sensitive subjects and in normal individuals. These studies found no relationship between chromium sensitivity and HLA typing (Roupe et al., 1979; Kapoor-Pillarisetti et al., 1981).

6.10   The Role of Noncutaneous Absorption in Cutaneous Hypersensitivity

Study of chromium absorption by noncutaneous routes has shed light on the mechanisms involved in chromium hypersensitivity of the skin, especially in its chronic forms. Orally administed trivalent chromium is only minimally absorbed through the gut ($\sim 0.5\%$), while intestinal absorption of hexavalent chromium occurs to a greater degree ($\sim 10\%$; Donaldson and Barreras, 1966; Chapter 2). It is known that gastric acidity promotes the conversion of ingested chromate to the trivalent state, thus reducing absorption. [No association has been noted between achlorhydria and chromium dermatitis! (Burrows, 1978).]

A number of studies have dramatically shown the effects of oral absorption of even small amounts of chromium in sensitive individuals. Within two hours of oral administration of 50 μg of potassium dichromate, five chromium-allergic subjects developed acute vesiculation of the palms (Fregert, 1965, cited in Cronin, 1980). Kaaber and Veien (1977) compared the effects of orally administered potassium dichromate and placebo in a double blind study in 31 chromium-sensitive patients with current dermatitis. Chromate produced exacerbation of the dermatitis in 11 individuals, while placebo was associated with aggravation of the skin condition of two subjects. The dose given was equivalent to 2.5 mg of chromium, regarded as within the range of ordinary daily intake in Danish diets. Goitre et al. (1982), using this oral chromium dose, produced exacerbation of chromium contact dermatitis in a construction worker.

No evidence is available concerning the benefits of reduction of dietary chromium in sensitive individuals. Such a strategy would seem difficult to carry out, given the ubiquitous occurrence of chromium in foods (usually in very small amounts).

As already indicated (Section 3.13), chrome-containing metal implants constitute a possible source of internal exposure which is presently of uncertain significance with respect to induction or elicitation of sensitivity.

Little information is available concerning the relationship between inhalational exposure to chromium and hypersensitivity manifestations in the skin. Shelley (1964) reported that inhalation of welding fumes by a chromium-sensitive individual resulted in repeated episodes of hand dermatitis. In another report, exposure to welding fumes was associated with urticaria and asthma in a railway worker. However, the allergen was not identified (Kaplan and Zeligman, 1963).

6.11   The Role of Interaction with Light

In some individuals light may interact with chromium in the production of hypersensitivity phenomena. Feuerman (1971) reported that Israeli housewives sensitive to chromates in detergents were at increased risk of developing dermatitis upon exposure to sunlight, relative to other dermatological patients without chromium allergy. Wahlberg and Wennersten (1977) demonstrated that short wave ultraviolet light exacerbated the reactivity to chromium in 12 of 25 chromium-allergic patients. Hannuksela et al. (1981) observed a statistically significant increase in prevalence of delayed contact hypersensitivity to chromium among two groups of patients with photodermatitis (14 and 110 subjects, respectively), than in 1,714 patients with other kinds of dermatitis.

6.12   Concomitant Sensitivities: Independent or Interdependent?

It remains unresolved whether positive reactions to nickel and/or cobalt occasionally observed in chromium-sensitive patients (Fregert and Rorsman,

1966; Cronin, 1980; also see Table 4) are attributable to cross-sensitivity or to independent simultaneously occurring sensitivity to these metals. In cement dermatitis, concomitant allergy to cobalt apparently occurs only when traces of cobalt are present in the cement (Polak, 1983). Thus, sensitization to cobalt and chromium may be taking place independently. Similarly, it is believed that concurrent exposure to both nickel and chromium compounds may account for the occurrence of positive patch tests to both metals simultaneously. However, sound research is lacking concerning possible cross-sensitivity in chromium allergy.

6.13   Desensitization, Tolerance, and Flare

Lastly, the phenomena of desensitization, tolerance, and flare reactions in chromium hypersensitivity are briefly considered. First, the meaning of these terms in immunological usage is outlined. Loss of the immune response to an antigen or hapten to which an individual has already been sensitized is "desensitization." In contrast, "tolerance" to the development of contact sensitivity can occur in nonsensitized (naive) subjects on exposure to the hapten. And finally, "flare reactions" may occur at healed sites of previous eczematous lesions or patch tests when the hapten is applied orally or intravenously.

Attempts have been made to permanently desensitize patients with cement dermatitis, by repeated epicutaneous administration of potassium dichromate (Pautritzel et al., 1962; Sourreil et al., 1964a,b). This approach has met with variable success. As noted previously (Section 6.10), oral administration of small amounts of potassium dichromate to chromium-sensitive patients produced exacerbation of previous inflammatory sites rather than desensitization (Kaaber and Veien, 1977). The failure to demonstrate desensitization consistently in human subjects is not surprising, given the results observed by Polak (1983) in studies with chromate-sensitized guinea pigs. Intravenous injection of doses of potassium dichromate of the order of the LD_{50}, accompanied by an epicutaneous challenge with the same compound within 24 hours, was required to induce permanent, complete desensitization. Toxic oral doses of potassium dichromate are also effective (Christensen et al., 1984). Lower intravenous doses or a delay in the associated epicutaneous application resulted in a temporary desensitization response only. The role of desensitization in humans exposed to chromium over a long period of time, such as in the workplace, is not known. On the other hand, Polak (1983) readily induced complete, permanent hapten-specific tolerance in guinea pigs at much lower doses than were required for permanent desensitization. Similarly, Vreeburg et al. (1984) found that simultaneous oral administration of nickel and chromium (as metal powders or salts) for six weeks induced a state of (partial) tolerance to both metals in guinea pigs. The possible extent of tolerance among workers exposed to chromium compounds is also unknown.

Polak (1983) has also extensively investigated the flare phenomenon in guinea pigs, after intravenous injection of the hapten. Flare-up reactions were elicited in chromate-sensitized animals after injection of either potassium dichromate or Cr(III) sulphate. Oral doses of potassium dichromate have also produced flare-up reactions (Christensen et al., 1984). Polak (1983) suggested that the mechanism involved was related to the Arthus phenomenon (antibody-mediated responses), although recent work by Parker et al. (1984) suggests it is cell-mediated like the primary response (a Type IV immune response; See Section 6.14). Oral administration of potassium dichromate has been used diagnostically to induce a flare at previously negative skin challenge sites and thus to reveal a "latent state" of chromium hypersensitivity (Schleiff, 1968, cited in Polak, 1983). However, this approach has not been evaluated further.

6.14   Immunological Mechanism of Chromium Contact Dermatitis

6.14.1   Recognition Phases of Contact Dermatitis

The following summary concerning the various phases of contact dermatitis serves as an introduction to Section 6.14.2 and is based on the reviews of this topic by Adams (1983) and Burrows (1983). Chromium contact dermatitis is a delayed reaction (delayed hypersensitivity; a Type IV cell-mediated immune response; Petersdorf et al., 1983). It flares up 48 hours after exposure to a secondary challenge. The development of allergic contact dermatitis occurs in four different phases: refractory, induction, elicitation, and persistence. In the refractory period, an individual is exposed to the allergen but remains unaffected. It is hypothesized that the hapten penetrates the skin (Fig. 1) and conjugates with an epidermal protein; during the induction phase the complex interacts with and is carried by T-lymphocytes to regional lymph nodes via dermal lymphatics (Fig. 2). In the lymph node, the T-lymphocytes become immunoblasts that divide

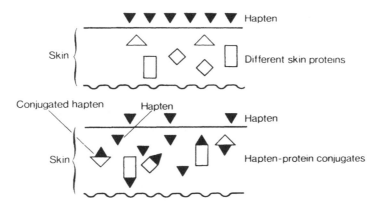

Figure 1. Antigen formation: an illustration of how low-molecular-weight substances, called haptens, can become immunogenic by combining with skin proteins subsequent to diffusion into the skin. Reproduced with permission from Polak (1983).

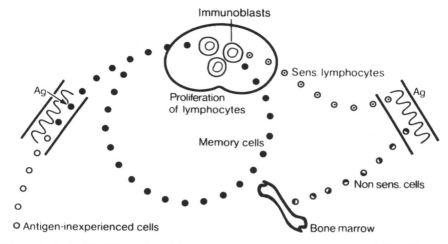

Figure 2. A simplified illustration of the events believed to occur in contact sensitivity. Hapten-protein conjugates are recognized by antigen inexperienced specific lymphocytes (T-cells). These cells migrate to the draining lymph node, where they proliferate and differentiate into memory and effector cells. Sensitized lymphocytes, after leaving the lymph node, circulate in the body. A new contact with the specific antigen elicits an inflammatory skin reaction with the help of nonsensitized cells, secondarily involved in the eliciting phase. Macrophage-like Langerhans' cells are thought to participate in the formation of the antigenic complex, and appear important in both the induction and elicitation phases of contact sensitivity. Reproduced with permission from Polak (1983).

into memory and effector cells. Memory cells retain the information of previous contact with the antigen and, upon recontact, these cells proliferate and produce a new population of sensitized small lymphocytes. Effector cells circulate in peripheral blood. During elicitation, which involves secondary challenge of the antigen, the hapten again penetrates the dermis, combines with a protein that is recognized by the effector cells, which comprise < 10% of the total lymphocyte population at the contact site. Upon contact, effector cells release chemotactic substances (e.g., Barr et al., 1984) that cause the inflammatory skin reaction by inducing vascular permeability, attraction of mononuclear cells, and retention and proliferation of cells within the area. Often, during both the induction and elicitation phases, macrophages or Langerhans' cells may be involved in processing and presentation of the antigen to the T-cells. Persistence involves continued presence of the specific effector cells capable of recognizing the hapten with the ability to produce inflammation.

6.14.2   *Working Hypothesis of Pathogenesis of Contact Sensitivity to Chromium*

A working hypothesis of chromium immunogenicity that unites the diverse observations reviewed in Sections 6.3 to 6.8 is as follows (Polak, 1983). Chromium in its hexavalent form readily penetrates the skin, whereas most trivalent chromium compounds traverse the epidermis relatively poorly, being readily complexed by skin proteins. Absorbed trivalent chromium tends to bind immunologically irrelevant proteins, and cross cell membranes poorly. Hexavalent chromium, on the other hand, is absorbed in greater concentrations, attains key membrane and intracellular sites, and undergoes reduction to the trivalent state which forms conjugates with proteins, constituting complete antigens (Fig. 1). Consistent with this hypothesis are reports that circulating antibodies may exist in sensitized animals against trivalent chromium but not against hexavalent chromium (Cohen, 1962; Mali et al., 1963; Novey et al., 1983), although such results are to be viewed with caution because of the inherent uncertainties and complexities in antibody characterization.

Antibodies do not play a primary role in chromium dermatitis which is mediated principally by peripheral lymphocytes (T-cells). A Cr(III)-protein complex is more likely the form that is recognized by antigen-inexperienced T-cells. Macrophage-like Langerhans' cells are thought to participate in the formation of the antigenic complex, and appear important in both the induction and elicitation phases of contact sensitivity (Fig. 2). Recent work by Siegenthaler et al. (1983) confirms that a common antigenic determinant is likely to be involved in chromium contact sensitivity, rather than independent entities corresponding to each valency state. As in earlier experiments (Section 6.3.2), guinea pigs sensitized with either trivalent chromium chloride or hexavalent potassium dichromate reacted in vivo and in vitro to both chromium salts. Reactivity to chromium in both valence states was retained after repeated restimulations with chromium of either valency. Interestingly, DNA synthesis could be induced in vitro by both chromate and Cr(III) salts in lymphocytes from patients with positive chromate patch tests (Al-Tawil et al., 1983). Thus, a common antigenic determinant is probable, and Siegenthaler et al. (1983) argued that this determinant is formed by chromium in the trivalent form.

From this model it is clear that the influences of many factors converge to yield clinically observable hypersensitivity manifestations. The nature of some of these manifestations will be strongly dependent on the clinical or laboratory maneuver performed to demonstrate them. Thus, the clinical signs may vary with route of antigen administration (Epstein, 1966), the test compound selected, the dose, or the type and composition (e.g., presence of emulsifiers; Rudzki et al., 1976) of the vehicle used (Skog and Wahlberg, 1969).

7. OCCUPATIONAL ASTHMA

In this section, evidence is reviewed concerning effects of occupational chromium exposure on reversible bronchoconstrictive disease, with a focus on hypersensitivity responses. A discussion of nonreversible airways disease and other pulmonary diseases occurs elsewhere in this volume (Chapter 19).

Given the well-documented irritant and sensitizing properties of chromium compounds, it is not surprising to encounter literature evidence for chromium-associated, occupational bronchial hypersensitivity. This evidence is, however, limited in extent, deriving predominantly from a number of case series studies. Despite the relatively uncommon reports of an association of chromium exposure with bronchial asthma, the etiological relation is strongly supported in most instances by consistent chronological patterns with regard to exposure, and by confirmatory responses to bronchodilator therapy. In these reports, the characteristic combination of (apparent) rarity, response to provocation, and reversibility (spontaneous or by treatment) suggest a hypersensitivity mechanism (Pepys, 1977; Petersdorf et al., 1983).

Both immediate (onset in minutes) and nonimmediate (onset in hours) asthmatic responses (Pepys, 1977) to chromium compounds are known. Williams (1969, 1973) reviewed and summarized the various occurrences reported up to 1969. Isolated cases of asthma from chromium compounds have been observed for occupational exposures to dichromates, to chromic acid, to chromite ore, to chromate paint pigments, and to welding fumes. A few specific examples will illustrate these case reports. Card (1935) precipitated asthmatic attacks in a chromium plater/polisher 2.5 to 5 hours after the intradermal injection of 0.004 g potassium dichromate or 0.004 g potassium chromate. A few years earlier, Joules (1932) had observed a positive response within one hour in a similar provocation in a metal plater. Marechal (1957) has reported four cases of asthma in association with exposure to chrome yellow, which contained lead and zinc chromates. A remaining report of chromium-associated asthma that deserves special mention concerns the response to chromic acid by ferrochromium workers (Broch, 1949; Lillingston, 1950). Two patients had violent asthmatic attacks to fumes from the factory furnace. Skin tests revealed that chromic acid was the offending substance.

In a nonatopic electroplating worker with a clinical picture consistent with occupational asthma, Novey et al. (1983) observed, on inhalation challenge by chromium sulphate, clear-cut depression of FEV_1 within 15 minutes of cessation of the challenge, with a return to prechallenge levels in 90 minutes. Interestingly, challenge with a nickel sulphate solution in the same patient produced a biphasic FEV_1 response, with the second depression maximal at four hours. Serological studies provided some evidence for circulating IgE antibodies to both chromium and nickel in this patient, although control tests for nonspecific contributions in the radioimmunoassays appear to have been inadequate.

Another recent report of detailed clinical assessment of chromium-associated asthma is that of Keskinen et al. (1980). A total of five stainless steel workers with work-related bronchospastic symptoms were documented, of which two were observed over a two-year period of nonexposure. Challenge by manual stainless steel welding yielded significant bronchial obstructive responses, whereas metal-inert gas welding and manual welding of mild steel were negative. This response was blocked by pretreatment with both disodium cromoglycate and with beclomethasone, but not with placebo. The main distinction in type of exposure between stainless steel and mild steel welding is that mild steel does not have chromium and nickel as constituents. A comparison of the composition of fumes generated by manual and inert gas welding of stainless steel has shown that manual welding generates considerably higher levels of water-soluble Cr(VI) (Stern, 1982).

The most comprehensive study of the role of chemicals (including chromate) in bronchial asthma is that of Popa et al. (1969), who studied 48 subjects with clear historical evidence of asthma occurrence in association with exposure to micromolecular substances having both irritating and sensitizing properties. Nine of these subjects had occupational exposure to dichromate (six tanners, one cement worker, one pottery worker, and one unclassified); however, all but two were also exposed to other possible respiratory sensitizers. In four of the chromate-exposed individuals, the inhalation challenge was positive. These workers were considered to have a nonspecific bronchial hyperreactivity, because of the failure to demonstrate sensitization by immunological tests. The diagnosis for the fifth worker was delayed hypersensitivity to chromate, for both the asthmatic and cutaneous responses. The remaining four patients could not be classified readily. They exhibited some but not all the features of allergic asthma. In addition to the person with the delayed responses, some of the other workers also exhibited delayed cutaneous sensitivity. Interestingly, concomitant contact sensitivity to chromate was also noted in some of the case reports cited previously (Williams, 1969; Keskinen et al., 1980).

Two additional epidemiological studies pertain to bronchial chromate hypersensitivity. Both were cross-sectional in design, and only utilized questionnaires to ascertain asthma. Among chromium platers, Royle (1975) found that 13.1%, compared to 9.8% of controls (p < 0.025), reported a history of symptoms compatible with asthma. While Langård (1980) observed a greater prevalence of wheeze and dyspnea among ferrochromium workers than among controls, members of the latter group were considerably younger than the exposed workers. The prevalence among the ferrochromium group was somewhat lower than that observed among ferrosilicon and maintenance workers in the same plant. Langård survey, perhaps in part because of methodological limitations, gives no indication that chromium bronchial hypersensitivity is more frequent among ferrochromium workers than the uncommon case reports suggest.

Finally, it should be noted that there is no evidence whatsoever in the literature to allow estimation of doses required for induction or elicitation of chromate-induced occupational asthma.

8. SUMMARY

Chromium is the second most common skin allergen in the general population, after nickel. In men it is the most frequent sensitizer. The prevalence of chromium dermatitis is the highest in construction workers using cement. Other occupational exposures associated with chromium sensitivity are numerous, of which chromium plating, tanning of leather, application of anti-corrosive agents, and printing are representative industrial activities. There is little evidence that the prevalence of chromium dermatitis correlates with dose. The consensus is that the outlook for persons with chromium dermatitis is one of continuing symptoms and disability. Avoidance strategies, including replacement of chromium-containing constituents and addition of suppressors, show promise for prevention and have been implemented in a few instances. Although respiratory sensitivity to chromium compounds is known, it occurs infrequently.

In terms of pathogenesis, chromium contact sensitivity of the skin is a characteristic cell-mediated phenomenon (delayed hypersensitivity). Hexavalent chromium compounds generally are more potent inducers and elicitors of sensitivity than trivalent compounds; however, the $Cr(III)$ ion very likely figures in the formation of the ultimate and common antigenic determinant. Concentration, solubility, pH, and presence or absence of reducing agents are significant influences on the likelihood and intensity of sensitivity associated with contact exposure to chromium compounds. There is no clear evidence that chromium dermal hypersensitivity is strongly genetically determined in humans. Sensitivity to light is a potentiating factor in some subjects. The existence of cross-sensitivity to other salts has not been excluded, but many "double allergies" could plausibly derive from the development of separate, independent sensitivity states. Oral ingestion of chromium can lead to skin reactions in sensitive subjects, but the overall significance of noncutaneous absorption of chromium is uncertain with respect to induction and elicitation of sensitivity, desensitization (loss of the immune response), and production of tolerance. Animal studies would suggest that the prospects for desensitizing interventions in humans are poor. It is not known whether chronic exposure to chromium results in desensitization in man. The induction of tolerance in nonsensitized (naive) subjects may be taking place to an uncertain extent in the community and in workplaces, and has been postulated to occur at much lower doses than desensitization.

ACKNOWLEDGMENTS

Financial support for the preparation of this review was received from the Occupational Health and Safety Division of the Ontario Ministry of Labour, and is gratefully acknowledged.

REFERENCES

Adams, R.M. (1983). *Occupational Skin Diseases*. Grune and Stratton, New York, pp. 1–26.

Adams, R.M., Fregert, S., Gruvberger, G., and Maibach, H.I. (1976). "Water solubility of zinc chromate primer paints used as antirust agents." *Contact Derm.* 2, 357–358.

Al-Tawil, N.G., Marcusson, J.A., and Möller, E. (1983). "Lymphocyte stimulation by trivalent and hexavalent chromium compounds in patients with chromium sensitivity." *Acta Derm. Venereol.* **63**, 296–303.

Anderson, F.E. (1960a). "Cement and oil dermatitis. The part played by chrome sensitivity." *Br. J. Dermatol.* **72**, 108–117.

Anderson, F.E. (1960b). "Biochemical experiments on the binding of chrome to the skin." *Br. J. Dermatol.* **72**, 149–157.

Angelini, G., Vena, G.A., and Meneghini, C.L. (1980). "Shoe contact dermatitis." *Contact Derm.* **6**, 279–283.

Anon. (1963). "Chrome ulceration of the nasal septum." *Br. Med. J.* **1**, 1364–1365.

Baer, R.L., Lipkin, G., Kanof, N.B., and Biondi, E. (1964). "Changing patterns of sensitivity to common contact allergens." *Arch. Dermatol.* **89**, 3–8.

Barr, R.M., Brain, S., Camp, R.D.R. et al. (1984). "Levels of arachidonic acid and its metabolites in the skin in human allergic and irritant dermatitis." *Br. J. Dermatol.* **111**, 23–28.

Bockendahl, H. (1954). "Chromnachweis und chromgehalt gefarbter kleiderstoffe." *Dermatol. Wochenschr.* **130**, 987–991 (cited in Polak, 1983).

Broch, C. (1949). "Bronchial asthma caused by chromium trioxide." *Nord. Med.* **41**, 996–997.

Burrows, D. (1972). "Prognosis in industrial dermatitis." *Br. J. Dermatol.* **87**, 145–148.

Burrows, D. (1978). "Chromium and the skin." *Br. J. Dermatol.* **99**, 587–595.

Burrows, D. (1981). "Chromium dermatitis in a tyre fitter." *Contact Derm.* **7**, 55–56.

Burrows, D. (1983). *Chromium: Metabolism and Toxicity.* CRC Press Inc., Boca Raton, FL.

Burrows, D. (1984). "The dichromate problem." *Int. J. Dermatol.* **23**, 215–220.

Burrows, D., and Calnan, C.D. (1965). "Cement dermatitis. II. Clinical aspects." *Trans. St. John's Dermatol. Soc.* **51**, 27–39.

Cairns, R.J., and Calnan, C.D. (1962). "Green tattoo reactions associated with cement dermatitis." *Br. J. Dermatol.* **74**, 288–294.

Calnan, C.D. (1960). "Cement dermatitis." *J. Occup. Med.* **2**, 15–22.

Calnan, C.D. (1978). "Chromate in coolant water of gramophone record presses." *Contact Derm.* **4**, 346–347.

Calnan, C.D., and Harman, R.R.M. (1961). "Studies in contact dermatitis. XIII. Diesel coolant chromate dermatitis." *Trans. St. John's Hosp. Dermatol. Soc.* **46**, 13–21.

Calnan, C.D., Fregert, S., and Pirila, V. (1969). "Cement additives." *Contact Derm. Newslett.* **6**, 112–113.

Card, W.I. (1935). "A case of asthma sensitivity to chromates." *Lancet* **229**, 1348–1349.

Carlsson, A.S., Magnusson, B., and Möller, H. (1980). "Metal sensitivity in patients with metal-to-plastic total hip arthroplasties." *Acta Orthop. Scand.* **51**, 57–62.

Christensen, O.B., Christensen, M.B., and Maibach, H.I. (1984). "Flare-up reactions and desensitization from oral dosing in chromate sensitive guinea pigs." *Contact Derm.* **10**, 277–279.

Cohen, H.A. (1962). "Experimental production of circulating antibodies to chromium." *J. Invest. Dermatol.* **38**, 13–20.

Connor, B. (1972). "Chromate dermatitis and paper manufacture." *Contact Derm. Newslett.* **11**, 265.

Covington, A.D., and Sykes, R.L. (1981). "Tannage with aluminum salts. Part III. Preliminary investigations of the interactions with polycarboxylic compounds." *J. Soc. Leather Techn. Chemists* **65**, 21–28.

Cronin, E. (1966). "Shoe dermatitis." *Br. J. Dermatol.* **78**, 617–625.

Cronin, E. (1980). *Contact Dermatitis.* Churchill Livingstone, Edinburgh, London, New York.

Czernilewski, A., Brykalski, D., and Depczyk, D. (1965). "Experimental investigations on penetration of radioactive chromium (^{51}Cr) through the skin." *Dermatologica* 131, 384–396.

Donaldson, R.M., and Barreras, R.F. (1966). "Intestinal absorption of trace quantities of chromium." *J. Lab. Clin. Med.* 68, 484–493.

Dooms-Goossens, A., Ceuterick, A., Vanmaele, N., and Degreef, H. (1980). "Follow-up study of patients with contact dermatitis caused by chromates, nickel, and cobalt." *Dermatologica* 160, 249–260.

Edmundson, W.F. (1951). "Chrome ulcers of the skin and nasal septum and their relation to patch testing." *J. Invest. Dermatol.* 17, 17–19.

Engel, H.O., and Calnan, C.D. (1963). "Chromate dermatitis from paint." *Br. J. Ind. Med.* 20, 192–198.

Epstein, S. (1966). "Detection of chromate sensitivity. Intradermal versus patch testing." *Ann. Allergy* 24, 68–72.

Feuerman, E.J. (1969). "Housewives' eczema and the role of chromates." *Acta Dermato-Vener.* 49, 288–293.

Feuerman, E.J. (1971). "Chromates as the cause of contact dermatitis in housewives." *Dermatologica* 143, 292–297.

Fischer, T., and Rystedt, I. (1985). "False-positive, follicular and irritant patch test reactions to metal salts." *Contact Derm.* 12, 93–98.

Fisher, A.A. (1976). "Blackjack disease and other chromate puzzles." *Cutis,* 18, 21–22, 35.

Fisher, A.A. (1984). "The role of patch testing in the management of dermatitides caused by orthopedic metallic prostheses." *Cutis* 33, 258, 264.

Förström, L., Pirilä, V., and Huju, P. (1969). "Rehabilitation of workers with cement eczema due to hypersensitivity in bichromate." *Scand. J. Rehab. Med.* 1, 95–100.

Fregert, S. (1961). "Chromate eczema and matches." *Acta Dermato-Vener.* 41, 433–442.

Fregert, S. (1962). "The chromium content of fuel ashes with reference to contact dermatitis." *Acta Dermato-Vener.* 42, 476–483.

Fregert, S. (1963). "Contact dermatitis due to chromate in foundry sand." *Acta Dermato-Vener.* 43, 477–479.

Fregert, S. (1965). "Sensitization to hexa- and trivalent chromium." *Proceedings of the Congress of the Hungarian Dermatological Society,* April, 1965, pp. 50–55 (cited in Cronin, 1980).

Fregert, S. (1971). "Remaining chromium in intracutaneous test sites." *Contact Derm. Newslett.* 10, 233.

Fregert, S. (1975). "Occupational dermatitis in a 10-year material." *Contact Derm.* 1, 96–107.

Fregert, S. (1981). "Chromium valencies and cement dermatitis." *Br. J. Dermatol.* 105 (Suppl 21), 7–9.

Fregert, S., and Gruvberger, B. (1970). "Chromium in leather." *Contact Derm. Newslett.* 8, 174.

Fregert, S., and Gruvberger, B. (1972). "Chemical properties of cement." *Berufsdermatosen* 20, 238–248.

Fregert, S., and Gruvberger, B. (1976a). "Chromate dermatitis from oil emulsion contaminated from zinc-galvanised iron plate." *Contact Derm.* 2, 121.

Fregert, S., and Gruvberger, B. (1976b). "Chromate dermatitis from zinc chromate used for marking textiles." *Contact Derm.* 2, 124.

Fregert, S., and Gruvberger, B. (1979). "Chromium in industrial leather gloves." *Contact Derm.* 5, 189.

Fregert, S., and Övrum, P. (1963). "Chromate in welding fumes with special reference to contact dermatitis." *Acta Dermato-Vener.* 43, 119–124.

Fregert, S., and Rorsman, H. (1964). "Allergy to trivalent chromium." *Arch. Dermatol.* 90, 4–6.

Fregert, S., and Rorsman, H. (1965). "Patch test reactions to basic chromium(III) sulphate." *Arch. Dermatol.* **91**, 233–234.

Fregert, S., and Rorsman, H. (1966). "Allergy to chromium, nickel and cobalt." *Acta Dermato-Vener.* **46**, 144–148.

Fregert, S., Gruvberger, B., and Heijer, A. (1970a). "Chromium dermatitis from galvanized sheets." *Berufsdermatosen* **18**, 254–260.

Fregert, S., Gruvberger, B., and Heijer, A. (1972). "Sensitization to chromium and cobalt in processing of sulphate pulp." *Acta Dermato-Vener.* **52**, 221–224.

Fregert, S., Gruvberger, B., Goransson, K., and Norman, S. (1978). "Allergic contact dermatitis from chromate in military textiles." *Contact Derm.* **4**, 223–224.

Fregert, S., Hjorth, N., and Gruvberger, B. (1970b). "Chromate in bridgetable felt." *Contact Derm. Newslett.* **8**, 173.

Fregert, S., Hjorth, N., Magnusson, B., et al. (1969). "Epidemiology of contact dermatitis." *Trans. St. John's Hosp. Derm. Soc.* **55**, 17–35.

Gafafer, W.M., Ed. (1953). *Health of Workers in Chromate Producing Industry: A Study.* U.S. Public Health Service, Division of Occupational Health, Pub. No. 192, Washington, D.C.

García-Pérez, A., Martín-Pascual, A., and Sánchez-Misiego, A. (1973). "Chrome content in bleaches and detergents. Its relationship to hand dermatitis in women." *Acta Dermato-Vener.* **53**, 353–358.

Geiser, J.D., and Girard, A. (1965). "Remarques sur les cas d'eczema au ciment observes a la clinique de dermato-venereologie de Lausanne de 1947 a 1961." *Dermatologica* **131**, 93–102.

Goitre, M., Bedello, P.G., and Cane, D. (1982). "Chromium dermatitis and oral administration of the metal." *Contact Derm.* **8**, 208–209.

Gray, S.J., and Sterling, K. (1950). "The tagging of red cells and plasma proteins with radioactive chromium." *J. Clin. Invest.* **29**, 1604–1613.

Gross, P.R., Katz, S.A., and Samitz, M.H. (1968). "Sensitization of guinea pigs to chromium salts." *J. Invest. Dermatol.* **50**, 424–427.

Guy, W.B. (1954). "Dermatologic problems in the railroad industry resulting from conversion to diesel power." *Arch. Dermatol. Syphilol.* **70**, 289–292.

Hall, A.F. (1944). "Occupational contact dermatitis among aircraft workers." *J.A.M.A.* **125**, 179–185.

Hammershoy, O. (1980). "Standard patch test results in 3,225 consecutive Danish patients from 1973–1977." *Contact Derm.* **6**, 263–268.

Hannuksela, M., Suhonen, R., and Förström, L. (1981). "Delayed contact allergies in patients with photosensitivity dermatitis." *Acta Dermato-Vener.* **61**, 303–306.

Harrison, P.V. (1979). "A postal survey of patients with nickel and chromate dermatitis." *Contact Derm.* **5**, 229–232.

Heise, H., and Mattheus, A. (1968). "Immunofluoreszenz-Untersuchungen bei verschiedenen Dermatosen. IV. Antigennachweis am positiven Kaliumbichromattest mit der direkten Immunofluoreszenz-Methode." *Arch. Klin. Exp. Dermatol.* **231**, 239–249 (cited in Polak, 1983).

Hellier, F.F. (1962). "Current problems in occupational dermatitis." *Proceedings of the XII International Congress of Dermatology.* Excerpta Medica Foundation, Washington, D.C., pp. 471–473 (cited in Langård, 1982).

Henning, H.F. (1972). "Chromium plating." *Ann. Occup. Hyg.* **15**, 93–100.

Hierholzer, S., Hierholzer, G., Sauer, K.H., and Paterson, R.S. (1984). "Increased corrosion of stainless steel implants in infected plated fractures." *Arch. Orthop. Trauma Surg.* **102**, 198–200.

Hilt, J. (1954). "La dermite du chrome hexavalent dans le cadre des dermites eczemateuses par sensibilisation aux metraux." *Dermatologica* **109**, 143–174.

Holz, H., Mappes, R., and Weidmann, G. (1961). "Chromatallergie bei bohrölekzem. Klinische beobachtungen, chemische untersuchungen." *Berufsdermatosen* 9, 113–122.

Høvding, G. (1970). "Cement Eczema and Chromium Allergy. An Epidemiological Investigation." Thesis. University of Bergen (cited in Cronin, 1980).

Hunziker, N., and Musso, E. (1960). "Apropos de l'eczema au ciment." *Dermatologica* 121, 205–212.

Huriez, C., Martin, P., and Lefebvre, M. (1975). "Sensitivity to dichromate in a milk analysis laboratory." *Contact Derm.* 1, 247–248.

Husain, S.L. (1977). "Contact dermatitis in the west of Scotland." *Contact Derm.* 3, 327–332.

Jaeger, H., and Pelloni, F. (1950). "Tests epicutanes aux bichromates, positifs dans l'eczema au ciment." *Dermatologica* 100, 207–216.

Jansen, L.H., and Berrens, L. (1968). "Sensitization and partial desensitization of guinea-pigs to trivalent and hexavalent chromium." *Dermatologica* 137, 65–73.

Johnston, A.J.M., and Calnan, C.D. (1958). "Cement dermatitis. I. Chemical aspects." *Trans. St. John's Hosp. Dermatol. Soc.* 41, 11–25.

Joules, H. (1932). "Asthma from sensitization to chromium." *Lancet* 223, 182–183.

Kaaber, K., and Veien, N.K. (1977). "The significance of chromate ingestion in patients allergic to chromate." *Acta Dermato-Vener.* 57, 321–323.

Kaplan, I., and Zeligman, I. (1963). "Urticaria and asthma from acetylene welding." *Arch. Dermatol.* 88, 188–189.

Kapoor-Pillarisetti, A., Mowbray, J.F., Brostoff, J., and Cronin, E.A. (1981). "HLA dependence of sensitivity to nickel and chromium." *Tissue Antigens* 17, 261–264.

Keskinen, H., Kalliomäki, P.-L., and Alanko, K. (1980). "Occupational asthma due to stainless steel welding fumes." *Clin. Allergy* 10, 151–159.

Kligman, A.M. (1966). "The identification of contact allergens by human assay. III. The maximization test: A procedure for screening and rating contact sensitizers." *J. Invest. Dermatol.* 47, 393–409.

Kresbach, H. (1967). "Untersuchungen zur aetiopatahogenese der kontaktekzeme. (2. Mitteilung) Chromallergie und kontaktekzema zur art und haufigkeit ursachlicher zusammenhange und zur klinik entsprechender krankheitsbilder." *Berufsdermatosen* 15, 317–335.

Krook, G., Fregert, S., and Gruvberger, B. (1977). "Chromate and cobalt eczema due to magnetic tapes." *Contact Derm.* 3, 60–61.

Lachapelle, J.M., and Tennstedt, D. (1979). "Epidemiological survey of occupational contact dermatitis of the hands in Belgium." *Contact Derm.* 5, 244–248.

Landsteiner, K., and Jacobs, E. (1935). "Studies on the sensitization of animals with simple chemical compounds." *J. Exp. Med.* 61, 643–657.

Langård, S. (1980). "A survey of respiratory symptoms and lung function in ferrochromium and ferrosilicon workers." *Int. Arch. Occup. Environ. Health* 46, 1–9.

Langård, S., and Hensten-Pettersen, A. (1981). "Chromium toxicology." In D.F. Williams, Ed., *Systemic Aspects of Biocompatibility.* Vol. 1. CRC Press, Boca Raton, pp. 144–161.

Levin, H.M., Brunner, M.J., and Rattner, H. (1959). "Lithographer's dermatitis." *J.A.M.A.* 169, 566–569.

Lidén, S., and Lundberg, E. (1979). "Penetration of chromium in intact human skin in vivo." *J. Invest. Dermatol.* 72, 42–45.

Lillingston, C. (1950). "Bronchial asthma caused by chromium trioxide (abstract of Broch, 1949)." *A.M.A. Arch. Ind. Hyg. Occup. Med.* 1, 588.

Magnus, I.A. (1958). "The conjugation of nickel, cobalt, hexavalent chromium and eosin with protein as shown by paper electrophoresis." *Acta Dermato-Vener.* 38, 20–31.

Magnusson, B., Blohm, S.-G., Fregert, S., et al. (1968). "Routine patch testing IV." *Acta Dermato-Vener.* **48**, 110–114.

Mali, J.W.H., van Kooten, W.J., and van Neer, F.C.J. (1963). "Some aspects of the behaviour of chromium compounds in the skin." *J. Invest. derm.* **41**, 111–122.

Mali, J.W.H., van Kooten, W.J., and van Neer, F.C.J. (1963). "Some aspects of the behaviour of chromium compounds in the skin." *J. Invest. derm.* **41**, 111–122.

Marechal, M.J. (1957). "Irritation et allergie respiratoires au jaune de chrome dans la peninture au pisctolet." *Arch. Mal. Prof. Med. Trav. Secur. Soc.* **18**, 284.

de Melo, J.F., Gjerdet, N.R., and Erichsen, E.S. (1983). "Metal release from cobalt-chromium partial dentures in the mouth." *Acta Odontol. Scand.* **41**, 71–74.

Moriearty, P.L., Pereira, C., and Guimaraes, N.A. (1978). "Contact dermatitis in Salvador, Brazil." *Contact Derm.* **4**, 185–189.

Morris, G.E. (1955). "Chromate dermatitis from chrome glue and other aspects of the chrome problem." *A.M.A. Arch. Ind. Health* **11**, 368–371.

Morris, G.E. (1958). "Chrome dermatitis: A study of the chemistry of shoe leather with particular reference to basic chromic sulfate." *Arch. Dermatol.* **78**, 612–618.

van Neer, F.C.J. (1963). "Sensitization of guinea pigs to chromium compounds." *Nature* **198**, 1013.

Newhouse, M.L. (1963). "A cause of chromate dermatitis among assemblers in an automobile factory." *Br. J. Ind. Med.* **20**, 199–203.

Nordlind, K., and Sandberg, G. (1983). "Leukocytes from patients allergic to chromium and nickel examined by the sealed capillary migration technique." *Int. Arch. Allergy Appl. Immunol.* **70**, 30–33.

North American Contact Dermatitis Group. (1973). "Epidemiology of contact dermatitis in North America: 1972." *Arch. Dermatol.* **108**, 537–540.

Novey, H.S., Habib, M., and Wells, I.D. (1983). "Asthma and IgE antibodies induced by chromium and nickel salts." *J. Allergy Clin. Immunol.* **72**, 407–412.

Olumide, Y.M. (1985). "Contact dermatitis in Nigeria." *Contact Derm.* **12**, 241–246.

Parker, D., Turk, J.L., and Drössler, K. (1984). "Comparison of DNFB and $K_2Cr_2O_7$ 'flare-up' reactions in guinea pigs." *Int. Arch. Allergy Appl. Immun.* **73**, 123–128.

Parkhurst, H.J. (1925). "Dermatosis industralis in a blue print worker due to chromium compounds." *Arch. Dermatol. Syphilol.* **12**, 253–256.

Pautritzel, R., Rivasseau, J., and Rivasseau-Coutant, A. (1962). "Le traitment des allergies cutanees de contact par epidermode-sensibilisation." *Med. Hyg.* **533**, 47.

Pedersen, N.B. (1977). "Chromate in a food laboratory." *Contact Derm.* **3**, 105.

Pedersen, N.B. (1982). "The effects of chromium on the skin." In S. Langård, Ed., *Biological and Environmental Aspects of Chromium, Topics in Environmental Health,* Vol. 5. Elsevier Biomedical Press, Amsterdam, pp. 249–275.

Pedersen, N.B., and Naversten, Y. (1973). "Disappearance of chromium(III) trichloride injected intracutaneously." *Acta Dermato-Vener.* **53**, 127–132.

Pedersen, N.B., Bertilsson, G., Fregert, S., et al. (1969). "Disappearance of chromium injected intracutaneously." *Int. Arch. Allerg.* **36**, 82–88.

Pedersen, N.B., Fregert, S., Naversten, Y., and Rorsman, H. (1970). "Patch testing and absorption of chromium." *Acta Dermato-Vener.* **50**, 431–434.

Peltonen, L. and Fräki, J. (1983). "Prevalence of dichromate sensitivity." *Contact Derm.* **9**, 190–194.

Pepys, J. (1977). "Clinical and therapeutic significance of patterns of allergic reactions of the lungs to extrinsic agents." *Am. Rev. Resp. Dis.* **116**, 573–588.

Perone, V.B., Moffitt, A.E., Jr., Possick, P.A. et al. (1974). "The chromium, cobalt, and nickel contents of American cement and their relationship to cement dermatitis." *Am. Ind. Hyg. Assoc. J.* **35**, 301–306.

Petersdorf, R.G., Adams, R.D., Braunwald, E., et al., Eds. (1983). *Harrison's Principles of Internal Medicine,* 10th ed. McGraw-Hill, New York, Chapters 50, 63, 87, and 273.

Pirilä, V. (1954). "On the role of chrome and other trace elements in cement eczema." *Acta Dermato-Vener.* **34**, 136–143.

Pirilä, V., and Kilpio, O. (1954). "On occupational dermatoses in Finland. A report of 1752 cases." *Acta Dermato-Vener.* **34**, 395–402.

Polak, L. (1983). "Immunology of chromium." In D. Burrows, Ed., *Chromium: Metabolism and Toxicity,* CRC Press, Inc., Boca Raton, FL, pp. 51–136.

Polak, L., Turk, J.L., and Frey, J.R. (1973). "Studies on contact hypersensitivity to chromium compounds." *Prog. Allergy* **17**, 145–226.

Popa, V., Teculescu, D., Stanescu, D., and Gavrilescu, N. (1969). "Bronchial asthma and asthmatic bronchitis determined by simple chemicals." *Dis. Chest* **56**, 395–404.

Rabeau, H., and Ukrainczyk, Mll. (1939). "Dermites des blanchisseuses. Role du chrome et du chlore (en France)." *Ann. de Dermatol. et de Syphiligr.* **10**, 656–680.

Rogers, S., and Burrows, D. (1975). "Contact dermatitis to chrome in milk testers." *Contact Derm.* **1**, 387–388.

Roupe, G., Rydberg, L., and Swanbeck, G. (1979). "HLA-antigens and contact hypersensitivity." *J. Invest. Dermatol.* **72**, 131–132.

Royle, H. (1975). "Toxicity of chromic acid in the chromium plating industry (2)." *Environ. Res.* **10**, 141–163.

Rudzki, E., and Czerwinska-Dihnz, I. (1977). "Sensitivity to dichromate in milk testers." *Contact Derm.* **3**, 107.

Rudzki, E., and Kozlowska, A. (1980). "Causes of chromate dermatitis in Poland." *Contact Derm.* **6**, 191–196.

Rudzki, E., Zakrzewski, Z., Prokopczyk, G., and Kozlowska, A. (1976). "Application of emulsifiers for the patch test. I. Patch test with potassium dichromate." *Dermatologica* **153**, 333–338.

Rudzki, E., Zakrzewski, Z., Prokopczyk, G., and Kozlowska, A. (1978). "Contact sensitivity to trivalent chromium compounds." *Dermatol. Beruf Umwelt* **26**, 83–85.

Rycroft, R. Doctorate of Medicine Thesis. Cambridge University (cited in Burrows, 1983).

Rycroft, R.J., and Calnan, C.D. (1977). "Chromate dermatitis from a boiler lining." *Contact Derm.* **3**, 198–200.

Samitz, M.H., and Gross, S. (1961). "Effects of hexavalent and trivalent chromium compounds on the skin." *Arch. Dermatol.* **84**, 404–409.

Samitz, M.H., and Katz, S. (1964). "A study of the chemical reactions between chromium and the skin." *J. Invest. Dermatol.* **43**, 35–43.

Samitz, M.H., and Shrager, J. (1966). "Patch test reactions to hexavalent and trivalent chromium compounds." *Arch. Dermatol.* **94**, 304–306.

Samitz, M.H., Katz, S., and Schrager, J.D. (1967). "Studies of the diffusion of chromium compounds through skin." *J. Invest. Dermatol.* **48**, 514–520.

Schleiff, P. (1968). "Provokation des Chromatekzems zu Testzwecken durch interne Chromzufuhr." *Hautarzt* **19**, 209–210 (cited in Polak, 1983).

Schwarz-Speck, M. (1968). "Experimentelle Sensibilisierung mit drei- und sechswertigem Chrom, XIII." *Congressees Internationalis Dermatologiae,* Munchen, 1967, Springer-Verlag, Berlin (cited in Polak, 1983).

Scutt, R.W. "Chromate sensitivity associated with tropical footwear in the Royal Navy." (1966). *Br. J. Dermatol.* **78**, 337–343.

Shanon, J. (1965). "Pseudo-atopic dermatitis." *Dermatologica* **131**, 176.

Shelley, W.B. (1964). "Chromium in welding fumes as cause of eczematous hand eruption." *J.A.M.A.* **189**, 772–773.

Shmunes, E., Katz, S.A., and Samitz, M.H. (1973). "Chromium-amino acid conjugates as elicitors in chromium-sensitized guinea pigs." *J. Invest. Dermatol.* **60**, 193–196.

Siegenthaler, U., Laine, A., and Polak, L. (1983). "Studies on contact sensitivity to chromium in the guinea pig. The role of valence in the formation of the antigenic determinant." *J. Invest. Dermatol.* **80**, 44–47.

Skog, E., and Wahlberg, J.E. (1969). "Patch testing with potassium dichromate in different vehicles." *Arch. Dermatol.* **99**, 697–700.

Sourreil, P., Fruchard, J., and Fruchard, J. (1964a). "Désensibilisation dans un eczéma du ciment." *Bull. Soc. Fr. Dermatol. Syphiligr.* **71**, 751.

Sourreil, P., Fruchard, J., and Fruchard, J. (1964b). "Nouvelle désensibilisation dans l'eczéma du ciment." *Bull. Soc. Fr. Dermatol. Syphiligr.* **71**, 752.

Spruit, D., and van Neer, F.C.J. (1966). "Penetration rate of Cr(III) and Cr(VI)." *Dermatologica* **132**, 179–182.

Spruit, D., and Malton, K.E. (1975). "Occupational cobalt and chromium dermatitis in an offset printing factory." *Dermatologica* **151**, 34–42.

Stern, R.M. (1982). "Chromium compounds: production and occupational exposure." In S. Langård, Ed., *Biological and Environmental Aspects of Chromium, Topics in Environmental Health, Vol. 5.* Elsevier Biomedical Press, Amsterdam, pp. 5–47.

Stevenson, C.J. (1975). "Fluorescence as a clue to contamination in TV workers." *Contact Derm.* **1**, 242.

Sykes, R.L., Hancock, R.A., and Orszulik, S.T. (1980). "Tannage with aluminum salts. Part II. Chemical basis of the reactions with polyphenols." *J. Soc. Leather Techn. Chemists* **64**, 32–37.

Tio, D. (1976). "A study on the clinical application of a direct leukocyte migration test in chromium contact allergy." *Br. J. Dermatol.* **94**, 65–70.

Vreeburg, K.J.J., de Groot, K., von Blomberg, M., and Scheper, R.J. (1984). "Induction of immunological tolerance by oral administration of nickel and chromium." *J. Dent. Res.* **63**, 124–128.
1.4-12.8. An experimental study in the guinea pig." *Dermatologica* **137**, 17–25.

Wahlberg, J.E. (1969). "Health-screening for occupational skin diseases in building workers." *Berufsdermatosen* **17**, 184–198.

Wahlberg, J.E., and Skog. E. (1965). "Percutaneous absorption of tri- and hexavalent chromium. A comparative investigation in the guinea pig by means of ^{51}Cr." *Arch. Dermatol.* **92**, 315–318.

Wahlberg, J.E., and Wennersten, G. (1977). "Light sensitivity and chromium dermatitis." *Br. J. Dermatol.* **97**, 411–416.

Weiler, K.-J. (1969). "Rezidivierendes chromkontaktekzem durch umgang mit chromstahl." *Berufsdermatosen* **17**, 316–320.

Weston, W.L., and Weston, J.A. (1984). "Allergic contact dermatitis in children." *Am. J. Dis. Child.* **138**, 932–936.

Williams, C.D., Jr. (1969). "Asthma related to chromium compounds. Report of two cases and review of the literature on chromate diseases." *North Carolina Med. J.* **30**, 482–491.

Williams, C.D., Jr. (1973). "Asthma and metals." *Ann. Internal Med.* **79**, 761–762.

Winston, J.R., and Walsh, E.N. (1951). "Chromate dermatitis in railroad employees working with diesel locomotives." *J.A.M.A.* **147**, 1133–1134.

Zelger, J. (1964). "Zur Klinik und pathogenere des chromatekzems." *Arch. Klin. Exp. Dermatol.* **218**, 498–542.

Ziegler, V., and Höhndorf, H. (1984). "Animal experiments with nickel-chromium-molybdenum implants." *Contact Derm.* **10**, 314.

OTHER HEALTH EFFECTS OF CHROMIUM COMPOUNDS

E. Nieboer

Department of Biochemistry and
Occupational Health Program
McMaster University
Hamilton, Ontario, Canada

A. Yassi

Occupational Health Program
Department of Community Health Sciences
University of Manitoba
Winnipeg, Manitoba, Canada

1. OVERVIEW

The risk of contracting cancer or developing hypersensitivity of the skin or respiratory system on exposure to chromium compounds has been reviewed in Chapters 17 and 18, respectively. Other debilitating effects associated with excessive exposure to chromium compounds have also been well documented in the literature, some of which are indeed more common than those mentioned (NIOSH, 1973, 1975; EPA, 1984). For example, skin ulcerations are the most common lesions in the occupational setting, and are experienced by tanners, electroplaters, and dichromate workers (Burrows, 1983). It is also well known that Cr(VI) compounds may cause acute, subacute, or chronic injuries to the nasal tissues of exposed workers, resulting in mucosal irritation, ulceration, and even nasal septal perforation (Bidstrup, 1983). As is reviewed briefly later, other nonallergic, inflammatory effects on the respiratory system that have been reported in association with chromium exposure include chronic rhinitis, chronic sinusitis, laryngitis, bronchitis, acute chemical pneumonitis, and even a suggestion of chromitotic pneumoconiosis. Conjunctivitis has also been attributed to chromium exposure, as well as olfactory disorders and dental abnormalities. In addition, isolated reports exist of inflammatory and ulcerative conditions of the gastrointestinal tract, and of liver and kidney damage. And, finally, there has been some concern expressed that chromium may have a deleterious effect on reproduction, possibly causing reduced fertility and birth defects. This, too, is reviewed in this concluding chapter on the health effects of chromium compounds.

2. IRRITANT EFFECTS ON SKIN AND NASAL TISSUES

2.1 Description and Prevalence

Ulcerations of the skin and nasal septum in workers exposed to Cr(VI) compounds have been recognized for well over a century. As early as 1827, Cumin described chrome ulcerous lesions of the skin, and in 1863, Bécourt and Chevallier reported ulcerous lesions of both skin and nasal tissues. NIOSH (1975) reviewed the dozens of reports in the literature in this area and concluded: "It may be stated unequivocally that chromium(VI) may cause skin ulcers, ulcers of the nasal mucosae and perforations of the nasal septum." Moreover, it is pointed out that chrome ulcers may appear anywhere on skin given sufficient contact with acidic or alkaline solution of hexavalent chromium. The most frequently reported sites are the fingernail root areas, skin folds over knuckles, finger webs, dorsum of hands, and forearms, because skin abrasions are more prone to occur at these locations and broken skin strongly facilitates chrome ulceration (Edmundson, 1951; Maloof, 1955; Burrows, 1984). Scars remain after healing. Maloof (1955) described these lesions as beginning as "painless papules or vesicles that eventually ulcerate and penetrate deeply into the tissue." He attributed the poor healing of the ulcers to tenacious deposits of Cr(III) with prolonged eroding action. Topical applications often include EDTA to aid both the reduction of hexavalent chromium to the trivalent form and the removal of the latter by chelation (Maloof, 1955; Burrows, 1983).

An annotation in the British Medical Journal 20 years ago (Anon, 1963) succinctly summarizes years of clinical experience with chrome ulceration of the nasal septum. Four well-defined stages are recognized (also see Bidstrup, 1983), progressing from nasal irritation, to ulceration and perforation of the septum. Initially, redness and irritation of the mucosa covering the septum occur, which are accompanied by rhinorrhea and followed by blanching and the development of adherent crusts. Nose-picking invites further contamination by chromates. After an interval that ranges from days to weeks, ulceration occurs. Subsequently, the septal cartilage becomes eroded, leaving a perforation of the septum. The perforation often increases in size, but neither the bony part of the septum nor the anterior and lower parts become involved. The shape of the nose is not affected. During the stage of active ulceration, slight bleeding is frequent. Commonly, within about three months the edges of the perforation are healed, providing it is not aggravated by picking. On healing, crusting and a tendency to "wet nose" may persist. Complications are rare and the condition does not give rise to disability. There is no evidence that malignant change occurs at the site of either nasal or skin ulceration caused by Cr(VI) compounds, and a loss of sense of smell appears to be infrequent. It has also been noted that there is no apparent relationship between the development of chrome ulcers and the prevalence of chromium sensitization (Burrows, 1983).

Given the irritant and corrosive nature of soluble, hexavalent chromium, it is not surprising that not only direct cutaneous contact but also inhalation of dust or mist may inflict damage. Chrome ulceration of the nasal septum may occur in any industry in which it is possible for hexavalent chromium to be deposited in this region of the nose. Indeed, the irritant effects of chromate on skin and nasal tissues have been reported in many countries for chromium platers (Kleinfeld and Rosso, 1965; Gomes, 1972; Henning, 1972; NIOSH, 1973, 1975; Royle, 1975a, b; Horiguchi, 1980), among chrome strippers (Mitchell, 1969), for galvanization shop employees (Petrzela et al., 1981), in individuals exposed to zinc chromate pigment (Langård and Norseth, 1979), in ferrochromium workers (Axelsson et al., 1980; Langård et al., 1980), and especially among chromate production workers (Edmundson, 1951; Mancuso, 1951; Gafafer, 1953; NIOSH, 1973, 1975; Krishna et al., 1976; Hayes et al., 1979; Watanabe and Fukuchi, 1981). Claims that skin and nasal ulcerations may also be associated with trivalent chromium exposure (Mancuso, 1951) have not been substantiated. As explained in Chapter 17, the chromate production plant studied by Mancuso was characterized by mixed exposure to both hexavalent and trivalent chromium. Langård and Norseth (1979) noted that in the case of tannery workers, the skin ulcers may be due to Cr(III) compounds (Maloof, 1955). However, as pointed out by Stern (1982), chromates are also common in tanneries as starting materials or in the "two-bath" tanning process (see Chapter 2).

The reported prevalence of skin and nasal ulcerations among workers exposed to hexavalent chromium is high. Among chromate production workers, a prevalence of nasal septal perforation near 60% has been consistently reported (Edmundson, 1951; Mancuso, 1951; Gafafer, 1953). Edmundson (1951) found that 70% of the workers examined had chrome skin ulcers, and 47% were inflicted with both the skin and nasal lesions. Occurrence among chrome platers was also substantial. In a study of 997 platers by Royle (1975b), the results of a questionnaire indicated that 17% of the workers had nose bleeding, 13% had nasal ulcers, 5% had nasal perforations, and 22% had skin ulcers. These percentages increased in a subgroup of 369 workers, who had been exposed for more than five years, to 16, 17, 9, and 32% respectively. In Britain, both skin and nasal septum ulcerations are notifiable. Although the number of notifications of skin ulcers have been declining since 1965, as many as 50 to 100 are still reported annually (Burrows, 1983). The majority of cases occur in the chromium plating industry (Royle, 1975a; Burrows, 1984).

2.2 Threshold Chromate Concentrations

As Langård and Norseth (1979) have concluded, the prevalence of chromium skin ulcers bears no simple relationship to atmospheric chromium concentrations, presumably because the development of these lesions depends on local deposition of the chromium compound on the skin and on the presence of any preexisting cutaneous lesions. Although most investigators stress the importance of individual work practices in preventing skin and nasal mucosal lesions (e.g., Anon, 1963; Bidstrup, 1983), it is acknowledged that the only effective way to prevent formation of nasal ulcers is to limit chromic acid mist from entering the workplace atmosphere (Henning, 1972). Methods of exhaust ventilation and spray suppressants, as well as of personal protection, are outlined in detail by Henning (1972) and Mitchell (1969).

Many of the investigators reporting chrome ulcers in association with exposure to mists of hexavalent chromium from plating tanks have provided atmospheric levels (Bloomfield and Blum, 1928; Kleinfeld and Rosso, 1965; Mitchell, 1969; Gomes, 1972; a number of studies cited in NAS, 1974 and NIOSH, 1975; Royle, 1975b; EPA, 1984). For example, Gomes (1972) reported that 86% of the workers examined had lesions due to chromic acid exposure, but noted that the TLV of 0.05 mg chromium m^{-3} was exceeded 50% of the time and extreme levels as high as 0.7 mg chromium m^{-3} had been noted periodically. At the mean Cr(VI) exposure levels listed in Table 3 of Chapter 17 (typically 0.03–0.30 mg chromium m^{-3}), Mancuso (1951) and Gafafer (1953) found the prevalence of skin and nasal septum ulcerations to be very high among chromate production workers, as already indicated.

Chromate-induced perforations of the nasal septum have often been considered to occur only when the atmospheric concentrations of chromic acid exceed 0.05 mg chromium m^{-3} (Bloomfield and Blum, 1928; NIOSH, 1975). This conclusion does not seem to be supported by more recent data. For example, Royle (1975b) reported that the risk of skin and intranasal ulcerations increased progressively with the duration of chromic acid exposure, which was shown in a single personal monitoring survey to be less than 0.015 mg chromium m^{-3} in 40 of 42 plants examined. Based on a questionnaire, he found that 22% of the platers had had skin ulcers, but when the duration of exposure exceeded five years, the incidence increased to 32%. Royle's results suggest that chromic acid at 0.015 mg chromium m^{-3} was sufficient to cause symptomatic nasal and skin ulcerations. Lucas and Kramkowski (1975) have found ulcerations at average levels of 0.002 mg chromium m^{-3} (range 0.001–0.010). Comparable findings are reviewed in NAS (1974). Mitchell's (1969) data suggest that a reduction of chromate exposure levels to 0.005 mg chromium m^{-3}, obtained by the installation of exhaust ventilation of chrome plating tanks (also see Kleinfeld and Rosso, 1965; NIOSH, 1975), prevented the occurrence of previously observed nasal lesions. In addition to chromic acid exposure, it is important to emphasize that settled, chromate-rich dust appears to be ubiquitous in electroplating plants, even when air levels are low (Royle, 1975b). This additional source of Cr(VI) may well contribute to increased skin contact. And finally, Langård and Norseth (1979) observed that while they had noted septal ulceration after two weeks and perforation after two months of exposure to zinc chromate dust at ∿ 0.3 mg chromium m^{-3}, they did not observe ulcerations or perforations in workers exposed to the same dust at 0.006–0.03 mg chromium m^{-3} for up to 18 months.

Within a retrospective mortality study of chromate production workers, Hayes et al. (1979) investigated by case-control pairs classified by the presence or absence of a history of lung cancer, the possible association of lung cancer with skin ulcers, nasal perforation, and dermatitis. Only a history of dermatitis was associated with lung cancer with a relative risk of 3.0 (p < 0.05). However, the occurrence of all three conditions within an individual was also highly associated with lung cancer history (relative risk of 6.0, p < 0.05). Hayes et al. (1979) expressed the view that this association between lung cancer and chromate-related morbid conditions likely indicates that greater exposure implies greater risk. However, they conclude: "Our knowledge of the epidemiology of these morbid conditions is, however, too scant to use them as direct indicators of cancer risk or as determinants for maximum allowable concentrations of carcinogens."

3. OTHER IRRITANT AND MISCELLANEOUS FINDINGS

In the extensive medical survey by Gafafer (1953) of chromate production workers, a number of other chromate-associated health effects were revealed. There was an excess of severe red throat, a yellowish discoloring of the teeth and tongue, and a higher percentage of workers with gingivitis and periodontitis (also see Gomes, 1972). Many of the inflammatory responses observed by Mancuso (1951) were not as prevalent (chronic chemical rhinitis, sinusitis and pharyngitis, laryngitis with hoarseness). Gastric distress (Mancuso, 1951; Lucas and Kramkowski, 1975), conjunctivitis (Gafafer, 1953; Gomes, 1972), and olfactory sense impairment (Mancuso, 1951; Watanabe and Fukuchi, 1981) have also been reported for workers exposed to hexavalent chromium.

4. BRONCHITIS, PNEUMONITIS, AND PNEUMOCONIOTIC PROCESSES

Occurrence of chromium-associated bronchitis is not well documented. Bronchitic symptoms have been reported for workers using dichromates (Williams, 1969) and chromate production workers (Capodaglio et al., 1975; Wieser et al., 1982). Evidence suggestive of bronchitis for electroplaters was provided by Royle (1975b), although assessment was by questionnaire only. Alderson et al. (1981) observed a total of 39 deaths associated with chronic bronchitis in chromate workers, which did not differ significantly from the number expected, although there was an excess (p < 0.001) in one of the three factories examined.

In 1939, Letterer described a "chromium-silicotic" lung in a sandblaster in a ferrochromium department. The diffuse enlargement of the alveolar septa and the chronic interstitial and alveolar pneumonia observed were believed by the author to be caused by inhalation of chrome dust. This finding was consistent with a previous report by Lukanin (1930) of pulmonary fibrosis in a chromate worker. Machle and Gregorius (1948) suggested the existence of a chromate-induced chemical pneumonitis from their survey of workers in chromate-producing plants, and Meyers (1950) reported two cases of acute chromic acid poisoning that resulted in severe and prolonged pulmonary disease. Mancuso (1951) detailed the case of a worker with chemical pneumonitis which was attributed to his work at a chromate plant. He and his colleagues also described several cases of spotty, moderately severe, but not nodular, pneumoconiosis which they called chromitotic pneumoconiosis (Mancuso, 1951; Mancuso and Hueper, 1951).

Sluis-Cremer and Du Toit (1968) have reported fine nodular pneumoconiosis in a few chromite miners in South Africa. They described the nodulation as finer in quality, though somewhat more radioopaque, than in simple coalminers' pneumoconiosis. However, these authors describe animal experiments that showed that chromite injected intratracheally into rats initially produced an acute inflammatory reaction, but that chromite deposits that accumulated and remained in the lungs were not fibrogenic. They also noted the study by Worth and Schiller (1955), in which intraperitoneal chromite caused an initial, mild inflammatory reaction in mice, but produced no long-term sequelae. They therefore concluded that pneumoconiosis in chromite miners is due to the deposition of radioopaque chromite dust in the tissues, but that the condition is benign and causes no fibrosis. Reports of pneumoconiosis in chromite miners have appeared in the U.S.S.R. as well (Andrievskaya and Mislavskaya, 1949).

Princi et al. (1962) described a pulmonary disease of progressive fibrosis characterized clinically by recurring bouts of acute pneumonitis, ventilatory impairment, as well as linear and nodular fibrosis in the chest x-ray. These pulmonary abnormalities were attributed to exposure to high concentrations of metallic (including chromium) silicide fumes. These authors suggested that this disease had occurred in workers of other ferroalloy plants, although it had not been previously recognized. Average air levels of total dust in the plant investigated were 15 ± 45 mg m^{-3} with the quartz content below 2%. The pathogenesis of this condition, which has since been noted by other investigators (Davies, 1974), is not yet clear.

The role of chromium in the progressive fibrosis described by Princi et al. (1962) has not been established, especially since the levels of total dust exposure were very high. For some of the other cases described previously, either trivalent or hexavalent chromium or both might have been determinants in the chromium-associated pneumoconiosis described (Langård and Norseth, 1979). However, the clinical significance of this is again unclear.

Finally, Bovet et al. (1977) investigated the spirometric alterations in chromate-exposed workers in the electroplating industry. They found that urinary chromium concentrations significantly correlated with reductions in FEV$_1$ (forced expiratory volume over one second) and FEF$_{25-75}$ (forced expiratory flow evaluated at 25–75% expiration of the vital capacity), but not with VC (vital capacity). Capodaglio et al. (1975) also demonstrated decrements in respiratory function among workers producing dichromates and chromic acid. By contrast, median vital capacity readings for chromate and nonchromate production workers showed practically no difference (Gafafer, 1953). The results obtained by Langård (1980) for ferrochromium workers suggest that the high level of total dust, rather than Cr(III) and Cr(VI) compounds per se, was responsible for the generalized obstructive lung disease observed. Examination of welders has also shown that there is a greater prevalence of small airway disease among shipyard manual metal-arc (MMA) welders of mild steel (no chromium exposure) than among stainless steel MMA welders [with exposure to Cr(VI)] (Kalliomäki et al., 1982).

In conclusion, chromium exposure is associated with a moderate risk of pulmonary hypersensitivity (Chapter 18), and heavy exposure to chromic acid may result in acute chemical pneumonitis. Otherwise, there is little evidence to suggest that long-term exposure to chromium compounds results in nonmalignant, chronic pulmonary disease.

5. GASTROINTESTINAL, HEPATIC, AND RENAL IMPAIRMENTS

Mancuso (1951) conducted a gastrointestinal x-ray examination of about 100 chromate workers. The findings (10% had evidence of ulcer formation and 6% of hypertrophic gastritis—compared to 5% in both categories in a control group of cement workers) indicated a highly probable relationship between occupational exposure and gastrointestinal pathology. Lucas and Kramkowski (1975) also reported that five of 11 "hard" chromium electroplaters showed evidence of gastric distress. Although gastrointestinal injury is not routinely reported in association with exposure to chromate, animal data are supportive. In experiments conducted by Nettesheim et al. (1971) in mice, not only did exposure to calcium chromate dust generate additional subepithelial connective tissue and flattened epithelium in the large bronchi, but also atrophy of the spleen and liver. Ulceration of the stomach and intestinal mucosae also occurred. Furthermore, Mosinger and Fiorentini (1954) observed acute gastroenteritis, as well as various lesions of the liver and kidneys, after the administration of potassium dichromate (2 mg chromium kg^{-1}) to various laboratory animals by a variety of routes.

Animal experiments illustrate that, generally speaking, hexavalent chromium compounds are more toxic than trivalent derivatives (NAS, 1974 and Tandon, 1982 provide extensive summaries of organ toxicities). Renal damage is most frequent, although necrotic changes in the liver have been reported. Isolated case reports of acute chromium poisonings confirm that both the liver and kidney are also susceptible to injury in man. A brief review of the human experience is provided as follows. A number of case reports indicate renal failure, liver failure, and death of humans following ingestion of high doses of potassium dichromate. In 1935, for example, Goldman and Karotkin published a case of a 25-year-old woman, who following ingestion of potassium dichromate, presented with gastrointestinal symptoms and renal and hepatic failure. The authors stated that this case was typical of the 69 cases previously reported. The evidence that acute renal failure follows ingestion of potassium dichromate is further exemplified by the clinical course of a young patient (Sharma et al., 1978). Oliguria lasted 12 days. A percutaneous renal biopsy on the eighteenth day showed a loss of tubular epithelial lining, interstitial edema, and evidence of regeneration. The patient made a complete recovery. Similarly, in a 22-year-old male patient who suffered renal failure as a result of chromic acid burns to 10% of his body surface area, kidney function returned to normal 28 days after the injury (Laitung and Earley, 1984).

In Chapter 2, a case study was mentioned in which cutaneous absorption of trivalent chromium culminated in acute renal failure and death. A man was accidentally immersed in hot acidic trivalent chromium sulphate solution but none was swallowed (Kelly et al., 1982). The clinical course was dominated by burns, intravascular hemolysis, and oliguric renal failure. Histological findings at necropsy revealed degeneration of liver, kidney, and small intestine tissues, as well as acute lung congestion and intravascular coagulation. The authors concluded that the observed tissue chromium distribution (high levels in kidney and spleen, but lower in muscle and none in brain) was comparable to that reported for hexavalent chromium poisonings and correlated with the damage sustained. It seems reasonable to conclude that trivalent chromium is the renal toxic agent, even in chromate toxicosis.

Perhaps the results of Franchini et al. (1978) are of greater relevance (also described in Mutti et al., 1979). Urinalysis suggested that occupational exposure to chromium may have caused subtle renal dysfunction. Of 99 workers exposed to chromium compounds, none of the 39 stainless steel welders had evidence of abnormalities, but eight of 36 workers using special chromium-alloy electrodes when welding had increased urinary levels of β-glucuronidase and three of these had acutal proteinuria. Among 24 chromium platers, nine had increased levels of β-glucuronidase, while four had evidence for proteinuria. More recently, Littorin et al. (1984) confirmed the absence of signs of kidney damage by similar tests in stainless steel MMA welders. Similarly, Pascale et al. (1952) suggested that subtle hepatic dysfunction may occur in chromate-exposed workers, based on the appearance of acute hepatitis in a chromium electroplater and asymptomatic hepatic dysfunction in four coworkers examined. In these coworkers, hepatic tests and liver biopsies showed mild to moderate abnormalities; all had nasal lesions typical of chromate exposure. At present, little can be said on the significance of these subtle kidney and liver changes, nor about the doses that may be expected to precipitate them.

On the basis of the evidence reviewed, it appears that chromate, at least in high doses, is capable of inducing gastrointestinal damage, as well as kidney and liver dysfunctions. The identification of hepatic impairment and mild renal damage in workers exposed to chromate suggest the existence of a potential health hazard that warrants further in-depth evaluation and study.

6. EMBRYOTOXICITY, TERATOGENICITY, AND OTHER DELETERIOUS EFFECTS ON REPRODUCTION

6.1 Synopsis

There is no human evidence at present that any form of chromium is associated with birth defects (Clarkson et al., 1985). However, the animal and other laboratory evidence reviewed in this section suggests that a cautious approach to the potential adverse effects of chromium compounds on development and reproduction is both judicious and warranted.

Experimental embryology and teratology provide evidence for, or are suggestive of, about 10 mechanisms by which causative agents can influence developing cells (Wilson, 1977). Agents identified as mutagenic or carcinogenic in man can often damage germ cells or embryo, resulting in reproductive deficits (Manson, 1978). In fact, it is estimated "that some 20–30% of human developmental errors can be attributed solely or primarily to mutation in a prior germ line" (Wilson, 1977). Genotoxicity may thus have some definite relationship to reproductive and developmental deficits. It was demonstrated in Chapters 16 and 17 that Cr(VI) compounds are mutagenic in bacteria, yeast, and cultured mammalian cells and carcinogenic in animals and man. Direct cytogenic effects in workers exposed to chromates have also been observed. Imreh and Radulescu (1982), for example, performed cytogenic analyses in 18 workers from a potassium dichromate producing plant and suitable controls. They found a significant increase in frequency of chromosome and chromatid aberrations in peripheral lymphocytes of workers with 20 years of exposure. In cultures of human lymphocytes, similar clastogenic effects were induced at concentrations of $K_2Cr_2O_7$ between 0.03 and 3.0 μg ml^{-1}. While sister-chromatid exchanges (SCEs) were produced in acute exposure in vitro, no significant increase was detected in vivo in the men with chronic occupational exposure. By contrast, Sarto et al. (1982, 1983) have detected both chromosomal aberrations and SCEs in lymphocytes of workers exposed to chromic acid. There is, therefore, no doubt that Cr(VI) compounds are genotoxic in man. However, no studies of abnormal fetal outcome have been undertaken in occupational settings. Consequently, the studies on fetal outcome in animals experimentally exposed to chromium are of considerable interest and these are now summarized.

6.2 Animal Studies

A number of animal and related studies have demonstrated that Cr(VI) compounds can be embryocidal and teratogenic (Matsumoto et al., 1976; Gale, 1978, 1982; Gale and Bunch, 1979; Gilani and Marano, 1979; Iijima et al., 1979). A comprehensive compilation of published studies is provided in the recent EPA Chromium Health Assessment Document (EPA, 1984). It has been shown, for example, that the frequency of a variety of malformations in the chick embryo at day 8 increases with the dose of Cr(VI) oxide injected into the egg air sac at doses of 1 to 25 μg chromium on days 0, 1, 2, or 3 of incubation (Gilani and Marano, 1979). Iijima and colleagues (1979), in a preliminary report on experiments with mice, found that Cr(VI) oxide injected subcutaneously at doses of 5–10 mg chromium kg^{-1} on day 8 of gestation had by day 18 induced both embryocidal and teratogenic effects. Fetal growth was also impaired. Matsumoto et al. (1976) and Iijima et al. (1983a) showed a positive dose-response relationship between amounts of Cr(III) chloride injected intraperitoneally on day 8 of gestation in pregnant mice (10–24 mg chromium kg^{-1}) and the incidence in live fetuses on day 18 of both external and skeletal abnormalities, as well as neural tube defects. Inhibition of fetal growth was also demonstrated. In a novel approach, Knudsen (1980) used a newly developed mammalian in vivo test for measuring genetic effects of chromates in the offspring of treated females. This test is dependent on transplacental transfer of the test compound, and assesses genetic alterations expressed as changes in fur pigmentation in the offspring. He found that intraperitoneal injection on day 8 to 10 of gestation of welding fume particles (total of 100 mg kg^{-1}) containing 3.6% hexavalent chromium, as well as administering potassium chromate (total of 2.7 mg chromium kg^{-1}), produced significant positive results.

A number of studies of the embryotoxicity and teratogenicity of Cr(VI) oxide in hamsters illustrates the complexities inherent in animal work. Firstly, Gale (1978) demonstrated a narrow dose-response range to Cr(VI) oxide adminstered intravenously to golden hamsters on day 8 of gestation: a 2.5 mg chromium kg^{-1}

dose exerted minimal effects, while doses of 3.8 and 5 mg chromium kg^{-1} were both embryolethal and teratogenic (cleft palate was the major malformation) and 7.5 mg chromium kg^{-1} killed the pregnant animals. Maternal morbidity did occur at all but the lowest dose. Secondly, the time of administration of the chromic acid was also a critical determinant of the embryotoxic and teratogenic responses (Gale and Bunch, 1979). And, finally, Gale (1982) demonstrated that different strains of hamsters did not exhibit the same degree of susceptibility to Cr(VI) oxide-induced embryotoxicity and teratogenicity. Three strains were susceptible, while three others were resistant to these toxic events. It is obvious from the foregoing experiments that extreme caution is necessary in attempting to extrapolate quantitatively animal results to humans. Wilson (1977) states: "the use of animal tests for evaluation of human risk will become more than empirical only when the degree of comparability of mechanisms between test animal and man is understood."

Various investigators (Newbold et al., 1979; Levis and Majone, 1979; and many others; see Chapter 16) have shown that survival and multiplication of cultured cells are affected in a dose-dependent manner by hexavalent salts, while it was considerably more difficult to demonstrate such effects with trivalent chromium. Jacquet and Draye (1982) confirmed that this is also the case in cultured mouse embryos by measuring the inhibition of blastocyst formation and hatching. Similarly, Danielsson et al. (1982) studied the effect of chromium salts on embryonic (chick limb bud mesenchymal) cells in vitro. Sodium dichromate inhibited cartilage formation (chondrogenesis) at 100 μg chromium l^{-1}, while Cr(III) chloride had no effect even at 15,000 μg chromium l^{-1}. Iijima et al. (1983b) have reported a correlation between developmental impairment observed in vitro in cultures of mouse blastocysts and egg cylinder embryos and SCE frequency in these units. Thus, again, cytogenetic damage appears implicated.

The variety of deleterious effects induced by both hexavalent and trivalent chromium on fertilization and development in sea urchins has recently been studied by Pagano et al. (1983). Sodium chromate was spermotoxic, suppressed mitosis, and impeded differentiation (even when treatment was of sperm or eggs before fertilization). Aside from causing acute spermotoxicity, it is of interest that trivalent chromium severely reduced sea urchin embryo hatchability despite its lack of demonstrable effect on mitotic activity and differentiation. Of note are also the findings of Billard and Roubaud (1985), who examined the effect of Cr(III) sulphate on rainbow trout gametes and fertilization. Spermotoxicity and reduced fertilization occurred at 5 μg chromium l^{-1}, while ova were not affected. Testicular degeneration and inhibition of spermatocyte production and testicular enzymes have been reported for intraperitoneally injected rabbits (Behari et al., 1978; EPA, 1984). Chromium(III) nitrate was more effective than potassium dichromate at doses of 2 mg/kg for three or six weeks.

In general, and as pointed out by IARC (1980) with respect to laboratory results such as those reviewed here, it is uncertain whether all the effects described represent direct insults on the embryo or fetus, or are a result of maternal toxicity. The issue of maternal toxicity is crucial in the testing of reproductive hazards. Often, the pertinent information is not reported by authors or is not considered in the interpretation of the results. Some effects must be considered inconsequential if toxicity of the dam occurs (e.g., embryolethality, retardation of ossification, reduced fetal weight, and an increase in nonspecific congenital malformations). Even for specific congenital malformations, a direct teratogenic effect on the fetus is not implicated unless the effect is rare in control animals and is produced at doses well below those toxic to the dam (Barlow and Sullivan, 1982).

6.3 Placental Transport

It is recognized that chromium may play an important role in the function and growth of the body (see Chapter 2; Guthrie, 1982). It is also suggested that if adequate amounts of chromium are not taken by the mother during pregnancy, deficiency may result, which is expected to worsen with increasing parity (Saner, 1981; Wallach and Verch, 1984). Apparently, chromium levels tend to be high in the newborn, then decline somewhat during the next decade or so (Schroeder et al., 1962). These authors also reported measurable concentrations of chromium in some, but not all, fetuses of laboratory animals immediately after birth, suggesting placental transport during pregnancy. However, as indicated in Chapter 2, analytical problems might invalidate these early observations.

Matsumoto et al. (1976) pointed out that intrauterine growth inhibition of specific chemical agents may reflect (singly or in combination): (1) an alteration of some factor in the maternal system that secondarily affects fetal growth, (2) blocking of the placental transfer of some essential material necessary for normal fetal growth, or (3) a direct effect upon specific embryonic or fetal tissue. The identification and quantification of chromium in the fetus would help to resolve this matter.

Placental transfer of inorganic chromium in either oxidation state was not observed in a number of early animal studies (Visek et al., 1953; Mertz et al., 1969). Mertz et al. (1969) showed that placental transport depends on the specific form in which chromium is administered. For example, intragastric administration on days 15 to 18 of gestation of ^{51}Cr(III)-labelled brewers' yeast to pregnant rats led to significant labelling of the newborn (20 ± 5% of the mother's level) compared to $\sim$ 1% for Cr(III) chloride administered similarly, but daily, during gestation (also see Chapter 2). By contrast, administration of Cr(III) acetate in the drinking water during gestation and intravenous injection of Cr(III) chloride at mating did not result in appreciable transfer into the young. On the basis of these studies, IARC (1980) concluded that inorganic chromium crosses the placenta only to a small extent, when administered to laboratory animals (regardless of oxidation state). The work of Matsumoto et al. (1976) with Cr(III) chloride supports this conclusion (administered subcutaneously to mice during the first 16 days of gestation), since little additional chromium over controls was detected in the fetal bodies on day 18. However in a follow-up study (Iijima et al., 1983a), the administration to mice by intraperitoneal injection of Cr(III) chloride (10 mg chromium kg^{-1}) on day 8 resulted in a steady decrease of label in the maternal blood over the first 24 hours and a concomitant increase of the label in the embryo. On a weight basis, the radioactivities of the maternal blood and embryo were similar at the end of the 24-hour observation period.

Recent studies have clarified the controversy concerning placental transport of chromium compounds. Polansky and Anderson (1980) showed that radio-labelled chromium, administered to rats as Cr(III) chloride or sodium chromate, was transported across the placenta to the unborn pups, independently of the mode of administration (whether intravenously, subcutaneously, or intra-peritoneally). Placental transport was time-dependent and peaked when the dose was administered shortly before birth. Administration of the labelled chromium more than 10 days before birth resulted in insignificant transfer. Similarly, Wallach and Verch (1984) found that the mean ^{51}Cr content per placentofetal unit in rats was $0.89 \pm 0.03\%$ of the dose injected intravenously on day 17 of gestation, as determined on day 20. The work of Danielsson et al. (1982) is even more helpful. They dosed mice by intravenous injection with radiolabelled sodium dichromate or Cr(III) chloride. Administration during the early organo-genetic period (day 8 to 11) showed radioactivity in all embryonic structures by autoradiography for Cr(VI) but not Cr(III). However, there was heavy accumulation in the visceral yolk-sac placenta for the Cr(III) injection. In mid-gestation (days 12 to 15), the fetal chromium levels were 10-fold higher for the administration of Cr(VI) compared to Cr(III). By contrast in late gestation (days 16 to 18), placental transfer for Cr(III) increased considerably. The relative fetal accumulation, Cr(VI) injection/Cr(III) injection, was now reduced to 5. The results of these studies clearly indicate that time of administration is a major determinant in placental transport experiments. This is not surprising since the placenta decreases in thickness and undergoes other structural and functional changes during gestation (Miller, 1983; Miller et al., 1983). It may therefore be concluded that placental transport of chromium does occur in rats and mice for both oxidation states. This conclusion is consistent with the evidence that Cr(III) and Cr(VI) compounds are embryotoxic and teratogenic in animals.

6.4 Concluding Remarks

In conclusion, there is some animal evidence that both Cr(III) and Cr(VI) compounds can be spermicidal, embryocidal, or teratogenic. There seems to be a wide range of susceptibilities in this regard among different species. Placental transport in rodents of chromium administered as Cr(III) or Cr(VI) has been demonstrated, but is strongly dependent on the time during gestation when the dose is given. The route of administration and chemical form are also deter-minative. The animal and related data reviewed suggest that concern about the potential adverse effects of chromium compounds on human reproduction and development is justified, even though there is no direct human evidence.

ACKNOWLEDGMENTS

Financial support for the preparation of this review was received from the Occupational Health and Safety Division of the Ontario Ministry of Labour, and is gratefully acknowledged.

REFERENCES

Alderson, M.R., Rattan, N.S., and Bidstrup, L. (1981). "Health of workmen in the chromate-producing industry in Britain." *Br. J. Ind. Med.* **38**, 117–124.

Andrievskaya, Z.M., and Mislavskaya, M.M. (1949). "Silicosis in a chromite mine and its prophylaxis." *Gig. i. Sanit.* **14**(5), 28–30 (cited in Sluis-Cremer and Du Toit, 1968).

Anon. (1963). "Chrome ulceration of the nasal septum." *Br. Med. J.* **1963**(1), 1364–1365.

Axelsson, G., Rylander, R., and Schmidt, A. (1980). "Mortality and incidence of tumours among ferrochromium workers." *Br. J. Ind. Med.* **37**, 121–127.

Barlow, S.M., and Sullivan, F.M. (1982). *Reproductive Hazards of Industrial Chemicals*, Academic Press, London, pp. 3–27.

Bécourt, M.M., and Chevallier, A. (1863). "Memoire sur les accidents qui atteignent les oeuvriers qui travaillent le bichromate de potass." *Ann. Hyg. Publique Med. Leg.* **20**, 83–95 (cited in Bidstrup, 1983).

Behari, J., Chandra, S.V., and Tandon, S.K. (1978). "Comparative toxicity of trivalent and hexavalent chromium to rabbits. III. Biochemical and histological changes in testicular tissue." *Acta Biol. Med. Germ.* **37**, 463–468.

Bidstrup, P.L. (1983). "Effects of chromium compounds on the respiratory system." In D. Burrows, Ed., *Chromium: Metabolism and Toxicity*, CRC Press, Inc., Boca Raton, FL, pp. 31–50.

Billard, R., and Roubaud, P. (1985). "The effect of metals and cyanide on fertilization in rainbow trout (*Salmo gairdneri*)." *Water Res.* **19**, 209–214.

Bloomfield, J.J., and Blum, W. (1928). "Health hazards in chromium plating." *Public Health Rep.* **43**, 2330–2351.

Bovet, P., Lob, M., and Grandjean, M. (1977). "Spirometric alterations in workers in the chromium electroplating industry." *Int. Arch. Occup. Environ. Health* **40**, 25–32.

Burrows, D. (1983). "Adverse chromate reactions on the skin." In D. Burrows, Ed., *Chromium: Metabolism and Toxicity*, CRC Press, Inc., Boca Raton, FL, pp. 137–163.

Burrows, D. (1984). "The dichromate problem." *Int. J. Dermatol.* **23**, 215–220.

Capodaglio, E., Catenacci, G., Pezzagno, G., et al. (1975). "Functional conditions of the respiratory system and work site pollution. II. Prevalence of respiratory function changes in workers employed in bichromate and chromic acid production." *Lavoro Umano,* **27**, 175–187 (Italian with English abstract).

Clarkson, T.W., Nordberg, G.F., and Sager, P.R. (1985). "Reproductive and developmental toxicity of metals." *Scand. J. Work Environ. Health* **11**, 145–154.

Cumin, W. (1827). "Remarks on the medicinal properties of madar, and on the effects of bichromate of potass on the human body." *Edinburgh Med. Surg. J.* **28**, 295 (cited in Bidstrup, 1983).

Danielsson, B.R.G., Hassoun, E., and Dencker, L. (1982). "Embryotoxicity of chromium: Distribution in pregnant mice and effects on embryonic cells *in vitro.*" *Arch. Toxicol.* **51**, 233–245.

Davies, J.C.A. (1974). "Inhalation hazards in the manufacture of silicon alloys." *Cent. Afr. J. Med.* **20**, 140–143.

Edmundson, W.F. (1951). "Chrome ulcers of the skin and nasal septum and their relation to patch testing." *J. Invest. Dermatol.* **17**, 17–19.

EPA. (1984). *Health Assessment Document for Chromium.* United States Environmental Protection Agency, Environmental Criteria and Assessment Office, Research Triangle Park, NC, EPA-600/8-83-014F.

Franchini, I., Mutti, A., Cavatorta, A., et al. (1978). "Nephrotoxicity of chromium. Remarks on an experimental and epidemiological investigation." *Contr. Nephrol.* **10**, 98–110.

Gafafer, W.M., Ed. (1953). *Health of Workers in Chromate Producing Industry: A Study.* U.S. Public Health Service, Division of Occupational Health, Pub. No. 192, Washington, D.C.

Gale, T.F. (1978). "Embryotoxic effects of chromium trioxide in hamsters." *Environ. Res.* **16**, 101–109.

Gale, T.F. (1982). "The embryotoxic response to maternal chromium trioxide exposure in different strains of hamsters." *Environ. Res.* **29**, 196–203.

Gale, T.F., and Bunch, J.D. (1979). "The effect of the time of administration of chromium trioxide on the embryotoxic response in hamsters." *Teratology* **19**, 81–86.

Gilani, S.H., and Marano, M. (1979). "Chromium poisoning and chick embryogenesis." *Environ. Res.* **19**, 427–431.

Goldman, M., and Karotkin, R.H. (1935). "Acute potassium bichromate poisoning." *Am. J. Med. Sci.* **189**, 400–403.

Gomes, E.R. (1972). "Incidence of chromium-induced lesions among electroplating workers in Brazil." *Ind. Med.* **41**(12), 21–25.

Guthrie, B.E. (1982). "The nutritional role of chromium." In S. Langård, Ed., *Biological and Environmental Aspects of Chromium, Topics in Environmental Health.* Vol. 5. Elsevier Biomedical Press, Amsterdam, pp. 117–148.

Hayes, R.B., Lilienfield, A.M., and Snell, L.M. (1979). "Mortality in chromium chemical production workers: A prospective study." *Int. J. Epidemiol.* **8**, 365–374.

Henning, H.F. (1972). "Chromium plating." *Ann. Occup. Hyg.* **15**, 93–100.

Horiguchi, S. (1980). "Industrial chromium poisoning in the small and medium sized electroplating factories, 1." *Sumitomo Bull. Ind. Health,* **33–37**.

IARC. (1980). *Chromium and Chromium Compounds. IARC Monogr. Eval. Carcinog. Risk Chem. Humans* **23**, 205–323.

Iijima, S., Matsumoto, N., and Lu, C.-C. (1983a). "Transfer of chromic chloride to embryonic mice and changes in the embryonic mouse neuroepithelium." *Toxicol.* **26**, 257–265.

Iijima, S., Shimizu, M., and Matsumoto, N. (1979). "Embryotoxic and fetotoxic effects of chromium trioxide in mice." *Teratol.* **20**, 152.

Iijima, S., Spindle, A., and Pedersen, R.A. (1983b). "Developmental and cytogenetic effects of potassium dichromate on mouse embryos in vitro." *Teratol.* **27**, 109–115.

Imreh, St., and Radulescu, D. (1982). "Cytogenetic effects of chromium in vivo and in vitro." *Mutat. Res.* **97**, 192–193.

Jacquet, P., and Draye, J.P. (1982). "Toxicity of chromium salts to cultured mouse embryos." *Toxicol. Lett.* **12**, 53–57.

Kalliomäki, P.-L., Kalliomäki, K., Korhonen, O., et al. (1982). "Respiratory status of stainless steel and mild steel welders." *Scand. J. Work Environ. Health* **8** (Suppl. 1), 117–121.

Kelly, W.F., Ackrill, P., Day, J.P., et al. (1982). "Cutaneous absorption of trivalent chromium: Tissue levels and treatment by exchange transfusion." *Br. J. Ind. Med.* **39**, 397–400.

Kleinfeld, M., and Rosso, A. (1965). "Ulcerations of the nasal septum due to inhalation of chromic acid mist." *Ind. Med. Surg.* **34**, 242–243.

Knudsen, I. (1980). "The mammalian spot test and its use for the testing of potential carcinogenicity of welding fume particles and hexavalent chromium." *Acta Pharmacol. Toxicol.* **47**, 66–70.

Krishna, G., Mathur, J.S., and Gupta, R.K. (1976). "Health hazard among chrome industry workers with special reference to nasal septum perforation." *Indian J. Med. Res.* **64**, 866–872.

Laitung, J.K.G., and Earley, M. (1984). "The role of surgery in chromic acid burns: Our experience with two patients." *Burns* **10**, 378–380.

Langård, S. (1980). "A survey of respiratory symptoms and lung function in ferrochromium and ferrosilicon workers." *Int. Arch. Occup. Environ. Health* **46**, 1–9.

Langård, S., Andersen, Aa., and Glyseth, B. (1980). "Incidence of cancer among ferrochromium and ferrosilicon workers." *Br. J. Ind. Med.* **37**, 114–120.

Langård, S., and Norseth, T. (1979). "Chromium." In L. Friberg, G.F. Nordberg, and V.B. Vouk, Eds., *Handbook on the Toxicology of Metals.* Elsevier/North-Holland Biomedical Press, Amsterdam, pp. 383–397.

Letterer, E. (1939). "Examination of a chromium-silicotic lung." *Arch. Gewerbepath* **9**, 496–508; abstracted in *J. Indust. Hyg. Toxicol.* **21**, 215 (1939).

Levis, A.G., and Majone, F. (1979). "Cytotoxic and clastogenic effects of soluble chromium compounds on mammalian cell cultures." *Br. J. Cancer* **40**, 523–533.

Littorin, M., Welinder, H., and Hultberg, B. (1984). "Kidney function in stainless steel welders." *Int. Arch. Occup. Environ. Health* **53**, 279–282.

Lucas, J.B., and Kramkowski, R.S. (1975). *Health Hazard Evaluation Report No. 74-87-221.* Health Hazard Evaluation Branch, DHEW, National Institute for Occupational Safety and Health, Cincinnati, OH.

Lukanin, W.P. (1930). "Zur pathologie der chromat-pneumokoniose." *Archiv Fur Hygiene* **104**, 166–174.

Machle, W., and Gregorius, F. (1948). "Cancer of the respiratory system in the United States chromate-producing industry." *Public Health Rep.* **63**, 1114–1127.

Maloof, C.C. (1955). "Use of edathamil calcium in treatment of chrome ulcers of the skin." *Arch. Ind. Health* **11**, 123–125.

Mancuso, T.F. (1951). "Occupational cancer and other health hazards in a chromate plant: A medical appraisal. II. Clinical and toxicologic aspects." *Ind. Med. Surg.* **20**, 393–407.

Mancuso, T.F., and Hueper, W.C. (1951). "Occupational cancer and other health hazards in a chromate plant: A medical appraisal. I. Lung cancers in chromate workers." *Ind. Med. Surg.* **20**, 358–363.

Manson, J.M. (1978). "Human and laboratory animal test systems available for detection of reproductive failure." *Prev. Med.* **7**, 322–331.

Matsumoto, N., Iijima, S., and Katsunuma, H. (1976). "Placental transfer of chromic chloride and its teratogenic potential in embryonic mice." *J. Toxicol. Sci.* **2**, 1–13.

Mertz, W., Roginski, E.E., Feldman, F.J., and Thurman, D.E. (1969). "Dependence of chromium transfer into the rat embryo on the chemical form." *J. Nutr.* **99**, 363–367.

Meyers, J.B. (1950). "Acute pulmonary complications following inhalation of chromic acid mist." *A.M.A. Arch. Ind. Hyg. Occup. Med.* **2**, 742–747.

Miller, R.K. (1983). "Perinatal toxicology: Its recognition and fundamentals." *Am. J. Ind. Med.* **4**, 205–244.

Miller, R.K., Ng, W.W., and Levin, A.A. (1983). "The placenta: Relevance to toxicology." In: T.W. Clarkson, G.F. Nordberg, and P.R. Sager, Eds., *Reproductive and Developmental Toxicity of Metals.* Plenum Press, New York, pp. 569–605.

Mitchell, A.J. (1969). "An unsuspected hazard of chrome stripping." *Trans. Soc. Occup. Med.* **19**, 128–130.

Mosinger, M., and Fiorentini, H. (1954). "Sur la pathologie due aux chromates. Première recherches expérimentales." *Arch. Mal. Prof. Med. Trav. Secur. Soc.* **15**, 187–199.

Mutti, A., Cavatorta, A., Pedroni, C., et al. (1979). "The role of chromium accumulation in the relationship between airborne and urinary chromium in welders." *Int. Arch. Occup. Environ. Health* **43**, 123–133.

NAS (National Academy of Sciences). (1974). *Chromium.* National Academy Press, Washington, D.C.

Nettesheim, P., Hanna, M.G., Jr., Doherty, D.G., et al. (1971). "Effect of calcium chromate dust, influenza virus, and 100 R whole-body X-radiation on the lung tumor incidence in mice." *J. Natl. Cancer Inst.* **47**, 1129–1144.

Newbold, R.F., Amos, J., and Connell, J.R. (1979). "The cytotoxic, mutagenic and clastogenic effects of chromium-containing compounds on mammalian cells in culture." *Mutat. Res.* **67**, 55–63.

NIOSH. (1973). *Criteria for a Recommended Standard: Occupational Exposure to Chromic Acid.* National Institute for Occupational Safety and Health, Cincinnati, OH, DHEW (NIOSH), Pub. No. 73-11021.

NIOSH. (1975). *Criteria for a Recommended Standard: Occupational Exposure to Chromium(VI).* National Institute for Occupational Safety and Health, Cincinnati, OH, DHEW (NIOSH), Pub. No. 76-129.

Pagano, G., Esposito, A., Bove, P., et al. (1983). "The effects of hexavalent and trivalent chromium on fertilization and development in sea urchins." *Environ. Res.* **30**, 442–452.

Pascale, L.R., Waldstein, S.S., Engbring, G., et al. (1952). "Chromium intoxication with special reference to hepatic injury." *J.A.M.A.* **149**, 1385–1389.

Petrzela, K., Bardon, J., Vit, M., and Szwarc, J. (1981). "Experience with the infection of upper respiratory pathways in the employees of a galvanization shop." *Prac. Lek.* **33**, 15–18.

Polansky, M.M., and Anderson, R. (1980). "Role of inorganic chromium in placental transport." *Fed. Proc.* **39**, 903.

Princi, F., Miller, L.H., Davis, A., and Cholak, J. (1962). "Pulmonary disease of ferroalloy workers." *J. Occup. Med.* **4**, 301–310.

Royle, H. (1975a). "Toxicity of chromic acid in the chromium plating industry (1)." *Environ. Res.* **10**, 39–53.

Royle, H. (1975b). "Toxicity of chromic acid in the chromium plating industry (2)." *Environ. Res.* **10**, 141–163.

Saner, G. (1981). "The effect of parity on maternal hair chromium concentration and the changes during pregnancy." *Am. J. Clin. Nutr.* **34**, 853–855.

Sarto, F., Cominato, I., Bianchi, V., and Levis, A.G. (1982). "Increased incidence of chromosomal aberrations and sister chromatid exchanges in workers exposed to chromic acid (CrO_3) in electroplating factories." *Carcinogenesis* **3**, 1011–1016.

Sarto, F., Cominato, I., Bianchi, V., and Levis, A.G. (1983). "Chromosomal damage in workers exposed to chromic acid (CrO_3)." *Mutat. Res.* **113**, 303–304.

Schroeder, H.A., Balassa, J.J., and Tipton, I.H. (1962). "Abnormal trace metals in man—chromium." *J. Chronic Dis.* **15**, 941–964.

Sharma, B.K., Singhal, P.C., and Chugh, K.S. (1978). "Intravascular haemolysis and acute renal failure following potassium dichromate poisoning." *Postgraduate Med. J.* **54**, 414–415.

Sluis-Cremer, G.K., and Du Toit, R.S.J. (1968). "Pneumoconiosis in chromite miners in South Africa." *Br. J. Ind. Med.* **25**, 63–67.

Stern, R.M. (1982). "Chromium compounds: Production and occupational exposure." In S. Langård, Ed., *Biological and Environmental Aspects of Chromium, Topics in Environmental Health.* Vol. 5. Elsevier Biomedical Press, Amsterdam, pp. 5–47.

Tandon, S.K. (1982). "Organ toxicity of chromium in animals." In S. Langård, Ed., *Biological and*

Environmental Aspects of Chromium, Topics in Environmental Health. Vol. 5. Elsevier Biomedical Press, Amsterdam, pp. 209–220.

Visek, W.J., Whitney, I.B., Kuhn, U.S.G., and Comar, C.L. (1953). "Metabolism of ^{51}Cr by animals as influenced by chemical state." *Proc. Soc. Exp. Biol. Med.* **84**, 610–615.

Wallach, S., and Verch, R.L. (1984). "Placental transport of chromium." *J. Am. Coll. Nutr.* **3**, 69–74.

Watanabe, S., and Fukuchi, Y. (1981). "Occupational impairment of the olfactory sense of chromate producing workers." *Jpn. J. Ind. Health* **23**, 606–611.

Wieser, O., Grünbacher, G., Prügger, F., et al. (1982). "Spastische bronchitis bei chromarbeitern." *Wien. Med. Wochenschr.* **132**, 59–62.

Williams, C.D., Jr. (1969). "Asthma related to chromium compounds. Report of two cases and review of the literature on chromate diseases." *North Carolina Med. J.* **30**, 482–491.

Wilson, J.G. (1977). "Current status of teratology." In J.G. Wilson, and F.C. Fraser, Eds., *Handbook of Teratology. Vol. 1. General Principles and Etiology.* Plenum Press, New York, pp. 47–74.

Worth, G., and Schiller, E. (1955). "Geshundheitsschadigungen durch chrom und seine verbindungen." *Arch. Gewerbepath. Gewerbehyg.* **13**, 673–686.

INDEX

Lightning Source UK Ltd.
Milton Keynes UK
UKOW07n1847121214

243017UK00001B/72/P